ADVANCES IN WELDING PROCESSES

Fourth International Conference

Harrogate — 9–11 May 1978

Volume 1 — Papers

Conference Technical Director
J.C.NEEDHAM

THE WELDING INSTITUTE

Abington Hall Abington Cambridge CB1 6AL

CONTENTS

Papers | *Page*

SESSION IV Control developments in arc and resistance welding processes

SESSION V High power density fusion welding

SESSION VI Techniques in special and heavy section fabrication

PAPER 4

The diffusion bonding of steels

D.S.Taylor, BSc, and G.Pollard, MA, PhD, MIM

Diffusion bonds have been produced in steels with carbon content up to 0.8%C in the temperature range 800°-$1000^{o}C$ using pressures up to $5Nmm^{-2}$ at times up to 30min. Resistance heating in a vacuum of 10^{-4} torr has been used. Tensile testpieces cut from the samples have given parent strength and ductility measurements but microstructural examination has shown only a single layer of small ferrite grains along the original interface. No recrystallisation of the austenite across the interface was evident in bonds giving satisfactory mechanical properties. In bonds between dissimilar steels a transition band about 1mm wide was present with a smooth transition in hardness. Experiments with artificially large voids showed the importance of bonding temperature if bonding times were to be kept below 1hr.

INTRODUCTION

Diffusion bonding has already been applied to an exotic range of products and materials: aerospace fabrications and the joining of superalloys are typical examples. The study of the mechanisms involved in producing diffusion bonds has been centred around titanium and its alloys[1-3] which can be bonded relatively easily over a wide range of conditions (since the oxide film becomes soluble in the matrix metal at bonding temperatures). Studies, particularly in the USA and the USSR, have shown that a wide range of materials can be diffusion bonded to a satisfactory standard without recourse to a high vacuum or long bonding times. Indeed Kazakov[3] claims that by using diffusion bonding it is possible to join together all crystalline materials including those which cannot be properly joined by fusion welding. Bartle[4] has also shown the wide applicability of the process. In the USSR impressive economic gains have been claimed for the fabrication, by diffusion bonding, of high throughput, low cost, steel articles.[5] It is therefore evident that in the right circumstances this bonding process can be usefully and economically applied to steels, especially those having a poor conventional weldability. The aims of the present work were to test the viability of diffusion bonding on a range of carbon steels using simple surface preparation techniques, readily achievable levels of vacuum, and acceptable process times. Resistance heating has been used because of its convenience in the laboratory for giving a rapid temperature response and its compatibility with a pneumatic ram loading system.

Mr Taylor is with ESAB Limited and Dr Pollard is Senior Lecturer, Department of Metallurgy, University of Leeds.

EXPERIMENTAL PROCEDURE

The experimental diffusion bonding equipment used for the work is shown, in schematic form, in Fig.1. The bonding pressure was applied to the specimen using a pneumatic ram fitted with a manostatic valve which ensured a constant loading force independent of any specimen expansion; a transducer provided accurate load setting and measurement. The specimens were resistance heated from a 5kW variable low voltage output system delivering 1000A at up to 5V and the interface temperature was monitored, using spot welded chromel-alumel thermocouples.
Both the bonding forces and the heating current were supplied by the water-cooled copper rams,

the bottom movable ram being sealed into a flexible stainless steel bellows. A vacuum of better than 10^{-4} torr was obtained using a single stage rotary pump backing on an oil diffusion pump.

The specimens were 28mm long and 11.28mm in diameter and diffusion bonded butt welds using both mild steel and plain carbon steel up to 0.8%C have been obtained. Dissimilar steel bonds within this range of materials have also been produced. The normal machined finish of the parted-off specimens was wet abraded to a 320 grit finish in a polishing holder to prevent rounding of the faying surfaces. This treatment gave a mean surface roughness Ra = 0.11μm (Ra is equivalent to CLA). The specimens were immersed in acetone for degreasing and then immediately positioned in pairs axially in the vacuum chamber. The interval between specimen preparation and bonding was kept to a minimum and was always less than 1hr.

The heating system has been found to be very flexible and the bonding temperatures were comfortably achieved within 1min. The temperature was maintained within ±5degC at the bonding temperature. After the required bonding time the specimens were cooled at an approximately normalising rate by reducing the heating current. The extent of specimen deformation was measured by the overall increase in specimen diameter.

The bonded specimens were either sectioned axially for metallographic examination and microhardness testing or machined into tensile testpieces. These testpieces were round proportional specimens of $12.5mm^{-2}$ cross-section and were tested on an Instron machine.

To follow the progress of the bonding — in particular the sequence of closing of surface voids — sets of specimens were prepared with an artificially rough surface. One of the faying surfaces was wet abraded to a superfinish using 600 grit paper and the mating surface was lathe finished. By the careful selection of the machining conditions it was possible to produce annular triangular grooves 0.17mm high with a peak-to-peak distance of 0.35mm. A section through a partially bonded specimen of this type is shown in Fig. 2.

The surface textures of the bonding surfaces have been measured with stylus profilometer techniques; the arithmetical mean deviation (Ra), the error of form, and the specimen axiality were examined. The 320 grit finish on mild steel produced an average Ra value of 0.11μm and the error of form was a single central hump across the specimen diameter 1μm high.

Table 1

Material, %	C	Si	Mn	S	P
Iron	0.006	0.30	0.47	0.004	0.006
078A18	0.18	0.27	0.78	0.058	0.013
059A24	0.24	0.26	0.59	0.056	0.014
083A45	0.45	0.26	0.83	0.032	0.038
058A57	0.57	0.32	0.58	0.039	0.016
078A81	0.81	0.32	0.78	0.030	0.005

RESULTS AND DISCUSSION

Butt diffusion bonds have been successfully produced on the range of carbon steels shown in Table 1, using bonding pressures of up to $5Nmm^{-2}$ in the temperature range 800° to 1000^oC for up to 30min.

Specimens of the same steel have been bonded as well as dissimilar steels. The metallographic examination of bonds made at 900^oC, $5Nmm^{-2}$ pressure, revealed complete bonding to have occurred and sections from such bonds are shown in Fig. 3. The structures obtained are typically normalised with ferrite grains outlining the original interface. This ferrite grain structure becomes increasingly discontinuous as the bonding time is increased. For those tests performed at 1000^oC the bonding interface quickly dispersed, but at the same time massive grain growth occurred in tests exceeding 15min and this resulted in a Widmanstätten ferrite structure being produced on cooling to room temperature. At 800^oC very little grain growth was observed but, although regions of high integrity bonding were observed metallographically, there remained unbonded regions along the interface after 30min under a load of $5Nmm^{-2}$.

Figure 4 shows the results of tensile tests on samples of the 0.24%C and the 0.57%C steels bonded at 900° and 1000^oC for times up to 30min. Failure was ductile in all of them and away from the bond interface in the parent metal, the parent strength being achieved within 5min. The ductilities achieved were at least equal to those obtained from unbonded parent metal although considerable variation in elongation values was observed, particularly with the 0.57%C material.

The microstructures obtained in bonds between dissimilar steels showed the same region of fine ferrite grains at the interface, and examples are shown in Fig. 5. To estimate the extent of the carbon interdiffusion between the two parent materials microhardness measurements were taken across the bond line. These profiles are shown in Fig. 6 and the total extent

Table 2

Parent material	Tensile strength, N/mm^2		Percentage elongation to failure	
Iron	398		25.38	
Mild steel	544		24.75	
0.24%C steel	503		28.25	
0.45%C steel	914		18.38	
0.57%C steel	981		15.88	
Diffusion bonding combination	**900°C**	**1000°C**	**900°C**	**1000°C**
Iron v. iron	491	454	29.50	27.00
Mild steel v. mild steel	523	554	32.50	27.00
0.24%C steel v. 0.24%C steel	515	529	33.75	28.63
0.45%C steel v. 0.45%C steel	627	846	18.75	17.38
0.57%C steel v. 0.57%C steel	922	966	24.00	18.63
Iron v. mild steel	495	433	29.25	19.63
Iron v. 0.24%C steel	487	472	29.13	25.50
Iron v. 0.45%C steel	491	476	22.00	24.00
Iron v. 0.57%C steel	490	479	23.25	24.63

of the diffusion zone was found to be slightly less than 1mm. The microstructure developed in the 0.57%C steel bonded to the iron is shown in Fig. 5a and a bond between the 0.81%C steel to iron is shown in Fig. 5b. The tensile test results for the various bond configurations are summarised in Table 2.

In all the tests the overall extent of deformation was measured across the diameter as this was found to be the most sensitive direction for the specimen aspect ratio used. At temperatures up to 900°C this barrelling was not more than 0.12mm for a 30min bonding time corresponding to about 1% deformation; at 1000°C the barrelling increased to about 0.35mm in 15min, i.e. about 3% strain. Because of the temperature distribution along the specimen length almost all the strain was limited to the central third of the specimens.

The removal of voids from along the weld interface is a critical stage in the production of satisfactory diffusion bonds.[6] The experiments using the artificially rough surfaces were designed to follow the closing sequence between the two faces. A bonding force of 1.5kN was used, this corresponding to a mean stress of $15Nmm^{-2}$ assuming full contact over the interface. The height of the remaining voids was measured from sectioned specimens and the results are shown in Fig. 7 for four test temperatures. The rate of closure is seen to be strongly temperature-dependent with closure under 1hr only for temperatures above 900°C. There was no evidence of oxide present in these voids presumably because of the low operating pressure and the activity of the carbon within the pores. The results so far available from these experiments do not enable the possible mechanisms of void closure to be identified. The values of activation energy determined are consistent with bulk diffusion but this does not imply a bulk diffusion process since high temperature creep and other mechanisms would show similar temperature-dependence. Since the metallographic studies are of the transformed pearlitic structure these studies also do not enable the mechanisms to be identified. It is indeed not clear that any diffusion flux is necessary to achieve satisfactory 'diffusion bonding'.

From the practical point of view it is not the relatively fine scale irregularities of the two bonding surfaces which will control the time required for final bonding but rather the longer wavelength texture. This surface shape is a function of the mode of preparation since turned specimens are usually humped in the centre and ground specimens have more variable deviations from flatness.

The main microstructural feature of all the tests other than at the highest temperature of 1000°C was the presence of a layer only one grain thick of ferrite. This ferrite will be nucleated from austenite grains and it is thought that at the bonding temperature these grains form a planar grain boundary along the interface with no significant grain growth or recrystallisation across the boundary. Previous workers[7, 8] have suggested that recrystallisation across the boundary is a prerequisite for forming strong bonds, but the current work does not indicate that this is necessary when a transformation product forms on cooling. The single layer of small ferrite grains does not impair the static mechanical properties of the bonds, but fatigue and fracture toughness tests have still to be carried out to see if there is any detrimental effect on the dynamic mechanical properties. However the structural variations across the bond are much less than those obtained by other welding processes.

CONCLUSIONS

As a result of the work reported here the following conclusions may be drawn:

1 Parent strength bonding together with parent ductility can be obtained in a range of carbon steels up to 0.8%C within 30min at 900°C. The loads required do not cause general yielding and the overall deformation is less than 1%. Stresses of $5Nmm^{-2}$ have produced satisfactory bonding in a vacuum of 10^{-4} torr

2 Recrystallisation and grain growth of the austenite across the bond interface is not necessary to produce good bonding and a single layer of ferrite grains is present along the original bond line. The microstructure across the bond is otherwise homogeneous

3 Bonding between dissimilar steels is possible with the same bonding conditions and a carbon diffusion band about 1mm wide has been produced at the interface with the conditions used in this work

4 Mechanical polishing on 320 grit followed by degreasing in acetone produced a satisfactory surface for bonding although overall surface flatness is considered a more important preparation

5 Bonding temperature is seen as the most important process variable for the achievement of complete bonding

ACKNOWLEDGEMENTS

The authors would like to thank the Science Research Council for provision of financial support for this work.

REFERENCES

1 KING, W.H. and OWCZARSKI, W.A. Welding J., 46 (7), 1967, 289s-98s.
2 KING, W.H. and OWCZARSKI, W.A. Welding J., 47 (10), 1968, 444s-50s.
3 KAZAKOV, N.R. Welding Prod., 14 (11), 1967, 83-6.
4 BARTLE, P.M. Welding J., 54 (11), 1975, 799-804.
5 USHAKOVA, S.E. Welding Prod., 10 (5), 1963, 34-6.
6 GARMONG, G., PATON, N.E., and ARGON, A.S. Metallurgical Trans., 6A (6), 1975, 1269-79.
7 McKEAG, D. and WILLIAMS, J.D. Procs of 3rd Int'l Conference 'Advances in Welding Processes', Harrogate, 7-9 May 1974, Paper 38, 263-71.
8 McKEAG, D. and WILLIAMS, J.D. 'Advanced Welding Technology'. Procs 2nd Int'l Symposium of Japan Welding Soc., Osaka, 25-27 August 1975, vol.1, Paper 4-(5), 227-32.

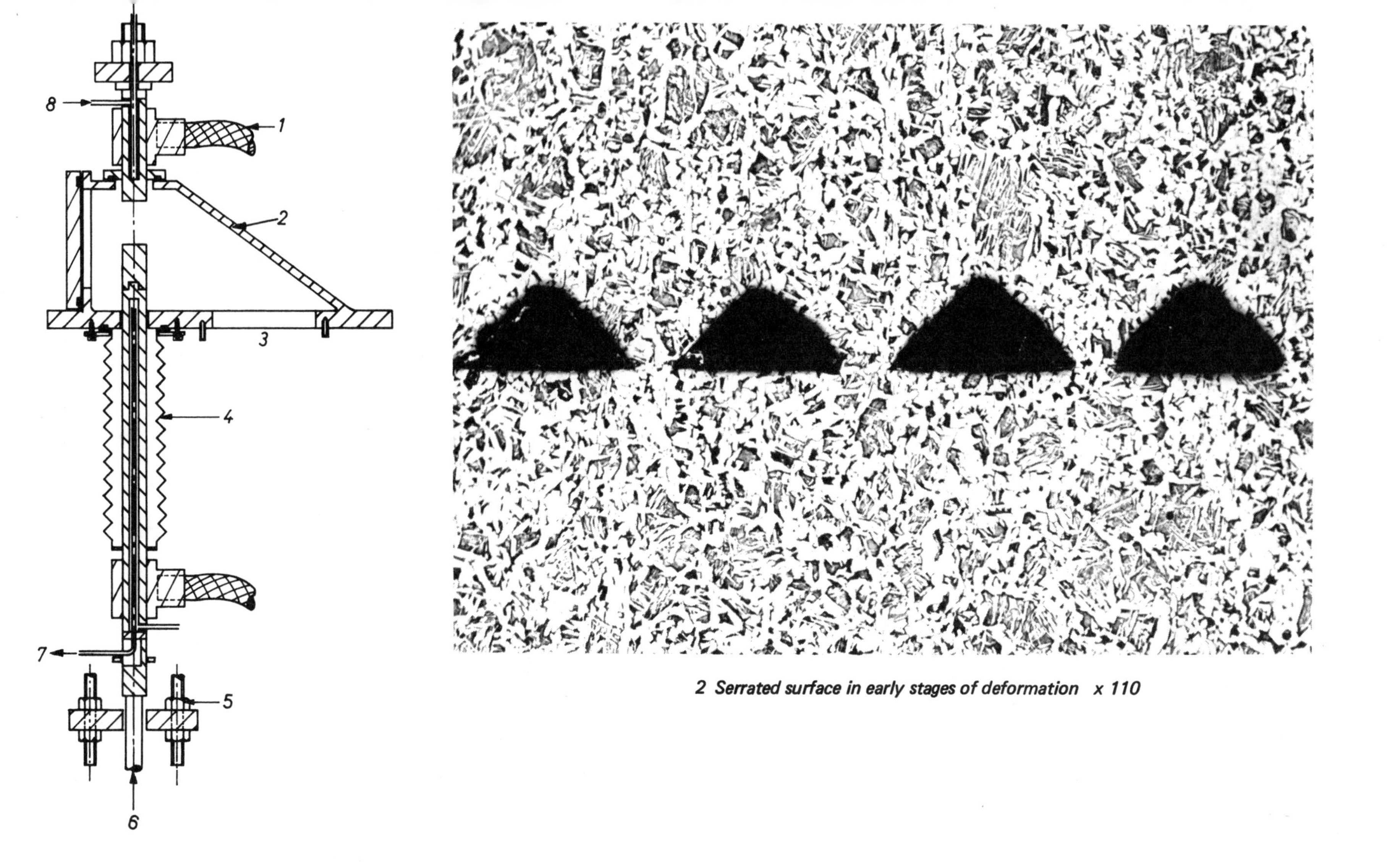

1 Schematic section through experimental rig.
1 – electrical connection to upper ram; 2 – vacuum chamber; 3 – diffusion pump; 4 – flexible bellows surrounding lower ram; 5 – stabilising bars; 6 – pneumatic ram; 7 – coolant out; 8 – coolant in

2 Serrated surface in early stages of deformation x 110

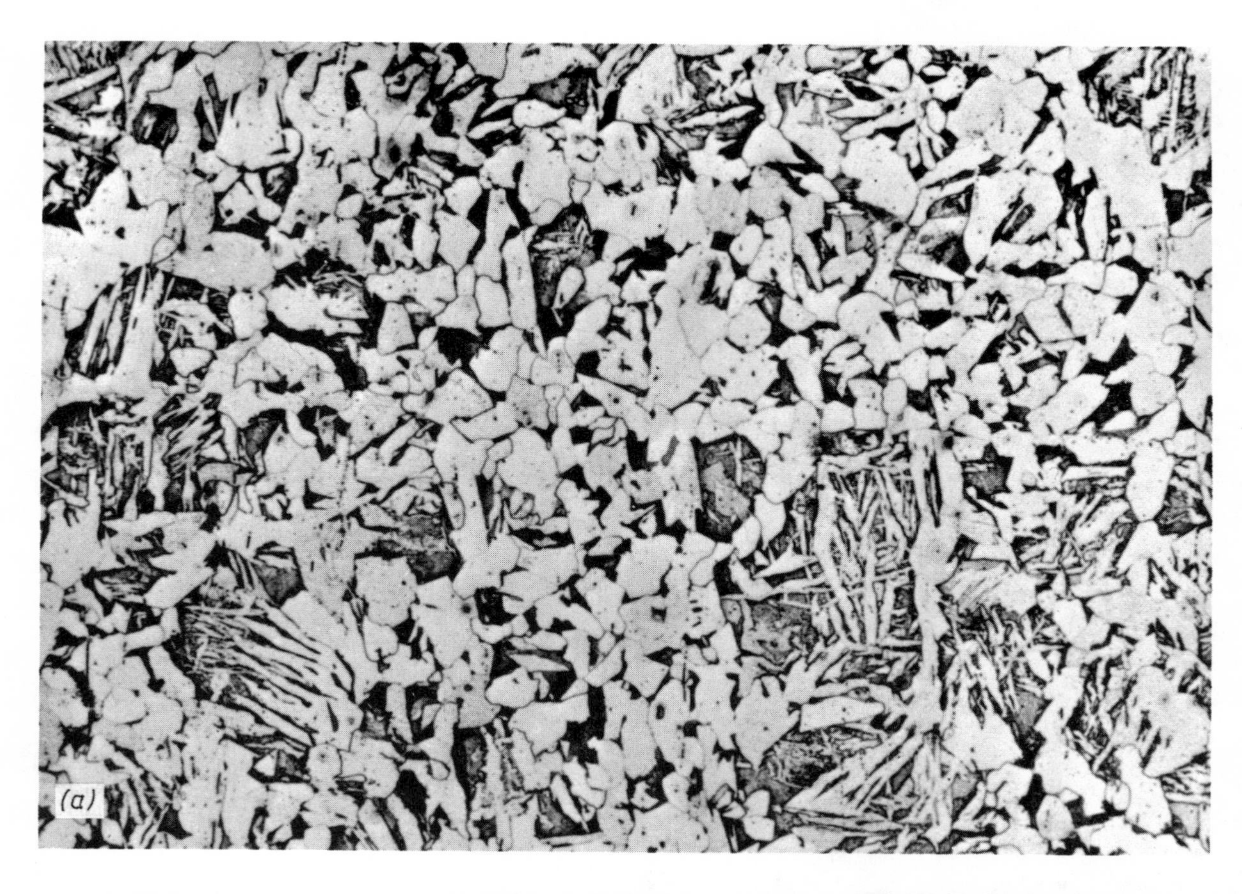

3 Microstructures of diffusion bond between: (a) mild steel specimens (x 275), (b) 0.45%C steel specimens (x 315)

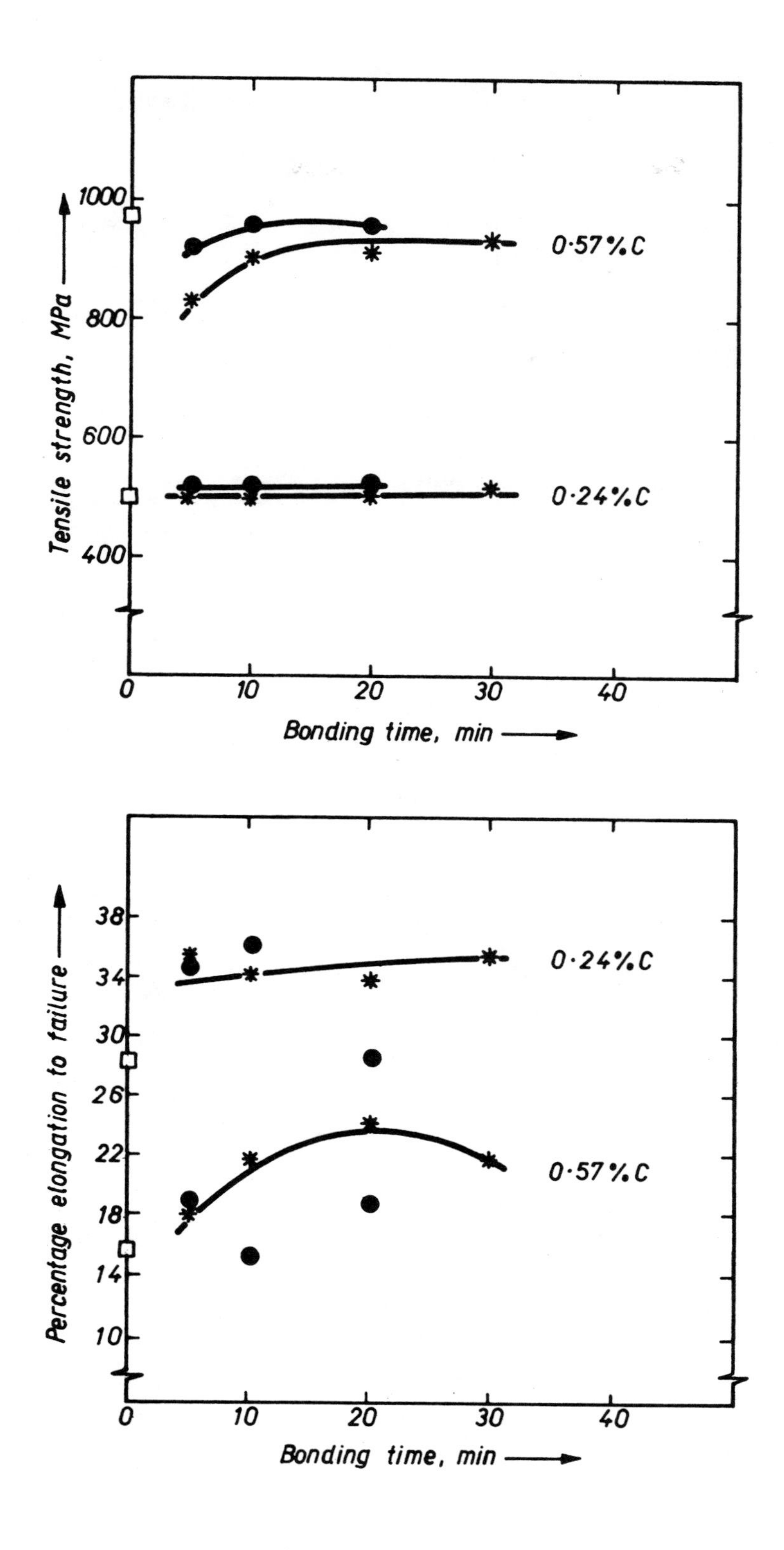

*4 Variation of tensile properties with time and temperature for diffusion bonds in 0.24%C and 0.57%C steels. ● – 1000°C; * – 900°C; □ – parent material*

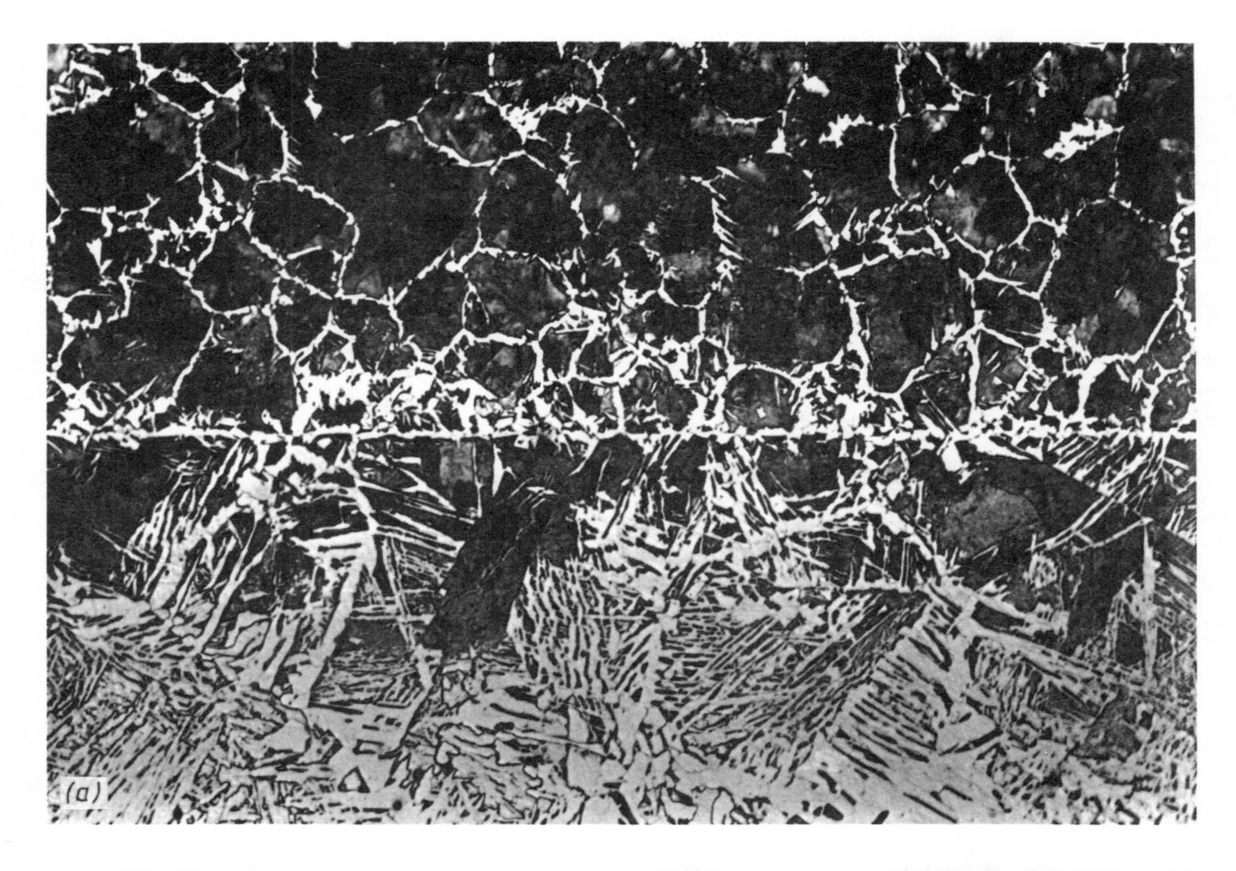

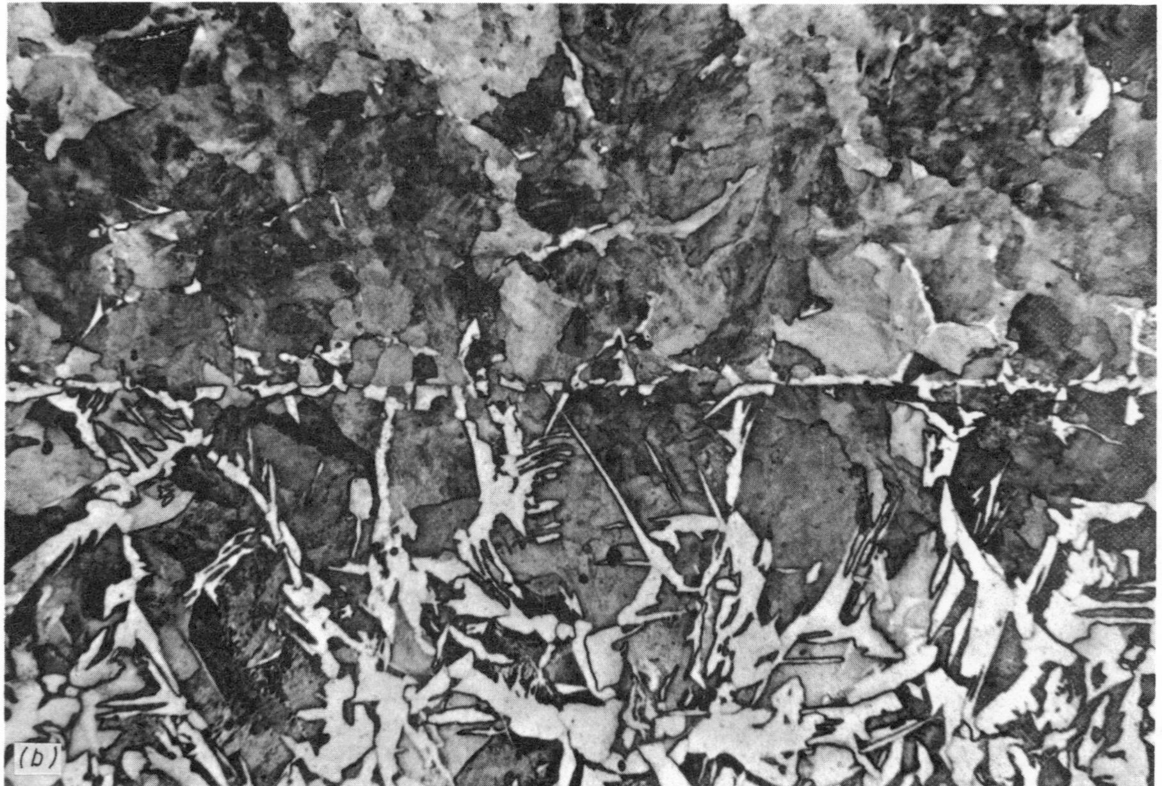

5 Microstructures of: (a) 0.57%C steel diffusion bonded to iron (x 130), (b) 0.81%C steel diffusion bonded to iron (x 275)

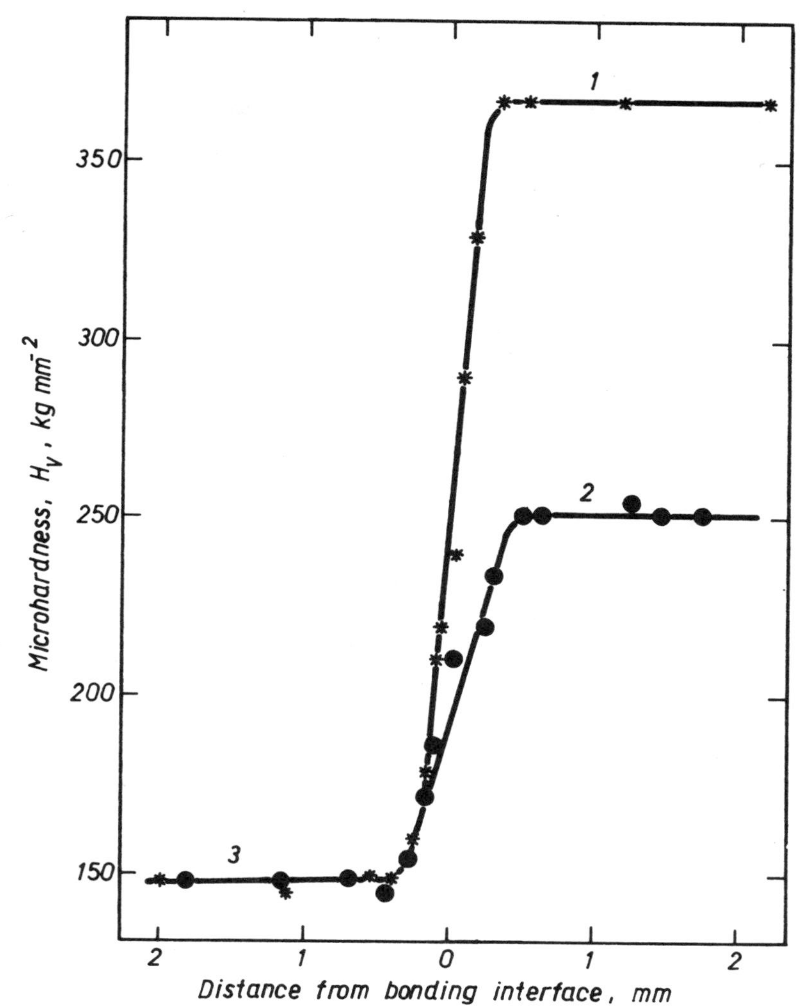

6 Microhardness profiles of interface regions of two iron-steel diffusion bonds. 1 – 0.81%C steel; 2 – 0.57%C steel; 3 – iron

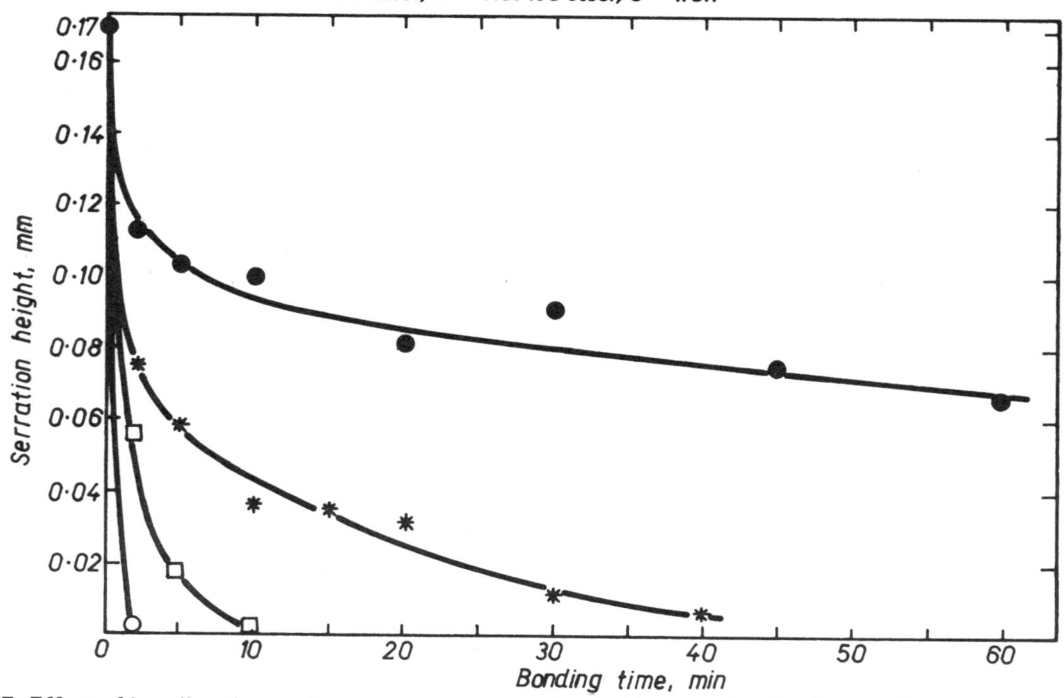

*7 Effect of bonding time and temperature on serration height. ● – 800°C; * – 900°C; □ – 950°C; ○ – 1000°C*

Intermetallics in solid-phase welds

E.R.Wallach, MA, MSc, PhD, MIM

The demand for bimetallic welds and opportunities for their use have expanded recently, especially since environmental and economic factors have assumed greater importance in the selection of materials for engineering applications. Accordingly, bimetallic solid-phase welding, i.e. the joining of two different metals without bulk fusion at their interface, is increasingly used to combine the advantages of one material with those of another. However, even when using solid-phase welding, intermetallic phases (generally brittle) may form at the interface between the two metals and may impose limitations on the performance of the joint. Thus knowledge of, for instance, the rates of nucleation and growth of such intermetallic phases is essential if welding processes and techniques are to be optimised or developed to minimise their formation. Research is in progress to study nucleation, diffusion, and the effect of intermetallic phases on solid-phase welds with the aim of correlating the results with the bonding processes and techniques.

This Paper describes the formation and nature of intermetallic phases, the factors controlling their growth, and their effects on the properties of aluminium copper roll-bonded welds. The results have implications for the manner in which, and the processes with which, bimetallic welds are fabricated.

INTRODUCTION

Bimetallic solid-phase welds are being used more frequently as materials are employed in more varied applications and in more hostile environments. Such welds are generally fabricated by processes such as pressure welding, ultrasonic welding, diffusion bonding, friction welding, and explosive bonding. These processes ideally prevent (or minimise) fusion at a joint interface and hence minimise intermetallic phase formation. However, if fabricated incorrectly, or used under adverse service conditions, interdiffusion may occur at the interface between the two parent metals in a weld, so causing brittle intermetallics to nucleate and grow with a consequent deterioration in weld properties.

The work outlined in this Paper is part of a larger research programme on solid-phase bonding, one objective of which is to provide information on the rates of nucleation and growth of intermetallics in bimetallic welds and on the effects of the intermetallics on joint properties. The results of the programme have practical implications for both optimising or assessing fabrication techniques and selecting

Dr Wallach, Assistant Lecturer, is in the Department of Metallurgy and Materials Science, University of Cambridge.

or assessing design criteria for the service use of specific bimetallic composites. This Paper emphasises these practical implications by summarising results from aluminium-copper welds. The aluminium-copper system has commercial importance, especially in electrical application, although the techniques described have more general application than for this system alone.

ALUMINIUM-COPPER SYSTEM

From the aluminium-copper equilibrium diagram,[1] Fig.1, five solid intermetallic phases are stable providing the temperature is kept below 548°C. The properties of these individual intermetallics are not well documented.* However, there is general agreement that the intermetallics are inherently brittle at homologous temperatures less than about $0.65T_m$[2-6] and cracks will propagate easily and rapidly. Thus it is imperative to minimise intermetallic formation both during weld fabrication and when the welds are in service.

If fabricating welds by pressure welding, e.g. roll bonding, local heating at the mating surfaces will be of very short duration and so minimal interdiffusion will occur, i.e. intermetallic formation will be negligible. However, when using a process such as diffusion bonding the times at relatively high temperatures may be long, and intermetallic nucleation and growth may be significant. Although a number of studies have been made of the kinetics of nucleation, interdiffusion and intermetallic growth for individual phases in aluminium-copper composites (solid-phase welds and conventional diffusion couples),[7-15] several phenomena were unresolved. These included:

1. Uncertainty whether all possible equilibrium intermetallic phases are nucleated, especially at temperatures less than 400°C
2. Doubt regarding the existence of claimed incubation periods for the onset of intermetallic nucleation
3. Little detailed data on the rates of growth of the intermetallic layers at temperatures less than 400°C
4. Insufficient information on the detrimental effects on weld properties caused by intermetallic growth

Accordingly, when developing procedures for fabricating bimetallic aluminium-copper welds, there were few guidelines for the selection of conditions which would ensure minimal intermetallic formation. One purpose of the work described in this Paper was to provide such guidelines. A second purpose of the work was to provide information on properties such as mechanical strength and electrical conductivity so that estimates could be made of the maximum tolerable thicknesses of intermetallic layers for specific applications.

EXPERIMENTAL PROGRAMME

Aluminium-copper-aluminium composites were fabricated by cold roll bonding. Using this process, atomic contact is ensured while interdiffusion is minimised. The composites were then sectioned and annealed in argon at atmospheric pressure during temperatures between 200° and 530°C for times up to 1036hr (3.7×10^6sec). After annealing, polished cross-sections through the weld interfaces were examined using optical and scanning electron microscopes. Concentration-distance curves through the interfaces were generated using electron probe microanalysis. From these curves the individual intermetallic phases could be identified and both rate constants and interdiffusion coefficients calculated.

With regard to properties, mechanical tests (shear, tensile, and impact) were carried out on samples (the designs of which are shown in Fig.2) which had been heat treated to produce intermetallic layers of different thicknesses. The surface morphologies of fractured specimens were examined by scanning electron microscopy and by X-ray diffraction (to identify the individual intermetallics on the fracture surfaces). Electrical resistance changes, as a result of intermetallic growth in the heat-treated welds, were determined by measuring the potential drops across an aluminium-copper interface when a known current was passed through an electron-beam welded sample, Fig.2b, before and after heat treatments. Details of the experimental techniques are available elsewhere.[16-18]

RESULTS

Intermetallic growth and identification

A typical optical micrograph of the cross-section through a heat-treated aluminium-copper

---

* Even in the systematic experimental investigation by Rabkin et al,[2] there is some doubt whether the techniques employed would have produced a single intermetallic rod, as claimed, rather than a rod consisting of several intermetallic phases. Hence the properties reported probably are not, as implied, for one intermetallic alone.

composite is shown in Fig. 3. The concentration-penetration curves obtained by electron probe microanalysis across the sample shown in Fig. 3 are presented in Fig. 4. To facilitate interpretation of such curves (potentially a major source of error), the aluminium-to-copper ratio (included in Fig. 4) was calculated since it clarifies the position of a boundary between adjacent phases. From such curves, intermetallic phase identification and widths of phases after different heat treatments were obtained.

Three particular points emerge from the results. Firstly, the intermetallic layers initially did not grow as continuous layers at low temperatures. In any particular composite nucleation and growth occurred in discrete regions, and only after growth had proceeded for a finite time (the length of which being temperature-dependent) was a continuous layer formed. This phenomenon is attributed to the mode of joining the two surfaces. Interface discontinuities and/or oxide 'blocks' remain distributed randomly along the interface after roll bonding.[19] Such artefacts inhibit atom transfer.

Secondly, there was not a sharp discontinuity in concentration gradient between the Cu_3Al_2 and Cu_9Al_4 phases. It is possible that, since the two phases have similar structures and lattice parameters,[1,20] a 'diffuse' rather than discrete interface between these phases is thermodynamically more probable. In other words, since a bimetallic weld is not representative of an equilibrium situation (owing to the presence of the large concentration gradient across the interface), there is no theoretical justification for assuming that an equilibrium diagram, such as Fig. 1, can be used to predict either the presence or compositions of individual phases.

Thirdly, as is clear from Figs 3 and 4, optically resolved layers do not correspond uniquely to individual intermetallic phases. Accordingly, caution is advocated when interpreting optical micrographs of interdiffusion effects in bimetallic welds. For instance, the incubation periods reported in earlier work[8-9] probably arose from the limited resolution of the optical techniques employed.

Calculation of rate constants and interdiffusion coefficients

Rate constants and interdiffusion coefficients are used respectively to describe the growth of individual intermetallic layers and rates of transport of elements through intermetallic layers in bimetallic diffusion couples. In transport phenomena, rate constants generally are derived for individual temperatures from the gradients of lines of

$$w^n = k\ t$$

where w is the width of an intermetallic layer (say)
t is the diffusion time
k is the rate constant for the particular temperature
n is the exponent and for volume diffusion is equal to 2

Thus knowledge of the values of rate constants at defined temperatures of interest allow predictions to be made, for heat treatments of set times, of both individual and the total intermetallic phase widths.

Interdiffusion coefficients also describe the rates of material transport. However, unlike rate constants which merely are the consequence of experimental data fitting, interdiffusion coefficients can under ideal conditions, e.g. in pure metals or when one mechanism is dominant, be related to actual physical atomic mechanisms of diffusion. Moreover in multiphase systems, such as bimetallic welds, allowance can be made for changes in partial molar volumes with concentration, concentration discontinuities at phase boundaries, etc. when calculating interdiffusion coefficients for the individual intermetallic phases.

Rate constants and 'average' interdiffusion coefficients were calculated for the various intermetallic phases by measuring individual phase widths and boundary concentrations from the concentration-penetration curves.[18] Several computer programs were written to facilitate handling of this data and to calculate the results using existing mathematical treatments.[21-23] It was further shown that the temperature-dependencies of both rate constants and interdiffusion coefficients could be described by the Arrhenius equation. This empirical relationship has the form

$$D = D_o \exp(-Q/RT)$$

where D is the diffusion coefficient (or rate constant k)
T is a particular temperature (^oK)
Q is an activation energy
D_o is a constant (or k_o)
R is the gas constant

Typical results for one intermetallic phase are shown in Fig. 5. From such graphs, the values of D_o (or k_o) and Q can be derived and predictions made of expected diffusion coefficients (or rate constants) at different temperatures. Using these, intermetallic phase widths for heat treatments of given times can be estimated.

The resulting values of activation energies and pre-exponential terms for temperatures of 400°C and above were in substantial (but not complete) agreement with previous work.[15] Moreover, at temperatures less than 400°C, changes in activation energies for the various intermetallic phases were indicated, e.g. change in gradient of line in Fig. 5.[18] A change in activation energy generally indicates a shift in emphasis from one diffusion mode to another. The two common diffusion modes in metal systems are volume diffusion and short-circuit diffusion (along crystal discontinuities such as grain boundaries, interphase boundaries, and dislocations). Although both modes operate simultaneously, volume diffusion predominates at high temperatures and short-circuit diffusion at low temperatures (below $\frac{1}{2}T_m$).

The preceding result has important practical implications. Firstly, it is appealing in research to measure diffusion rates at high temperatures so that the times for diffusion experiments are practical, i.e. not too long. Then, using the techniques outlined previously, predictions can be made of expected diffusion distances or intermetallic phase thicknesses by extrapolation from the high temperature rates. However, it is clear that if there is a change in diffusion mode at lower temperatures (as found in the aluminium-copper system) and if this is not realised, such predictions may be disastrously overoptimistic. Secondly, the change in diffusion mode may be of importance when considering the atomic mechanisms of processes such as diffusion bonding. Knowledge of the dominant diffusion process may assist in elucidating the rate-controlling mechanisms. In turn, such information could help to improve bonding procedures and extend the scope of a particular joining technique to new bimetallic systems.

Mechanical properties

The results from the mechanical tests (tensile, shear, and impact) are shown together in Fig. 6.[17] It is clear that the impact properties of welds are extremely poor as the intermetallic layer thickens even though the corresponding tensile and shear strengths remain adequate. Thus to assess or describe weld properties using the results of tensile or shear tests can be very misleading. Several other points also emerge.

Firstly, the results confirm those of Winkle[14] which also showed that the critical intermetallic thickness for weld embrittlement was approximately 2μm. In the present work a layer thickness of 2μm reduced the impact strength from 28 to $2MJm^{-2}$. A contribution to the low impact strengths could also result from the existence of discontinuities in the interface region, such as oxide blocks.

Secondly, the differences in tensile and impact values are not contradictory. For instance, Weiss and Hazlett[24] noted that relatively high tensile strengths were obtained even when testing poorly bonded friction welds or welds with brittle intermetallic layers. It was postulated that a triaxial state of stress existed in the weaker metal adjacent to the weld interface. Thus, the yield stress of the metal in the interface region was higher than that for the bulk material. A similar explanation had been put forward[25] earlier to account for enhanced strengths in thin brazed and soldered joints.

Thirdly, the ratios of tensile to shear ultimate strengths at specific intermetallic thicknesses are found to range between 1.5 (low thicknesses) to 2.0 (thicknesses around 40.0μm). These ultimate strength ratios are similar to theoretical values for the predicted onset of plastic yielding which can be calculated using the von Mises (value of 1.73) and Tresca (value of 2.0) yield criteria.[26] This implies that little plastic deformation occurs prior to fracture and that the experimental ultimate strengths are not much greater than the yield strengths of the welds, i.e. confirmation of brittle failure. In addition, ductile polycrystalline materials tend to obey the von Mises criterion whereas brittle materials tend to obey the Tresca criterion. Thus, the increase in the experimental ratio (of tensile to shear strengths) as the intermetallic layer thickens is consistent with the increased dominance of the brittle layers over the ductile parent metals.

Fourthly, it was shown using X-ray diffraction that the welds invariably failed between the $CuAl_2$ and CuAl intermetallics for layer thicknesses up to 25.0μm. Above this, failure was predominantly transgranular through the CuAl layer. The implication of these results is of greater importance than merely providing a description of the fracture path since the results have shown the welds to be very susceptible to 'instantaneous' loading (impact tests). Bimetallic welds are frequently subjected to rapidly changing temperatures in service environments, i.e. are thermally cycled. On thermal cycling the two metals repeatedly expand and contract by different amounts because of differences in their coefficients of thermal expansion and this can give rise to thermal stresses which may be large and instantaneous. In fact, in studies of the thermal cycling of fibre-reinforced composites, yielding and cracking in the vicinity of the interfaces were produced by thermal strains.[27-28] Thus thermal effects may well be of considerable significance in certain applications. Research is continuing in this area.

Electrical properties

From measurements of the voltage drops (for known currents) across intermetallic layers of different known thicknesses and cross-sectional areas, the value of the resistivity of the overall intermetallic layer in aluminium-copper heat-treated welds was found to be approximately 13$\mu\Omega$cm.[16] Since, for most practical applications, the intermetallic layer thickness must be kept below two microns to maintain adequate impact strength, it is unlikely that the resistance alone of such a thin layer will result in a bimetallic weld being unacceptable for reasons of poor conductivity. However, it was shown when measuring the resistivity that even a thin intermetallic layer of higher resistance than the parent metals can have a marked effect on the actual current path through a bimetallic joint. The effect will depend on the original joint geometry but may cause an additional and significant change in the overall joint resistance. An analogue method was used to demonstrate this phenomenon.*

Typical results of the analogue studies are shown in Fig.7 and are discussed in detail elsewhere.[16] It is clear that the path of least resistance is followed for the bimetallic analogue, i.e. the current prefers to flow through the middle low-resistance layer in Fig.7b. As is shown by the sharp bending of the equipotential lines at the bottom left and top right corners of the middle layer, most of the current passes through a small percentage of the total available interfacial area. When an intermetallic layer of higher resistance than either parent metal is present, Fig.7c, the current path alters and a larger interfacial area is used. Thus, the overall resistance of such a joint increases more than would be expected from consideration of the resistance of the intermetallic layer alone.

*Graphitised electrical analogue paper models were used and the position of lines of equipotential were plotted when constant currents flowed through these analogues.[16] Since the current path is perpendicular to the lines of equipotential, the analogues indicate the probable current path. The plotted equipotential lines are directly proportional to resistance. Therefore, the results are semi-quantitative and the numbers associated with each line in Fig.7 are percentages of the total resistance across each analogue. The total resistances are included in the Figure.

SUMMARY AND CONCLUSION

There are two approaches (at least) for obtaining satisfactory joints between dissimilar metals. The first is to vary the bonding parameters, perhaps trying a multitude of conditions until a weld is formed which will meet the specified standard. Traditionally this method has been proven, particularly in autogenous welds. Often, suprisingly quickly, suitable bonding parameters can be defined. However, the method essentially is a 'hit-miss' approach. Firstly, there is no guarantee that conditions will be found and considerable time and effort may be expended. Secondly, the method does not necessarily build up the detailed understanding which subsequently can be used to make predictions for bonding new systems.

The second method is to understand the problems likely to be encountered when bonding dissimilar metals, define these closely, and then adjust bonding conditions to avoid or minimise them. When bonding dissimilar metals this second approach is to be recommended. Accordingly this Paper has attempted to outline potential problems stemming from intermetallic phase growth in bimetallic welds. The results clearly show the detrimental influence of the intermetallic phases. In particular, the total width of such phases at a bimetallic interface must be strictly controlled if adequate mechanical strength is to be maintained. Thus data must be available on rates of growth of the intermetallic phases to define conditions (times and temperatures) which can be used, say in diffusion bonding, to ensure minimal intermetallic growth and so guarantee joints with adequate properties. Such data may readily be obtained using the techniques described.

The same data can be used to make predictions of expected service behaviour of joints under defined operating conditions, i.e. assessments can be made of the suitability of proposed bimetallic welds for a specific application. It should be emphasised, however, that care must be taken both when deriving and using the data. Firstly, as has been seen, the dominant diffusion mechanism may change with temperature with a consequent change in intermetallic growth rate. This could result in overoptimistic extrapolations from high temperature data to low temperature predictions. Secondly, results from the incorrect selection of a test method could be misleading, e.g. tensile versus impact results. Thirdly, an intermetallic layer might have an effect on properties not immediately apparent from the results measured, e.g. effect an overall joint resistance because of changes in current path. Nonetheless, data derived in the manner outlined in this Paper can considerably facilitate the fabrication of bimetallic welds and their assessment for service applications.

ACKNOWLEDGEMENTS

The author would like to thank Professor R.W.K. Honeycombe for providing laboratory facilities and Professor G.J. Davies for useful discussions and encouragement. The author is also grateful to Alcan International Research Centre, Kingston, Ontario, Canada, and, subsequently, the Science Research Council for financially supporting the research programmes of which this work was a part.

REFERENCES

1 HANSEN, M. and ANDERKO, K. 'Constitution of binary alloys', 2nd ed. New York, McGraw-Hill, 1958, 85.
2 RABKIN, D.M. et al. Soviet Powder Met. Metal. Ceram., 8, 1970, 695.
3 VOL, A.E. 'Handbook of binary metallic systems — structure and properties', Vol.1, 1959, 81. (Translated by Israel Translation Program for Scientific Translation Service, Jerusalem, 1966.)
4 PETTY, E.R. J. Inst. Metals, 89, 1960-61, 343-9.
5 HAASEN, P. 'Physical metallurgy' (Ed. R.W. Cahn). North Holland, 1970, 1070.
6 DEY, B.N. and TYSON, W.R. Physica Status Solidi (A), Applied Research, 9 (1), 1972, 215-21.
7 RYABOV, V.R. et al. Welding Production, 15 (4), 1968, 16-22.
8 LARIKOV, L.N. et al. Akad. Nauk. Ukrain. SSR Metallofizika, 28, 1969, 5.
9 RABKIN, D.M. et al. Welding Production, 16 (5), 1969, 21-6.
10 RABKIN, D.M. et al. Automatic Welding, 22 (2), 1969, 20-27.
11 TRUTNEV, V.V. et al. Welding Production, 18 (1), 1971, 23-5.
12 RYBOV, V.R. and LOZOVSKAYA, A.V. Schweisstechnik (Berlin), 22 (6), 1972, 250-53.
13 KOZLOV, Yu I. and ITIN, V.I. Sov. Powder Metall. Met. Ceram., 12, 1973, 454.
14 WINKLE, R.V. English Electric Nelson Research Laboratories, Beaconhill, Stafford, Report nos NC u 3154, 1968, and Nc y 110, 1970.
15 FUNAMIZU, Y. and WATANABE, K. Trans Japan Inst. Metals, 12, 1971, 147.
16 WALLACH, E.R. and DAVIES, G.J. Metals Sci., 11 (3), 1977, 97-103.
17 WALLACH, E.R. and DAVIES, G.J. Metals Tech., 4 (4), 1977, 183-90.
18 WALLACH, E.R. and DAVIES, G.J. Acta Met., submitted.
19 CANTALEJOS, N.A. and CUSMINSKY, G. J. Inst. Metals, 100 (1), 1972, 20-23.
20 BRADLEY, A.J. et al. J. Inst. Metals, 63, 1938, 149-62.
21 HEUMANN, T. Z. Chem. Phys., 201, 1952, 168.
22 WAGNER, C. Acta Met., 17 (2), 1969, 99-107.
23 WALLACH, E.R. Scripta Met., 11, 1977, 361.
24 WEISS, H.D. and HAZLETT, T.H. Paper 66-MET-8 presented at ASME Metals Eng. Conference, Cleveland, Ohio, 1966.
25 GRIFFITHS, J.R. and CHARLES, J.A. Metals Sci., 2, May 1968, 89-92.
26 COTTRELL, A.H. 'The mechanical properties of matter'. New York, Wiley, 1964, 312.
27 CHAWLA, K.K. Metallography, 6 (2), April 1973, 155-69.
28 CHAWLA, K.K. and METZGER, M. J. Mat. Sci., 7 (1), 1972, 34-9.

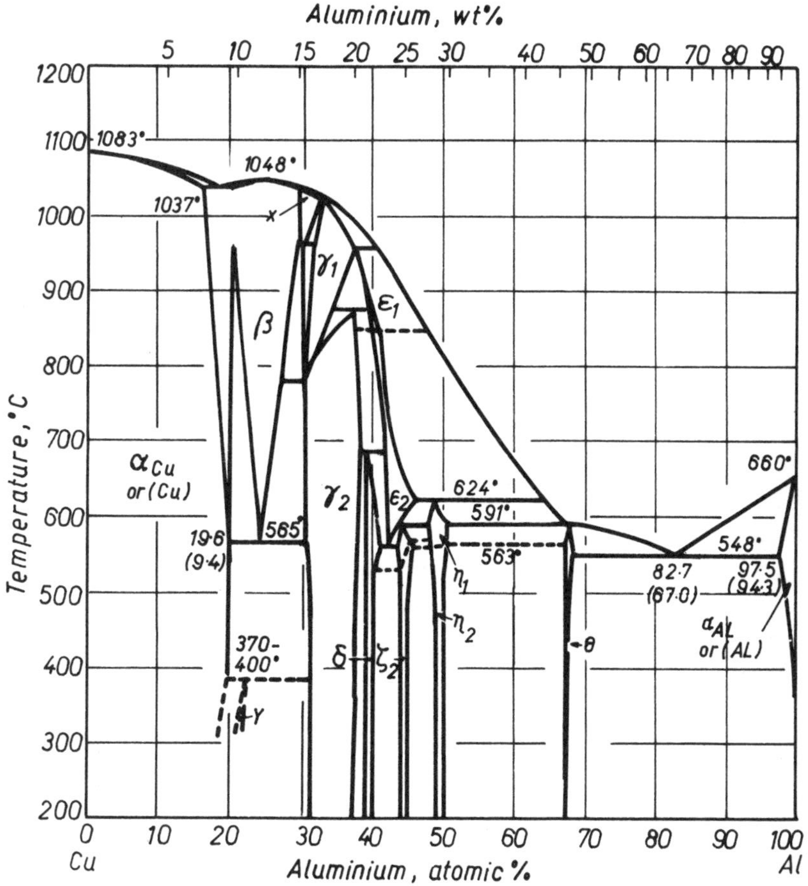

1 *Aluminium-copper equilibrium diagram (after Hansen and Anderko[1])*

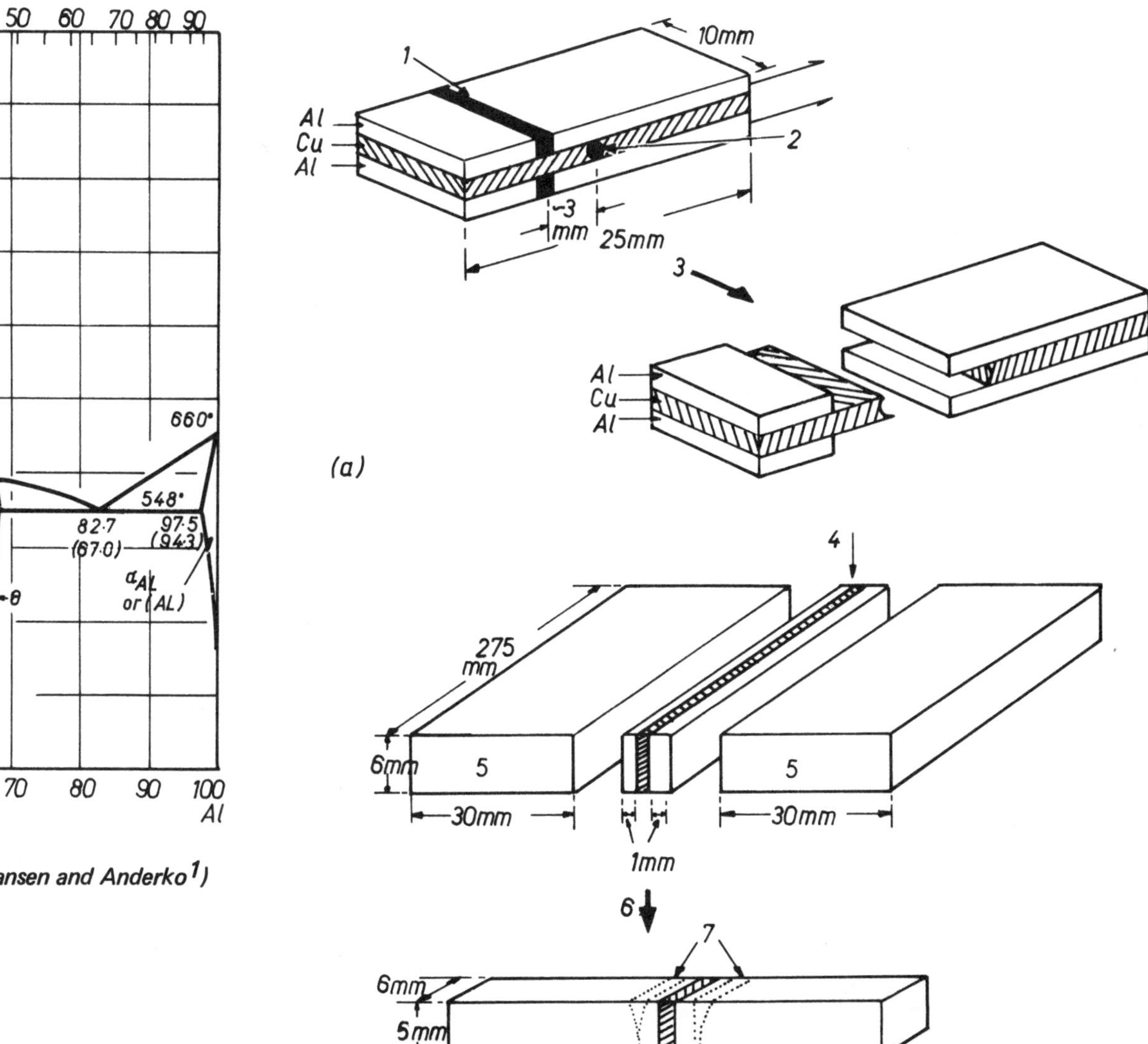

2 *Specimens for mechanical tests: (a) design for shear test, (b) design for electron-beam welded tensile and impact tests. 1 — milled slits; 2 — drill hole; 3 — after testing; 4 — machined strip of roll-bond composite; 5 — high strength aluminium; 6 — after welding and machining; 7 — electron-beam welds*

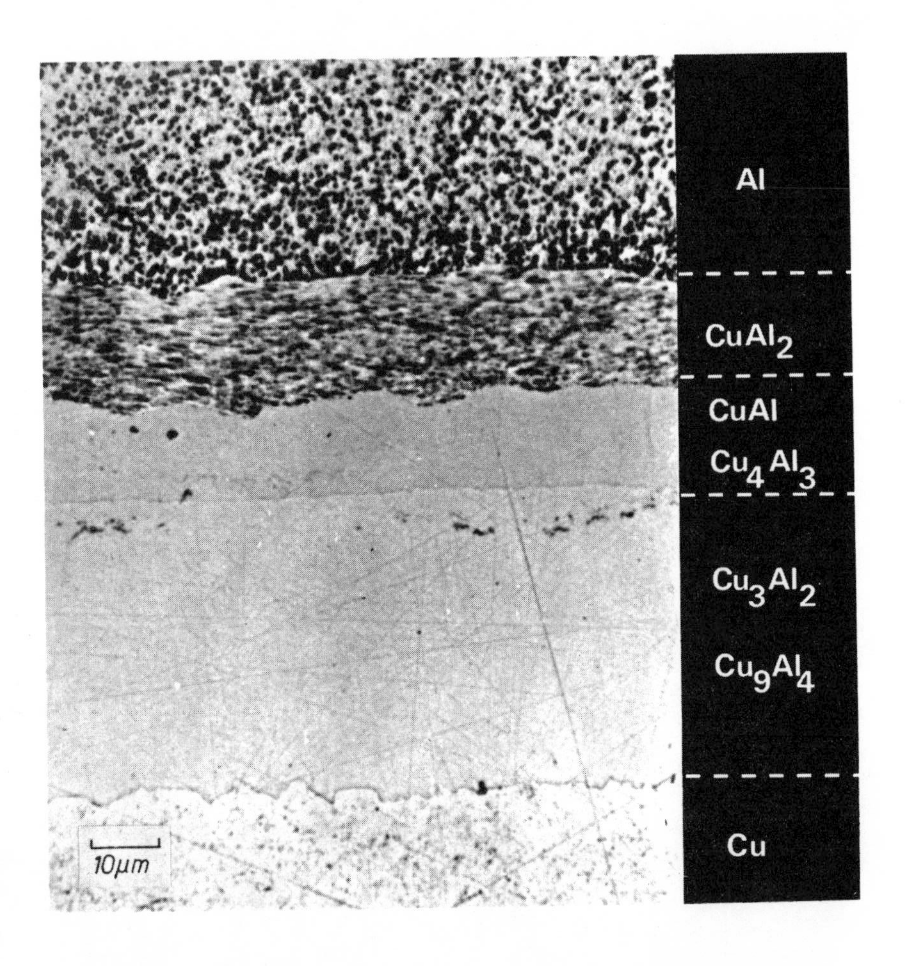

3 Optical micrograph of interface of polished aluminium-copper roll-bonded weld annealed for 48hr at 475°C. Etched in NaOH and $FeCl_3$ solutions

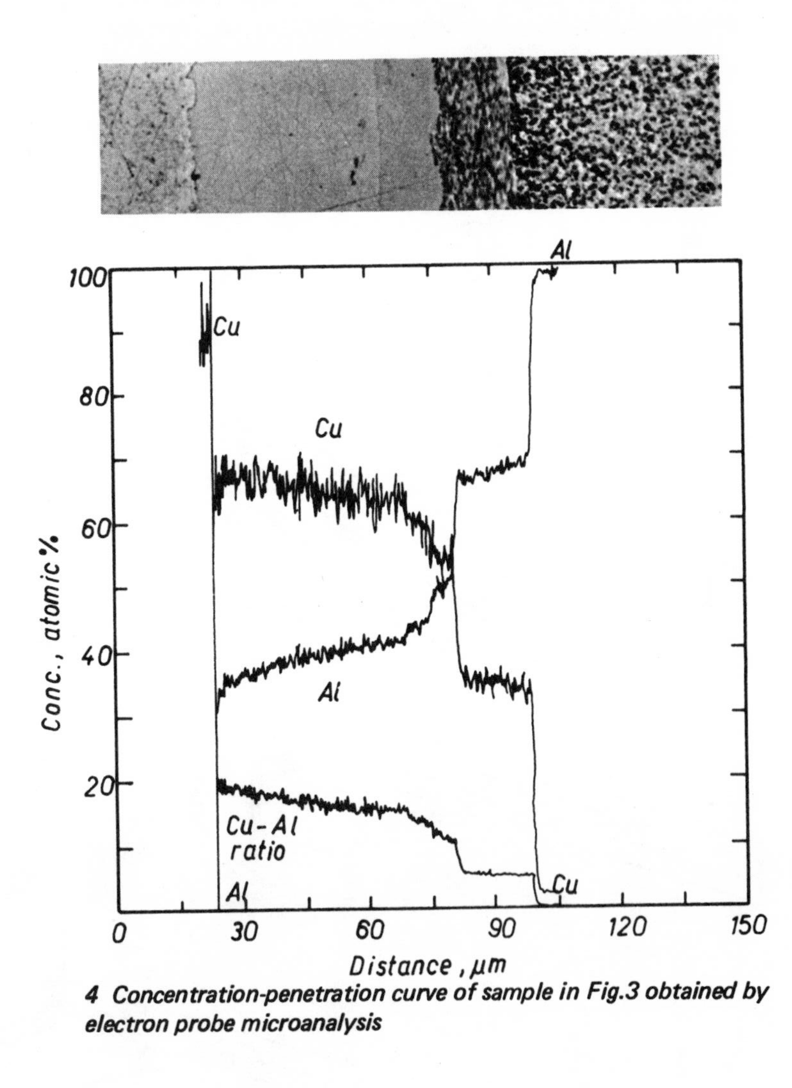

4 Concentration-penetration curve of sample in Fig.3 obtained by electron probe microanalysis

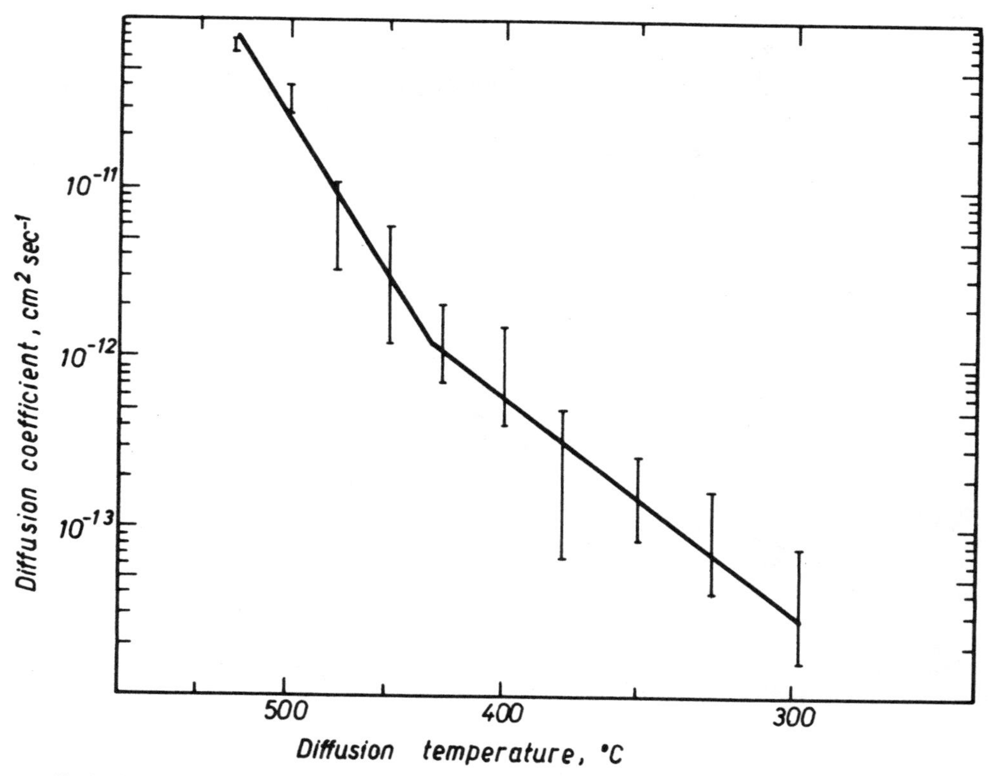

5 Arrhenius plot of average interdiffusion coefficient for Cu_4Al_3 intermetallic phase

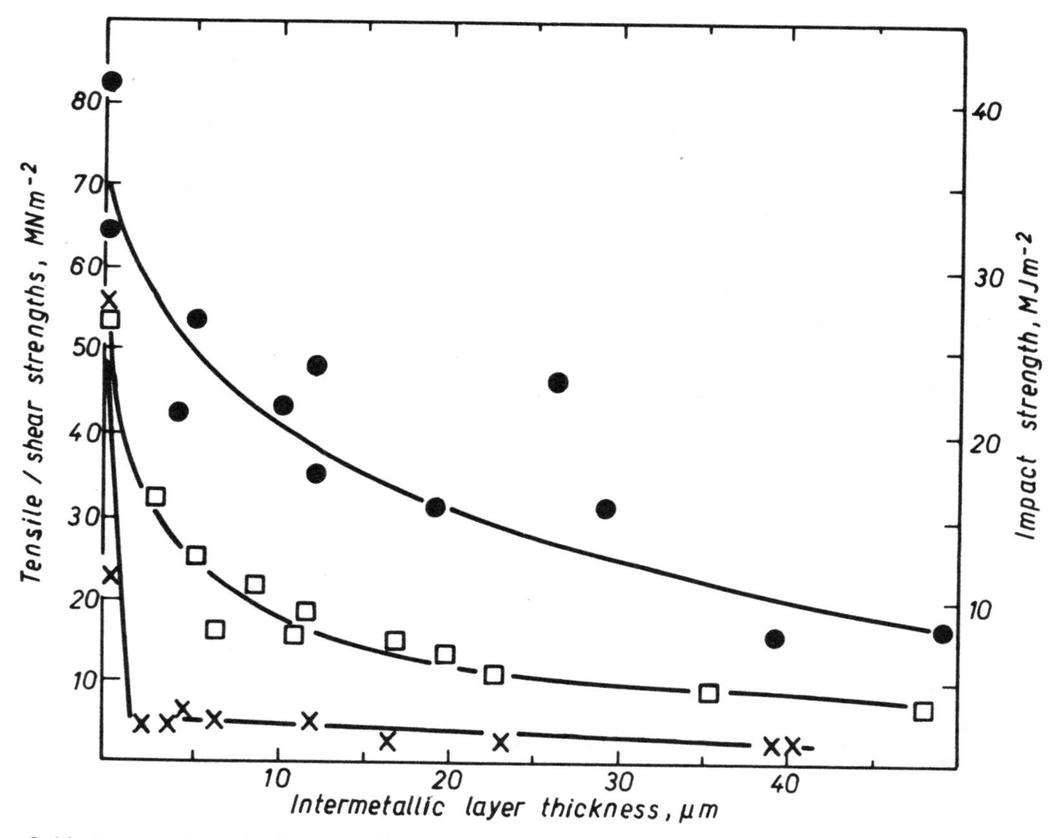

6 Variations of tensile (●), shear (□), and impact (X) strengths with intermetallic thickness for aluminium-copper solid-phase welds

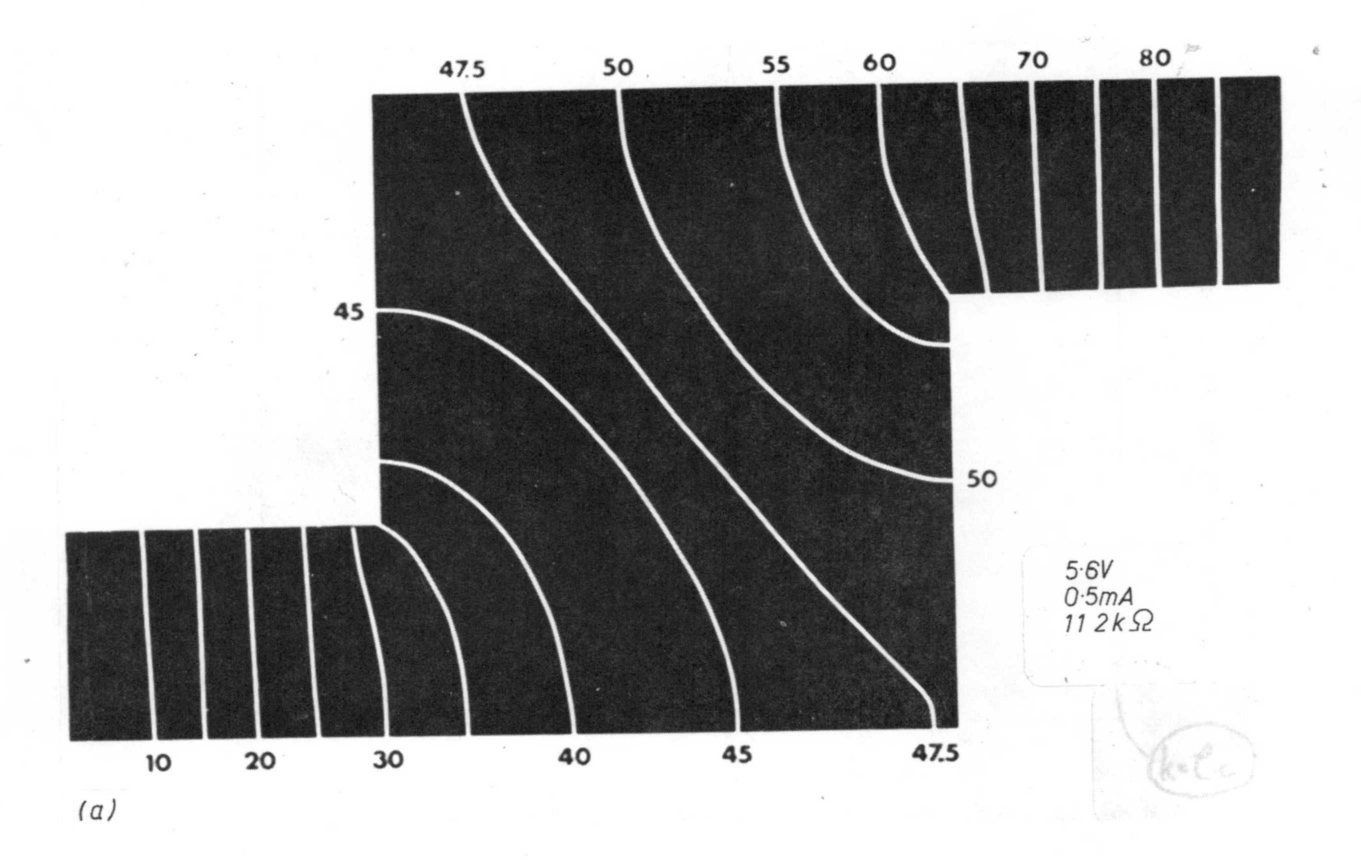

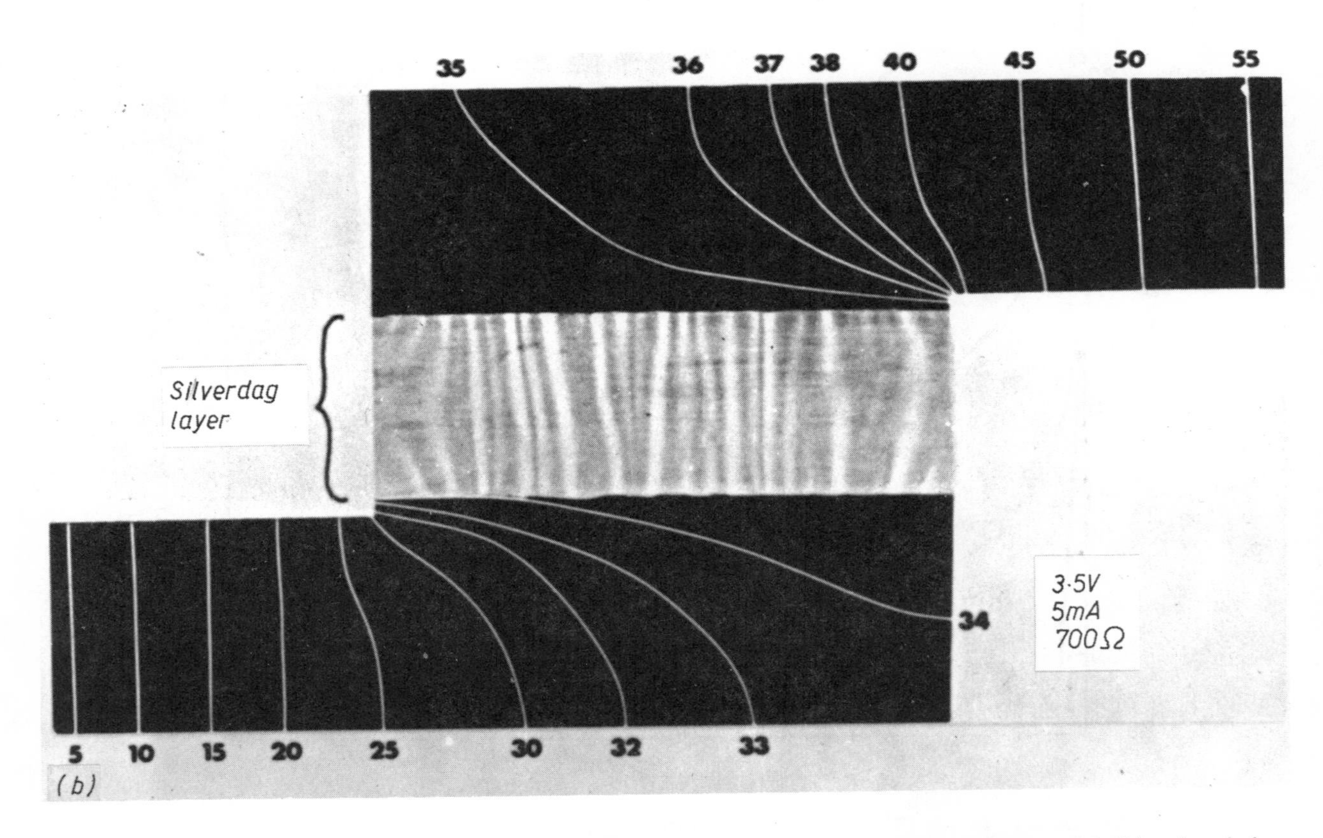

7 Distribution of equipotential lines on analogues of bimetallic welds: (a) homogeneous material, (b) as-bonded material without intermetallic layer

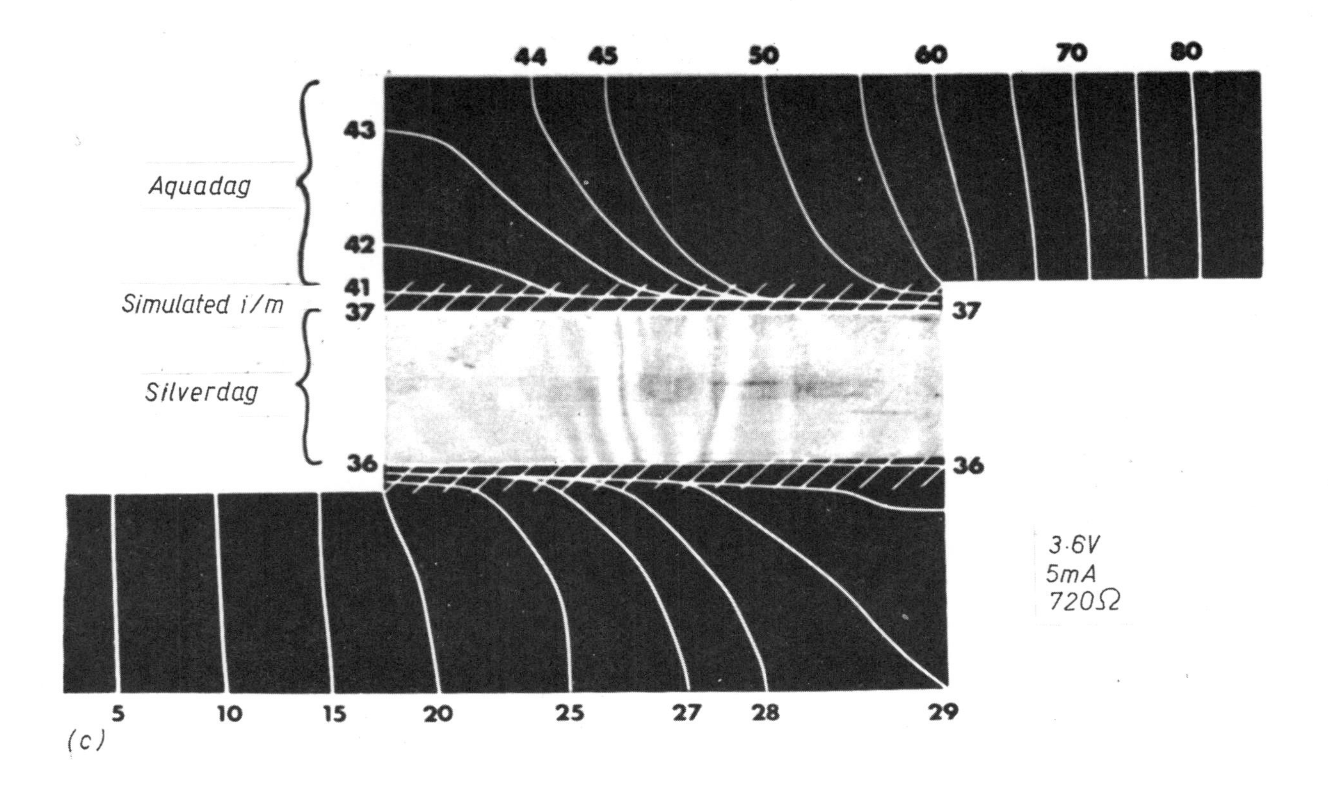

7 Distribution of equipotential lines on analogues of bimetallic welds: (c) and (b) with high resistance intermetallic layer

Friction welding dissimilar metals

T.J.Jessop, BSc(Eng), ACGI, MWeldI, E.D.Nicholas, BSc, MIM, MWeldI, and W.O. Dinsdale

Friction welding can be used to join many dissimilar metal combinations which are difficult to weld by other processes. Although the friction welding of similar metal combinations, particularly mild steel, has been given considerable attention there is little general information on dissimilar metals.

The dearth of information on the process parameters for dissimilar metal friction welds is not, however, the only area of ignorance, as data on the mechanical properties of the joints are also very sparse. This subject is further complicated by the fact that in many combinations, for example pure aluminium and stainless steel, the ultimate tensile strength of one of the materials is lower than that of the joint. This makes an accurate assessment of the bond strength difficult and reliance has had to be placed on a qualitative assessment from a hammer bend test.

This Paper outlines the development of a quantitative mechanical test and shows that an old, but probably unfamiliar, technique—the shear test—is eminently suitable to determine bond quality. The results of a parameter optimisation programme, using the shear test, for stainless steel/pure aluminium, stainless steel/copper, and stainless steel/aluminium alloy $\frac{1}{2}$ Si$\frac{1}{2}$Mg is presented and the effects of parameter variations on bond strength and metallurgical features discussed.

INTRODUCTION

The thermal energy generated when two surfaces are rubbed together under an applied pressure can be utilised to join dissimilar metal combinations in the solid phase. By careful control of the welding procedure sound, high strength bonds can be produced with metal combinations which are difficult if not impossible to weld by other processes.

Mr Jessop, Group Leader, NDT Research, Mr Nicholas, Section Leader, and Mr Dinsdale, Senior Research Engineer, Friction Welding, Process Operation and Control Department, are all at The Welding Institute.

The applications for such joints are wide and numerous, although for the more diverse dissimilar metal joints most uses can be placed in the category of electrical connector or transition piece. Dissimilar metal friction welds can be characterised as hard/hard or hard/soft combinations which do or do not form intermetallic compounds. For example, stainless steel/low alloy steel is a hard/hard combination which does not form intermetallics. Applications of this combination are found in pumps and shafts working in corrosive environments. Stainless steel/titanium is an example of a hard/hard combination which does form intermetallics and is used in nuclear power plant instrumentation rigs. Steel/copper is a hard/soft combination without intermetallics which finds applications of the transition piece type in furnace equipment. Hard/soft combinations which form intermetallics are typified by aluminium (or aluminium alloy)/ferrous materials. This type of joint is potentially the most difficult to weld since the combined features of wide dissimilarity in strength and brittle intermetallic formation can present particular problems.

Industrial interest in joints between aluminium (or aluminium alloy) and ferrous materials generally has been considerable over the last few years. Such joints have found applications in: chemical plant for the production of chlorine gas (electrical coupling) and aluminium (electrical connector), nuclear power plant for pipeline construction (transition piece), and cryogenic plant for welding pipework to vessels (transition piece).[1] Despite the increasing use of dissimilar metal friction welds research of a general nature has been rather limited and, therefore, three combinations, stainless steel/pure aluminium (E1C), stainless steel/aluminium alloy (HE9), and stainless steel/copper were chosen for detailed study, which is the subject of this presentation.

The value of simple mechanical tests for dissimilar metal combinations has always been dubious as the strength of the bond is often greater than that of the softer material being joined. Therefore it was necessary before embarking on a research programme to establish a quantitative mechanical test method to assess weld strength. A brief literature survey[2-5] indicated that tensile, torsion, bend, impact, and fatigue tests had all been used in the past, but it was clear that severe limitations existed. An initial programme of work on the currently used tests served to assess the difficulties involved so that a simple quantitative test could be developed which would eliminate or substantially reduce these difficulties. Following this, the new 'shear' test was used systematically to study the effect of varying both machine controlled variables and externally controlled factors (such as surface finish and contamination of the parts to be joined) on weld strength of the dissimilar metal combinations. This enabled optimum conditions and tolerance ranges to be determined as well as identifying any problem areas.

SHEAR TEST DEVELOPMENT

A method had to be established which would compel failure to take place only at the weld interface and, in the context of friction welding round bars, this indicated some form of shearing action. A jig was therefore designed to test 25mm diameter bars in which one end of the the specimen was clamped and a shear load applied to the other by means of a vertical sliding member. The interface could be easily set at the shear line by eye. Load was applied by situating the jig in an 'Instron' testing machine, Fig.1, so that the load/displacement characteristics during the test could be recorded.

This technique was fully evaluated using the stainless steel to pure aluminium combination. It was found that 'pure' shear was never obtained and a certain degree of bending always took place. This was minimised if the aluminium end of the specimen was clamped. A comparison of maximum shear load values with hammer bend properties showed that similar values were obtained (~45kN) over a wide variation in hammer bend qualitative properties. However, good correlation with the hammer bend properties was obtained when the shear energy to failure, i.e. area under the load/displacement curve, Fig.2a, was measured. Values ranged from 250J for a weld exhibiting poor bend properties to 580J for a weld exhibiting good bend properties. The fracture faces were characterised by a thin film of aluminium (~1mm thick) adhering to the stainless steel. The area of film present also showed good correlation with the hammer test results ranging from 40 to 100% of the total cross-section, Fig.2b.

An impact version of the shear test using drop weight equipment and appropriate instrumentation exhibited similar results to the static test.

In view of these encouraging results the static shear test was used in conjunction with the qualitative hammer bend test in a parameter optimisation programme on the three dissimilar combinations mentioned above. Also, metallurgical studies were undertaken to supplement the mechanical tests.

WELDING TRIALS

Equipment and materials

The Welding Institute's continuous drive friction welding machine rated at 110kN axial thrust and 15kW electric drive transmission was used. The machine was instrumented to monitor and record the major welding parameters, i.e. rotational speed, torque, welding force, and axial displacement. A brief description of the

Table 1 Optimum operating parameters

Material combination	Pure aluminium/ stainless steel	Copper/ stainless steel	HE9 alloy/ stainless steel
Stickout from collet, mm	20	25	25
Speed, rev/min	1300	1500	975
Friction force, kN	4.0	10	9.0
Forge force, kN	23	32	25
Burnoff, mm	3.4	9.4	4.0
Brake pressure	Zero	Zero	Zero
Film area on shear fracture face, %	100	100	80
Average shear strength, J	580	670	395

instrumentation methods used is schematically represented together with a typical instrumentation record in Fig.3. Bar to bar specimens 25mm diameter by 100mm long were welded in the following combinations:

(a) stainless steel to pure aluminium (BS 1476-E1C)
(b) stainless steel to aluminium alloy (HE9 0.5%Mg, 0.5%Si)
(c) stainless steel to copper (hard drawn, tough pitch)

The stainless steel used throughout was fully austenitic to BS 970,321S20 (En58B).

Procedure

The faces to be welded were dry turned to a very fine finish, typically 0.5-1.5μm, and care was taken to avoid contamination during loading. In all instances the softer materials, which deform severely during welding, were held in a steel collet in the rotating chuck. Before each weld the specimens were butted together statically under the friction load to settle them against their backstops and zero the burnoff control.

The major welding variables of rotational speed, burnoff (which controls heating duration), friction and forge forces, as well as arrest time (variable braking effort) were investigated for each of the combinations to determine the nominally optimum conditions, Table 1, which provided acceptable bend and shear test results.

The machine-controlled parameters were then systematically varied about these optimum settings to determine the tolerance limits within which bond strength is still acceptable. In the pure aluminium/stainless steel combination the friction and forge forces were independently changed, but were varied together for the other two combinations. The external variables of surface roughness and surface contamination were also studied. For the former this involved the use of saw cut surfaces (on either component) and for the latter either cutting fluid during turning or by contaminating the surfaces with finger grease. These variables were examined in turn at otherwise optimum conditions for the principal parameters.

Analysis of instrumentation records established the effects of machine-controlled parameters and surface condition on welding torque, burnoff rate, forge displacement, initial conditioning time, and total weld time.

Weld quality

Table 2 summarises the tolerance ranges for the machine-controlled parameters within which adquate weld quality can be maintained, together with comments on the effects of surface condition and contamination, for the three metal combination concerned. The effect of each variable is discussed separately below.

Rotational speed

The relationships between weld strength (shear energy) and rotational speed for each of the metal combinations are illustrated in Fig.4. For the copper/steel combination varying the speed from 975 to 2190rev/min has little effect on weld quality. This is perhaps to be expected since no intermetallic compound forms between these materials. Therefore the joint interface can tolerate the relatively wide variation in

Table 2 Weld strength tolerance to variations in welding conditions

Variable under study / Material being joined to stainless steel	Pure aluminium	Copper	HE9 aluminium alloy
Speed	±40%	±33%	±10%
Friction force	+40% -20%	±33%	±25%
Forge force	+75% -60%		
Burnoff	+75% -60%	±25%	±25%
Surface roughness of stainless steel (saw cut)	Intolerant (43%)*	Intolerant (40%)	Intolerant (75%)
Surface roughness of other metal (saw cut)	Intolerant (17%)	Tolerant (0%)	Intolerant (42%)
Surface contamination (oil or grease)	Intolerant (36%)	Tolerant (4%)	Intolerant (60%)

* Reduction in strength

interfacial temperature which results at the extremes of welding speed. A satisfactory operating range (between 750 and 1800rev/min) also exists for pure aluminium/stainless steel, but in the aluminium alloy/stainless steel welds speed appears to be relatively critical as joint strength can be maintained only at nominally 975rev/min, with a tolerance, say, of ± 10%. Metallurgical studies of both combinations showed that intermetallic compound formation had taken place at the medium and higher speeds (975rev/min upwards), but was absent at the low speed condition (500rev/min). There was no evidence of lack of bond at the low speed which would otherwise suggest that inadequate metal scavenging at the rubbing surfaces combined with insufficient heat generation was responsible for poor quality. The deterioration in bond strength at high welding speeds can probably be explained by the higher surface velocities resulting in increased interface temperatures. Diffusion is time- and temperature-dependent and therefore the welding conditions prevailing at high speeds promote the growth of the iron/aluminium intermetallic compound. When the weld interfaces from optimum and high speed pure aluminium/stainless steel friction welds are compared, Fig.5, it is clear that a substantially thicker layer is produced at the higher speed. This intermetallic layer has very poor mechanical properties, thus an embrittled weld of low strength results.

The difference in the optimum shear strength levels between pure aluminium and alloy welds (580 and 400J respectively) can be explained by the different effects produced as a result of the thermomechanical conditions that prevailed throughout the welding sequence. Work hardening of the pure aluminium adjacent to the weld interface occurs, thus increasing strength, but for the HE9 alloy softening takes place from the parent hardness level value of 80 to 50HV either because both overage or return to the solution-treated condition (or both effects together).

Welding force

From Table 2 it can be seen that substantial changes in both friction and forge forces when varied together or independently have very little effect on weld strength for the three combinations. In the pure aluminium only, when the forge force is lowered to a level approaching 90% below the optimum there is a significant reduction in quality. At these low forge force levels there is presumably insufficient consolidation at the weld interface.

Burnoff (heating duration)

Over the range of values studied for each of the metal combinations very little effect on weld quality was observed. With the pure aluminium/steel combination a reduction in joint strength (30%) was noted only when burnoff was set to zero. This is to be expected since insufficient time was available to generate the necessary thermal and mechanical conditions.

Arrest time

Decreasing the arrest times from a nominal value of 0.45 to 0.15sec by increasing the brake pressure from zero to 34.5 bar did not affect weld strength for the pure aluminium and copper/stainless steel combinations. However, a reduction in shear energy of 16% from the optimum value was found for the HE9/stainless welds. This strength loss may be related to the effective reduction in the rotary forging effect which manifests itself by a significant drop in the measured forge displacement.

Surface condition

The effects of surface roughness and presence of organic contamination on bond strength are presented in Fig.6. Clearly, the use of a saw cut stainless steel surface drastically reduced the weld properties for all the metal combinations under investigation. Shear energy strength reductions of 78, 45, and 40% from optimum values were rated for the alloy, pure aluminium, and copper/stainless steel joints respectively. In the copper/stainless combination metallurgical examination showed that unbonded regions were present at the weld interface, although no clear features indicative of poor strength were noted for the other combinations. It is reasonable, however, to assume that, because no deformation of the stainless surface takes place, the inherent roughness of saw cut profile modifies the scavenging and thermomechanical effects of the softer deforming metals. Consequently inadequate welding conditions prevail which restrict good interfacial adhesion. When welding smooth turned stainless steel to saw cut pure aluminium and HE9 alloy there was a less significant reduction in weld strengths, and with saw cut copper there was no loss of strength. Metallurgical examination of the aluminium welds revealed no satisfactory explanation for the lower strengths.

Introducing small amounts of contaminants, whether in the form of finger grease or machining lubricant, to the rubbing surfaces had a marked effect on weld quality, Fig.6, when welding both aluminium materials. In contrast no problems were observed with copper/stainless steel combinations. It is thought that the difference in behaviour is a result of the higher temperatures generated (approximately 1000°C for copper as opposed to 600°C for the aluminium materials) burning off the grease during the friction stage. If the soft materials of the three combinations are arranged in ascending order of their solidus temperatures, i.e. HE9 alloy (580°C), pure aluminium (660°C), tough pitch copper (1080°C), it is evident that the effect of contaminants decreases with this increasing temperature. For example, grease reduced the shear strength of HE9/stainless steel joints by 78%, the pure aluminium/stainless steel joints by 40%, and had no noticeable effect on the copper/stainless steel combinations.

Process characteristics

Table 3 shows the values of the various monitored parameters at the respective optimum condition for each combination and indicates the effect on the same process characteristic of increasing each of the machine-controlled. parameters in turn.

In the first part of Table 3 the most dominant feature is the higher values of torque for the stainless steel/copper combination and it would seem that the torque increases with an increase in hardness and conductivity of the material being welded. The high torque is also associated with higher burnoff rate.

The increases in speed, force (friction and forge), and burnoff length resulted in changes in process characteristics (see Table 3) which have been noted also for similar metal combinations.[6] The use of rougher faying surfaces resulted in a predictable increase in initial peak torque with little effect on other characteristics, and the use of surface contaminants was typified by 'lubrication' effects, i.e. lower initial torque, increased conditioning time, and a reduced burnoff rate.

There is nothing in the measured process characteristics which would be regarded as unusual and, indeed, the trends observed were to be expected from experience with similar metal combinations.

General metallurgical observations

Selected welds from each of the metal combinations were sectioned for metallurgical examination. In the pure aluminium/stainless

Table 3 Process characteristics: normal values and effect of variation in welding conditions

	Measured process characteristic	Peak torque, Nm	Equilibrium torque, Nm	Arrest torque, Nm	Conditioning time, sec	Burnoff rate, mm/sec	Forge displacement, mm	Total weld time, sec
Values measured at optimum conditions	Pure aluminium	100	40	80	1.5	1.0	6.8	5.2
	Copper	150	50	160	1.5	2.2	7.0	6.5
	HE9 aluminium alloy	130	35	40	1.0	1.1	5.0	5.5
General effect on all combinations of increasing each variable in turn	Speed	Decrease	Decrease	Decrease	Decrease	General decrease	Slight increase	Decrease
	Friction force	Increase	Increase	Increase	Decrease	Increase	Slight increase	Decrease
	Forge force	-	-	Increase	-	-	Increase	-
	Burnoff	-	Slight decrease	No change	-	No change	Slight Increase	Increase
	Surface roughness on stainless steel	Decrease	← LITTLE CHANGE →					
	Surface roughness on other metal	Slight Decrease	← LITTLE CHANGE →					
	Surface contaminant (oil or grease)	Decrease	No change	No change	Increase	Decrease	Slight increase	Increase

steel joint, tapered sections (5° angle) were necessary to magnify the weld interface features (for example, see Fig.5). Polishing such specimens required considerable care to prevent the introduction of a 'step' at the weld line which would make microscopic examination difficult. Very little information could be gained from low power magnification, apart from the obvious feature that only the softer metal in the combination deformed to form a 'flash' collar. A conventional plane section of an aluminium/stainless steel weld made at the normal condition is shown in Fig.7. This Figure gives a general indication of the weld shape and also indicates the flow profile of aluminium which affects metal up to 6mm back from the interface.

With respect to the copper/stainless steel interface, very few distinct features were noted. For almost all the welds examined over a wide range of conditions no lack of adhesion was observed at the weld line. The only indication of unsoundness was seen for the saw cut stainless steel welds. Some of the sections when etched in acidified $FeCl_3$ solution revealed a heavily worked zone characterised by a fine-grain structure adjacent to the weld line. The metallurgical evidence suggests that the copper/stainless steel welds are solid phase, with no interdiffusion of elements. This agrees with Scott[7] who confirmed his observations by electron microprobe analyses.

In marked contrast, both the pure and alloy aluminium/stainless steel combinations exhibited distinct intermetallic layers, which suggested that considerable interdiffusion of elements had taken place.

Figure 5 shows a taper section of a specimen welded at the normal condition which highlights the presence of a grey interfacial film of aluminium/iron intermetallic varying in thickness. The thickness of the film was measured at 1.0mm intervals across the interface and the result is plotted in Fig.8. This pattern is typical of all welds sectioned with the greatest thickness always occurring at the midradius, and with a central region devoid of intermetallic compound. The average film thickness at this, the normal condition, was 0.2μm. The pattern indicates that the hottest areas are at midradius despite the fact that at the periphery the relative velocity is greater and more heating would be expected. Changes in intermetallic thickness with welding conditions were generally consistent with diffusion theory in that higher heat conditions gave thicker layers, e.g. increasing speed, see Fig.5b.

A plot of average film thickness v. average shear strength is presented in Fig.9. Some correlation between the two is indicated and there appears to be an initial value of film thickness of 1.0μm, above which lower strength welds are obtained. This has also been found by other workers.[8] The only instances of failure stemming from this excessive formation of intermetallic compound was at the high speed condition, which was also characterised by a shiny fracture surface. A satisfactory bonding condition is normally associated with an intermetallic compound thickness in the range 0.2-1.0μm. If the thickness is below this range there is a tendency for weaker welds to be obtained, although the data, Fig.9, indicate that in certain circumstances (notably at high friction forces) high strength welds can be obtained with an almost complete absence of intermetallic compound.

CONCLUDING REMARKS

Existing mechanical test methods were found to be inadequate for the evaluation of 25mm diameter bar, pure aluminium/stainless steel friction welds although the hammer bend test could be used to make qualitative comparisons.

A simple static shear test was developed in which fracture is compelled to occur at the interface. A reliable indication of weld strength is obtained either by measuring the energy to failure, i.e. the area under the load displacement curves or the area of aluminium film adhering to the stainless steel fracture face.

The static shear test has been used successfully on a procedure optimisation programme for various dissimilar combinations: stainless steel to pure aluminium, aluminium alloy HE9, and copper respectively. All combinations were adequately tolerant to variations in the machine-controlled parameters in some degree. Pure aluminium was the most tolerant followed in turn by copper and aluminium alloy. However with both aluminium combinations the effect of using rough or contaminated surfaces had a drastic effect of weld quality. Copper/stainless steel weld quality was also reduced with rough surfaces but the introduction of surface contaminants could be tolerated.

On sectioning at the interface of both aluminium combinations Fe/Al intermetallic layers were observed and the thickness of the layer increased with higher heat input to the joint, i.e. faster speeds and higher burnoff rates. Metallurgical examination of the interface was hampered by preferential grinding of the softer material when polishing.

Some correlation between intermetallic thickness and weld strength for the pure aluminium/stainless steel welds was noted. High strength welds normally exhibited an intermetallic film in the 0.2 to 1.0μm thickness range.

ACKNOWLEDGEMENT

The authors would like to thank Dr K. Johnson for his cooperation and contribution made in discussion.

REFERENCES

1 BAKER, B.H. and LUCEY, J.A. 'Fabrication of cryogenic pressure vessels'. Brit.Welding J., 15 (5), 1968, 205-13.

2 ELLIS, C.R.G. and NICHOLAS, E.D. 'Determination of a procedure for making friction welds between electrical grade aluminium and tough pitch copper'. Welding Inst. Report P62/73, September 1973. Also in Welding Research Int'l, 5 (1), 1975, 1-32.

3 BROWN, E. and BROWN, I.J. 'The joining of copper to aluminium by friction welding'. UKAEA Research Group Report, 1969.

4 TRUTNEV, V.V. et al. 'Friction welding of steel 35 to aluminium alloy VAD-1'. Welding Production, 20 (3), 1973, 34-6.

5 SUGIYAMA, Y. et al. 'Friction welding

aluminium to copper'. Sumitomo Light Metal Technical Report, 9 (3), 1968, 158-68.

6 ELLIS, C.R.G. 'Continuous drive friction welding of mild steel'. Welding J., 51 (4), 1972, 183s-97s.

7 SCOTT, M.H. Private communication.

8 McEWAN, K.J.B. and MILNER, D.R. 'Pressure welding of dissimilar metals'. Brit.Welding J., 9 (7), 1962, 406-20.

1 Shear test setup. 1 — sliding crosshead (acting in compression); 2 — sliding plunger; 3 — specimen; 4 — load cell

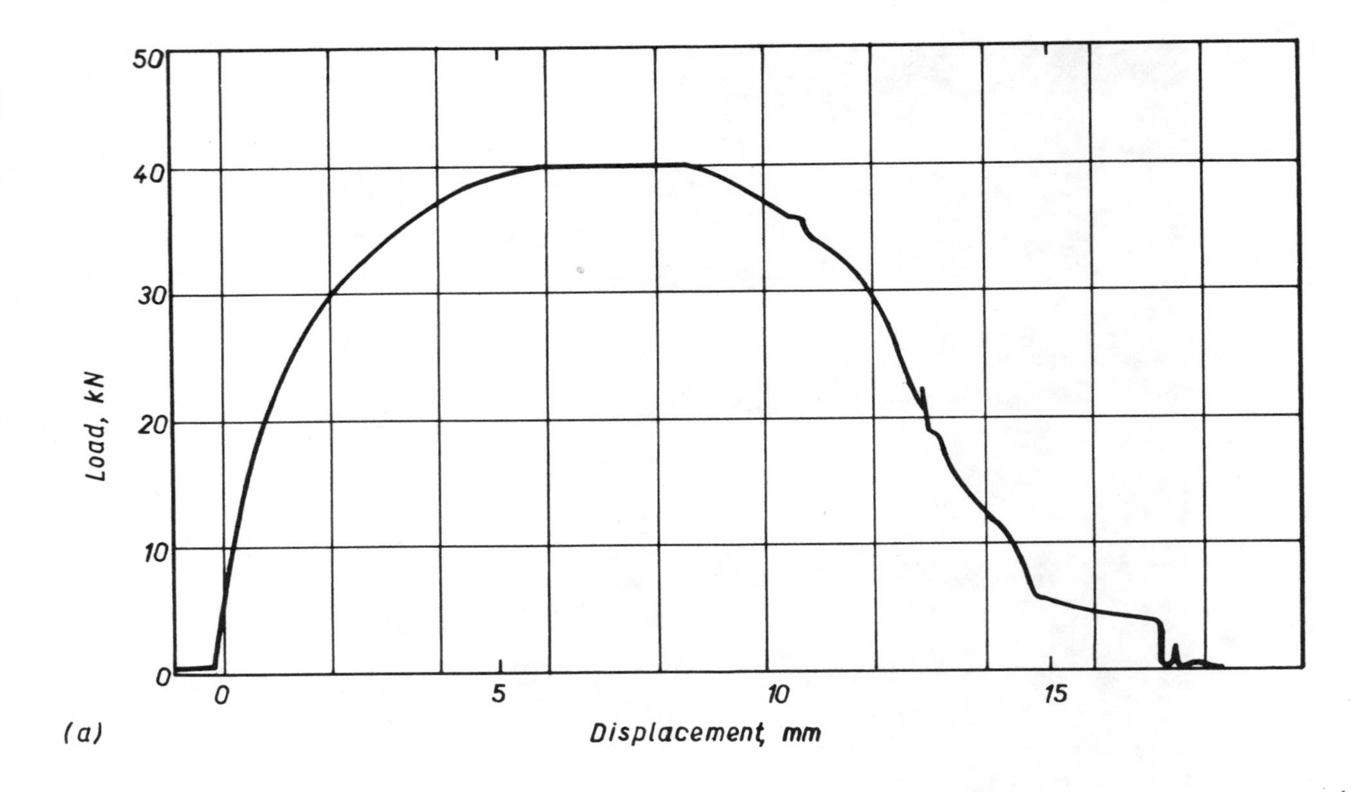

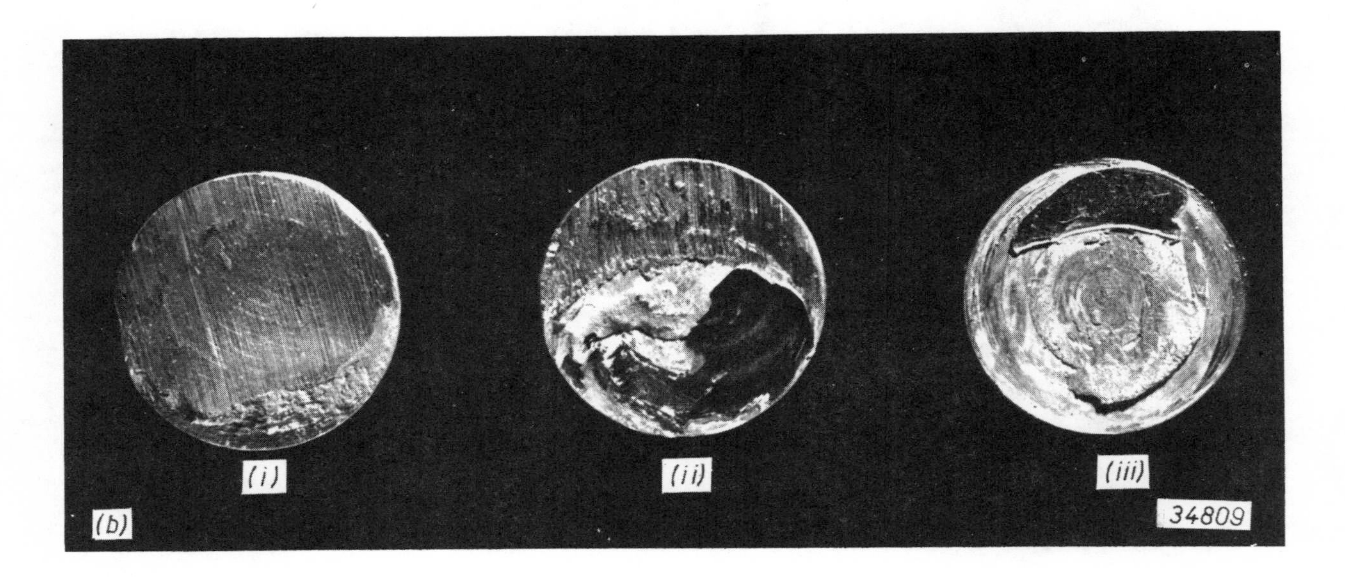

2 Shear test characteristics for aluminium/stainless steel: (a) typical load/displacement record for shear test, (b) fracture faces showing amount of aluminium adhering to stainless steel:

(i) 100% film area — strong
(ii) approximately 60% film area ↓
(iii) approximately 30% film area — weak

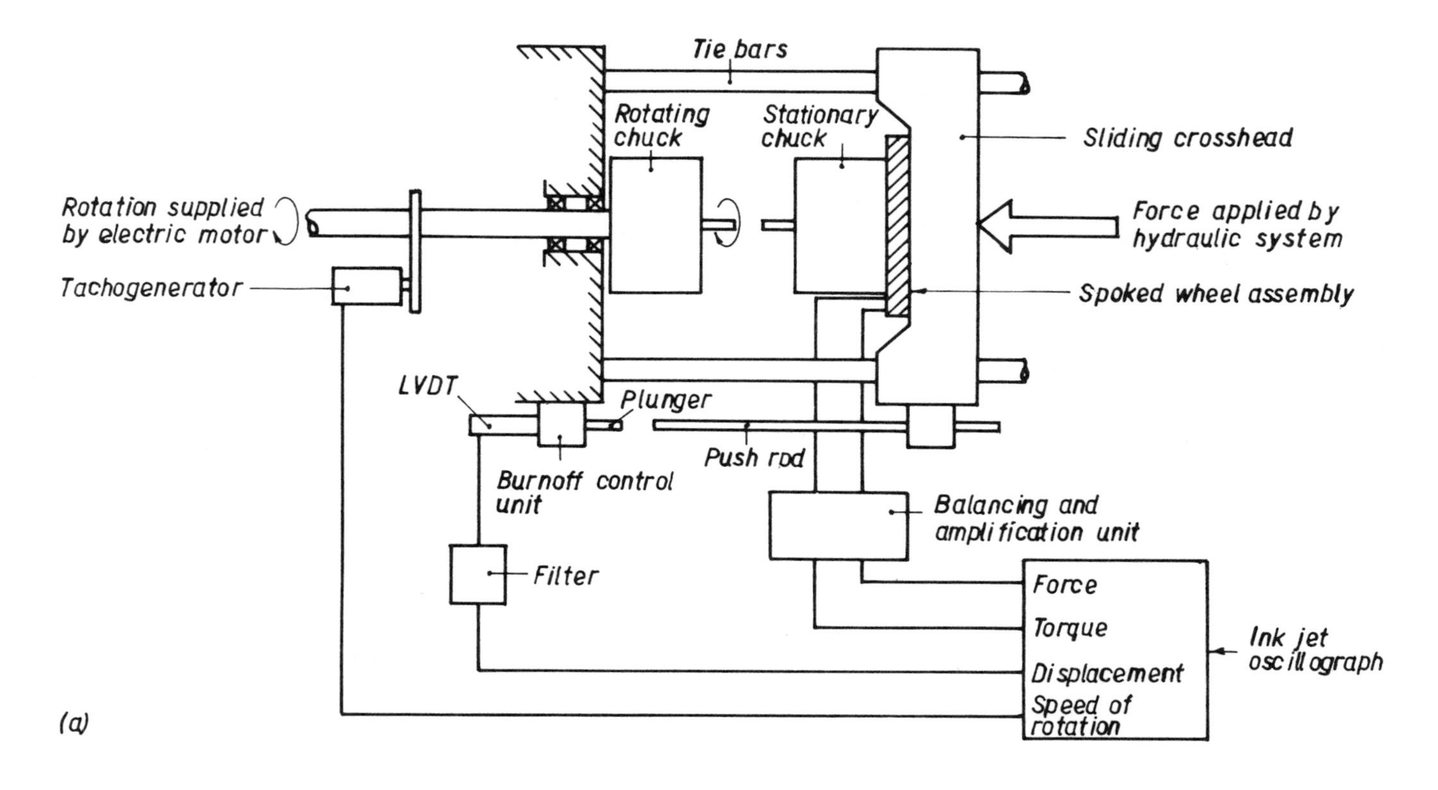

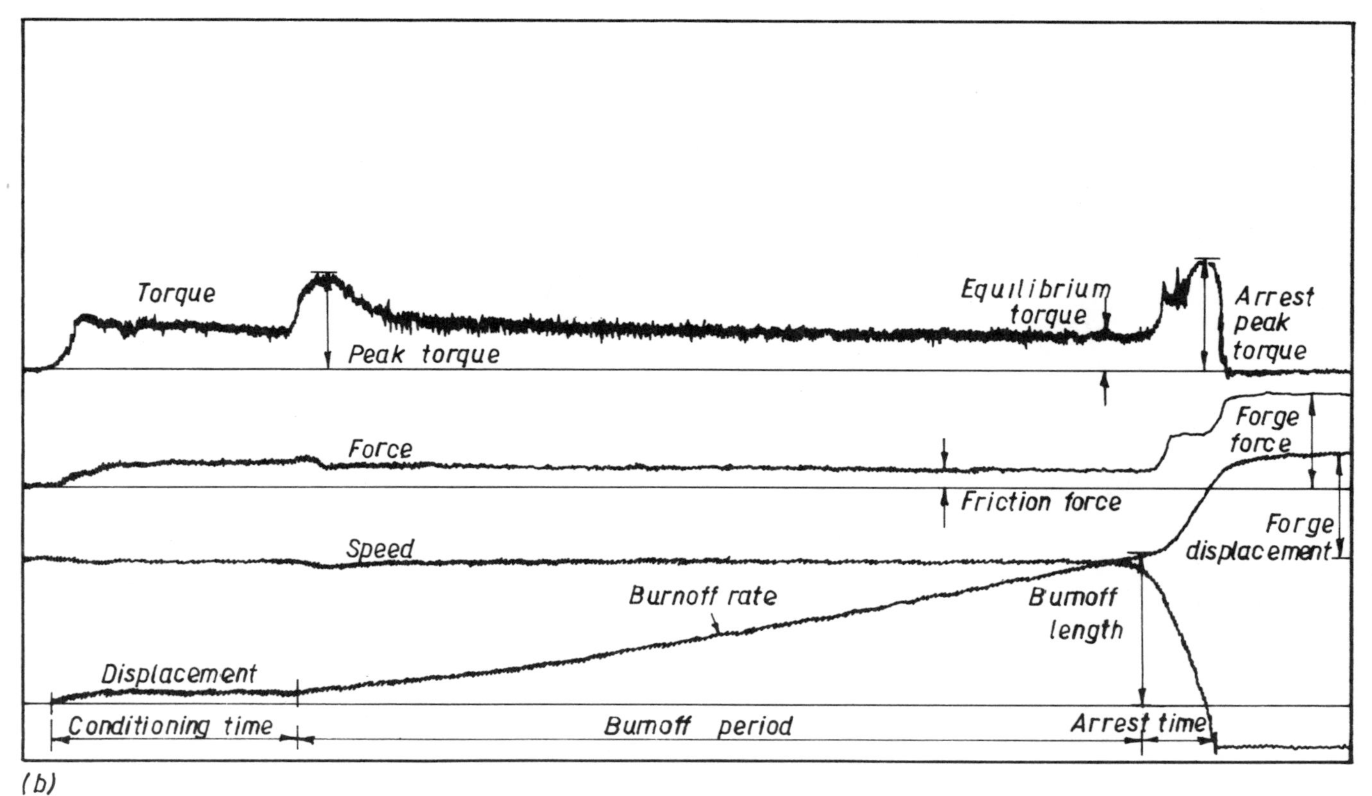

3 Instrumentation techniques and definition of parameters measured: (a) schematic arrangement for instrumentation, (b) typical oscillograph record (copper/stainless steel)

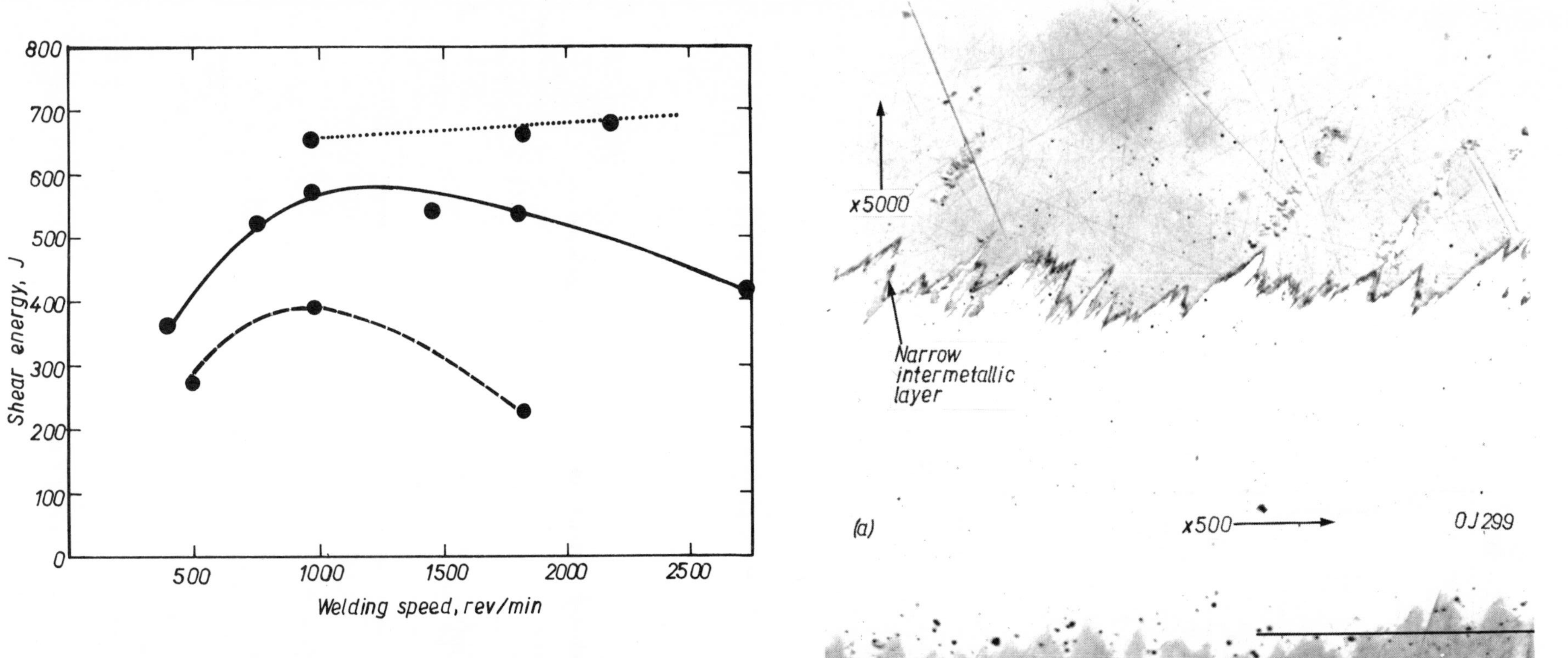

4 Effect of welding speed on weld shear strength for all combinations. .... copper/stainless steel; —— pure aluminium/stainless steel; - - - - aluminium alloy/stainless steel

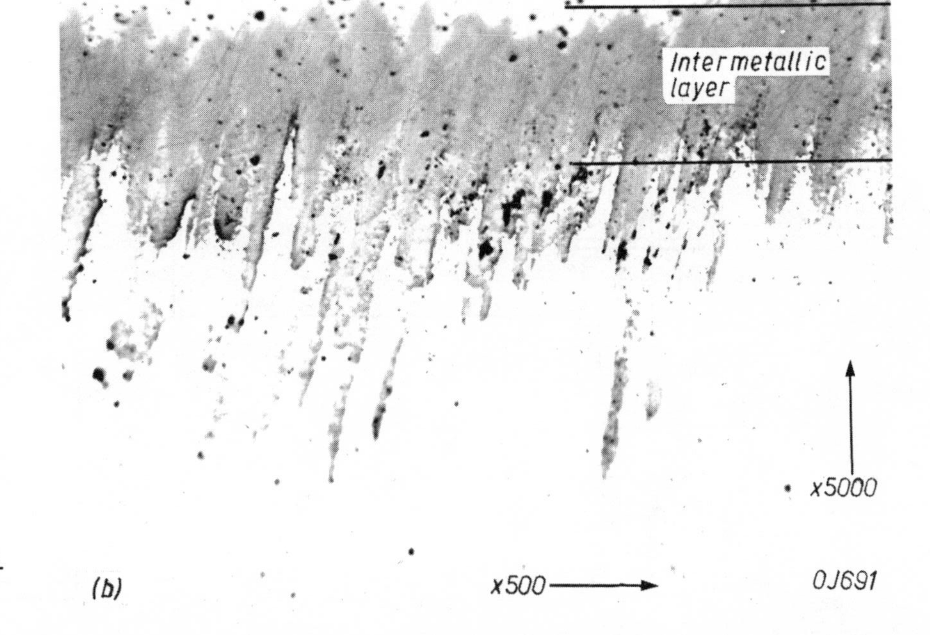

5 Taper microsection of pure aluminium/stainless steel welds: (a) good condition (975rev/min), (b) high rotational speed (2400rev/min)

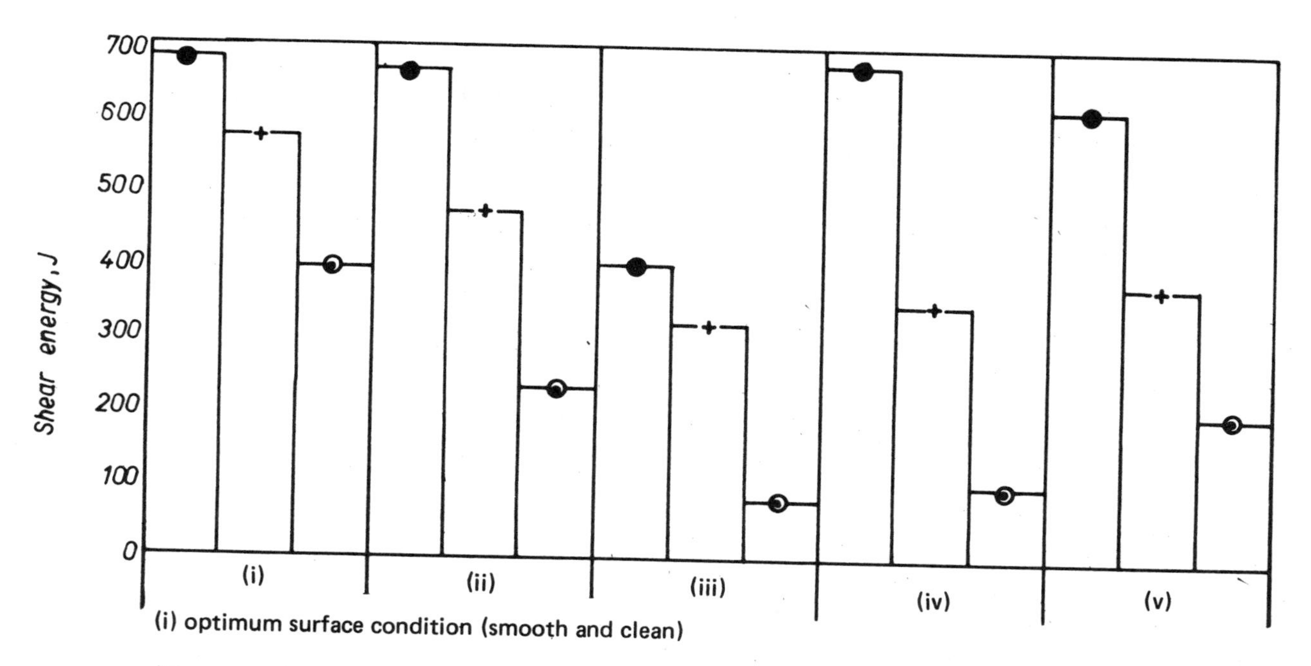

(i) optimum surface condition (smooth and clean)

(ii) saw cut Cu, Al, and HE9 alloy

(iii) saw cut stainless steel

(iv) finger grease

(v) machining lubricant

● – copper/stainless steel

+ – pure aluminium/stainless steel

ʘ – aluminium alloy/stainless steel

6 Effect of surface condition on weld shear strength for all combinations

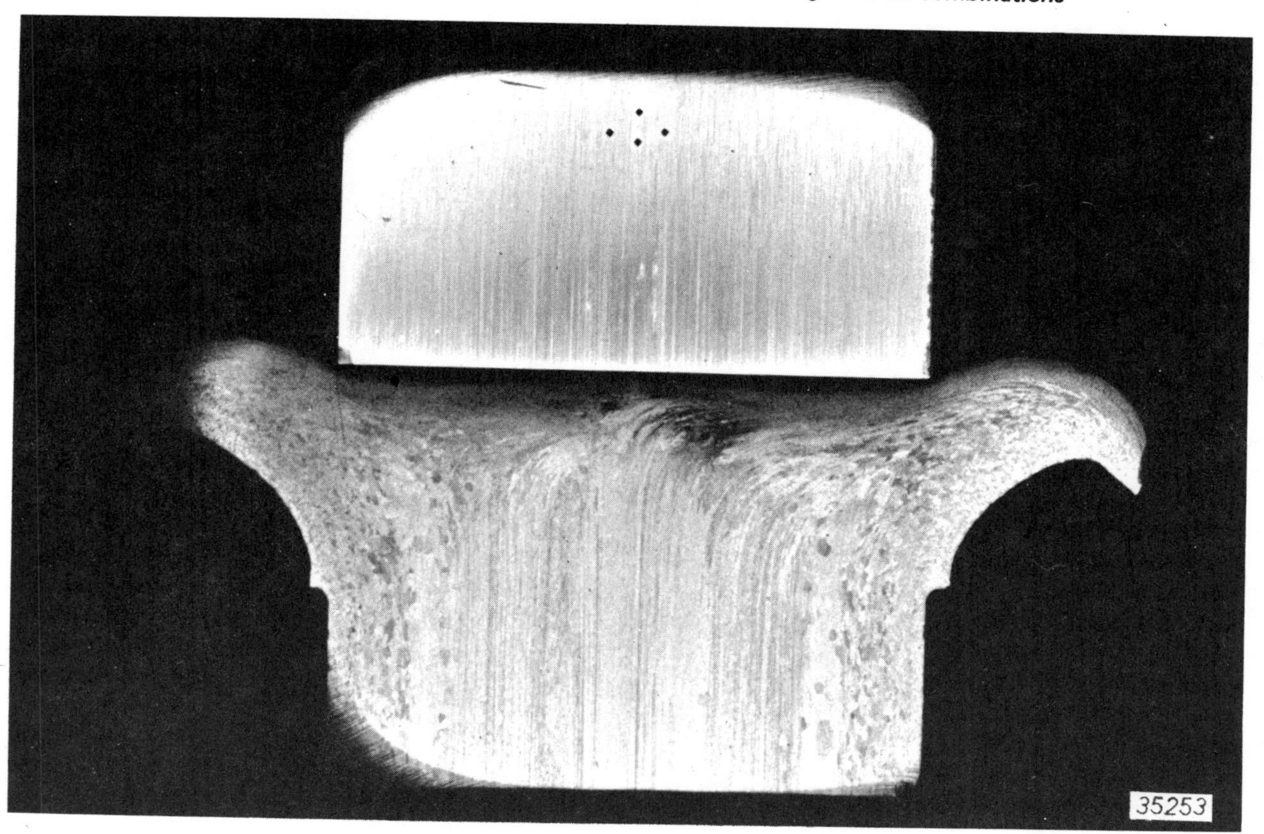

7 General configuration of pure aluminium/stainless steel friction weld (section etched in sodium hydroxide solution followed by acidified ferric chloride) x3

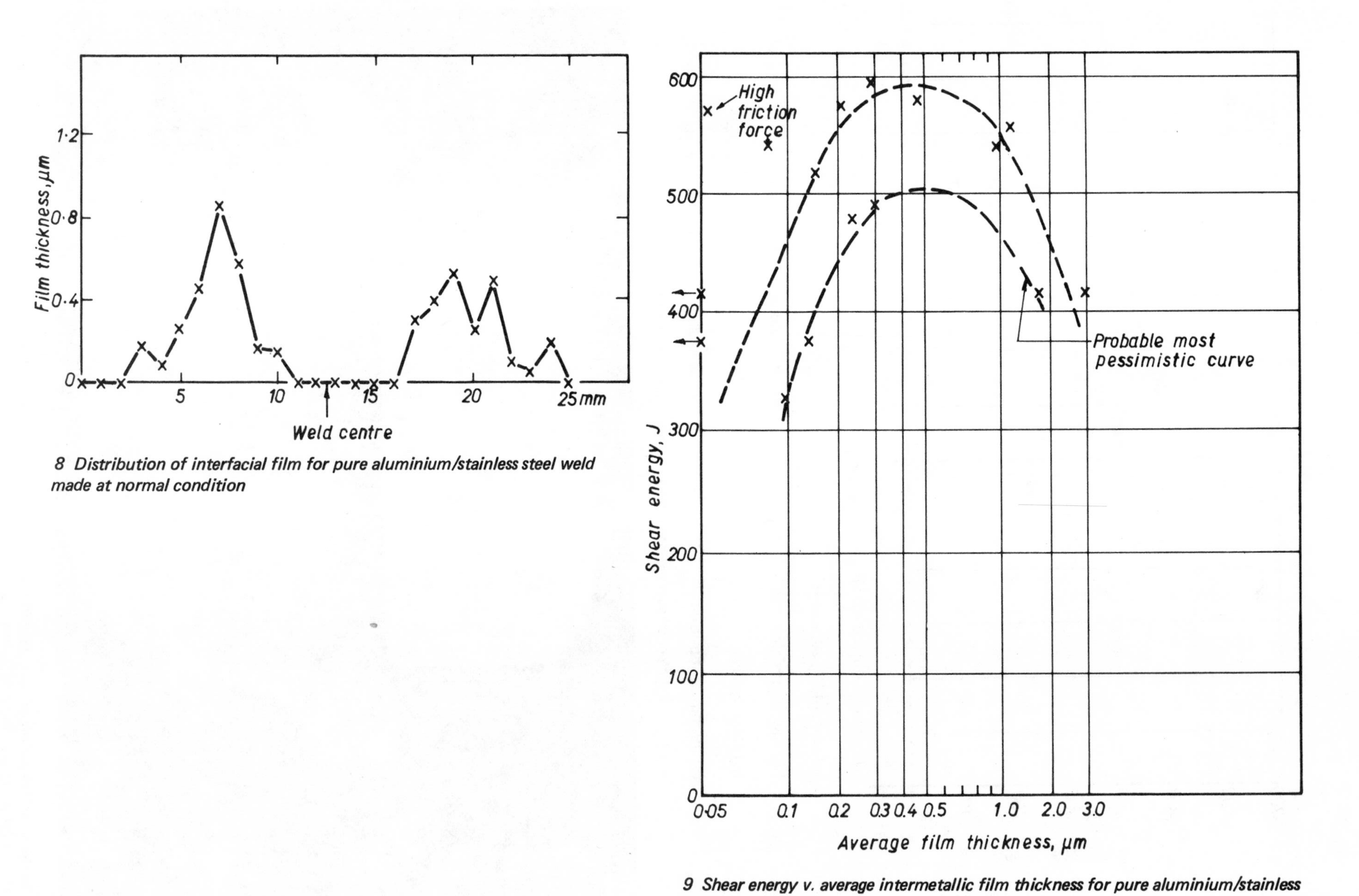

8 Distribution of interfacial film for pure aluminium/stainless steel weld made at normal condition

9 Shear energy v. average intermetallic film thickness for pure aluminium/stainless steel combination (range of welding conditions)

Radial friction welding

E.D.Nicholas, BSc, MIM, MWeldI, and R.H.Lilly

INTRODUCTION

Welding by friction is now a well-established process for joining a wide variety of metal combinations and component geometries. The method has found application in fields as divergent as the motor and nuclear industries.

The thermomechanical conditions required for a conventional friction weld are derived from relative motion between two parts aligned on a common axis which are butted together while an axial thrust is applied. At least one part has to be rotated, and axial movement during the heating and forging stages takes place as material is extruded out from the rubbing surfaces to form a collar. These features apply both to Continuous Drive and Stored Energy (Inertia) friction welding systems. However, the process has not been so successfully exploited for joining long hollow sections where restriction in the bore cannot be tolerated. Unfortunately ovality, and lack of concentricity and straightness, of parts further aggravate the situation. Also there are engineering aspects, such as transmission and load reaction, which make the rotation of long hollow sections problematic.

These difficulties have been recognised for some time and to some extent can be overcome by using a rotating insert between long stationary parts. However the difficulties in rotating an intermediate length of hollow section, and applying uniform and consistent welding forces to both weld interfaces, present extra problems for the machine designer. Furthermore, because two internal flash metal collars are formed, additional machining of the bore will be necessary to avoid restriction.

A new approach termed Radial Friction Welding* was put forward in 1975 by The Welding Institute which provided a means to overcome these limitations. In this arrangement a relatively simple solid ring is rotated between suitably prepared pipe ends while it is subjected to radial compressive loading to generate the welding force and friction heat. The development of this new welding approach is briefly reviewed. Two methods of ring compression and rotation were examined and the effect of ring and pipe geometries investigated to determine the preferred approach for production requirements. Finally potential applications for this technique are indicated.

WELDING METHOD

The simplest arrangement for radial friction welding is schematically presented in Fig.1. Two pipes with bevelled ends are butted together and clamped securely to ensure they cannot rotate or move axially apart. A mandrel is located in the bore, where the weld is to be made, to support the pipe walls and prevent penetration of flash metal formed during the weld sequence. A solid ring of suitably compatible material is positioned within the apex provided by the machined pipe ends. The ring is more sharply bevelled to promote metal flow from the base of the weld preparation as well as to reduce the initial high torque (power) demand normally associated with the beginning of the friction cycle when cold surfaces come into contact.

To generate the thermomechanical conditions necessary for friction welding the ring is rotated while subjected to uniform radial compression. To terminate the weld sequence, ring rotation is arrested while the radial compressive load is maintained or increased to consolidate the joint. This method can also be applied with relative ease to the attachment of collars to solid shafts. Alternatively, by changing the mode of ring deformation to that developed by expansion, rings can be inserted into hollow cylindrical bodies, or indeed the method can be used for pipe welding where access to the outside is restricted.

For this new friction welding concept to be feasible three main requirements had to be satisfied. The first and probably the most

Mr Nicholas, Section Leader, and Mr Lilly, Senior Welding Engineer, are both in the Friction Welding Section of the Process Operation and Control Department of The Welding Institute.

* Patent applied for.

important was to devise a method for applying uniform radial compression to the welding ring without causing major distortion. The second important feature was to develop a welding arrangement which would rotate the ring while it was subjected to radial compression. Finally for pipes it was also necessary to find a suitable material which could be used for the internal mandrel support, that is, good strength at elevated temperature combined with relatively poor thermal conductivity to reduce rapid cooling of the weld region.

EXPLORATORY EQUIPMENT

Initially, methods were devised both to radially compress and impart drive to the solid ring which could be interfaced with a conventional friction welding machine. One of the simplest but effective means of inducing uniform radial compression was to force a premachined tapered solid ring, Fig.2a, into a reduction housing with a matching taper. To assist transmission of rotational drive, grooves were machined in the ring which would engage on splines added to the reduction taper housing, Fig.2b. The unit designed for 49mm OD x 29mm ID rings relied upon axial movement through the taper housing for compression. Experiments indicated that an axial load of 100kN was needed to induce 5.5mm reduction in the bore diameter without distortion. Although this unit operated successfully certain shortcomings were recognised:

(a) requirement for axial movement of the ring through the taper housing
(b) machining drive grooves in the ring, and
(c) complex ring geometry with large drive segments

Consequently circumferentially located individual jaws (similar to multijaw collet chucks) was considered. The welding arrangement illustrated in Fig.3b was constructed to evaluate the practicality of compressing and driving the ring without relying on splines.

Basically the drive unit comprised an arrangement of jaws (twelve) capable of rotation about a centre mandrel on to which the pipes were fitted. The mandrel also provided bore support using a sleeve of heat-resistant metal. The jaw tips were provided with contact splines which indent on to the plain external surface of the ring to ensure positive drive. To activate the chuck the latter fitted a taper housing with splines (similar to that shown in Fig.2b) to impart both rotary drive and radial movement to the jaws. It was found that adequate drive and radial compression could be imparted to the simple ring, Fig.2a.

WELDING TRIALS

Procedure

Both of the above units were evaluated using a friction welding machine rated at 110kN axial thrust, 15kW electric drive transmission. The reduction housing was fitted taper outwards to the machine's rotating spindle, and the mating bodies were attached to the nonrotating but axially moving carriage.

Welding trials to assess feasibility and determine acceptable bore support materials, and to establish conditions giving full penetration, were carried out using mild steel pipes with specially machined rings. In the sliding ring, Fig.2, a 49mm OD by 29mm ID ring was used for 33mm OD by 25mm ID pipes, but for the multijaw arrangement the pipe and ring dimensions were 38mm OD x 25mm ID and 50mm OD x 27mm ID respectively. The pipes were prepared with an included angle of 100°-110°, with 0.4-1.5mm root face, and were clamped firmly prior to welding. A flow control valve was incorporated in the machine's hydraulic circuit to maintain an essentially linear displacement rate in the taper housing and the frictional heating duration was controlled by a preset axial displacement. Allowance was made both for displacement resulting from initial cold collapse of the ring and for that during welding. To avoid possible damage to the welding units and backing sleeve no increased force was applied at the arrest or forge stage.

Sliding ring arrangement

Initial trials were undertaken at 2190rev/min to establish the operating characteristics of the arrangement involving direct movement of the ring through the taper housing, as in Fig.2.

This demonstrated that the overall performance of the equipment was adequate for radial friction welding and that it was free from heavy radial vibration. However the stainless steel backing sleeve did not provide sufficient internal support and both the pipe bore and sleeve suffered permanent deformation. Consequently other materials were tried and a nickel-based turbine casting alloy, commercially termed PK24, was found to provide the ideal combination of strength and low conductivity properties.

From these initial trials the following ring and pipe preparations and burnoff settings were standardised in an investigation of the effects of welding speed, axial displacement rate, and braking effort:

Welding ring bore	28mm
Pipe included angle	100°
Root face height	0.4mm
Axial burnoff length	9mm

A typical instrumentation record for a radial friction weld is illustrated in Fig.4. The applied axial force increases to a maximum value as the cold ring is radially compressed to contact the pipe ends. It then decreases as the friction energy raises the ring temperature. The maximum force levels noted varied from 34-62kN, reflecting differences in the heat generation characteristics from one weld to another. The maximum level of 62kN is substantially lower than the 100kN required to compress the ring when cold. At the moment of contact the welding torque rises to a first peak before dropping to a steady level which is maintained until the end of the heating sequence when it rises again to a high level during ring deceleration.

Axial movement of the ring down the taper is fairly linear during cold ring reduction and welding. The axial displacement rate was varied from 0.6-1.5mm/sec, equating to radial burnoff rates of 0.1-0.25mm/sec, which are lower than those normally expected for conventional friction welding of mild steel. However, it must be stressed that considerable caution was being exercised at this stage to obviate damage to the welding equipment. Higher burnoff rates could be employed on more robust equipment.

A typical weld, when internally pressure tested in the as-welded condition, did not fail up to the maximum pressure available from the test ring. Furthermore reducing the wall thickness from the outside to 2mm caused failure in the parent metal, well away from the weld region at a pressure of 550 bar, Fig.5. No significant change in penetration leading to 100° of the pipe wall was observed at the three welding speeds used, i.e. 2190, 1460, or 975rev/min, or with changes in deceleration characteristics caused by varying the braking effort. However, separating the pipe ends by approximately 0.8mm prior to welding caused a significant change leading to complete penetration of the tube wall, but lack of bond was then found in the peripheral regions. This was attributed to insufficient radial reduction of the ring and restriction of metal flow. These problems probably could have been overcome by simple modifications to the welding apparatus.

Although no great emphasis was placed upon acceptable weld quality at this stage some welds were subjected to bend tests. Longitudinal segments, 6mm wide, were bent around a 2t former (t = wall thickness of pipe) to establish bend ductility of the peripheral and bore regions. Although bend angles of 5°-45° were noted before failure occurred, one of the weld interfaces exhibited 'blued' oxidised regions which suggest that the metal flow characteristics needed to be improved.

Multijaw arrangement

Similar success was achieved when welding with the multijaw assembly, Fig.3. The distortion of the ring was found to be uniform, and the provision of small splines in the tip jaws resulted in adequate indentation into the ring to maintain transmission of power without slip. The final external appearance of such a ring, Fig.6a, is shown together with a section of the weld, Fig.6b. The indents are clearly visible on the external surface of the ring, as well as the flash collars formed by metal being displaced from the rubbing surfaces. Closer examination of the weld zone revealed that sound defect-free weld interfaces were achieved above the root face where a partial hot pressure bond was formed which contained small islands of retained oxide.

Machine development

Following the encouraging results obtained from this relatively short evaluation programme, a grant was obtained from the Mechanical Engineering and Machine Tool Requirements Board to develop a research prototype radial friction welding machine and evaluate its capabilities. A machine to weld 50mm OD x 6mm wall thickness pipes has been constructed, Fig.7, and is being used to determine the effect of ring and pipe geometries, and welding parameters, on the static and dynamic mechanical properties of the resultant welds. The weld section shown in Fig.8a illustrates the capability of this machine.

In parallel with this development an experimental rig, similar to the aforementioned multijaw unit, was built to weld nominally 110mm OD x 10mm wall thickness pipe. The performance of this rig when interfaced with a 1000kN axial thrust, 270kW power, machine was most encouraging. After a few procedural trials to prove satisfactory operation, it was just a short step to produce pipe welds, Fig.8b, which were characterised by satisfactory metallurgical features consistent with high strength. The section also shows that no step has taken place into the bore. The bore was in fact supported by a special segmented plug which presented a virtually solid surface with only fine radial gaps. Preliminary bend tests indicated that greater than 90° bend angles could be obtained without failure whether the coupons were bent towards or away from the bore. Tensile tests with machined specimens revealed that failure occurred in the parent metal well away from the joint. Fatigue tests were also undertaken with these larger diameter joints with the collar both left attached and machined off. The results showed substantially increased fatigue life when compared with conventional pipe friction and arc welds. Also Charpy V notch tests gave satisfactory results with energy absorbed values of 71J (average of three tests) at $-10^{o}C$.

POTENTIAL APPLICATION AREAS

Pipelines

Off shore

Numerous economic and technological advantages can be realised if a 'one-shot' welding process is viable for offshore pipe laying. As a result of the relatively short welding sequence (under 15sec for 110mm OD pipe) higher production rates could be achieved, thus making greatest use of the 'weather window' in regions where adverse weather conditions prevail. It is significant that the process can be operated in almost any plane from horizontal to vertical thus permitting pipe laying in varying water depths. Perhaps one of the most important features of using such a process is that, once welding conditions have been optimised consistent with high integrity welds, reproducible weld quality can be expected without relying upon the skill of an operator. In-process monitoring of the major welding parameters together with ultrasonic inspection can be applied to ensure quality assurance.

On shore

Similar benefits to the aforementioned can also be expected when operating radial friction welding on shore, although the terrain where pipe laying is required may produce severe limitations. Problems can also occur where bends and protruding attachments etc. make it impossible to slide a welding head along to the next weld location. However, a significant modification — that of 'split' ting' the ring drive/compression unit — would certainly overcome these latter problems and thus extend the scope of application in this area.

Double jointing

To reduce the number of welds that have to be made at site, particularly with respect to pipe laying off shore, two pipe lengths are welded in a workshop environment. This application provides an ideal situation since pipes can be easily presented to the welding machine at the start of a production line. No problems are envisaged in removing the backing mandrel necessary to prevent bore restriction or in machining off the metal collars and excess ring material if required.

Collars and sleeves to shafts

When there is a requirement for an increased diameter to be provided on a shaft, it is usually produced by machining away a large part of the bar or forging from which the shaft is to be made. This quite obviously results in increased costs incurred in machining time and metal loss. Therefore using the compression approach it is now possible to weld on collars, Fig.9a, or sleeves requiring only a final machining operation to provide the correct geometry. This procedure affords additional choice of material which will yield optimum properties at the location when the shaft is in operation.

Inserts

The expansion technique used to provide ring deformation offers the possibility of inserting rings, Fig.9b, or sleeves of like or dissimilar materials into hollow cylindrical bodies which may improve either wear resistance, thermal stability, or stiffness. A good example is the positioning of a thermally stable metal in the combustion bowl lip zone of engine piston crowns. Elimination of thermal erosion by this approach should increase engine efficiency and reduce the production of toxic waste.

Repair

Considerable attention is being devoted to finding a satisfactory procedure for repairing damaged pipelines, particularly in hostile environments such as those found at the ocean depths. Many welding methods are at present under review to solve the problem. Explosive welding and arc welding in hyperbaric chambers to name but two are actively being pursued. With the advance of radial friction welding it is now reasonable to consider using this process. Many difficulties are recognised, none more so than those associated with supporting the internal bore to prevent pipe deformation and flash penetration. What is needed to resolve this particular problem is the development of a segmented plug and or plug of specially prepared material which will break down into small component parts on completion of welding, thus allowing easy removal using conventional methods. For this application the ideal rotation compression unit would require to be split, so that it can be recovered from the pipeline. However, it is not unreasonable to consider the welding head as a throw-away item in view of the very high repair costs.

Noncompatible materials

Experience with the friction welding process has shown that a far wider range of dissimilar material combinations can be successfully joined than by other available welding processes. However, there are certain combinations such as zirconium alloys/stainless steel and aluminium alloys/steel which cannot be satisfactorily welded. When environmental circumstances allow, a compatible intermediate metal such as pure aluminium has been used to effect a sound joint of adequate mechanical properties. However by adopting this approach two friction welds have to be made which generally involve considerable machining. Now the compatible material can be radially friction welded to the otherwise incompatible metal combination thus yielding economic benefits.

CONCLUSIONS

The work carried out to date confirms that radial friction welding can provide high strength solid phase welds particularly when joining hollow steel sections. The engineering requirements to drive and compress a steel ring have been demonstrated on a prototype machine. It is considered that this machine can be translated without undue difficulty into a site welding system. By using special bore support mandrels internal flash can be eliminated.

ACKNOWLEDGEMENTS

The authors would like to express their gratitude to Welding Institute personnel, in particular D. Wilson and Dr K. Johnson who have cooperated on this project and also to the Mechanical Engineering Machine Tool Requirements Board for financial support for the current development programme.

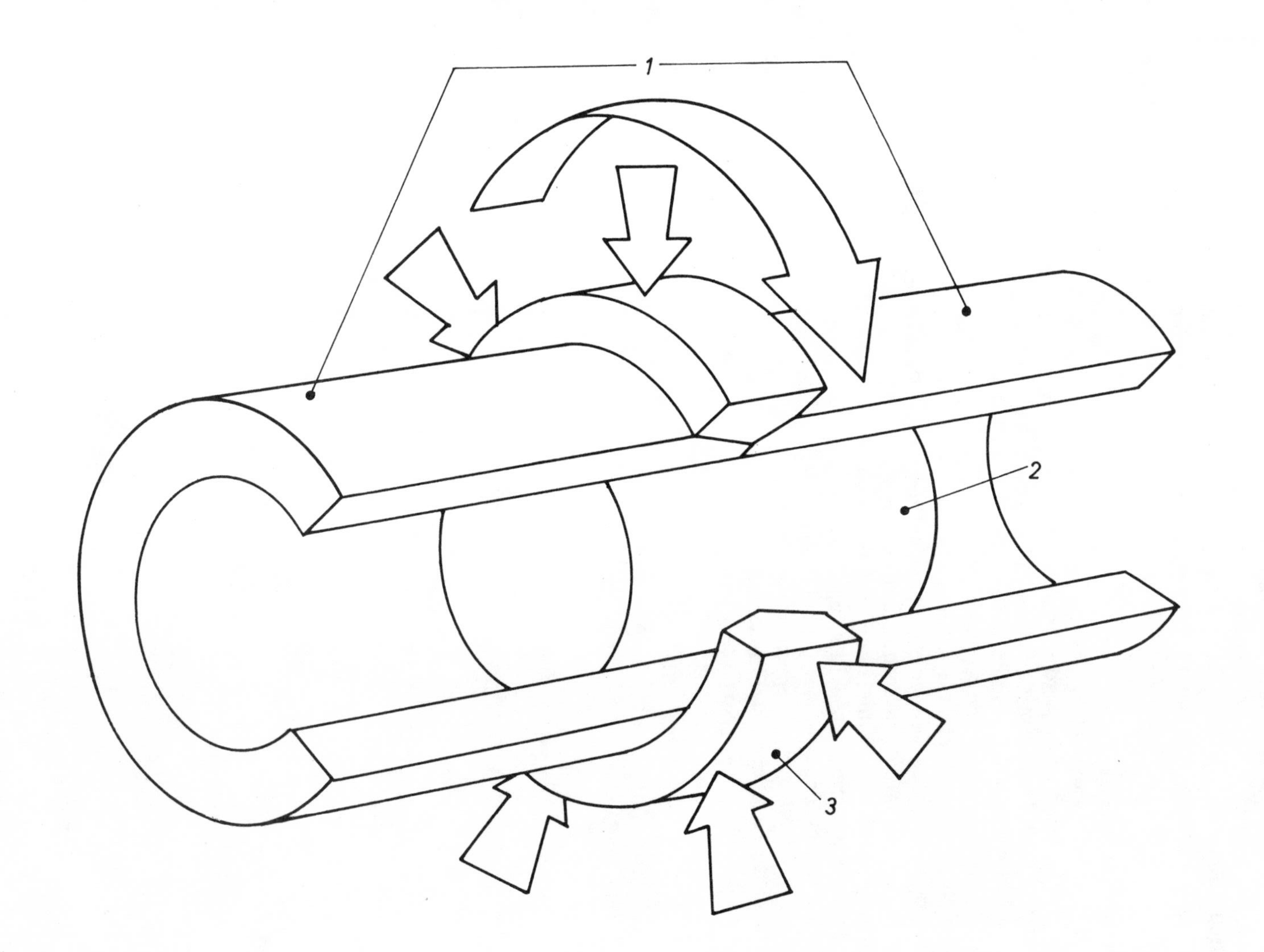

1 Schematic arrangement for radial friction welding. 1 – stationary clamped pipes; 2 – expanding plug mandrel; 3 – integral ring rotated and compressed radially

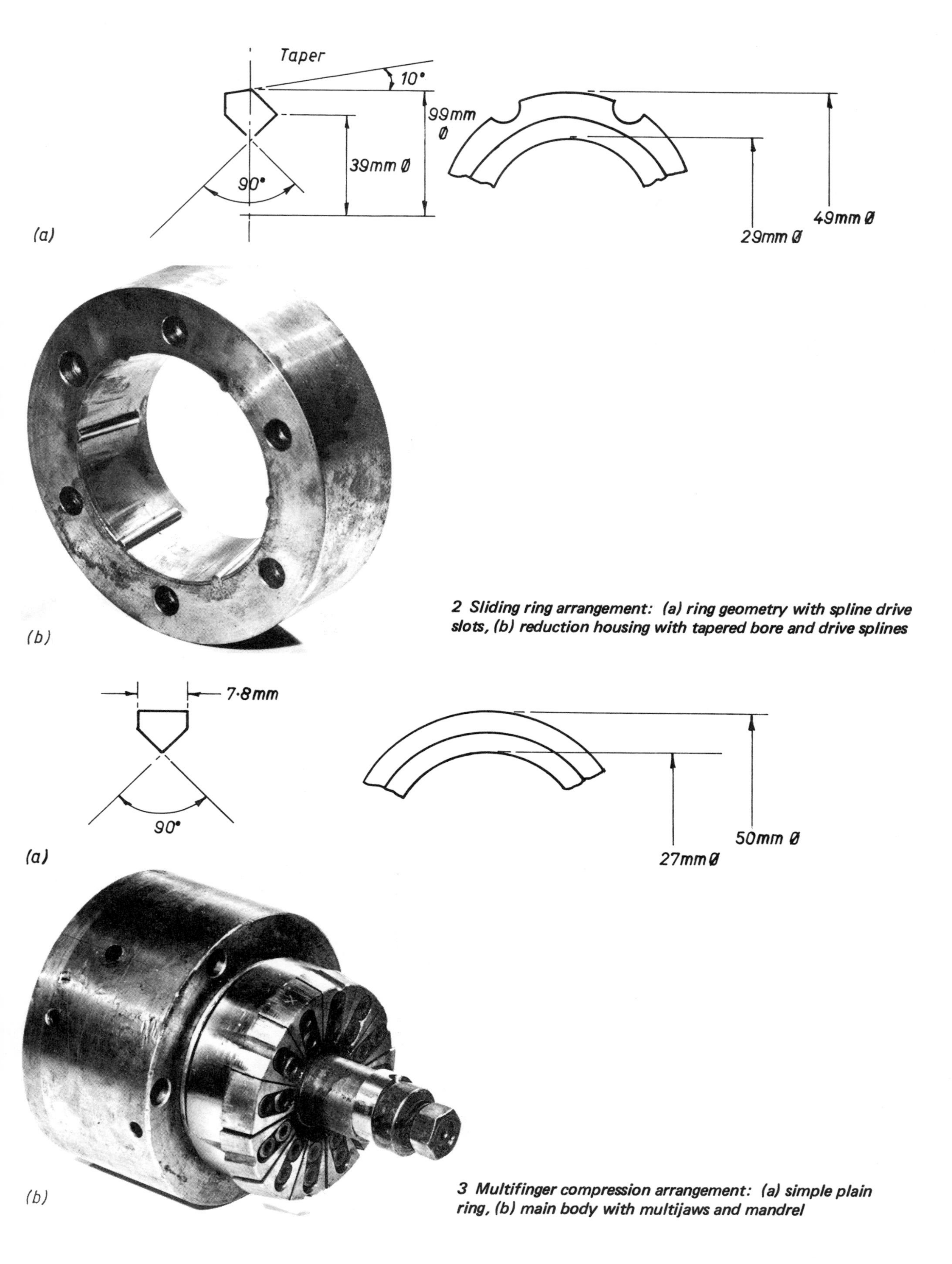

2 Sliding ring arrangement: (a) ring geometry with spline drive slots, (b) reduction housing with tapered bore and drive splines

3 Multifinger compression arrangement: (a) simple plain ring, (b) main body with multijaws and mandrel

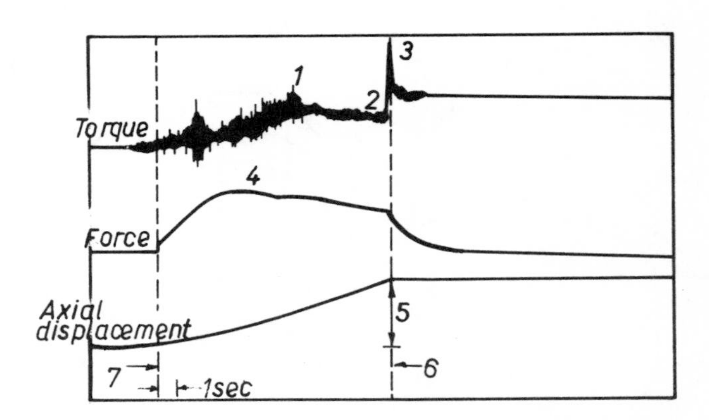

4 Typical instrumentation record for radial friction weld. 1 — initial peak, 110Nm; 2 — equilibrium, 80Nm; 3 — final peak, 280Nm; 4 — maximum, 37kN; 5 — burnoff setting, 9mm; 6 — end of weld; 7 — weld start

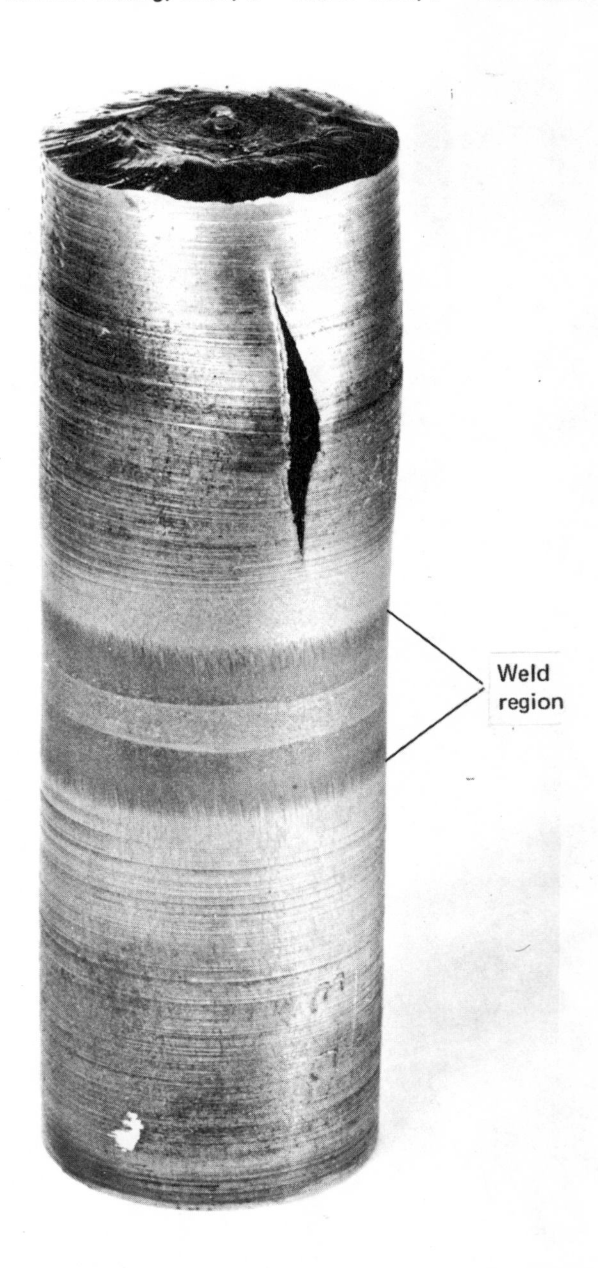

5 Burst pressure tested radial friction weld; note failure in parent pipe away from weld region

(a)

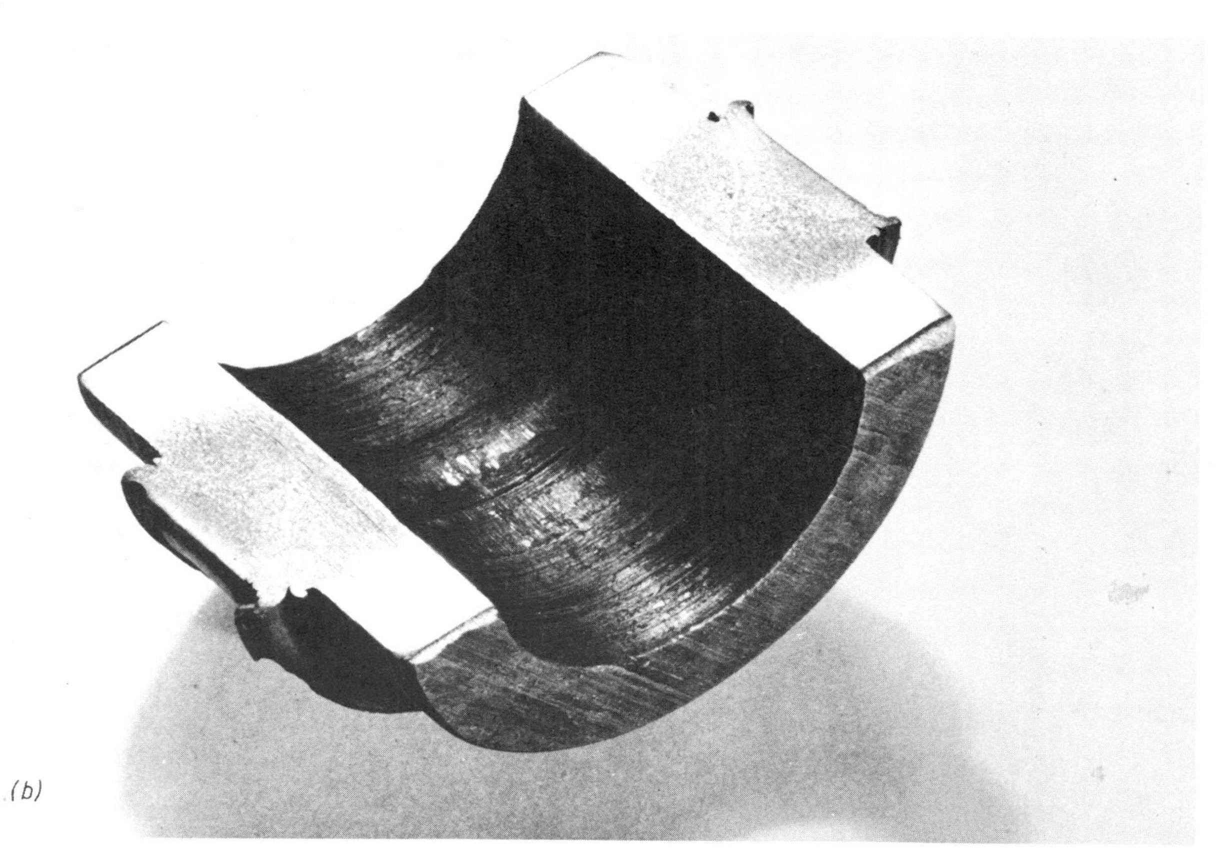

(b)

6 Radial friction welded pipe using multijaw arrangement: (a) external ring profile, (b) weld section

7 Radial friction welding machine for 50mm OD pipe

(a)

(b)

8 Sections through radial friction pipe welds: (a) 50mm OD x 6mm wall, (b) 110mm OD x 10mm wall

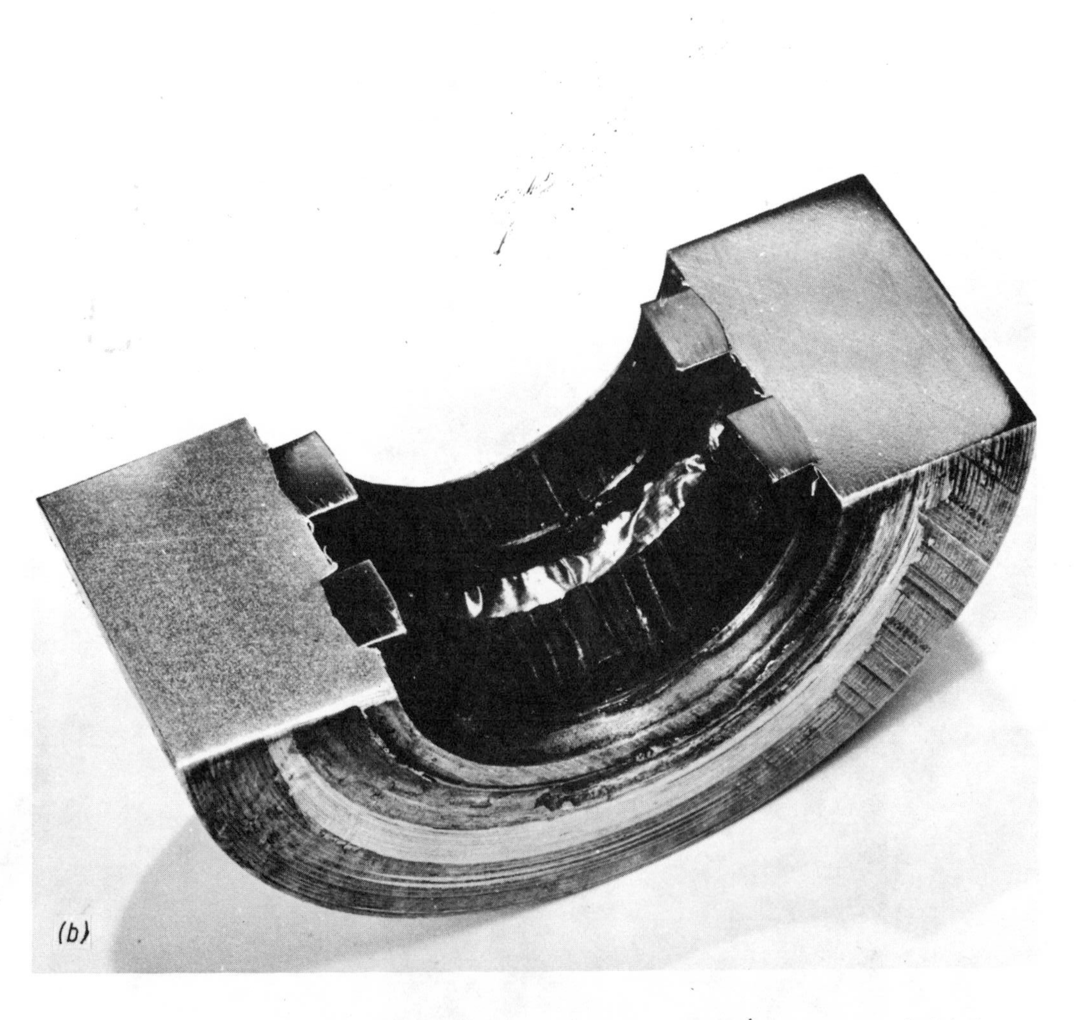

9 Further applications for radial friction welding: (a) collars to shafts (compression), (b) ring inserts in cylindrical body (expansion)

Explosive welding tubes into cladded tubeplates

P.E.G.Williams, BSc, PhD, CEng, MIMechE, MWeldI

The experimental work reported here on explosive welding of tubes into cladded tubeplates is an extension of work previously carried out and reported on welding tubes to tubeplates. Brief reviews of the explosive welding parameters controlling the process, explosive cladding work, and the explosive welding of tubes to tubeplates are presented. It is noted that there is little information on cladding thin flyer plates over a large area and also on welding tubes into cladded tubeplates.

Experimental work is reported on the cladding of thin flyer plates using multiple support points and the region of no-bonding in the vicinity of these supports is noted. The success of welding tubes into cladded tubeplates is reported and problems of deformation in the weakest material in tube — cladding plate — combinations are outlined.

INTRODUCTION

Explosive welding results in a pressure weld which is generated by the oblique collision of two metal components at a very high velocity. An explosive charge is placed in contact with one or both of the two components. On detonation a pressure pulse is produced and imparted to the components, which then accelerate to a very high terminal velocity over a short distance. On impact, oxide layers on the surfaces of the components are removed and virgin surfaces are exposed enabling bonding to take place.

The general setup for explosive welding is shown in Fig.1a. The flyer plate and base plate are initially set apart by a distance known as the stand-off gap. The explosive is separated from the upper surface of the flyer plate by a thin buffer plate, e.g. rubber, which protects the metal surface from the burning explosive gases. Detonation in this setup takes place from one edge and, as illustrated in Fig.1b, the detonation front propagates along the plate ahead of the impact region. The collapsing mechanism of the flyer plate on to the base plate has been widely reported and, therefore, it is intended only to establish the equations which relate the most important parameters. Figure 1b shows the position of the flyer plate at a time, Δt, after detonation. The dynamic bend angle, β, of the flyer plate relative to the base plate is assumed to remain constant throughout the action and the velocity of the flyer plate, V_p (and hence its components, V_f, the velocity of the flyer plate relative to the point of impact, and, V_w, the velocity of the collision point relative to the base plate), is also assumed to remain constant.

Resolving these components of velocity as in Fig.1c it can be seen that

$$V_p = V_d \tan \beta \text{ or } \tan \beta = V_p/V_d \qquad [1]$$

$$V_f = V_p \sin \beta \qquad [2]$$

and $V_w = V_d$ [3]

Dr Williams is a Senior Lecturer, School of Mechanical and Industrial Engineering, Ulster College, The Northern Ireland Polytechnic, Newtownabbey.

where V_d is the detonation velocity of the explosive material. However, if an oblique setup is used, Fig. 2a, the collapse that takes place is as shown in Fig. 2b. The initial angle of inclination, α, will have to be considered and eqs 1, 2, and 3 can be rewritten as

$$\tan \beta = \frac{V_p \cos\alpha}{V_w - V_p \sin \alpha} \quad [4]$$

$$V_f = \frac{V_p \cos\alpha}{\sin \beta} \quad [5]$$

$$V_w = \frac{V_d V_p}{V_p \cos\alpha + V_d \sin\alpha} \quad [6]$$

The oblique setup, Fig. 2, is necessary only when the dynamic bend angle, β, becomes less than 4°. This occurs when high detonation velocity explosives are used with the parallel setup for welding.

Low detonation velocity explosive — 1500 to 4000m/sec — are preferred for many applications for the following reasons:

(a) the detonation velocity can be varied by varying the thickness of explosive

(b) spalling as a result of reflected high pressure tensile waves in flyer and base components is less likely to occur, and

(c) either oblique or parallel setups can be used

PRACTICAL EXPERIENCE

Welding parameters

From the practical viewpoint the investigation of the parameters which control the process is of paramount importance. The results of experiments performed and reported by Wylie et al[1] highlighted and confirmed the proposal by Crossland and Williams[2] that the parameters V_w, V_f, β, and impact energy have limiting values for welding. The limiting conditions of these parameters and others, which have since been developed, are:

(a) $\frac{V_w}{V_{sp}}$ and $\frac{V_f}{V_{sf}}$ should be less than 1.25 where V_{sp} is the sonic velocity in the base component and V_{sf} is the sonic velocity in the flyer component

(b) the dynamic bend angle, β, should exceed a limiting value below which welding does not take place regardless of the amount of explosive. The exact limiting value has been difficult to determine because β not only varies with detonation velocity of the charge, particularly when using low detonation velocity explosive, but also with other parameters which must exceed limiting values. However an angle of 4° or above is generally acceptable

(c) there is a minimum impact velocity below which welding will not occur, as reported by Williams et al.[3] The minimum impact velocity is related to the strength of the materials to be welded and in a combination of dissimilar materials the minimum velocity is determined by the material with the greater strength. It is generally accepted that impact velocity is associated with the pressure at impact but no information is available on experimental work which gives typical impact pressures

(d) above the minimum impact velocity or impact pressure there is a minimum impact energy which is given by the following expression

$$E = \tfrac{1}{2} \rho t V_{fr}^2 \quad [7]$$

where ρ is the density of the flyer component, t is the flyer component thickness, and V_{fr} is the velocity at impact of the flyer component

(e) the stand-off gap or initial clearance between flyer and base component, although previously mentioned but not confirmed in the above equations, should exceed half the thickness of the flyer component for thicknesses greater than 1.2mm. For thicknesses less than 1.2mm a stand-off gap greater than the wall thickness of the component should be used otherwise the terminal velocity of the flyer component will not be achieved before impact

Plate cladding

The welding of two different materials over a large area has proved to be technically and commercially viable. In the UK, Nobel's Explosive Company Ltd, Stevenston, Scotland, clad plates on a commercial scale. The flyer plate sizes vary from 9 to 15mm in thickness and can be as large as 4 x 4m. The setup for welding these very large plates is shown in Fig. 3 and it can be seen that the detonation point is located at a corner. The detonator is inserted into a booster charge of high detonation velocity explosive which ensure spontaneous detonation in the through thickness of the explosive. Welding does not occur immediately below the booster charge, as indicated in Fig. 3b, and edge effects are known to prevent complete welding throughout the surface. It is therefore, necessary to trim the cladded plate before finishing to size by rolling and machining or just machining, depending on customers' requirements.

In practice there are few applications that require a clad plate thickness greater than 3.2mm. In heat exchangers, tubeplates having a clad plate thickness of 3.2mm or less may suffice but this depends on the wall thickness of tube to be used. If the tube has a wall thickness of 1.6mm the need for a clad plate thickness greater than 1.6mm is questionable. Similarly, if the wall thickness is less than 1.6mm the cladding thickness could also be less but in practice the clad plate thickness may not be determined by the tube wall thickness. Moreover to provide a continuous system of the same material on one side in a heat exchanger the tubes must be bonded to both materials in a bimetal cladded tubeplate. If the tube walls are very thin this may present difficulties in achieving a bond in the clad plate of the same thickness.

The existing procedure to obtain a thinly cladded tubeplate is by rolling thicker cladded plates and then performing two operations before the final size is achieved, i.e. rolling and machining to size. It is possible to reduce the total number of operations to two if plates were fabricated to their final composite thickness by explosively cladding a flyer plate of required thickness to a much thicker base plate of finished size. This involves explosively cladding thin flyer plates, which is, in principle, well established since research work on the development of the explosive welding process has involved welding relatively small specimens of thin flyer plate to thin or thick base plates, as in Wylie et al.[1] Because the specimens were small, sagging of the flyer plate was not considered a problem but with much larger plates sagging or bending of the flyer plate cannot be ignored.

Crossland and Bahrani reported on the sagging of large thin flyer plates when the plates were supported at their edges or corners, and then suggested floating the flyer plate on a cushion of air to achieve uniformity of standoff. However, this technique creates problems of sealing the air between the flyer and base plate which would present particular difficulties on site.

One of the parameters that must be correct to be able to achieve an acceptable bond in any welding setup is the impact velocity, which must exceed some minimum value. Shribman and Crossland[4] measured the velocity of collapsing plates and produced empirical relations of plate velocity against mass ratios of explosive and flyer plate for various explosive substances. Williams et al[3] reported on the results of similar experiments for expanding tubes and later Williams[5] reported that the minimum velocity for welding mild steel to mild steel was between 220 and 320m/sec, copper to copper between 180 and 210m/sec, and stainless steel to stainless steel between 305 and 405m/sec. Although these results were obtained from tube to tubeplate welding experiments, several researchers have confirmed that the values are similar in plates for the same material combinations.

Tube to tubeplate welding

Crossland et al[6] explicitly mentioned explosive welding tubes to tubeplates in 1967, and subsequently the subject has been widely reported by many workers and experimental work has been carried out contiguously with plate cladding work to ascertain important welding conditions. Yorkshire Imperial Metals Ltd, Leeds, pioneers of commercialising explosive welding tubes to tubeplates use the technique shown in Fig. 4a and have successfully welded many material combinations which would otherwise have been impossible with conventional fusion welding techniques.

The parallel welding setup shown in Fig. 4b is an alternative method which is widely accepted and uses a low detonation velocity explosive charge. Internal detonation of the charge, as in both setups shown, ensures that, if welding is achieved before the front edge, the problems of stress corrosion cracking stemming from a crevice will be eliminated at the front edge. Another advantage of explosive welding is the ability to produce long (deep) welds between tube and tubeplate, whereas fusion welding can weld only a depth equal to the wall thickness of the tube. Remoteness of tubes in heat exchangers has caused problems for the nuclear power industry when leaks occur, but with the development of the explosive welding process in the area of tube plugging the ability to weld plugs into tubes at a distance has been simplified. Bahrani et al[7] have reported extensively on their achievements in this field.

TUBE TO CLAD TUBESHEET FABRICATION

Experimental setup

The explosive used was ammonia nitrate/TNT/ aluminium flake mixture, better known in the UK as Trimonite 1. The detonation velocity of the explosive varies with thickness ranging from 1800m/sec for a thickness of 6.4mm to 3600m/sec for 25mm. In the plate cladding tests 150mm diameter plates were used with the charge detonated at the centre. The clearance between flyer and base plates was achieved by using four 10mm diameter x 3.2mm thick stubs located so that they coincided with the positions where holes would be drilled for tubes. Seven-hole arrays with a PCD of 35mm, as shown in Fig. 5a, were adopted for the hole configuration in the clad plates. The parallel welding arrangement, Fig. 4b, was used which was achieved by first drilling holes equal to the OD of the tubes and then counterboring to a depth of 40mm, i.e. approximately half the tubeplate thickness. The counterbored hole created the clearance between the tube and hole, and the gap varied depending on the wall thickness of the tube, i.e. 1.6mm clearance for wall thicknesses up to 1.6mm and 3.2mm clearance for tube walls between 1.6 and 3.2mm thick. Tubes of 25mm OD in three different materials were welded one at a time into the tubeplates to a depth of approximately 25mm. All tests were evaluated by ultrasonic, peel testing, and microscopic examination.

EXPERIMENTAL RESULTS

Plate cladding tests

It has already been pointed out that, if explosive welding is to be achieved between flyer and base components, it is necessary to have a uniform clearance that is sufficiently large to allow the flyer component to achieve its terminal velocity before impact. To satisfy this condition with large, thin flyer plates many areas of support over the surface are necessary because of the lack of rigidity of the plate. To determine the effect of multiple supports on the weld area that can be achieved, a preliminary test was conducted using 1.6mm thick copper flyer plate and a four-point support arrangement similar to that shown in Fig. 5b. The studs were 22.5mm in diameter and their thickness equalled that of the flyer plate. An impact velocity of 320m/sec giving a specific kinetic energy of 1.46J/mm^2 was achieved. The welded area was then examined using an ultrasonic flaw detector with the results as shown in Fig. 6a. It can be seen from the detail section, Fig. 6b, that welding began 3.18mm from the edge of the centre stud, which represents a no-bond region of 28.86mm in diameter. The area of no-bonding around the outer three studs was similar, with a maximum no-bond width of 4.3mm around one edge, Fig. 6a. With welding ceasing at 1.1mm on the near or opposite edge, this represents a maximum no-bond region of 27.9mm diameter. Clearly the diameter of the studs has to be reduced otherwise a no-bond region would exist outside the 25.4mm diameter holes.

Based on this result flyer plates 1.6 and 3.2mm thick in stainless steel and copper were welded to a 75mm thick mild steel base plate using 10mm diameter studs. The experimental setup is shown in Fig. 5b together with the bonding zones as indicated in Fig. 7 resulting from the conditions listed in Table 1.

Comparing the weld areas in copper clad and stainless steel clad plates with the same clad plate thickness it can be seen that in both a slightly larger area is bonded when copper is the flyer plate. This is because of the difference in energy absorbed in deformation during the collapse of the flyer plates. Copper, having a lower dynamic yield stress than stainless steel, flows more easily around the studs and will be in the correct position for welding at an earlier stage. Comparing flyer plates of the same materials with different thicknesses it can be seen from Fig. 7a and b that the thinner plate started welding earlier and that consequently the overall weld area is larger.

Table 1

Flyer plate details	V_p, m/sec	Specific kinetic energy, J/mm^2	Remarks
Stainless steel, 1.6mm thick	320	0.73	Weld area shown, Fig. 7a
Stainless steel, 3.2mm thick	330	1.55	Weld area shown, Fig. 7b
Copper, 1.6mm thick	380	0.92	Weld area shown, Fig. 7c
Copper, 3.2mm thick	380	1.84	Weld area shown, Fig. 7d

Table 2

Test no.	Tube details	V_t, m/sec	Specific kinetic energy, J/mm	Remarks
1-central hole	Mild steel, 3.2mm wall	200	34.8	No weld
2	Mild steel, 1.6mm wall	310	44.7	No weld
3	Mild steel, 0.8mm wall	480	55.5	Weld throughout length; copper severely deformed
4	Stainless steel, 1.6mm wall	310	44.7	No weld
5	Stainless steel, 0.8mm wall	492	58.4	Weld throughout length; copper severely deformed
6*	Copper, 1.6mm wall	310	51.1	Weld in copper but small crevice at front edge, Fig. 10a
7	Copper, 0.8mm wall	480	62.7	Weld throughout length

* Explosive charge detonated outside the tubeplate

Table 3

Test no.	Tube details	V_t, m/sec	Specific kinetic energy, J/mm	Remarks
1-central hole	Mild steel, 3.2mm wall	180	28.2	No weld
2	Mild steel, 1.6mm wall	310	44.7	No weld in copper
3	Mild steel, 0.8mm wall	480	55.5	Weld throughout length, Fig. 8a
4	Stainless steel, 1.6mm wall	310	44.7	No weld
5	Stainless steel, 0.8mm wall	490	57.8	Weld throughout length
6	Copper, 1.6mm wall	330	47.9	Weld throughout length, Fig. 8a
7	Copper, 0.8mm wall	480	62.7	Weld throughout length, Fig. 8a

Tube to tubeplate welding tests

Copper-clad base

The experimental data for the two thicknesses of flyer plate — 1.6 and 3.2mm — are summarised in Tables 2 and 3 for tubes of mild steel, stainless steel, and copper.

Stainless steel clad base

Similarly, the experimental data for the two thicknesses of flyer plate — 1.6 and 3.2mm — are summarised in Tables 4 and 5 respectively for tube materials mild steel, stainless steel, and copper.

The 35mm pitch between the central and surrounding six holes ensured that deformation in the material between the holes was kept to a minimum during welding. In all instances the deformation was internal, except for one test, Table 2, no. 6, for which the explosive charge was detonated external to the plate.

It was observed, and can be seen in Figs 8 and 9 for copper and stainless clad base plates, that welds were achieved between tubes and cladded tubeplates when the maximum velocity for welding had been exceeded for the combination. For example, in the test, Table 3, no. 5, for the trimetal combination (consisting of mild steel base plate, copper cladding plate, and stainless steel tube) a velocity of 490m/sec produced a satisfactory weld between tube and plate materials. An impact velocity of 490m/sec is considerably higher than the recorded minimum velocity to weld a copper-copper combination (210m/sec), and higher than that to weld mild steel to mild steel (320m/sec). However it is only slightly higher than the minimum recorded to weld stainless steel to stainless steel (405m/sec).

Table 4

Test no.	Tube details	V_t, m/sec	Specific kinetic energy, J/mm	Remarks
1-central hole	Mild steel, 0.8mm wall	320	24.6	No weld at front edge
2	Mild steel, 1.6mm wall	420	82.2	Weld throughout length, Fig. 9a
3	Mild steel, 1.6mm wall	540	70.3	Weld throughout length
4	Stainless steel, 0.8mm wall	540	70.3	Weld throughout length
5	Stainless steel, 0.8mm wall	490	57.8	Weld throughout length, Fig. 9b
6	Stainless steel, 0.8mm wall	390	36.6	No weld
7	Copper, 1.6mm wall	360	69	No weld

Table 5

Test no.	Tube details	V_t, m/sec	Specific kinetic energy, J/mm	Remarks
1-central hole	Mild steel, 1.6mm wall	340	53.8	No weld
2	Mild steel, 1.6mm wall	400	74.6	Weld throughout length
3	Mild steel, 0.8mm wall	380	34.8	Weld throughout length
4	Stainless steel, 0.8mm wall	410	45.6	Weld with front crevice
5	Stainless steel, 0.8mm wall	500	60.3	Weld throughout length, Fig. 10b
6	Stainless steel, 1.6mm wall	340	53.8	No weld
7	Copper, 1.6mm wall	420	85	No weld in stainless steel

This satisfies one of the conditions for welding: that, in a dissimilar material combination, the minimum velocity is determined by the material with the greatest strength.

However the high impact velocity and consequently the kinetic energy can seriously affect the deformation of the weakest material. Thus Fig. 8a and b shows the weld regions of a copper tube welded to copper clad mild steel tubeplate. Here the impact velocity of the tube exceeded the minimum velocity necessary to weld mild steel, but in achieving this velocity the tube was so severely deformed at the interface between copper and mild steel plates that its thickness was considerably reduced. The severe shearing action experienced by the tube at the interface can also be seen in detail in Fig. 8b.

A further example of tube thinning and severe deformation in the weakest material was found when a mild steel tube was welded to a copper clad mild steel tubeplate, Fig. 8c. Although this latter combination of materials (and those involving mild steel, stainless steel, and copper) would have little use in a heat exchanger because of deformation and corrosion problems, it does provide useful information on the limitations of certain combinations.

DISCUSSION

Plate cladding

It is to be pointed out that the 10mm diameter studs used to create the clearance between flyer and base plates in these tests, Fig. 7, are much larger than necessary. Smaller studs could be used and a greater weld area would be achieved. Also, it is to be emphasised that location of the supports is important and knowledge of the layout of the holes is essential before cladding.

The problem of edge effects or edge instability has been experienced in welding large, thick flyer plates. It is generally assumed that this effect stems from supporting the flyer plates at their edges, and detonation of the charge at an edge which produces an 'open loop' detonation front. This problem is eliminated by using thin flyer plates and detonating the charge at the centre. Different conditions then occur, all of which contribute to the overall elimination of the effect. Firstly, the distance the detonation front travels is reduced by half. Secondly, the propagation of the weld is continually interrupted by the supporting studs and the size to which the waves can grow is limited. Finally, detonating the explosive charge at the centre produces a closed detonation loop which moves radially outwards in an unbroken path. An example of the latter is shown in the detail in Fig. 6b, where it can be seen that there is no variation in amplitude and wavelength of the waves over a distance of 65mm. Nevertheless further tests on larger diameter plates will be necessary to confirm this result and also to support the closed loop proposal.

Another important feature of using thin flyer plates is that less explosive is required, and hence the detonation velocity associated with the thickness of charge can result in a dynamic bend angle well above the limiting angle for welding, whereas, in thick flyer plates and consequently thick charges, bend angles tend to approach their limiting value.

Crevice formation in tube to clad tubeplate welding

It is generally accepted that crevices between tubes and tubeplate, particularly at the outside edge, cause stress corrosion cracking. There is an absence of crevices more often if the explosive charge is detonated from inside the tubeplate. Only one test was conducted using an external detonation point, resulting in a crevice at the front edge, as can be seen in Fig. 10a. This should be compared with Fig. 10b which shows a crevice-free joint between stainless steel and stainless steel when detonation was internal, as illustrated in Fig. 4b. The ability to weld tubes across the interface of a thinly cladded tubeplate with the minimum of distortion to the tubes and clad material adds a new dimension to the application of the explosive welding process.

CONCLUSIONS

A technique of multiple supports for creating clearance between thin flyer plates and base plates has been outlined and information has been presented on the area welded when the plates are explosively impacted. It has been shown that the technique is suitable for cladding plates for use as tubeplates. Information has been presented on the deformation of tubes and on quality of the welds when tubes are explosively welded into holes in plates which have been clad with thin plates. It has been shown that the controlling parameter for high quality welds is the velocity at impact between tubes and clad plate and the value of minimum impact velocity is determined by the strength of the stronger material.

ACKNOWLEDGEMENTS

The author would like to thank Professor B. Crossland for permission to use the facilities of The Queen's University of Belfast for the experimental work.

Thanks are also due to Mr George Stewart and Mr Stuart Beggs of the Ulster College, The Northern Ireland Polytechnic, Newtownabbey, for the support provided by the technician staff.

This work would not have been possible without the financial support given by the Ulster College, The Northern Ireland Polytechnic.

REFERENCES

1 WYLIE, H.K., WILLIAMS, P.E.G., and and CROSSLAND, B. 'An experimental investigation of explosive welding parameters'. Procs Conference 'Use of Explosive Energy in Manufacturing Metallic Materials of New Properties of use in the Chemical Industry'. Marianske Lazne, 1970, 45-70.

2 CROSSLAND, B. and WILLIAMS, J.D. 'Welding parameters for explosive cladding'. Procs Conference 'Advances in Welding Processes', Harrogate, 14-16 April 1970. Abington, Welding Inst., 1971, 78-82; discussion 252-5.

3 WILLIAMS, P.E.G., WYLIE, H.K., and CROSSLAND, B. 'Welding parameters for

explosively welding nonferrous tubes to tubeplates'. Procs Conference 'Welding and Fabrication of Nonferrous Metals', Eastbourne, 2-3 May 1972. Abington, Welding Inst., 1972, 108-21.

4 SHRIBMAN, V. and CROSSLAND, B. 'Experimental investigation of the velocity of the flyer plate in explosive welding'. Procs 2nd Conference of the Centre for High Energy Forming, Denver, Colorado, 1969, P.7.3.1.

5 WILLIAMS, P.E.G. 'Explosive welding of tubes to tubeplates'. PhD thesis, The Queen's University of Belfast, December 1972.

6 CROSSLAND, B. et al. 'Explosive welding of tubes to tubeplates'. Welding and Metal Fab., 35 (3), 1967, 88-94.

7 BAHRANI, A.S., HALLIBURTON, R.F., and CROSSLAND, B. 'The explosive plugging of heat exchangers'. Procs Int'l Conference 'Welding Research Related to Power Plant', CEGB Marchwood, Southampton, 17-21 September 1972, 617-33.

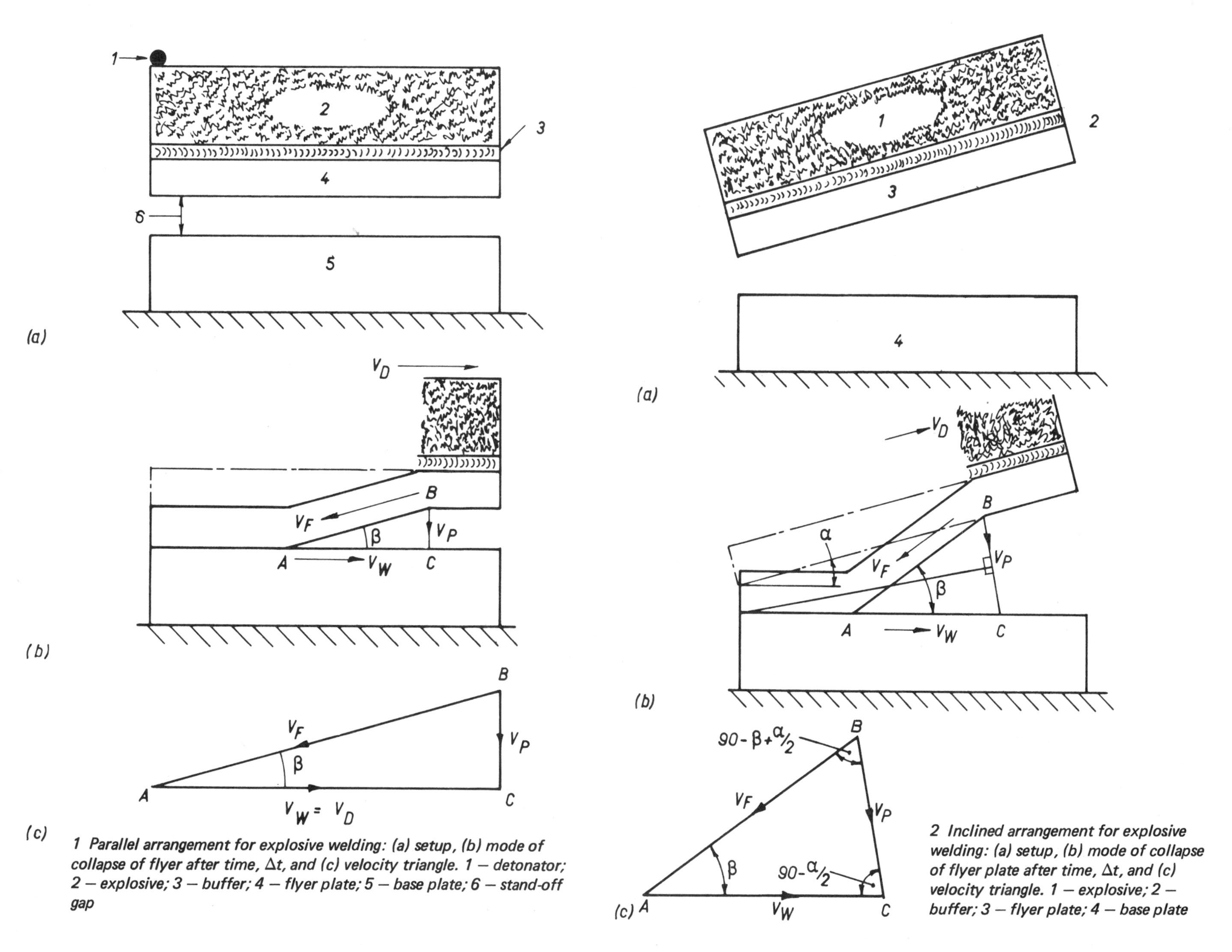

1 Parallel arrangement for explosive welding: (a) setup, (b) mode of collapse of flyer after time, Δt, and (c) velocity triangle. 1 – detonator; 2 – explosive; 3 – buffer; 4 – flyer plate; 5 – base plate; 6 – stand-off gap

2 Inclined arrangement for explosive welding: (a) setup, (b) mode of collapse of flyer plate after time, Δt, and (c) velocity triangle. 1 – explosive; 2 – buffer; 3 – flyer plate; 4 – base plate

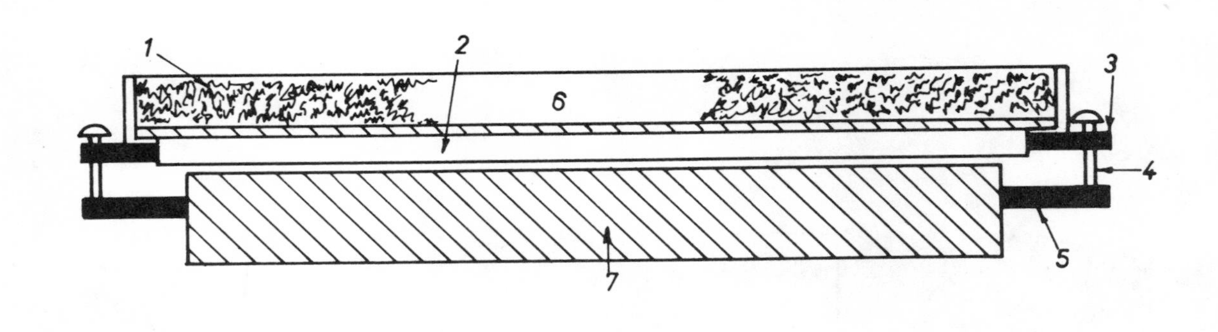

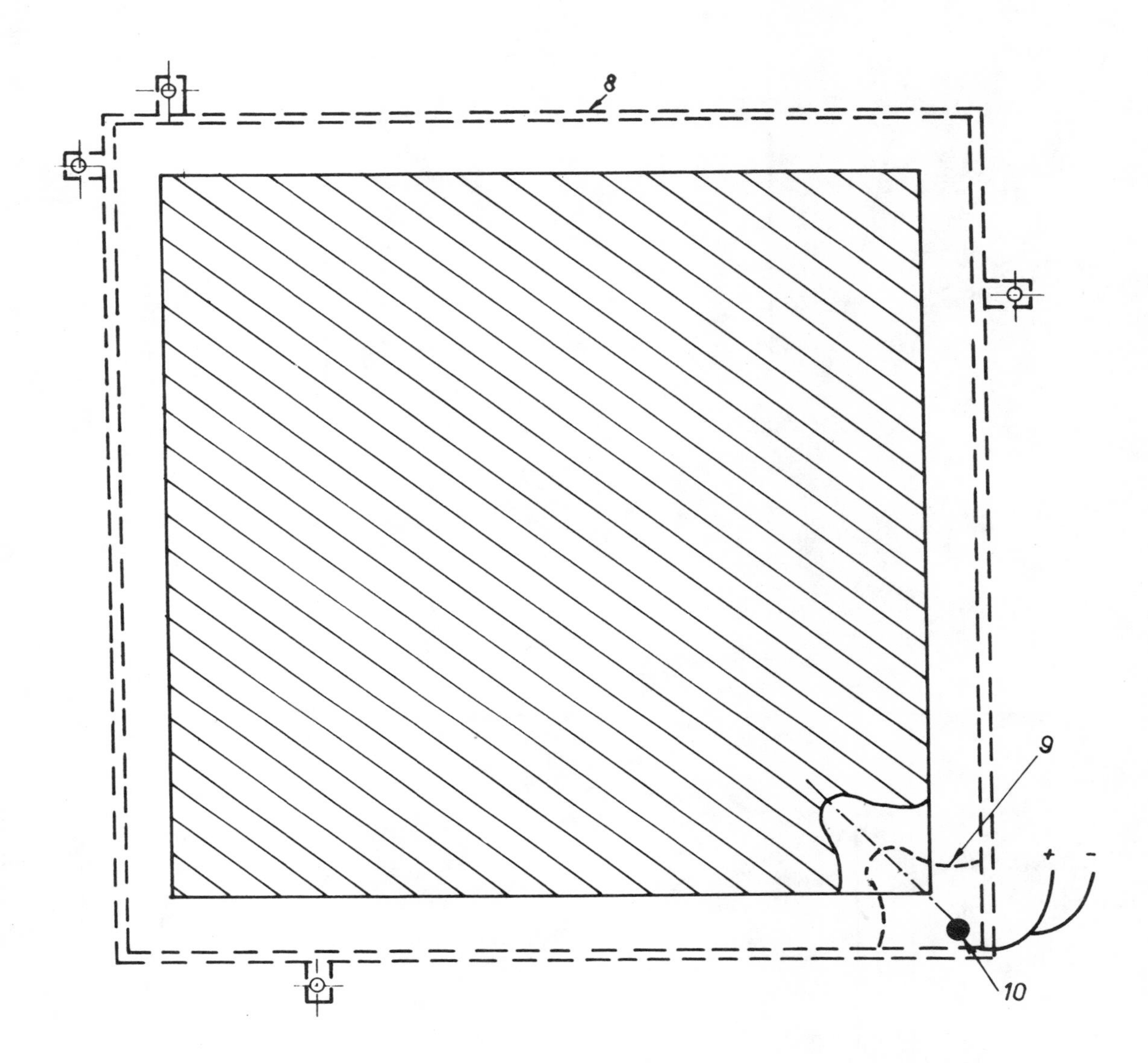

3 Arrangement for cladding with thick flyer plates of large area: (a) welding setup, (b) plan view showing bonded area (shaded). 1 — buffer; 2 — flyer plate; 3 — four flyer plate lugs; 4 — screws; 5 — four parent plate lugs; 6 — explosive charge; 7 — parent plate; 8 — Z section containment frame; 9 — Metabel booster charge; 10 — detonator

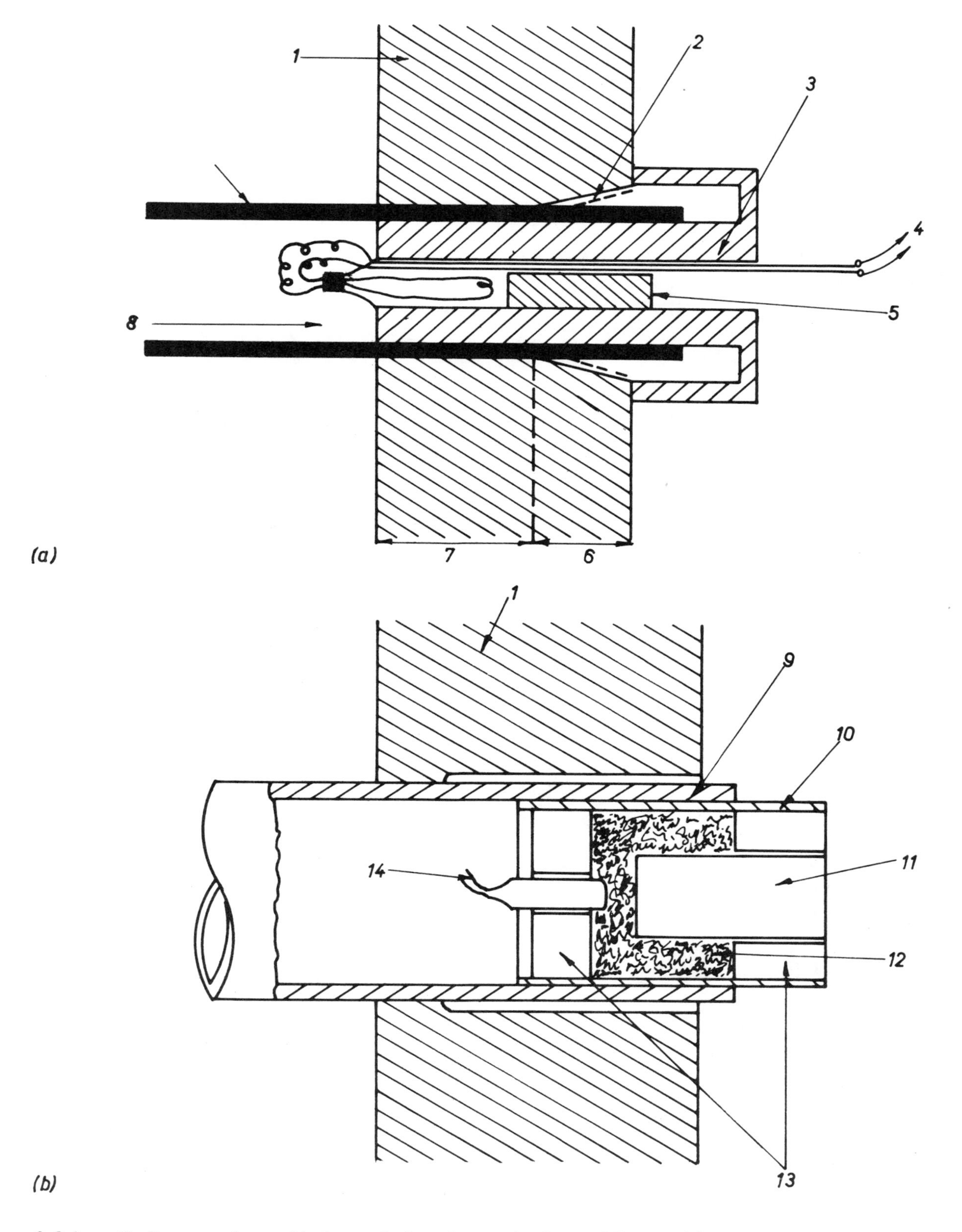

4 Schematic diagrams of assembly for explosive tube to tubeplate welding by: (a) YIMpact process, (b) parallel tube welding technique. 1 — tubeplate; 2 — position of tube after welding; 3 — polythene insert; 4 — detonator wires to exploder; 5 — base charge of detonator; 6 — profiled section of tubeplate; 7 — parallel section of tubeplate hole; 8 — direction of detonation; 9 — tube; 10 — polythene buffer; 11 — polythene spigot; 12 — Trimonite powder explosive; 13 — polythene endpieces; 14 — detonator

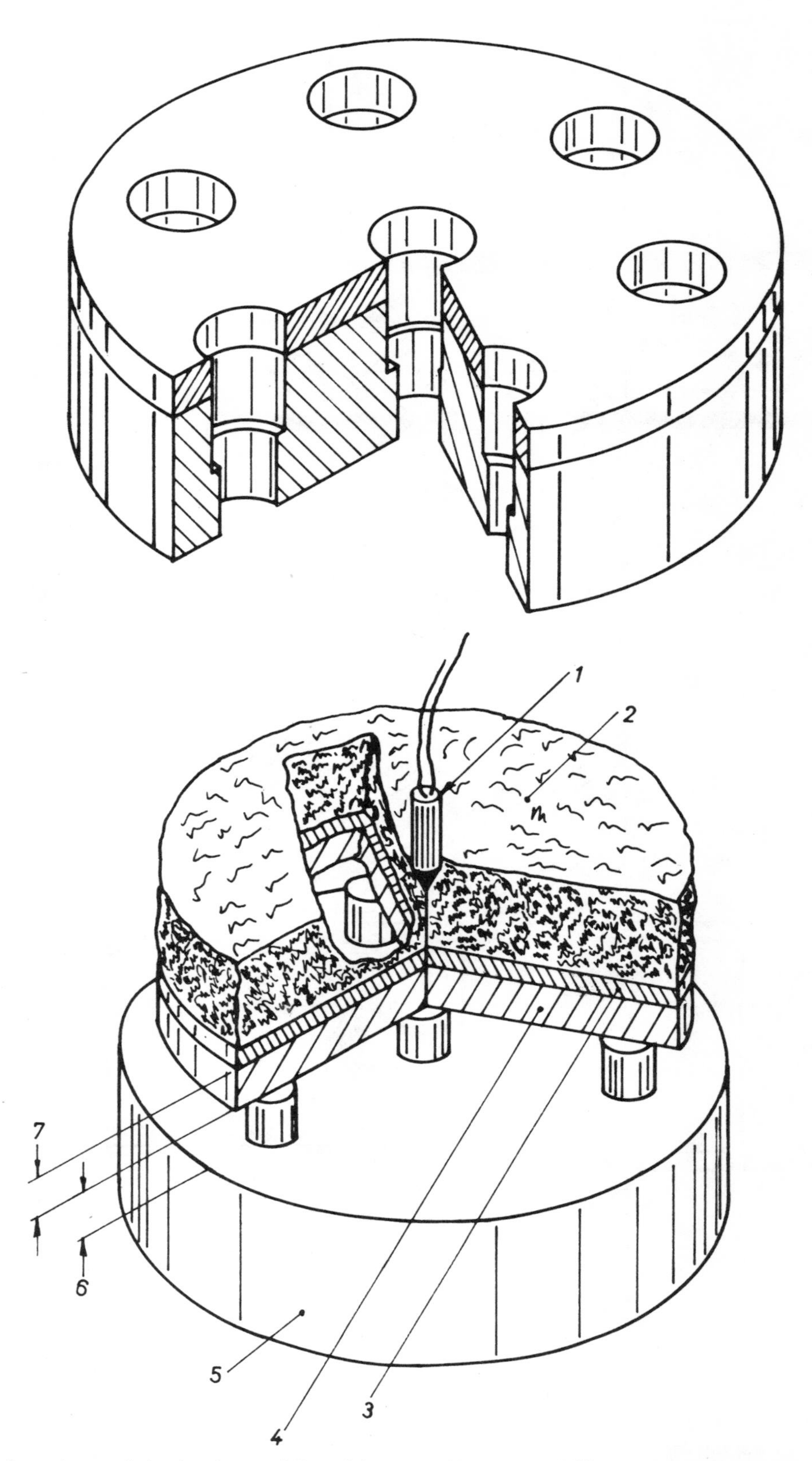

5 Arrangement for tube to clad tubeplate welding: (a) seven-hole array in 150mm diameter plate with six holes equally spaced on PCD, as finished prior to tube welding, (b) experimental setup for thin plate cladding (showing four support studs equally spaced on PCD); not to scale. 1 — detonator; 2 — explosive; 3 — buffer; 4 — flyer plate; 5 — base plate; 6 — clearance; 7 — varying thickness

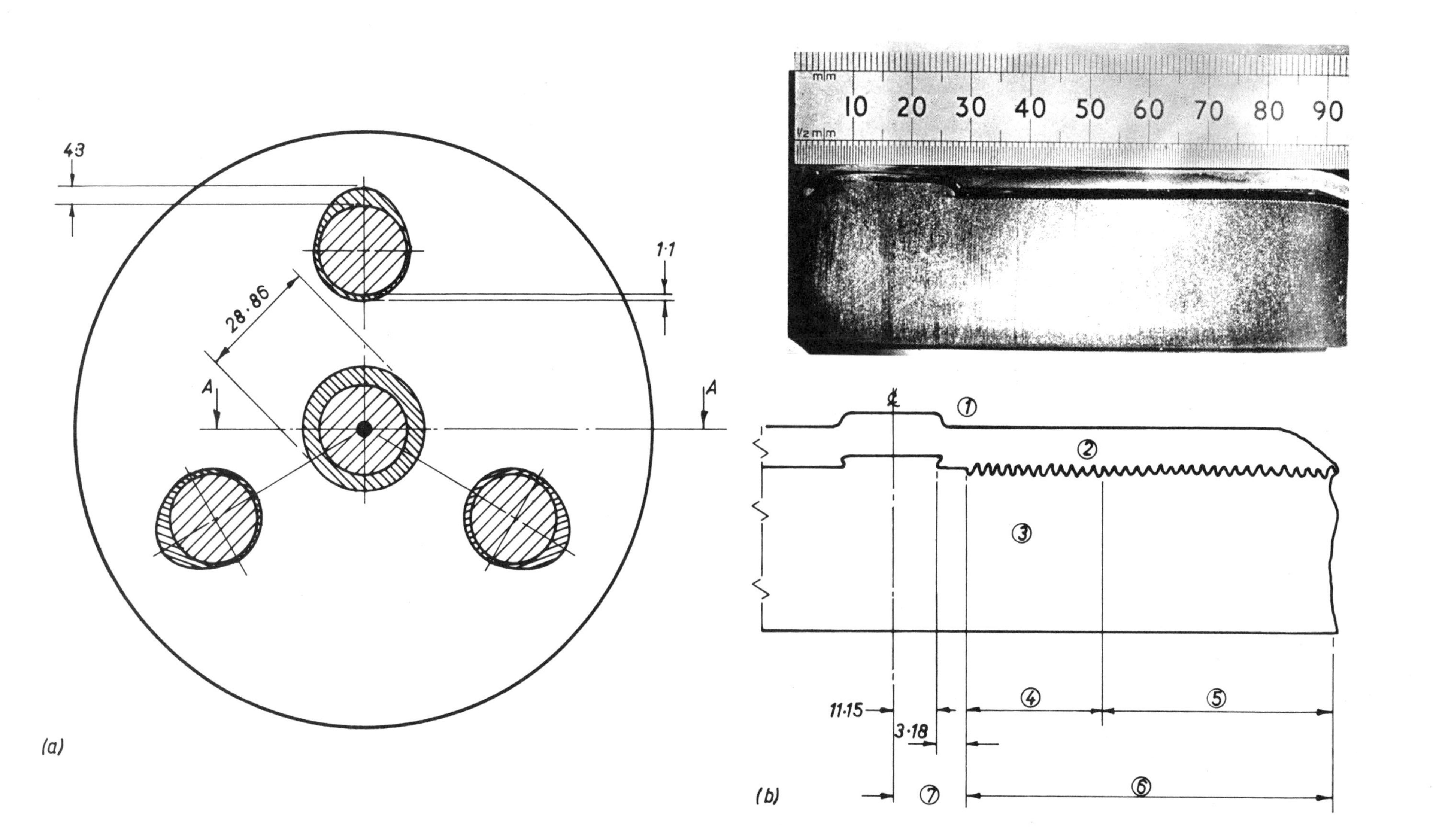

6 Weld of 1.6mm thick copper-mild steel clad using 22.5mm diameter supports: (a) diagram of weld area, shaded zone – unwelded area; ● – detonation point; surface area – 17 670mm^2; weld area – 15 842mm^2, (b) detail showing welded regions and uniformity of interfacial waves over welded length; ① – 22.5mm Ø stud; ② – flyer plate; ③ – base plate; ④ – thirty-eight waves in 20mm; ⑤ – seventy-one waves in 36mm; ⑥ – bonding; ⑦ – no bond. Dimensions in millimetres

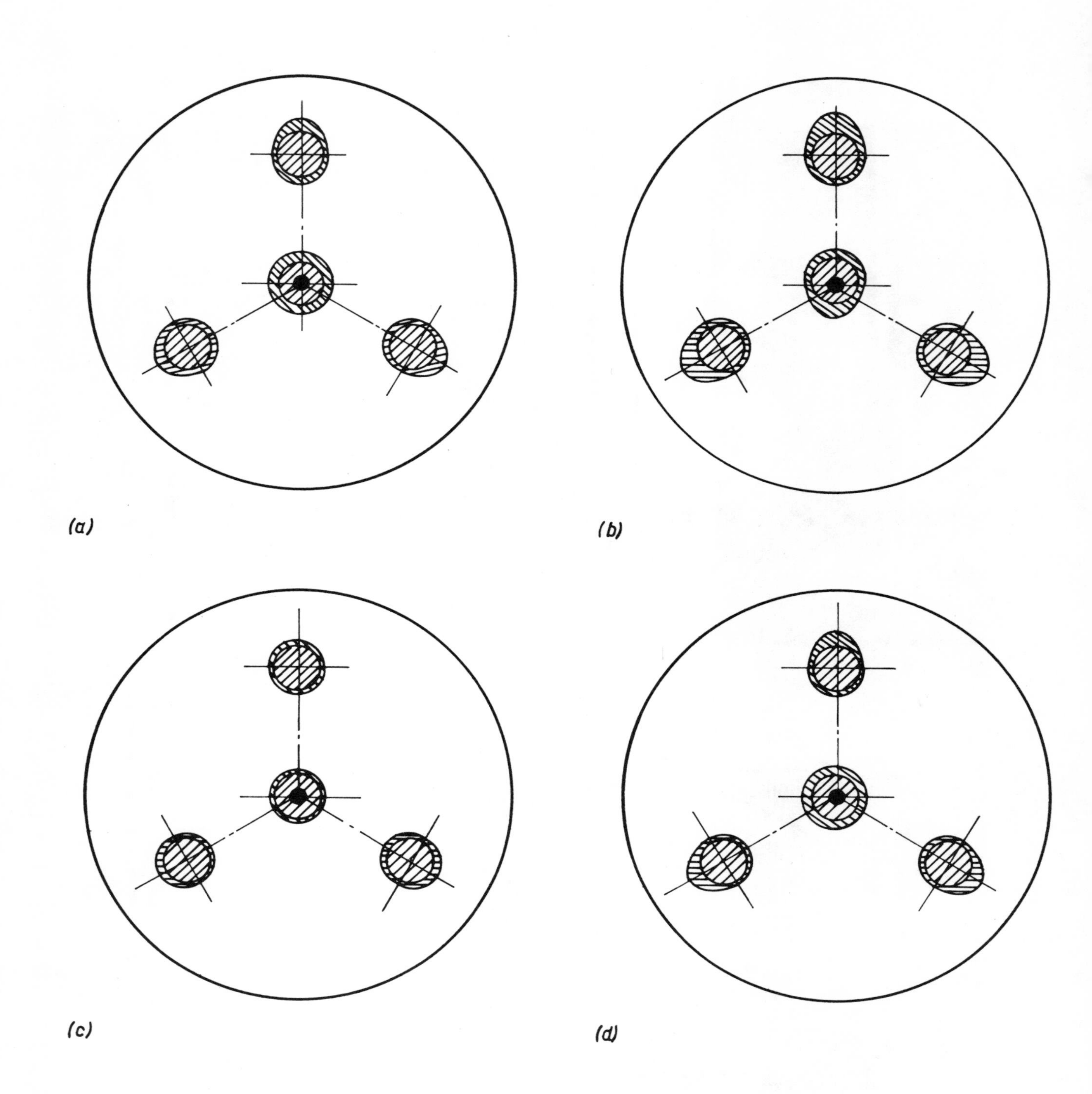

7 Schematic diagram of weld area for thin plate clad: (a) 1.6mm thick stainless steel-mild steel clad; weld area — 16 900mm^2, (b) 3.2mm thick stainless steel-mild steel clad; weld area — 15 820mm^2, (c) 1.6mm thick copper-mild steel clad; weld area — 15 900mm^2, and (d) 3.2mm thick copper-mild steel clad; weld area — 15 400mm^2. ● — detonation points; shaded zones — unwelded areas

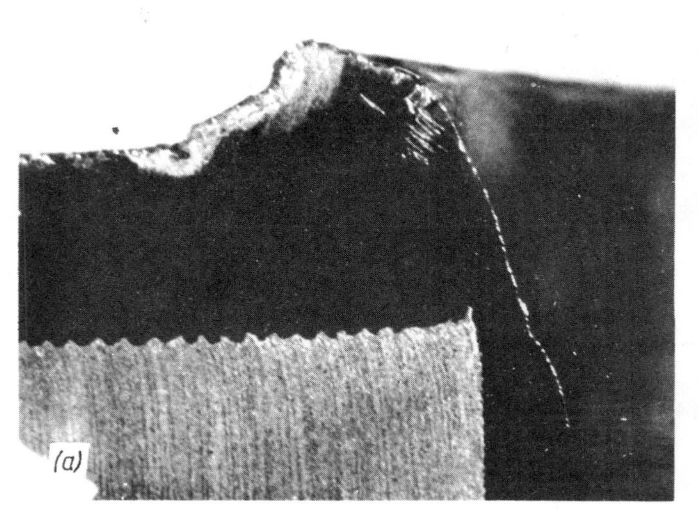

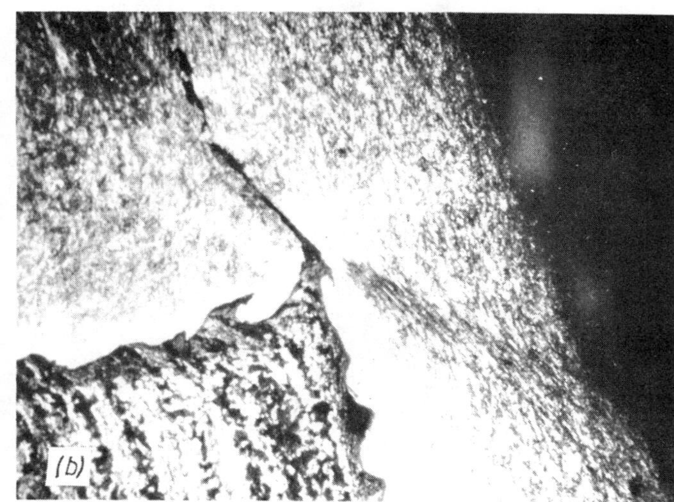

8 Tube welds to copper clad tubeplate: (a) copper tube (x 60), (b) detail of deformation in copper tube (x 100), and (c) mild steel tube (x 60)

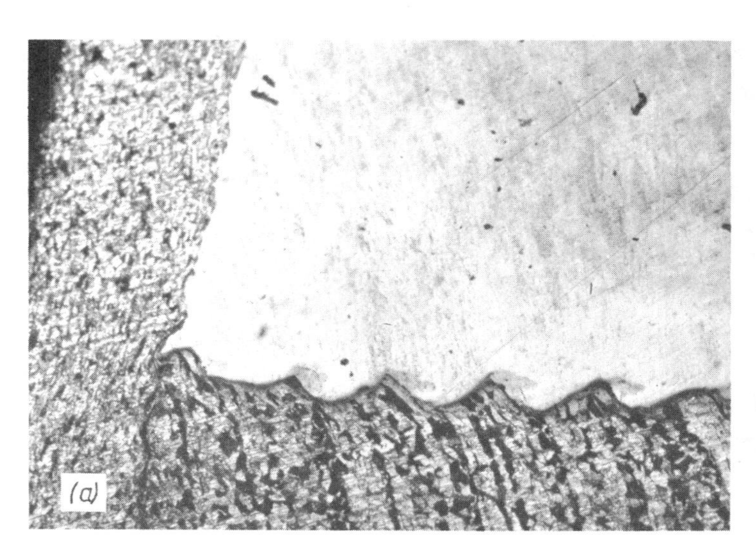

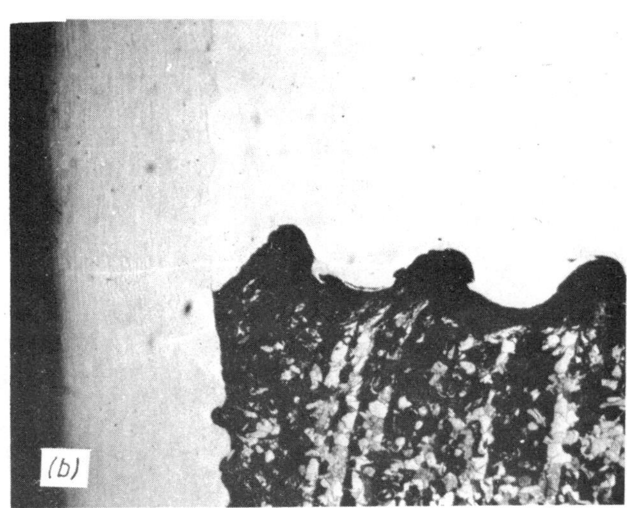

9 Tube welds to stainless steel clad tubeplate: (a) mild steel tube, (b) stainless steel tube. x100

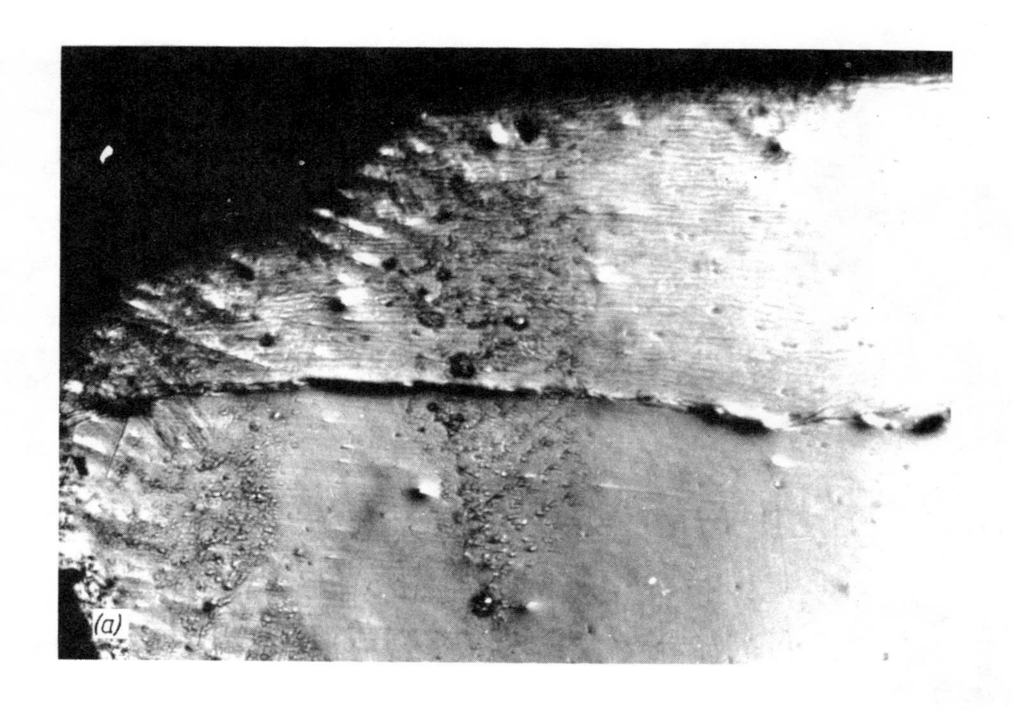

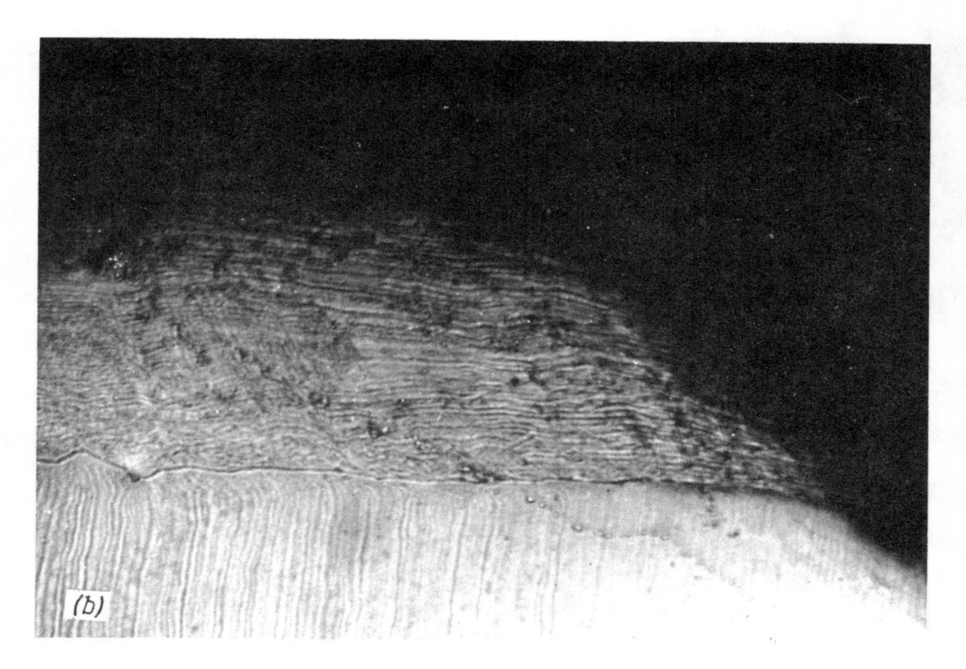

10 Front edge detail of tube welds: (a) copper tube to copper clad tubeplate with external detonation; note crevice, (b) stainless steel tube welded to stainless steel clad tubeplate with internal detonation; note no crevice. x400

PAPER 6

The influence of electrode extension and cold filler wire additions on the deposition rates and properties of submerged-arc welds

W.J.F.Thomas and R.L.Apps, BSc, PhD, FIM, MWeldI

There is a developing interest in increasing the productivity of submerged-arc welding, but it is also important to maintain weld metal properties. Two methods of increasing weld deposition rates have been investigated.

Deposition rates were established for electrode negative submerged-arc welding with a basic agglomerated flux, utilising extended electrode stickouts or cold wire additions to the weld pool. Both methods were capable of increasing deposition rates by significant amounts (up to 37% in the present work) with no, or marginal, increase in heat input.

The mechanical properties of weld deposits produced at three different electrode extensions and with two different amounts of cold wire additions have been determined. Only comparatively minor variations in mechanical properties were observed and these were partially attributable to changes in weld metal composition.

It is concluded that these two methods of increasing deposition rates (without major changes in the welding process) are worth further investigation.

INTRODUCTION

Since the introduction of submerged-arc (SA) welding in 1935 there has been a continuing interest in the increase of productivity by means of increased deposition rates. Once the significance of weld metal structure and properties in relation to joint performance was appreciated, the need to maintain adequate joint properties has been an opposing requirement which has often restricted deposition rates and hence productivity. Furthermore, in the present economic climate fabricators have been concerned with increasing productivity not only without deterioration in weld properties, but also without the use of expensive additional equipment.

The initial means of increasing deposition rates was to raise the welding current[1] with consequent increase in arc energy (heat input), leading to coarse-grained weld metal and heat-affected zone structures and hence poor notch toughness. The introduction of multiwire SA welding gave a further marked increase in deposition rate[2] but no improvement in properties.

With mounting awareness of the need to keep arc energy low, alternative methods of increasing deposition rates have been tried. The most successful method has been hot wire welding[3] in which a second small diameter wire, resistance heated by current from a

Mr Thomas is with The Unit Inspection Company, and Professor Apps is Professor of Welding Technology at Cranfield Institute of Technology.

second power source, is fed into the weld pool. Resistance heating is much more efficient than arc heating, and increases in deposition rates from 50 to 100% can be achieved with little increase in total energy expenditure and without impairment of weld metal properties. The disadvantages of the hot wire addition technique are the extra setting up time needed, the reduction of weld penetration, and the cost of the additional equipment.

Alternative methods of increasing deposition rates appear to have received far less attention, although they are simpler than hot wire additions and can often be achieved without the purchase of extra equipment. Such methods include the use of:

(a) extended electrode extensions or stickout to increase resistance heating of the electrode
(b) cold wire additions made in a similar manner to hot wire but without additional resistance heating, and
(c) metal powder or chopped wire additions to the weld preparation

All these alternative techniques aim at using the welding energy more efficiently, to produce higher deposition rates with little or no increase in energy expended.

Increased electrode stickout can certainly increase deposition rates by 25-50% by making use of I^2R heating and several workers[4-6] have recommended it, although industrial application in the UK appears limited. The problems appear to be the need to use an insulated guide to control the longer extension and a reduction in penetration (of about 10%) which could require a change in edge preparation and joint fitup. Additionally, if excessive electrode extensions are used the wire overheats leading to electrode pulsation, arc instability, and stubbing.

Cold wire additions have been shown to be feasible using both solid and flux-cored wires[5,7] without deterioration in weld properties. The technique, however, does not appear to have been applied industrially. Again, the disadvantages of the method appear to be the need for additional equipment and more careful setup, together with a reduction in penetration.

Metal powder and chopped wire additions have been reported[5,6,8] as being able to increase deposition rates by up to 70%. The techniques are claimed to give smoother fusion, improved weld bead appearance, and reduced penetration and dilution. Information on properties is limited although one paper states[8] that there is no impairment of properties.

EXPERIMENTAL PROCEDURE

Equipment

All the equipment used was conventional. A standard controlled arc SA welding unit was used together with a 1200A drooping characteristic rectifier. Testplates were traversed beneath the fixed welding head on a movable carriage.

For long electrode stickout an insulated electrode extension guide was constructed from asbestos blocks and 3mm thick mild steel sheet, Fig.1. Equipment for cold wire additions was assembled by attaching a small MIG-welding torch with its own wire feed motor to the SA welding head, Fig.2.

Consumables

Welding was carried out on 25mm thick steel plate to BS 4360, Grade 43A, and backing strips in similar materials were also used. An agglomerated basic flux (Basicity 3.0) together with 4.0mm diameter 1.5%Mn SA welding wire were used for all welding. Where cold wire additions were required a 1.2mm diameter wire with 1.4%Mn was used.

Welding

The tests to determine weld deposition and flux consumption rates, together with weld bead dimensions, were carried out by bead-on-plate runs. The weld preparation and assembly, Fig.3, were used to produce welds for the determination of mechanical properties. All testplates were clamped in a heavy jig during welding to minimise distortion.

Bead-on-plate tests were made over a range of welding parameters and from these tests 600A (DC, electrode negative), at 38V, and 380mm/min welding speed were chosen as optimum. With these conditions electrode extensions up to 160mm were examined. Extensions of 140mm and greater were found to give pulsing and arc instability, and it was concluded that 120mm was the maximum practical extension for 4.0mm wire. Similarly, cold wire additions up to 73% of the SA welding deposition were made in bead-on-plate tests but it was considered that an addition of 35-40% was the maximum that could be achieved consistently in practice.

For property determination and metallographic examination welded testplates were produced in three series, two to examine the influence of electrode extension and one that of cold wire additions. The welding conditions

Table 1 The influence of electrode extension on deposition rate and flux consumption

Series	Electrode extension, mm	Electrode polarity	Energy, kJ/mm	Operating voltage	Current, A	Deposition rate, kg/hr	Flux consumption, kg/hr
1	40	Negative	3.60	38V head	600	11.05	11.88
	80		3.60			12.49	10.62
	120		3.60			15.12	10.26
2	40	Negative	3.32	33V arc	600	11.66	10.10
	80		3.41			13.28	10.12
	120		3.60			15.12	
3	40	Positive	3.24	36V head	600	8.70	10.45
	80		3.24			10.75	11.20
	120		3.24			12.00	12.00
	160		3.24			14.10	12.90

were standardised to those given above for two of the series: Series 1 (electrode extension with constant welding head voltage) and Series 4 (cold wire addition). For a further series, Series 2, the voltage between the welding head and the testplate was varied to maintain a constant arc voltage, i.e. arc length. Thus electrode extension was studied under two conditions:

1 With a constant head voltage of 38V in which the arc voltage (arc length) decreased as the electrode extension increased (Series 1)

2 With a constant arc voltage so that the total head voltage increased from 33 to 38V as the electrode extension was increased from 40 to 120mm (Series 2)

Examination and testing

Each weld was visually examined and radiographed and found to be generally free from defects. Subsequently welds were machined to give all-weld-metal tensile testpieces (Hounsfield no. 13 size). For Charpy V notch impact tests, each specimen was taken transverse to the welding direction 4mm from the top weld surface and notched normal to the plate surface in the weld metal at a position equidistant from the fusion boundary and the weld centre line. This notch location was chosen rather than the weld centre line, since the latter consisted of almost wholly refined weld metal which was considered unrepresentative of most practical situations. Further sections of each weld were used for metallographic examination and chemical analysis.

EXPERIMENTAL RESULTS

The welding conditions were chosen so that the welding operation could be performed satisfactorily to give good weld bead profiles and sound weld metal. The effect of increasing electrode extension is shown in Table 1 for a constant head voltage (Series 1) and a constant arc voltage (Series 2). For comparison the effect of electrode extension with DC electrode positive is also included (Series 3) for similar, though not identical, welding parameters and energy expenditure.[9] For a fixed energy expenditure an increase in electrode extension from 40 to 120mm gave 37% increase in deposition rate for both negative and positive polarity. With a fixed arc voltage the increase was somewhat less even though energy expenditure increased by 7%. For negative polarity, flux consumption decreased by 13% as electrode extension was increased with constant head voltage (Series 1) but remained virtually static for a constant arc voltage (Series 2). Surprisingly the results for positive polarity (Series 3) indicated a 15% increase in flux consumption with the same increase in stickout. Similar increases in deposition rate were also obtained with cold wire additions, Table 2, with only a minor increase (4%) in flux consumption.

The variation in weld bead dimensions with increased electrode stickout and cold wire additions is shown in Table 3. The results are generally inconclusive but reinforcement tends to increase with increase in electrode extension, especially with positive polarity, and cold wire additions, while penetration decreases. However, the variations are barely outside the errors of measurement and fluctuation along a weld bead.

Table 2 The influence of cold wire additions on deposition rate and flux consumption

Proportion of cold wire addition, % of deposited weld metal	Total deposition rate, kg/hr	Flux consumption, kg/hr
0	11.05	12.63
13	12.50	13.18
37	15.10	13.18

Notes: Electrode polarity: Negative
Current: 600A
Head voltage: 38V
Electrode extension: 40mm
Travel speed: 380mm/min

The results of tensile and Charpy tests are given in Table 4, and Charpy transition curves are presented in Figs 4 and 5. Increased electrode extension at a constant head voltage (and energy) resulted in an increase in yield (12%) and tensile (5%) strengths, but with inconsistent changes in Charpy behaviour. With constant arc voltage (Series 2) changes in strength were small and inconsistent, and the Charpy transition behaviour deteriorated slightly with increased electrode extension and energy expenditure. Cold wire additions gave the most marked increase in yield (20%) and tensile (10%) strength without significant loss in ductility and some improvement in Charpy behaviour, Fig. 5.

Macroexamination failed to reveal any defects in the weld. The main feature was a reduction in the number of runs to complete each weld as deposition rate was increased, with resultant variation in the degree of weld metal refinement. Figure 6 indicates the influence of cold wire additions on macrostructure. The microstructures were typical of multirun welds deposited with a CMn steel wire and basic flux. The 'as-deposited' weld metal showed acicular ferrite within a columnar structure delineated by proeutectoid ferrite, whereas the refined weld metal consisted of fine equiaxed ferrite with significant amounts of a dark etching microconstituent. The microstructure remained essentially the same for all electrode extensions and cold wire additions. Inclusion content was also examined but the amounts were low and did not vary significantly. Attempts to correlate inclusion content with Charpy upper shelf energy and Mn:Si ratio gave no consistent relationship.

DISCUSSION

No practical difficulties were found in increasing electrode extension to 120mm and it is surprising that this method of increasing productivity has not been pursued more actively. In any welding process optimum operating

Table 3 The influence of electrode extension and cold wire additions on weld bead dimensions

Series	Electrode polarity	Energy, kJ/mm	Electrode extension, mm	Cold wire addition, %	Weld bead dimensions, mm		
					Width	Reinforcement	Penetration
1	Negative	3.60	40	0	24.0	4.0	5.0
			80		22.3	4.3	4.8
			120		22.0	5.0	5.3
			150		22.5	5.8	4.5
2	Negative	3.32	40	0	22.0	4.0	5.5
		3.41	80		22.5	5.0	6.0
		3.60	120		22.0	5.0	5.3
3	Positive	3.24	40	0	21.9	3.3	9.3
			80		23.2	3.8	8.3
			120		22.3	4.1	7.4
			160		22.9	4.5	6.8
4	Negative	3.60	40	0	22.8	4.0	5.5
				13	22.3	4.5	5.0
				31	21.5	5.3	4.3
				48	22.8	5.5	3.5

Table 4 Weld metal properties

Series	Energy, kJ/mm	Electrode extension, mm	Cold wire addition, %	Tensile strength, N/mm^2	Yield stress, N/mm^2	Elongation, %	Reduction of area, %	Hardness, HV	Charpy V notch tests	
									CV at -20°C, J	Temp. for CV of 40J, ^{o}C
1	3.60	40	0	547	419	34	70	175	46	-23
		80		564	430	32	66	178	31	-11
		120		584	471	32	70	181	54	-31
2	3.32	40	0	590	484	32	66	193	75	-43
	3.41	80		604	507	31	66	199	60	-33
	3.60	120		584	471	32	70	181	54	-31
4	3.60	40	0	547	419	34	70	175	46	-23
			13	591	497	32	63	187	87	-46
			37	601	504	30	68	199	76	-39

parameters must be established carefully, but deposition rates can be substantially increased with little change in energy expenditure, while bead size and shape is little altered although penetration can be reduced. The minor changes in weld bead dimensions should cause no surprise since the only change in Series 1 is arc voltage, which is known to have only a minor influence on bead dimensions.[10] For Series 2 arc voltage was kept constant so that no change would be expected. For negative polarity flux consumption decreased with increasing stickout for Series 1 and was unchanged for Series 2, which would be expected in relation to the arc length changes in these series. The results quoted by Kho[9] for DC + SA welding (Series 3) do show a more significant change in weld bead dimensions with increase in electrode extension, in particular a 20% decrease in penetration. This discrepancy between negative and positive polarities may be merely experimental error, but it is more likely to be related to changes in plasma jet forces which are much stronger in DC + welding. Cold wire additions have little effect on weld bead width, but the energy available is used to melt more filler metal and less parent metal with a consequent fall in penetration and increase in reinforcement.

The strength, ductility, and notch impact properties of all the weld deposits were satisfactory and, although many more results are needed to indicate or confirm trends, revealed nothing to deter the use of either increased electrode extensions or cold wire additions. Increased stickout for constant energy expenditure (varying arc voltage) indicated a significant increase in strength with only marginal loss in ductility, but Charpy results had no obvious trend. These trends may stem from a more effective use of available energy leading to lower average weld pool temperatures and higher cooling rates.

With a constant arc length, i.e. increasing energy expenditure, no change in tensile properties was found as the electrode extension was increased, but there was a deterioration in Charpy behaviour. The addition of cold wire significantly improved strength for a slight loss of ductility, and Charpy behaviour was also improved. Although the energy available was used more efficiently there was no obvious change in grain size or microstructure.

The limited results achieved in this investigation do nothing to discourage the use of increased electrode extension or cold wire addition. The resulting improvement in efficiency does not exact a penalty in weld bead shape although weld penetration effects merit more attention. Properties are also satisfactory although more results are needed.

CONCLUSIONS

1 With 4.0mm diameter wire electrode extensions up to 120mm can be used in SA welding to give significant increases in deposition rate

2 Yield and tensile strength increase with electrode extension at constant welding energy

3 Yield and tensile strength showed little variation with electrode extension when the arc length was kept constant

4 Charpy V notch impact behaviour varied with electrode extension but no trend was established

5 Cold wire additions can be used to increase deposition rates

6 Cold wire additions improve tensile strength and Charpy impact behaviour

REFERENCES

1 OUTCALT, F.G. and KEIR, J.M. 'Survey of welding and cutting in ship construction'. Welding J., 21 (1), 1942, 5-15.

2 McMAHON, B.P. and DUFFY, F.H. 'Multipower submerged-arc welding'. Procs Conference 'Advances in Welding Processes', Harrogate, 14-16 April 1970. Abington, Welding Inst., 1970, 51.

3 LEE, L.F. 'Hot wire methods of surfacing and submerged-arc welding'. Metal Constr., 4 (11), 1972, 408-12.

4 HINKEL, J.E. 'Long stickout welding — a practical way to increase deposition rates'. Welding J., 47 (11), 1968, 869-74.

5 MALLOY, J. 'High deposition techniques in submerged-arc welding'. MSc thesis, CIT, 1975.

6 REYNOLDS, D.E.H. 'High deposition rate submerged-arc welding', Chapter 7 'Submerged-arc welding'. Abington, Welding Inst., 1978.

7 ANDERSON, O., BAGGERUD, A., and THAULOW, C. 'The influence of cold wire and fluoride additions on weld metal toughness in submerged-arc welding'. Welding and Metal Fab., 43 (11), 1975, 704-708.

8 HAECK, R.J. de. 'A new high efficiency welding procedure using finely divided filler metal'. Welding J., 50 (10), 1971, 689-700.

9 KHO, T.C. 'High deposition rate submerged-arc welding — welding parameters'. MSc thesis, CIT, 1976.

10 RENWICK, B.G. and PATCHETT, B.M. 'Operating characteristics of the submerged-arc process'. Welding J., 55 (3), 1976, 69s-76s.

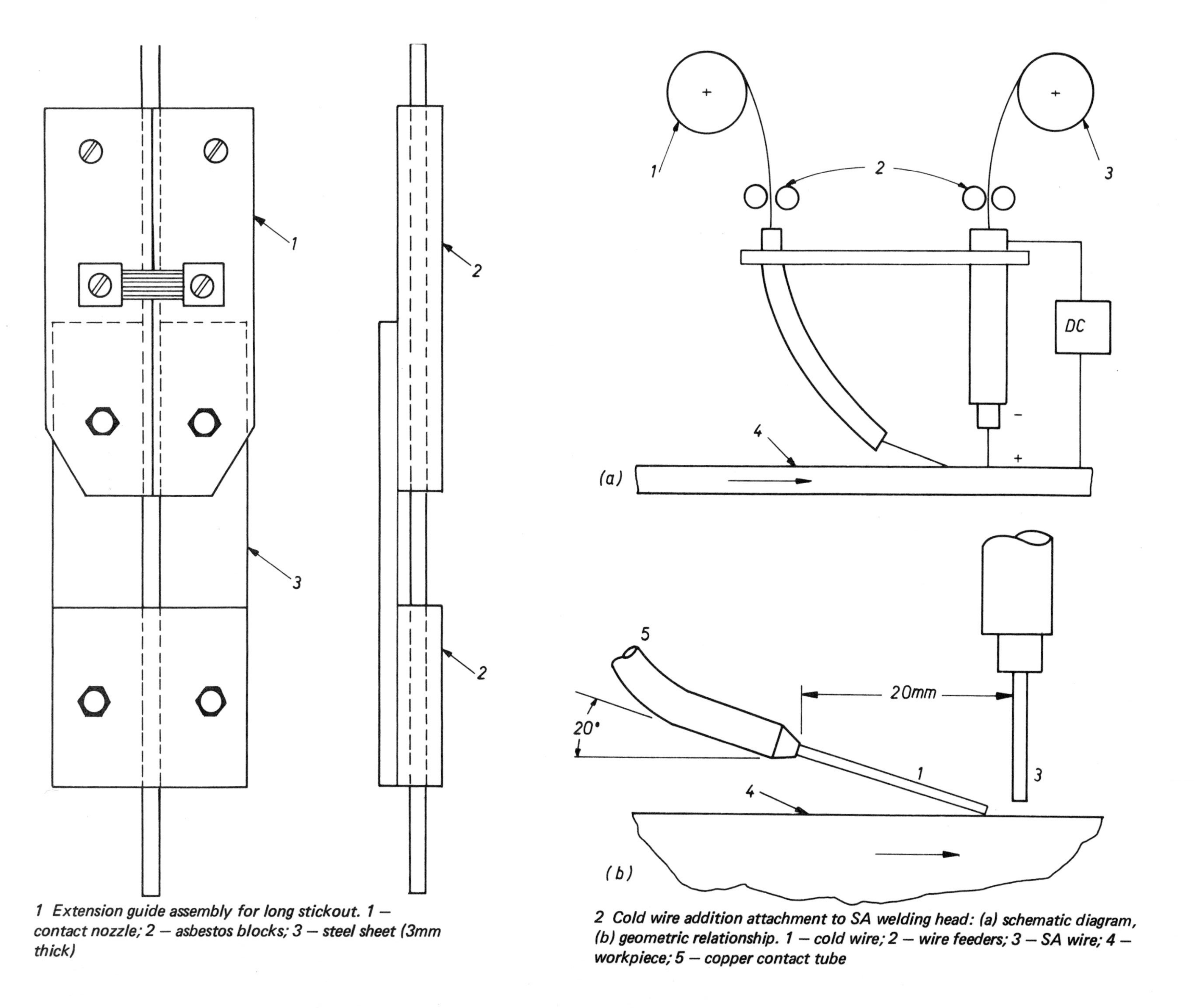

1 Extension guide assembly for long stickout. 1 — contact nozzle; 2 — asbestos blocks; 3 — steel sheet (3mm thick)

2 Cold wire addition attachment to SA welding head: (a) schematic diagram, (b) geometric relationship. 1 — cold wire; 2 — wire feeders; 3 — SA wire; 4 — workpiece; 5 — copper contact tube

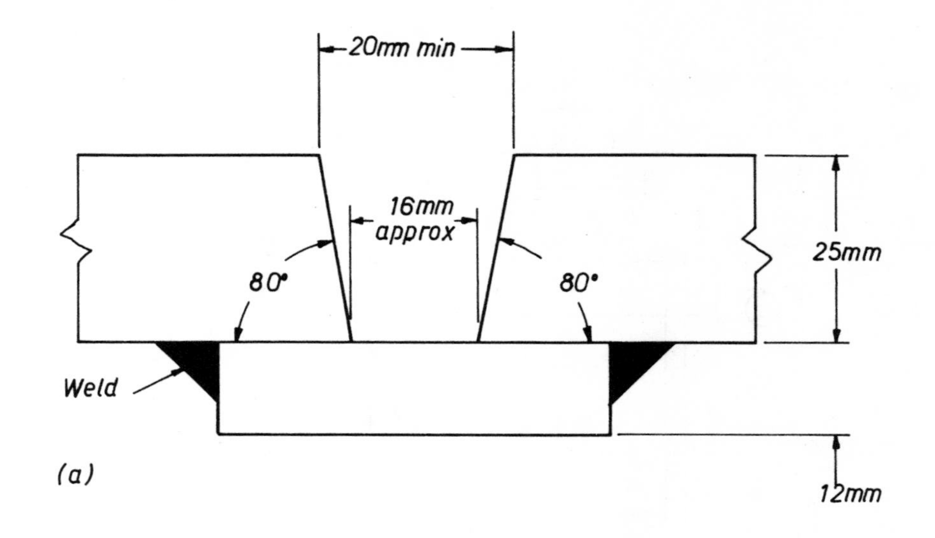

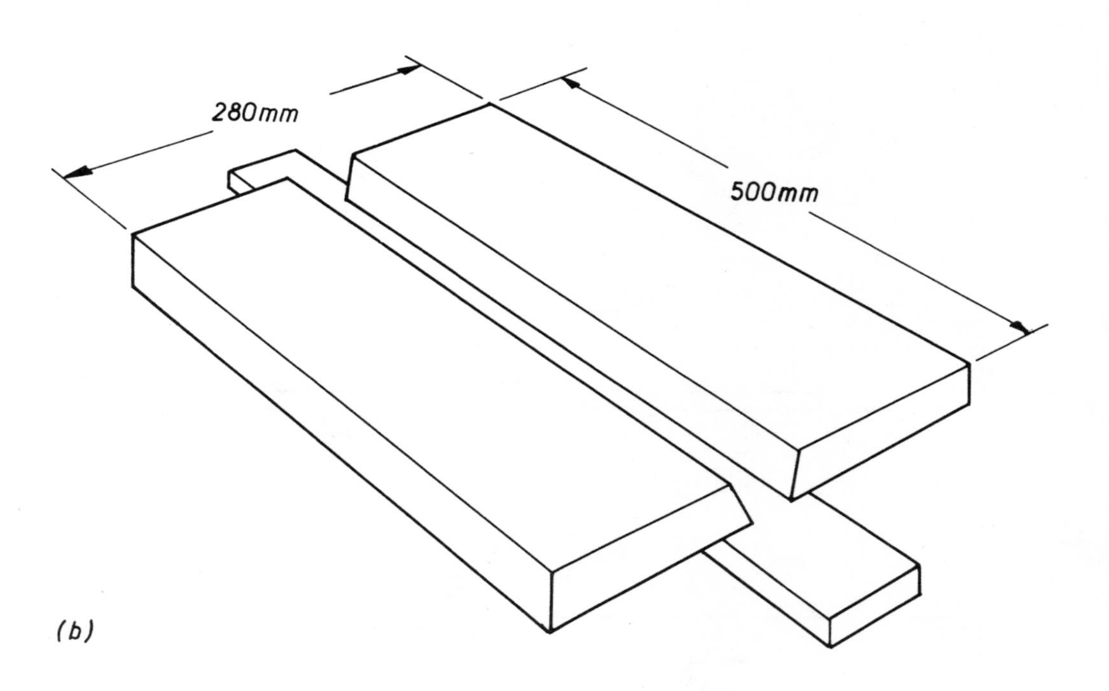

3 Weld preparation and testplate assembly: (a) preparation, (b) assembly

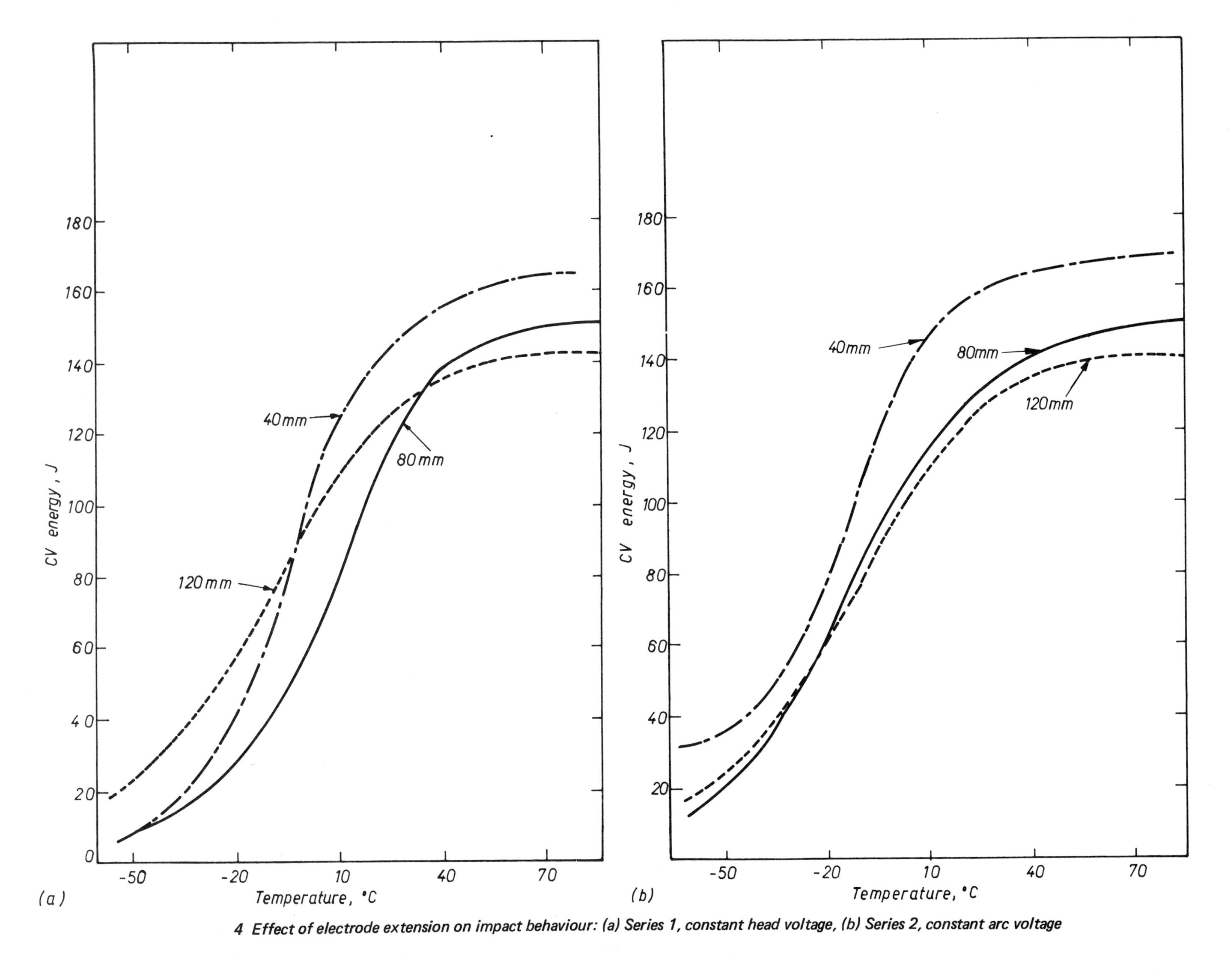

4 Effect of electrode extension on impact behaviour: (a) Series 1, constant head voltage, (b) Series 2, constant arc voltage

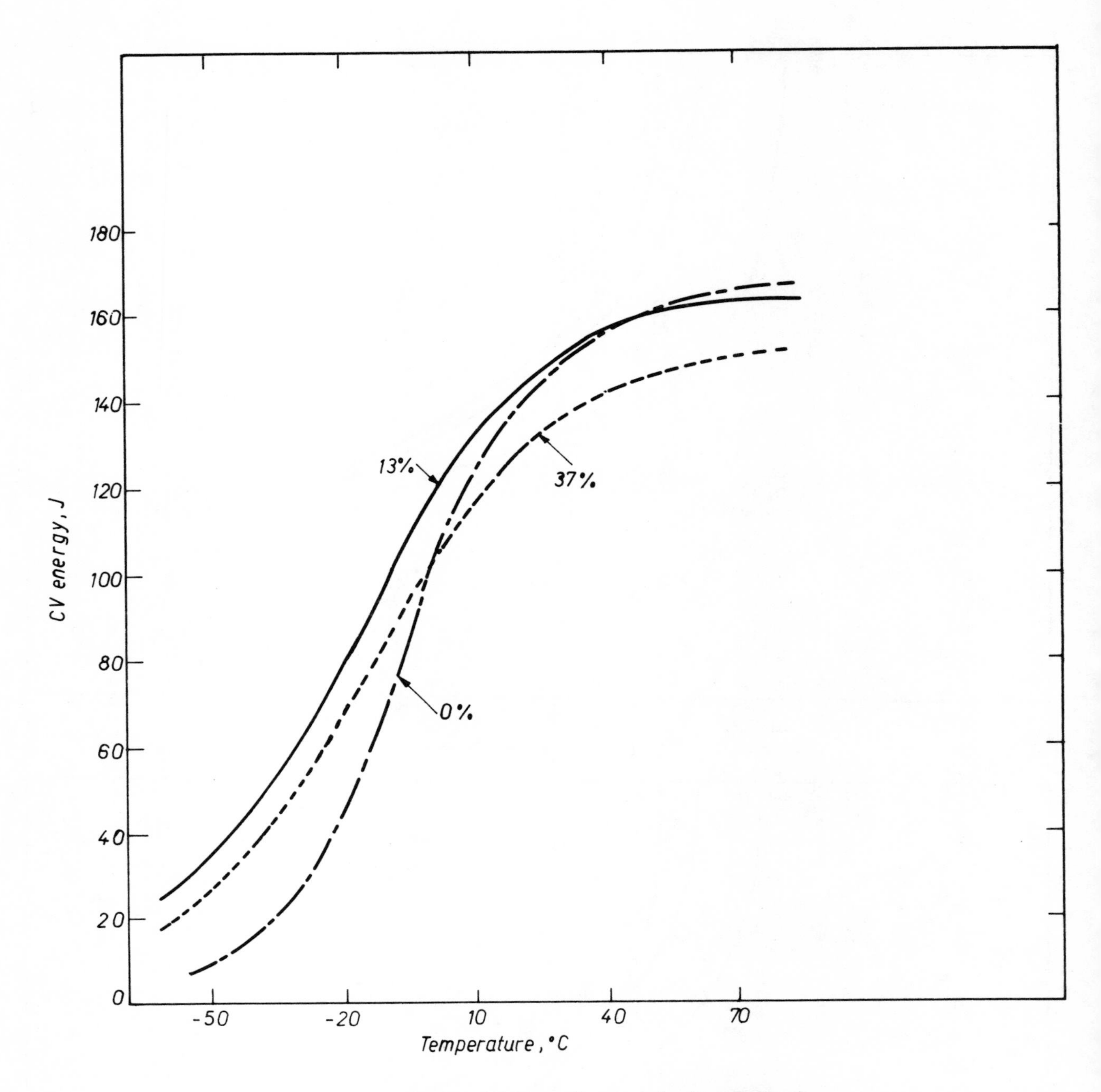

5 Effect of cold wire addition on impact behaviour, Series 4

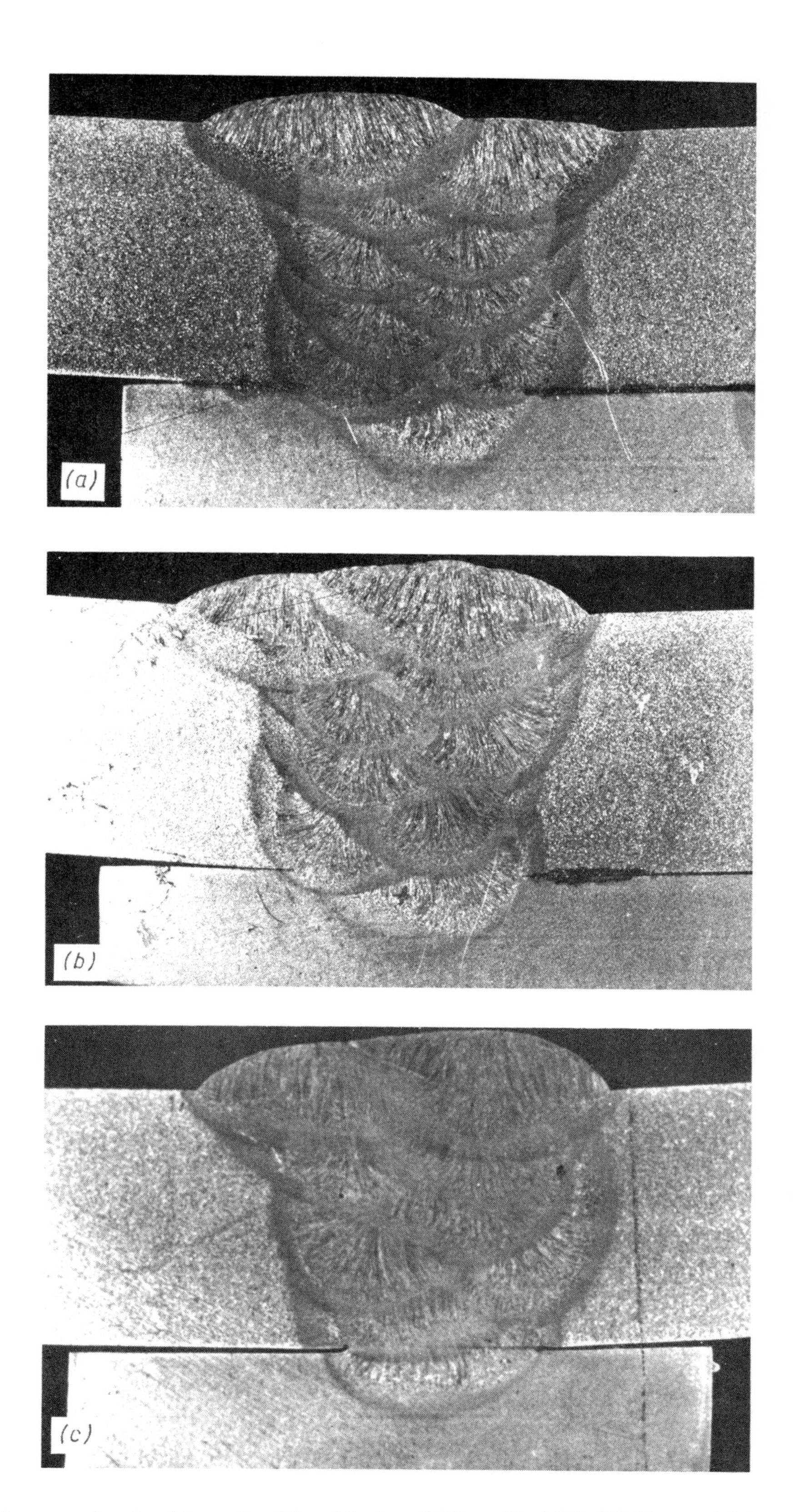

6 Macrographs of welds made with cold wire additions, %: (a) 0, (b) 13, and (c) 37

PAPER 43

Dilution and energy input control in multipower submerged-arc welding

S.B.Jones, BSc, PhD

The variation of metal deposition rate with welding current and electrode extension has been measured for the submerged-arc welding of CMn steel using DC electrode positive (DC+), DC electrode negative (DC-), and AC. The influence of subsidiary variables including wire diameter, arc voltage, flux, and equipment type was also considered to obtain a range of basic deposition data for the process.

This information was supplied in a series of welding trials designed to improve control in multipower submerged-arc welding. Conventional multipower welding procedures were modified using high deposition rate techniques with suitable alteration to joint preparations and welding conditions, in efforts to reduce heat input and dilution. Results indicate that high deposition methods give added flexibility to multipower submerged-arc systems, and may in principle be used to control energy input and weld metal dilution, especially in one- and two-pass applications.

INTRODUCTION

From the earliest development stages it was recognised that the efficiency of the submerged-arc (SA) welding process could be improved by increasing metal deposition rates. A wide variety of suitable techniques was rapidly developed, which have been known for many years.[1] Nevertheless, exploitation has been limited and only recently have demands for improved economy, together with developments to consumables, awakened widespread interest in the production use of high deposition methods.

The methods available may be divided into three categories:

1 Techniques which make use of the inherent qualities of the SA system, e.g. alteration of current type and polarity or the use of long contact tip to workpiece distances.[2,3] These methods are among the cheapest and simplest to apply, as they do not involve extensive modification to welding plant

2 Methods which stabilise the arc or weld pool to enable operation at increased overall current levels, such as parallel wire welding and the large range of multipower welding systems.[4,5] These are an effective means of improving deposition rate, since the latter primarily depends on the welding current

3 Methods which add supplementary filler material to the weld pool to improve deposition and process control, including direct addition of resistance-heated or

Dr Jones is a Senior Research Engineer in the Production and Economics Department of The Welding Institute.

cold wires to the weld pool, use of flux containing metal powder, or the addition of powdered, granulated, or otherwise divided metal to the joint preparation before welding.[6,7] These techniques utilise excess arc heating to melt extra filler material while still effecting full joint fusion

The individual deposition methods outlined above may be used singly or in combination to give a large number of SA process variations offering a wide range of welding characteristics. Recent publications in this field[8,9] demonstrate the variety of process routes which are being actively developed for production use.

Most of these approaches to high deposition rate welding have the principal objective of increasing joint completion rates, either by enabling joint filling in fewer passes or allowing increased welding speeds. This objective is of clear significance, especially in situations where the SA welding process determines overall production rates. In such instances improvements in process efficiency can be equivalent to an increase in overall plant capacity.

In addition to this straightforward means of exploiting increased deposition rates, there is evidence that high deposition methods increase overall process flexibility. Improvements in solidification cracking resistance have been reported for single-wire welding,[3] and increased deposition at restricted nominal energy input has been sought, and obtained, in multipass welding.[10]

The improvement obtained in process flexibility and control would appear to be especially appropriate to multipower welding systems. Situations where these systems are employed usually involve high production rates which place particular demands on the soundness and mechanical properties of the welds. Problems with weld and heat-affected zone (HAZ) quality can limit production capacity or impose uneconomic limits on the choice of consumables or even parent materials.

The present investigation is concerned with the application of simple high deposition methods (involving extended contact tip to workpiece distances) to multipower SA welding systems to determine whether these techniques offered potential improvements in process control compared with conventional multipower SA practice.

EXPERIMENTAL APPROACH

Welds were made using a tractor-mounted multipower SA unit, Fig.1. This was fitted with three welding heads: one, powered by a flat characteristic 1400A capacity transformer-rectifier, for DC operation and two powered by Scott-connected 1500A capacity transformers for AC welding.

The influence of power source characteristic on deposition rate was assessed for DC operation using an alternative welding head powered by a 900A capacity drooping characteristic transformer-rectifier.

Welding current and voltage levels were monitored using moving coil chart recorders, and dial gauges. An ink jet oscilloscope was used for more detailed electrical records. Elapsed wire feed was measured using digital meters maintained in contact with the electrode wires.

Bright drawn mild steel in 19.0 and 25.4mm thickness was used as testplate material throughout the welding trials. Butt and fillet welds were made between plates of 800 x 150mm with preparations as described later. Test panels were heavily tack welded at each end and rigidly clamped to a heavy steel welding bench.

Used in this project were CMn steel wires of 3.2, 4.0, and 5.0mm diameter to S3 (1.5%Mn) specification. A variety of fluxes was used in the course of the work, most of the welding being carried out with a basic agglomerated carbonate-bearing flux. Other fluxes considered included fused acid manganese silicate, fused basic rutile, and agglomerated neutral alumina types.

HIGH DEPOSITION RATE STUDIES

Basic deposition rate measurements were made from bead-on-plate welding trials with 4mm diameter electrodes at nozzle-plate distances of 30, 90, 120, and 150mm using DC electrode +ve (DC+), DC electrode -ve (DC-), and AC. The welding speed was kept constant at 500mm/min throughout and welding currents were varied from 400-1200A in 200A steps. Welding voltages were increased with electrode extension from 30V at 30mm to 35, 39, and 42V at extensions of 90, 120, and 150mm to allow for the voltage drop as a result of resistance heating.

In each test welding was continued for 60sec, the amount of electrode consumed being determined from the digital wire feed meter attached to the wire. Results of these trials gave the relationship between metal deposition rate and welding current for DC+, DC-, and AC over the range of electrode extensions studied, Fig.2.

At electrode extensions above 90mm resistance heating made the end of the electrode excessively flexible and mobile, and seam tracking was difficult. Therefore, to control the heated electrode extension, an insulated copper electrode support was used which comprised an outer copper body (which could be slid over the normal contact tip assembly) to adjust the amount of extension between the contact tip and workpiece. The body of the support was insulated from the electrode by a wide bore recrystallised alumina tube, with a dummy contact tip made from a replaceable length of 5mm bore silica tube held in a split Tufnol collet, Fig. 3.

With electrode extensions of 120 and 150mm arc starting was occasionally difficult because the extended electrode tended to overheat and melt before an arc was established. This problem was avoided by running the extended electrode as the middle or trail arc of a triple arc system. With this arrangement the extended electrode was fed directly into a molten weld pool, and arc starting was straightforward. The system could be operated in either hot wire or multiple arc mode by changing the voltage on the extended electrode.

From Fig. 2 it is evident that welding with DC- polarity and increased electrode extension is capable of giving deposition rates considerably greater than those obtainable with DC+ or AC. The maximum AC, at which stable operation was possible, decreased with increasing electrode extension, thus limiting the use of this technique on AC, Fig. 2c. The reason for this limitation was not definitely identified but appeared to be associated with arc stability.

The influence of wire diameter on metal deposition rate

Results of welding trials made with 3.2 and 5.0mm wires at electrode extensions of 30 and 150mm using DC+, DC-, and AC currents are shown in Fig. 4. The same overall trends, observed with 4.0mm wire, were repeated at the other two wire sizes although AC operation at long electrode extension appeared more stable with a 5.0mm wire. Comparison of results for the wire diameters studied demonstrates the influence of resistance heating in increasing metal deposition rate as wire diameters are reduced. Even in a tandem arc system the 3.2mm diameter electrode was usable only over a very limited current range at 150mm extension, and even then gave signs of instability which would preclude such use in practice. The 5.0mm diameter wire also showed a restricted usefulness at long electrode extension with DC- operation, and stable arc behaviour could be established only within a relatively narrow current range.

These results suggest that 4.0mm diameter wires give the greatest flexibility of operation at long electrode extension, especially over the current ranges commonly used in single and multipower SA welding.

The influence of arc voltage on metal deposition

The relationship between metal deposition rate and welding current was investigated for welds made with 4.0mm diameter wire and agglomerated basic carbonate-bearing flux over a range of arc voltages. No influence of arc voltage upon metal deposition rate could be detected, and no arc instabilities were encountered except at the extreme of the current/voltage range. The shape and penetration profile of the resulting weld beads on the other hand was markedly affected. These results would suggest that normal procedures for selecting voltage for multiarc systems should remain valid when high deposition techniques are used, provided allowance is made for voltage drop in the electrode extension.

The influence of flux type on metal deposition

A series of deposition rate determinations was made using 4mm diameter electrodes and DC+ and DC- with agglomerated basic carbonate-bearing, fused acid manganese silicate, and agglomerated neutral alumina-based fluxes. No influence of flux type on metal deposition could be detected at currents below 700A, but welds made with the fused rutile-based flux showed a slightly increased deposition tendency at currents above this level.

The influence of power source characteristic on metal deposition

Direct current power source characteristic did not appear to influence deposition behaviour up to 900A, the capacity of the drooping characteristic rectifier. Oscilloscope records of welding current and voltage did, however, reveal pronounced differences in the short-term stability of current and voltage levels. These were in line with established power source/arc behaviour.[11]

THE GENERAL APPROACH TO PROCESS CONTROL WITH HIGH DEPOSITION TECHNIQUES

There is often scope using conventional welding procedures for more efficient control of weld metal composition by choice of welding parameters and joint preparations. Even so, the level to which energy input and weld metal dilution can be reduced using these techniques is limited by the need to maintain sidewall fusion and produce defect-free welds. This is especially true of root runs, or two-pass welds, where weld metal dilution is high and danger of lack of fusion or solidification cracking is greatest.

Beyond the optimisation of conventional welding techniques, one approach to energy input and weld metal dilution control may be made by increasing metal deposition rates without enhancing penetration capability. This will enable production of weld beads containing lower proportions of parent material, and these should in turn be less susceptible to solidification cracking and other metallurgical problems.[12] Weld metal properties will therefore be more directly dependent on the composition of the consumables involved. This approach is shown schematically in Fig. 5 for butt and fillet welding situations, where low deposition, high penetration, techniques are replaced by high deposition, low penetration, conditions. In one- and two-pass butt welds this transition involves use of a larger joint preparation section, but fillet welds rely on increased deposition to produce an equivalent weld size without incurring so much parent metal fusion.

Figure 6 outlines the main approaches which may be followed to control weld metal dilution and energy input when replacing conventional welding methods by high deposition welding techniques. The two curves shown relate deposition rate and welding current for two process variants, one of which has a higher characteristic deposition rate than the other. Three main approaches may be identified from Fig. 6:

(a) equal deposition
Deposition rates obtained at high currents with the conventional welding process can be achieved at lower currents with the high deposition technique. A reduction in welding current, and nominal heat input, is therefore possible while maintaining joint filling capability. In the example shown, a weld made at 750A using the high deposition method, Y, will involve a lower heat input and have a lower dilution than a weld made at 1000A using the conventional method, X. This arises since the penetration capability of method Y at 750A will be lower than method X at the higher current, although the deposited volumes are equal

(b) equal current
If the two process variants are compared at similar welding currents, a change to a high deposition technique may be seen to increase joint-filling capability (and reduce dilution levels) considerably without increasing energy input. The change in penetration capability between the two techniques will depend on the high deposition method used, but is not necessarily drastic

(c) intermediate currents
Between these extremes there is considerable scope for using high deposition techniques at currents between the equal deposition and equal current points. This allows both an increase in deposition rate and reduction in energy input while still reducing penetration capability

The scope for controlling weld metal dilution and heat using high deposition methods would appear to be especially great in the multipower welding systems. In these systems the functions of penetration, joint filling, and bead profiling of the weld may be divided between the arcs involved. Figure 7 illustrates this process for a triple arc system in which the lead arc is primarily responsible for penetration, the second arc adds to penetration and controls joint filling, and the trail arc smooths the in-plate and surface profiles of the weld bead. Penetration and deposition characteristics of multiarc systems may therefore be controlled independently to an extent which depends on the makeup of the multiarc system concerned.

With a multipower system several high deposition rate methods may be used simultaneously. Even when the basic system has been established, the range of control approaches outlined in Fig. 6 may be used in combination to fulfil the various arc functions. Whatever arrangement is chosen, the objective of dilution reduction in two-pass butt welding may be reached by increasing joint volume and reducing root face measurements. In fillet welds it will be sufficient to reduce penetration while fulfilling the leg length and throat thickness criteria.

EXAMPLES OF THE USE OF HIGH DEPOSITION RATE TECHNIQUES FOR CONTROLLING WELD METAL DILUTION AND ENERGY INPUT

The following examples are intended to illustrate possible ways of using high deposition rate methods for reducing weld metal dilution levels and controlling energy input. The procedures shown do not represent a final

Table 1 Welding conditions and weld bead data for two-pass butt welds in 25.4mm plate

Current type and polarity			Current, A	Voltage, V	Electrode extension, mm	Welding speed, m/min	Nominal energy input, kJ/mm	Penetration, % plate thickness	Weld metal dilution, %
(a) conventional procedure: 75° symmetrical double V preparation 7mm root face									
	Lead	DC+	750	30	30				
Side 1	Middle	AC	1000	37	30	1.0	5.3	52	59
	Trail	AC	700	40	30				
	Lead	DC+	1150	30	30				
Side 2	Middle	AC	1000	37	30	1.0	6.0	66	61
	Trail	AC	700	40	30				
(b) revised procedure: 75° symmetrical double V preparation 4mm root face									
	Lead	AC	750	30	30				
Side 1	Middle	DC-	700	36	150	1.0	4.5	48	31
	Trail	AC	700	40	30				
	Lead	AC	1200	30	30				
Side 2	Middle	DC-	700	36	150	1.0	5.4	68	38
	Trail	AC	700	40	30				

Table 2 Analyses of parent plate, wire, and welds from triple wire two-pass butt welding trails on 25.4mm plate

Item	C	S	P	Si	Mn	Ni	Cr	Mo	V	Cu	Nb	Ti	Al	B	Pb	Sn
Parent plate	0.23	0.031	0.022	0.02	0.99	0.03	<0.01	<0.01	<0.01	0.04	<0.005	<0.01	<0.005	<0.001	<0.01	<0.01
4.0mm S3 wire	0.14	0.012	0.011	0.23	1.45	0.03	0.05	0.01	<0.01	0.10	<0.005	<0.01	0.008	<0.001	<0.01	<0.01
DC+ ACAC weld																
1st pass	0.19	0.016	0.017	0.10	1.10	0.03	0.02	<0.01	<0.01	0.06	<0.005	<0.01	0.005	<0.001	<0.01	<0.01
2nd pass	0.18	0.017	0.017	0.10	1.10	0.03	0.02	<0.01	<0.01	0.06	<0.005	<0.01	0.006	<0.001	<0.01	<0.01
ACDC- (150mm) AC weld																
1st pass	0.16	0.018	0.017	0.11	1.09	0.04	0.02	<0.01	<0.01	0.07	<0.005	<0.01	<0.005	<0.001	<0.01	<0.01
2nd pass	0.16	0.019	0.017	0.11	1.12	0.04	0.02	<0.01	<0.01	0.08	<0.005	<0.01	<0.005	<0.001	<0.01	<0.01

stage in procedure development and bead shapes would have to be modified further for production use. It is considered that this would be a relatively straightforward operation in procedural trials.

Butt welds in 25.4mm steel plate

A conventional DC+ ACAC two-pass welding procedure was used as a basis for welding developments in 25.4mm plate. The electrode configuration shown in Fig. 8a and the welding conditions shown in the first part of Table 1 were used to produce a DC+ ACAC test panel at 1.0m/min. Figure 9a shows a typical section through the resulting weld, which had a maximum dilution of 61% calculated from area measurements on transverse joint sections.

As a first step in the modification of this procedure to lower weld metal dilution, the root face of the preparation was decreased from 7 to 4mm, without altering the joint angle. An ACDC- AC welding technique, Table 1, was chosen as an alternative to the conventional DC+ ACAC power connection, advantage being taken of the enhanced deposition rate of the DC- middle arc to increase the deposition capability of the system as a whole. The increased deposition rate required by the enlarged volume of the modified joint preparation could have been obtained by the use of a high current on the DC- middle arc, but it was thought more worthwhile to attempt an overall decrease in energy input. A 150mm electrode extension was therefore applied to the DC- welding head, both AC heads being set with normal 30mm electrode extensions. Comparison of the weld beads produced by the two techniques Table 1, Fig. 9a and b, revealed that use of the ACDC- AC method had reduced the maximum dilution level from 61 to 38% while reducing the maximum nominal energy input by 10% to 5.4kJ/mm. Analyses of parent plate, wire, and the resulting welds made using conventional and improved techniques are shown in Table 2. The presence of reduced dilution in the ACDC- AC weld is indicated by the change in carbon level between the two welds.

Butt welds in 19.0mm steel plate

The conventional procedure used as a basis for butt welding trials in 19.0mm plate was taken from an industrial application and involved use of 5.0mm diameter electrodes for two-pass DC+ ACAC welding. Table 3 and Fig. 10a show the welding conditions and resulting weld bead produced with the conventional DC+ ACAC technique, using the electrode configuration shown in Fig. 8b.

Table 3 Welding conditions and weld bead data for butt welds in 19.0mm plate

Current type and polarity			Current, A	Voltage, V	Electrode extension, mm	Welding speed, m/min	Nominal energy input, kJ/mm	Penetration, % plate thickness	Weld metal dilution, %
(a) conventional procedure: 75° symmetrical double V preparation 8mm root face									
Side 1	Lead	DC+	710	33	35				
	Middle	AC	800	32	35	1.2	4.5	45	57
	Trail	AC	1000	40	35				
Side 2	Lead	DC+	980	34	35				
	Middle	AC	1000	35	35	1.1	5.8	57	65
	Trail	AC	1000	38	35				
(b) revised procedure : 75° symmetrical double V preparation 6mm root face									
Side 1	Lead	DC+	900	36	30			Pass 1 56	Pass 1 58
Side 2	Middle	AC	700	33	120	1.1	4.4	Pass 11 56	Pass 11 57
	Trail	AC	800	32	90				

The symmetrical 75° double V preparation with 8mm root face was modified by reducing the root face to 6mm, thus bringing the preparation into line with those used for single arc welding. In view of the relatively small change in preparation area involved, it was decided that the limited capability of extended electrode AC welding should be investigated. Conditions were selected accordingly using AC middle and trail arcs with 120 and 90mm extension respectively, Table 3, without altering the basic electrode configuration of Fig. 8b. A section through the resulting weld is shown in Fig. 10b.

Since there is only a small change in joint preparation, the reduction in dilution by use of the higher deposition process was small, amounting to only 7%, Table 3. The maximum nominal energy input was, however, reduced by 24% to 4.4J/mm.

The principal advantage of this particular type of procedure modification probably lies in directions other than those of straightforward dilution or energy input control. Use of a preparation suitable for either multiarc or single arc SA welding could give a useful degree of flexibility in production, and the use of identical welding procedures for first and second passes gives obvious production advantages.

DISCUSSION

The work covered by this Paper has outlined the deposition increases attainable with only a small number of the high deposition SA techniques available. It has also shown that these techniques are particularly applicable to multipower SA welding systems. In view of the large number of variables involved in the SA process, the present results cannot be considered comprehensive and do not necessarily represent the optimum joint qualities possible with the techniques considered. It is, however, hoped that they demonstrate the potential of these methods, and may encourage further work to improve the efficiency and flexibility of the process.

The pattern of metal deposition rate with respect to variation in current type and polarity is in line with previously reported work, but the present results represent a coherent body of information generated from the same experimental base. The influence of subsidiary variables including flux, equipment type, and arc voltage does not appear to be marked within the range of deposition considered. For this reason the use of the basic deposition data in determining conditions for industrial multipower systems appears justified.

Weld metal dilution control using high deposition rate systems is likely to be most necessary, and most effective, where dilution levels with normal techniques are high and parent materials are high in C and S. These methods may therefore be particularly effective for the control of solidification cracking in root runs, one- and two-pass butt welding situations, and fillet joints. Where dilution levels are lower, e.g. multipass filling runs, advantage may be taken of the improved deposition efficiency of these methods.

As mentioned earlier, the particular advantage of polarity changes and use of long electrode extensions is that only small changes are required to existing equipment and there is no need for additional consumables.

It must, however, be remembered that any technique which increases weld metal deposition rate without increasing penetration capability may, in principle, be used to control energy input and weld metal dilution. Methods which may be of use in this context include additions of powdered or granular metal,[7] hot or cold auxiliary wires, and parallel wires or series arc connections. The relative merits of such techniques in multipower situations remain to be determined both from the process and metallurgical points of view, but they should be borne in mind when heat input or dilution control are considered.

SUMMARY AND CONCLUSIONS

This work has outlined the possibilities of improving the efficiency and flexibility of the multipower SA process by improving metal deposition rates.

The influence of welding current type, polarity, and electrode extension on SA weld metal deposition rates has been established over a range of welding currents using 3.2, 4.0, and 5.0mm diameter wires and a basic carbonate-bearing flux. The effect of power source characteristic, arc voltage, and flux type have been assessed for a 4.0mm diameter wire.

Improved deposition methods have been applied to multipower SA systems to control weld metal dilution and energy input. Much work clearly remains to be done before the picture is fully coherent, both within the techniques studied and in the wider context of improved deposition rate methods. The following conclusions may be drawn at this stage.

1 High deposition methods are particularly applicable to multipower systems. They have some promise for controlling nominal energy input levels and weld metal dilution, and hence susceptibility to metallurgical defects such as solidification cracking. They may also help to alleviate difficulties with mechanical properties where these stem from excessive heat input or pickup of undesirable alloying elements from the parent plate

2 Techniques using high deposition rate methods for dilution control are most applicable to root runs, one- and two-pass butt welding, and fillet joints

3 In low dilution welds the composition of consumables, and their effect on weld pool composition, will assume a more direct significance than in high dilution welds

4 Alternative methods of increasing metal deposition rates may also be of use in the control of heat input and weld metal dilution. Use of current type, polarity, and electrode extensions for increasing deposition rates has the advantage of requiring little modification to existing equipment and does not involve the use of new or additional consumables

ACKNOWLEDGEMENTS

The author wishes to acknowledge the help of Messrs J. Clough, J. Haugh, and C. Robson in carrying out the experimental work, and colleagues at The Welding Institute for many useful discussions.

REFERENCES

1 KNIGHT, D. E. 'Multiple electrode welding by "Unionmelt" process'. Welding J., 33 (4), 1954, 303-12.
2 REYNOLDS, D. E. H. 'Decreasing welding costs in heavy fabrication, Pt 1'. Welding and Metal Fab., 42 (3), 1974, 94-8.
3 BERKHOUT, C. F. and AVEST, F. J. ter. 'A technological consideration of submerged-arc welding with a relatively large current-carrying portion of the wire' (in Dutch). Lastechniek, 38 (8), 1972, 191-8. English version in WI Translation no. 392.
4 HINKEL, J. E. and FORSTHOEFEL, F. W. 'High current density submerged-arc welding with twin electrodes'. Welding J., 55 (3), 1976, 175-80.
5 GRONBECK, I. 'Submerged-arc welding

with a multielectrode system'. Svetsaren, (2), 1973, 8-12.

6 SAENGER, J.F. 'Hot wire — a new dimension in arc welding'. Welding and Metal Fab., 39 (6), 1971, 227-34.

7 TROYER, R.W. and MIKURAK, J. 'High deposition submerged-arc welding with iron powder joint fill'. Welding J., 53 (8), 1974, 494-504.

8 REYNOLDS, D.E.H. 'High deposition rate submerged-arc welding', Chapter 7 in 'Submerged-arc welding'. Abington, Welding Inst., 1978.

9 MARSHALL, V.G. 'The submerged-arc process in general fabrication', Chapter 10 in 'Submerged-arc welding'. Abington, Welding Inst., 1978.

10 WOLFENDEN, T. 'Some applications of submerged-arc welding in shipbuilding'. Paper presented at Welding Inst. Submerged-arc Welding Technical Group, London, 1973.

11 UTTRACHI, G.D. 'A new DC power source for submerged-arc welding'. Welding J., 49 (12), 1970, 913-17.

12 BAILEY, N. and JONES, S.B. 'Cracking of ferritic steels during submerged-arc welding'. Abington, Welding Inst., 1977.

1 Multipower SA welding equipment. 1 — flux reservoir; 2 — control unit; 3 — wire feed units; 4 — workpiece

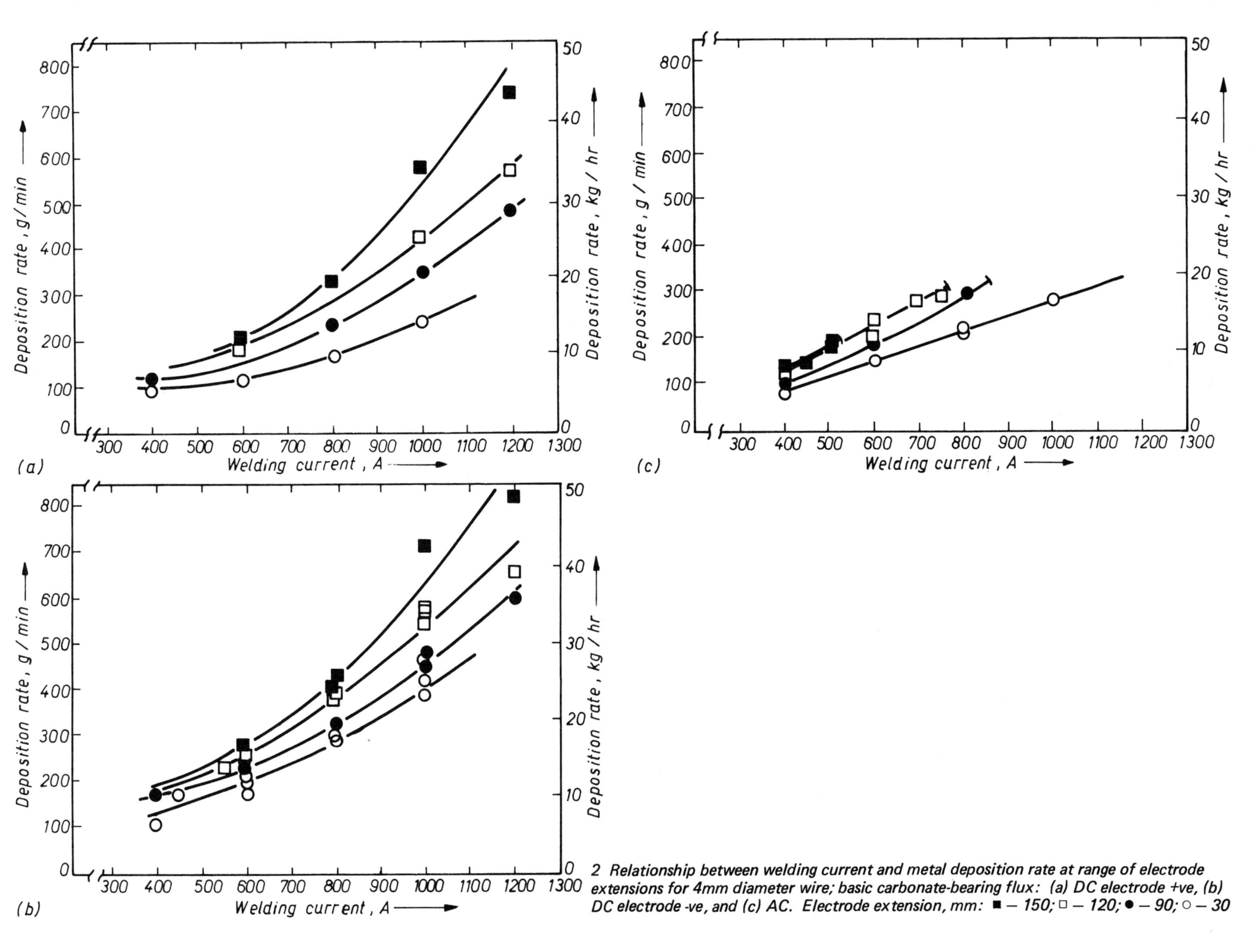

2 Relationship between welding current and metal deposition rate at range of electrode extensions for 4mm diameter wire; basic carbonate-bearing flux: (a) DC electrode +ve, (b) DC electrode -ve, and (c) AC. Electrode extension, mm: ■ – 150; □ – 120; ● – 90; ○ – 30

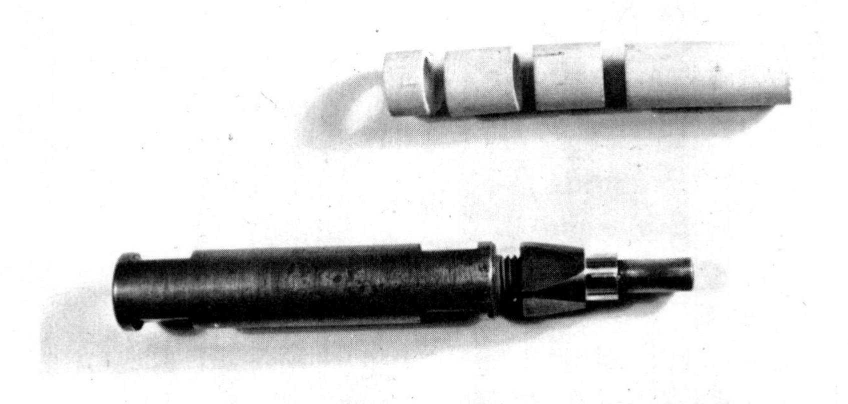

34782

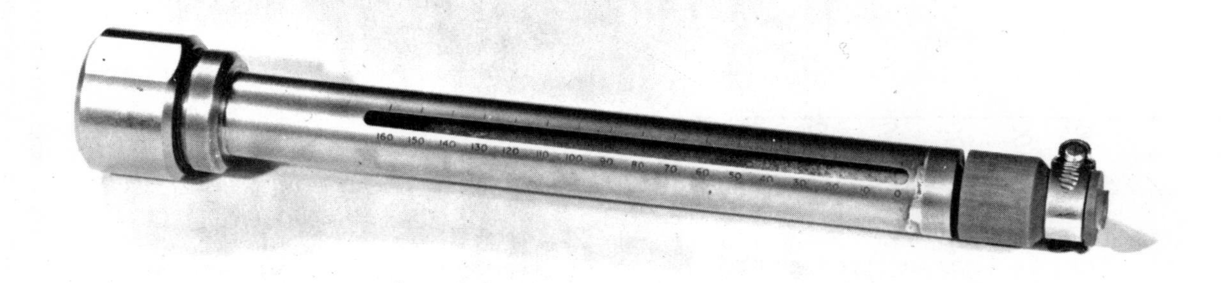

36528

3 Insulated electrode support: (top) internal alumina insulating sleeves; (centre) conventional contact tip assembly; (bottom) insulated electrode support with replaceable insulating silica tip

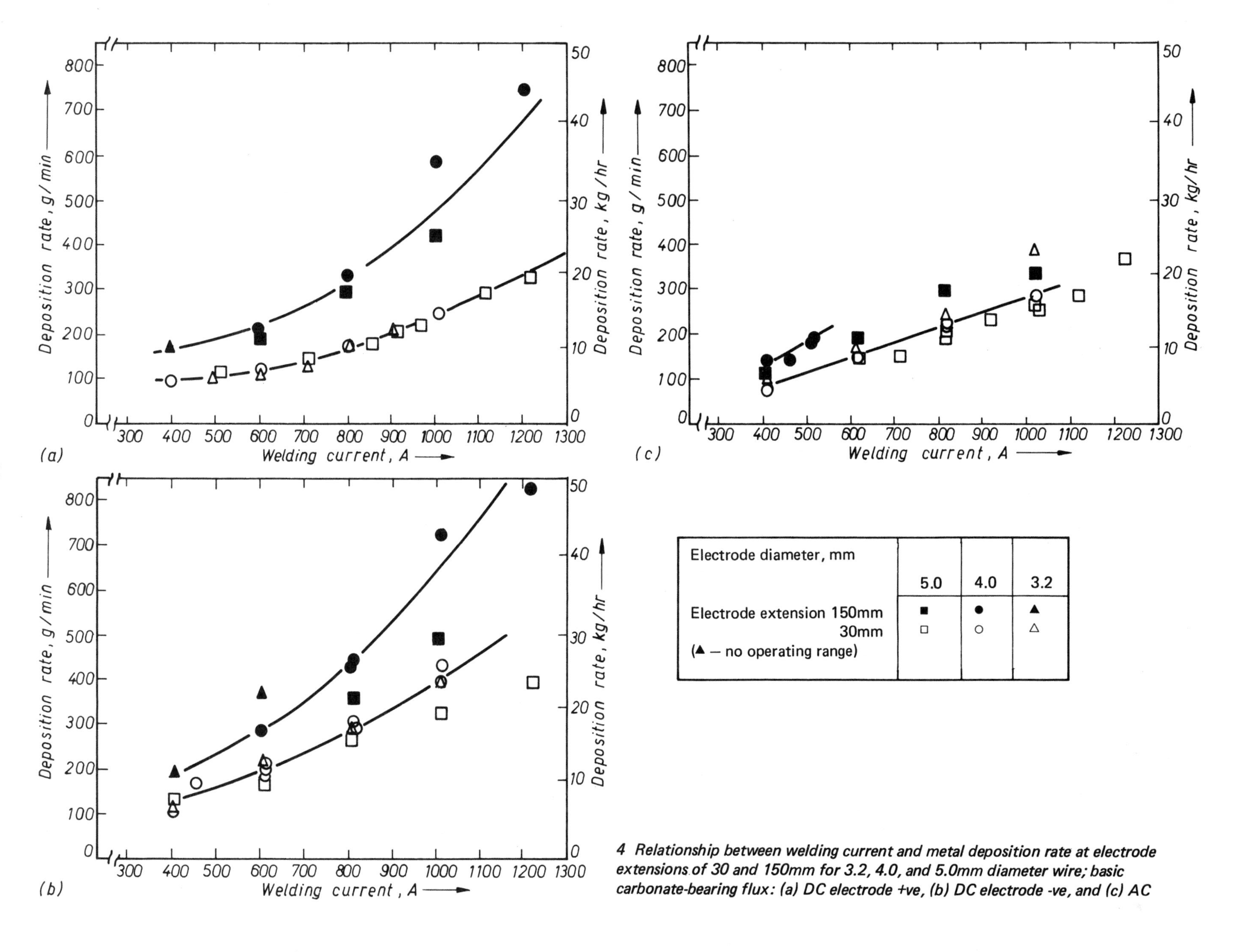

4 Relationship between welding current and metal deposition rate at electrode extensions of 30 and 150mm for 3.2, 4.0, and 5.0mm diameter wire; basic carbonate-bearing flux: (a) DC electrode +ve, (b) DC electrode -ve, and (c) AC

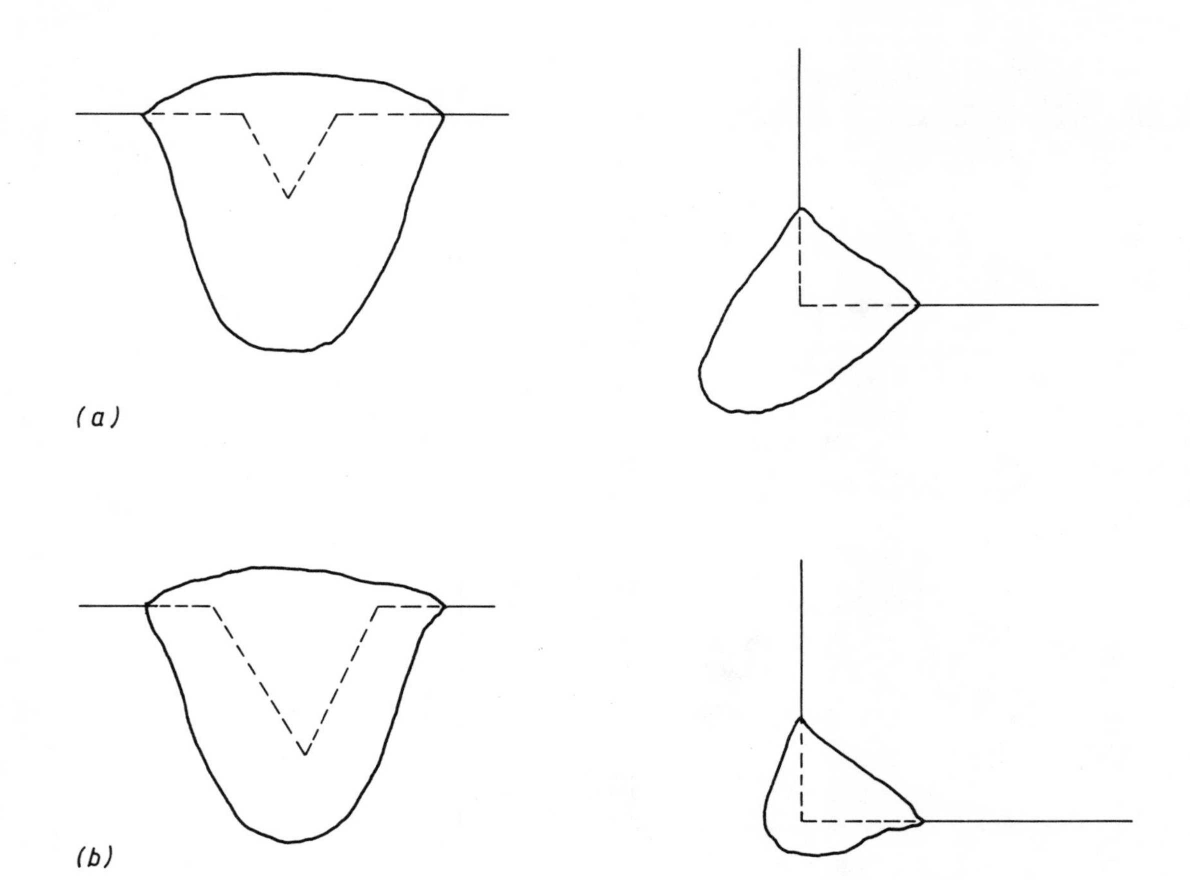

5 Variation of weld bead shape and dilution for low and high deposition rate welding conditions (schematic); weld beads made with: (a) high penetration, low deposition, welding conditions, (b) low penetration, high deposition, welding conditions

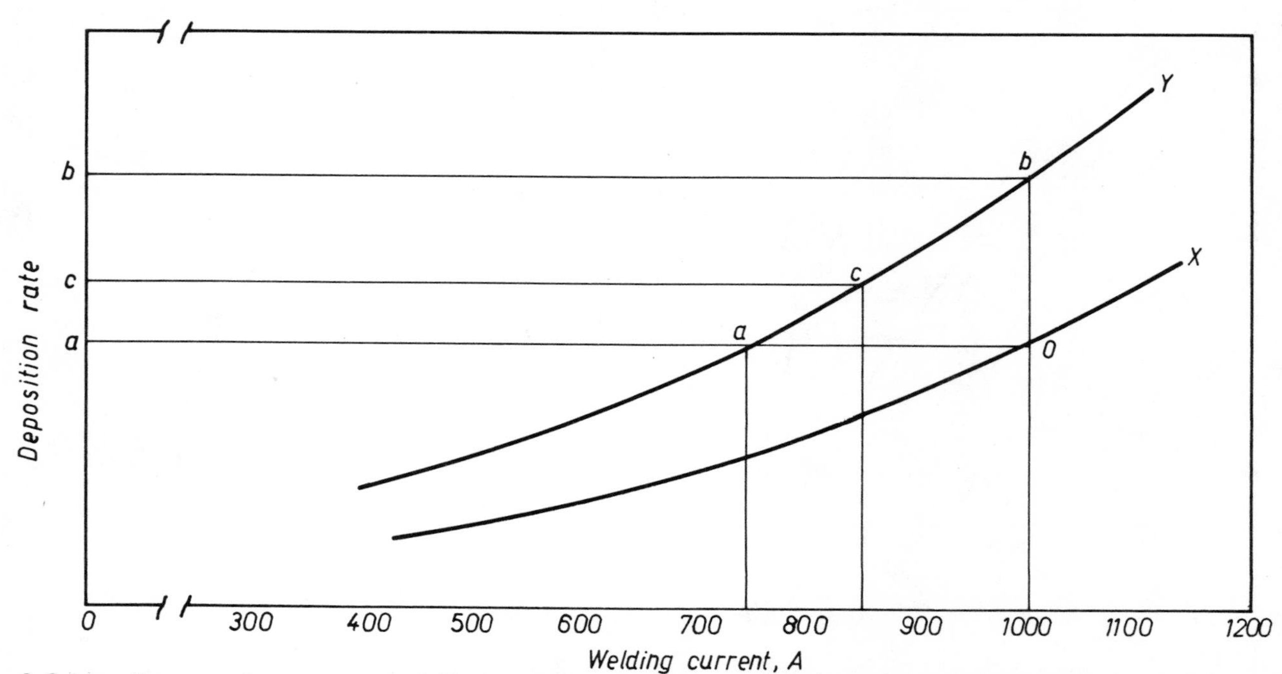

6 Schematic approaches to control of dilution and heat input using high deposition welding techniques (O — original welding condition) by: (a) direct reduction in welding current; deposition constant, (b) increase in deposition rate; welding current constant, and (c) reducing welding current and increasing deposition rate

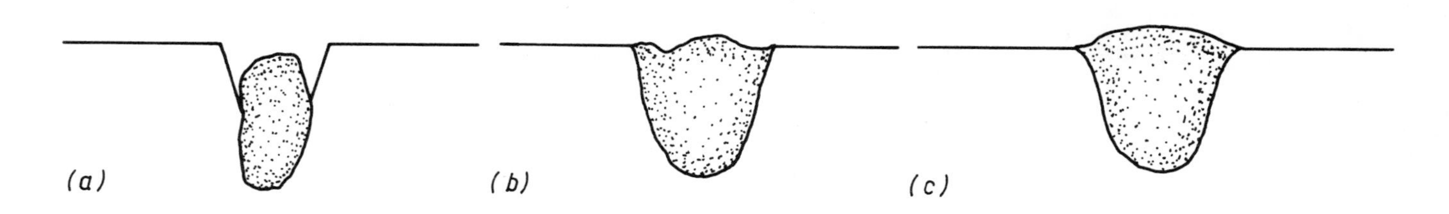

7 Arc functions in multipower SA welding (schematic): (a) lead arc weld bead shape, (b) lead/middle arc weld bead shape, and (c) final weld bead shape

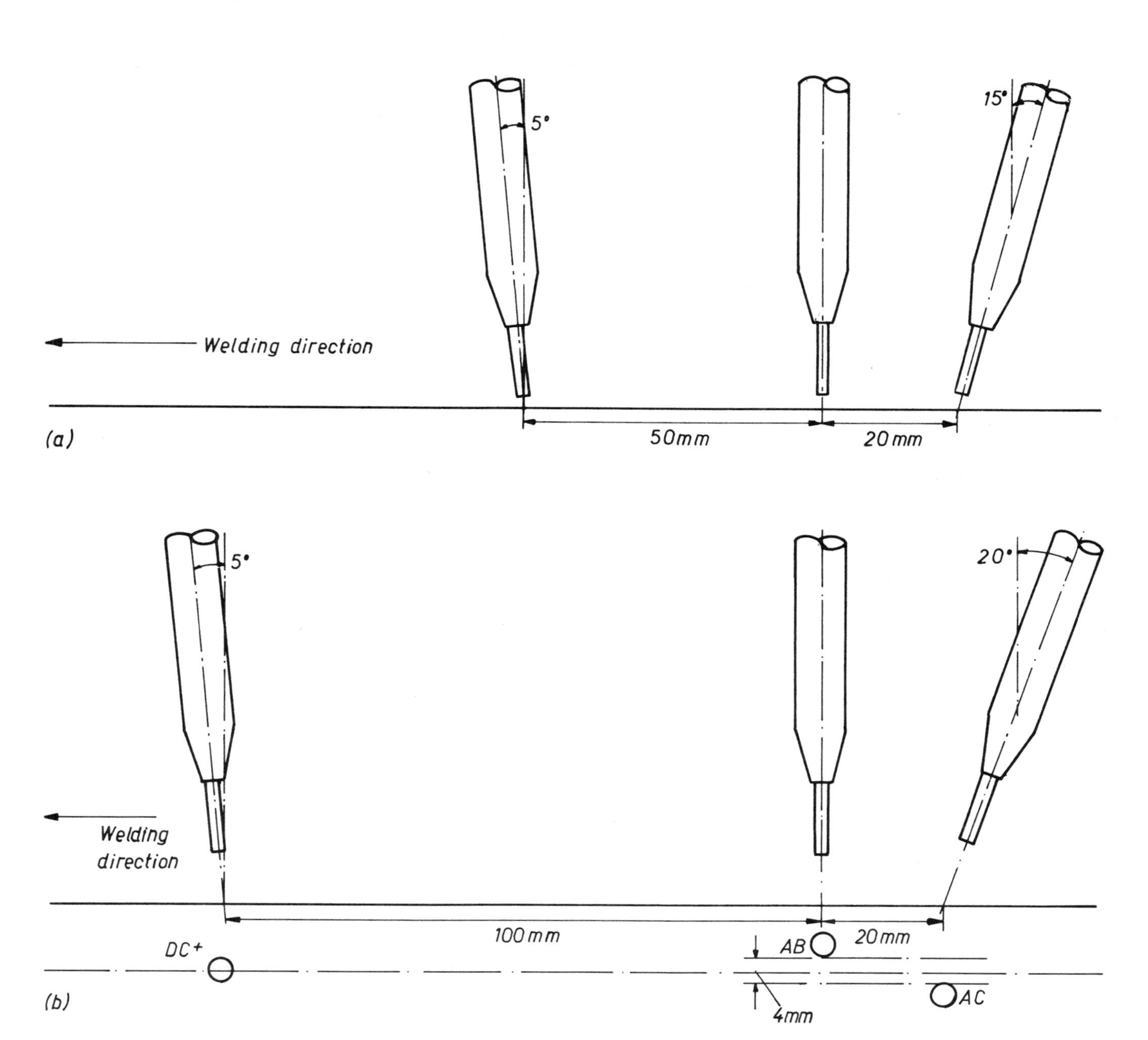

8 Electrode configurations for: (a) DC+ ACAC and ACDC- AC welding of 25mm plate, (b) DC+ ACAC welding of 19mm plate (electrodes were brought into line along weld seam for extended electrode welding)

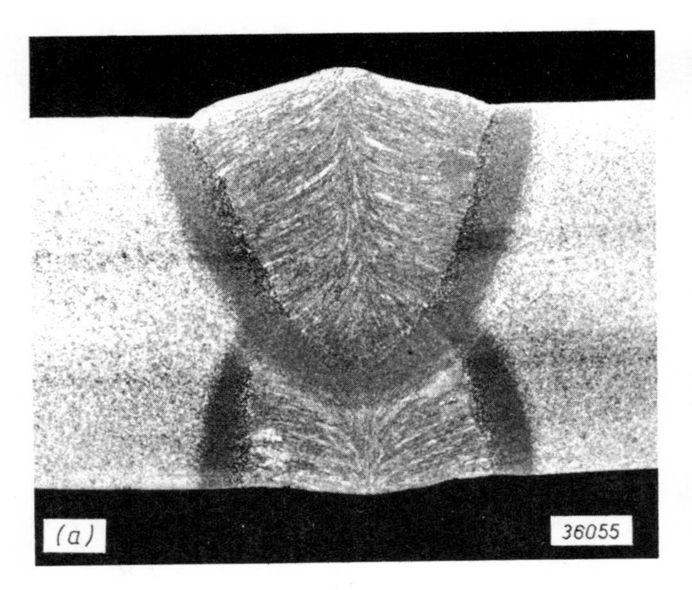

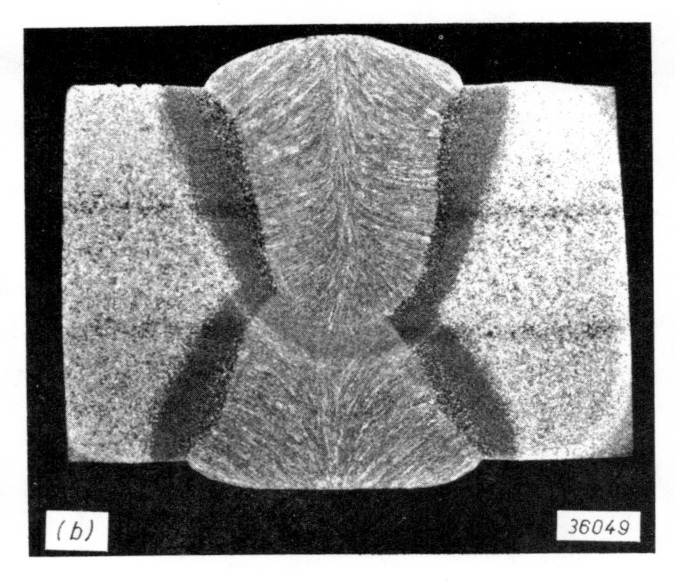

9 Triple electrode SA welds in 25mm plate

Welding approach	(a) DC+ ACAC	(b) ACDC- (150mm extension) AC
Nominal energy input, maximum, kJ/mm	6.0	5.4
Dilution, %	61	38

(for other procedural details see Tables 1 and 2)

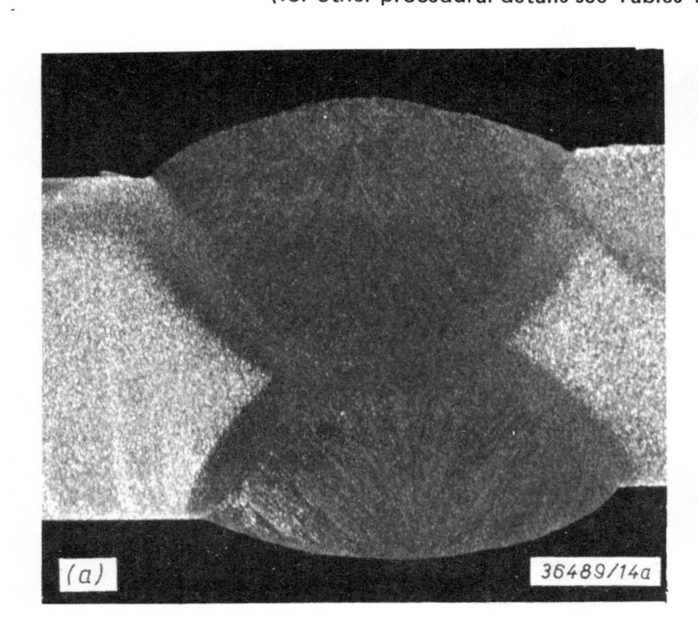

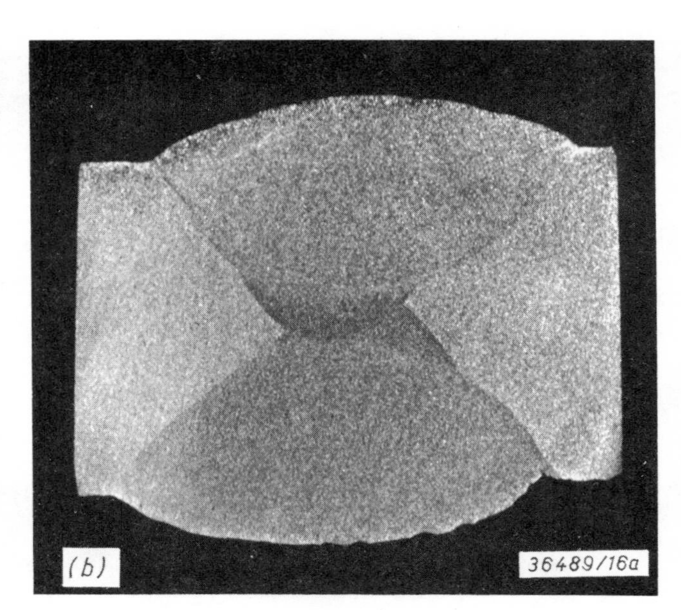

10 Triple electrode SA welds in 19mm plate

Welding approach	(a) DC+ ACAC	(b) DC+ AC (120mm extension) AC (90mm extension)
Nominal energy input, maximum, kJ/mm	5.8	4.4
Dilution, %	65	58

(for other procedural details see Table 3)

Simul arc welding process

S. Kimura, K.Iio, and K.Yamada

A new system has been developed called the 'Simul arc welding process' which provides double-sided welding of butt joints without turning over the plate. This avoids some of the deficiencies of previous welding methods, such as welding each side separately or the one-sided welding process, but preserves some of their good points. This new method carries out the welding operation simultaneously on both sides of the plate by a combination of flat welding and overhead submerged-arc welding. However it is not merely a combination, it is characterised by the feature that the welding heat produced by the leading flat weld is utilised by the overhead weld that trails it. Consequently the penetration of the overhead weld is distinctly increased, as compared with that of the other welding methods, and as a result this process has the following advantages:

1 Much wider range of application in I-shaped groove joints
2 Reduction of weld heat input
3 Decrease of welding distortion in thin plate welding

The tolerance range for the root gap, and difference in thickness, is about the same as those for the previous methods. Nevertheless this process shows great promise in practical application.

INTRODUCTION

In Japan the butt welding techniques that have been widely employed by shipbuilders and bridge constructors are welding each side separately and the one-sided welding method. Both of these approaches, however, have deficiencies. The two-sided method requires the steel plate to be turned over to complete the weld. Therefore this method cannot be applied in such examples as a big block butt joint, in which the routine of reversal often means a very dangerous operation. One-sided welding was developed to avoid this limitation by utilising tandem electrodes for welding as well as special backing material on the reverse side of the steel plate. This produced an underbead by the leading electrode and the surface bead by the trailing electrode to complete the entire welding operation from one side only.

Even with this method, however, new problems were experienced such as the backing equipment required to obtain the underbead in a panel line. To operate a panel line system smoothly the backing apparatus has to be installed to meet the welding needs of the longest joint expected. In other words, this tends to become a very large-scale installation itself. The second problem is that, since the backing material acts like a mould, a defective bead can easily arise on the reverse side. The third problem is that since

The authors are in the Development Department, General Technical Office for Development and Manufacture, Welding Electrode Division, Kobe Steel Limited, Fujisawa, Japan.

this approach requires a Y-shaped groove with a small root face the heat input increases for the larger cross-section joints. Consequently, the quality of the heat-affected zone often deteriorates in thick plate, and with thin plates there is a tendency to distortion by the welding heat.

To overcome these deficiencies an automatic double-sided butt joint method was developed in which the welding operation is carried out simultaneously on the surface and on the reverse side of plate.

BASIC PRINCIPLES

One possible approach to double-sided welding without turning over the plate, i.e. one which combines an overhead and flat welding process, could be that the reverse side of the plate is welded first, and then the rest of the joint is completed by flat welding on the top side. However, if the welding current is increased in overhead welding the weld pool drips out. Accordingly, the welding current cannot be increased indiscriminately and the efficiency of the overhead weld is much less than that of the flat process. Besides, the penetration achieved in this manner can be very shallow, and lack of fusion easily occurs between the overhead and flat beads. Alternatively, the latter can burn through the underbead and therefore the penetration must be carefully and skilfully controlled.

Achievement of deeper underbead penetration was thought to be the key to the problem and efforts were directed to obtaining deeper penetration in the overhead weld by, for example, preheating the steel plates, which is known to produce both greater penetration and a smoother bead edge even in high speed welding. It was decided to apply this approach to the overhead welding operation bearing in mind that, in ordinary double-sided welding, the reverse side of the steel plate is considerably heated by the weld. The temperature reaches several hundred degrees centigrade especially around the groove. The point is that, if the flat welding process is operated prior to the trailing overhead weld, by making use of welding heat from the former an overhead weld can be obtained without additional heating apparatus. Figure 1 illustrates a typical relationship between distance from the flat welding position and the temperature on the reverse side of the plate in the vicinity of the groove. Here a flat submerged-arc weld was executed so that there was an unwelded zone left on the reverse side of the steel plate. As shown the temperature rises to over 700° on the back face in the region between 100 and 400mm behind the flat welding electrode. Needless to say the welding conditions and changing plate thickness cause differences in this relationship. However the important point is that if the overhead welding operation is synchronised to the flat weld in an appropriate position behind it a preheat of over 700°C is always obtained. The general concept of this arrangement is presented diagrammatically in Fig.2. As expected with this preheating device the penetration becomes twice as deep as that of ordinary welding methods. Based on this technique various combinations of overhead and flat welding operations were studied.

PROCESS DEVELOPMENT

First stage: TIG-welding

In overhead welding any method that generates spatter leads to malfunction of the gas shield in the long run because of the deterioration of the gas cup. This interrupts the operation and requires cleaning or replacement of nozzles. To avoid such inconveniences a spatterless technique was examined: TIG-welding. Since no backing material is employed the only factor that sustains the molten metal is surface tension. This means that when the depth of penetration is greater than 5mm drop through of molten metal occurs. Thus, to eliminate lack of fusion between the welded zone of the overhead and flat beads, the depth of unwelded material left by the leading flat welding operation must be kept within 5mm. Practically, this leads to the problem of burnthrough and thus methods involving backing had to be examined.

Second stage: TIG-welding with sliding copper shoe

With copper backing it is possible in principle to prevent drop through of the welded steel, even when the penetration is greater than 5mm. However, as the backing has to be slid along it is very difficult to obtain close and constant adherence to the underside of the plate, which creates problems of escape of the molten metal. It is impossible to consistently produce a solid overhead bead by this method, and any failure to mould the weld metal into the groove of the backing plate leads to piping defects in the upper regions of the welded zone.

Third stage: overhead submerged-arc welding

Since it is also difficult to obtain a satisfactory underbead using a solid sliding backing, attempts were made to maintain the bead by a powder technique. Since TIG-welding, which employs a nonconsumable electrode, is not suited for this a consumable electrode system was investigated, i.e. submerged-arc welding where the flux for the powder that keeps the molten metal from falling also serves to shield the welded part. In other words, overhead welding is achieved by supplying flux (which is stored in advance in a hopper) and pressing it against the reverse side of the plate from underneath by a screw-feeder device, together with a consumable electrode which feeds through the axis of the screw. The molten metal does not drop out even with deep penetration, because it is contained by the flux, and moreover close adherence to the under-surface of the steel plate can be achieved without difficulty. This method results in better, non-defective beads, compared with using a solid sliding backing, and is the key to double-sided welding without requiring reversal of the plate.

WELDING APPARATUS

Synchronisation between overhead and flat electrodes

The overall installation is shown schematically in Fig.3 where the equipment above the steel plate is a conventional flat submerged-arc welding system as widely employed. Under the plate is the overhead submerged-arc welding apparatus that travels simultaneously with the normal flat submerged-arc system. The distance, L, between the welding electrodes, above and below the steel plate, is determined by the temperature relationship for the underside of the plate, and, as was indicated in Fig.1, is usually kept within 120 to 200mm range. As for welding power supplies an AC drooping characteristic system is used for the flat weld, whereas a DC constant potential source is employed for overhead welding, as the stability of the arc is very important here.

As explained above the method of Simul arc welding relies on utilising the welding heat produced by the flat weld in completing the overhead submerged-arc weld. Thus, both carriages, above and below, have to travel in synchronism. For this it is sufficient to use a digitiser or pulse motor to sense the number of revolutions of the driving motor of one carriage, and to ensure that the same number is obtained on the other carriage. Good welding results have been obtained with a constant distance between electrodes being held, even for long joints, since heavy pressures (as for sliding backing) which could interfere with the travel speed are not required.

Seam tracking for overhead weld

The most widely used system for detecting the weld line is a roller running in the groove, which operates limit switches to drive the tracking servo motor. Although this produces good results when applied to a groove, it cannot be used in the absence of a deep profile such as in Simul arc welding where, with the deep overhead penetration, the I-shaped groove has a considerable root face.

However a tracking device based on an air detector, Fig.4a, can be applied to the I-shaped groove, and this was investigated further. In this system compressed air passes through a small hole like a whirlpool, so that if there is a solid object without an opening in the vicinity of the top of the jet hole the pressure, P_o, of the air inside the central core is lower than outside. But if there is an opening like a weld line, outside air flows in and, P_1, the air pressure inside becomes greater than that without an opening, P_o. A test of the precision obtainable with a detector, based on the above principle, was made with the I-shaped groove, Fig.4b. This indicates that the degree of precision is within ±1.5mm, which in this instance is acceptable.

WELDING PERFORMANCE

Effect of plate thickness on operating conditions

Figure 5 presents an overall view of the Simul arc welding setup and Table 1 indicates the preferred operating conditions for 6 to 25mm thick steel plates. Since the Simul arc process produces deep penetration in the overhead weld, I-shaped grooves can be used in up to 19mm thick plate. Regarding heat input, even for 25mm thick plate the necessary input is less than 10kJ/mm, which represents 30 to 40% decrease compared with one-sided welding. So far, adequate performance has been assured up to 25mm thick plate, but it is expected that even thicker plates can be accommodated by changing welding conditions on the top surface or by increasing the number of electrodes. Figure 6a presents one example of the bead appearance produced by overhead welding, while Fig.6b and c shows examples of the macro-structure of the cross-section. As a general guide Table 2 gives the chemical composition of weld metal for one example and similarly Table 3 gives some data on tensile and elongation strengths as well as Charpy impact value.

Effect of differences in joint geometry

In production welding the joint has often to be made where there is a difference in thickness of the steel plate. In testing the applicability of the new method it was found that a satisfactory

Table 1 Welding conditions

Plate thickness, mm	Preparation of edge	d, mm	L, mm	Welding Position	Welding current, A	Welding voltage, V	Welding speed, mm/min	Heat input, kJ/mm
6	I	--	120	Flat Overhead	480 250	35 25	600	1.68 0.60
12	I	--	150	Flat Overhead	900 350	37 27	600	3.07 0.95
19	I	30	200	Flat	L: 1200 T: 900	35 45	700	4.2 4.05
				Overhead	350	27		0.81
25	50° 12	30	200	Flat	L: 1250 T: 1000	35 45	600	4.38 4.50
				Overhead	350	27		0.95

Welding material

Flat : US-29 (4.8ϕ, 4.8ϕ) x PFH-42

Overhead : USS-43 (2.0ϕ) x PFI-43s

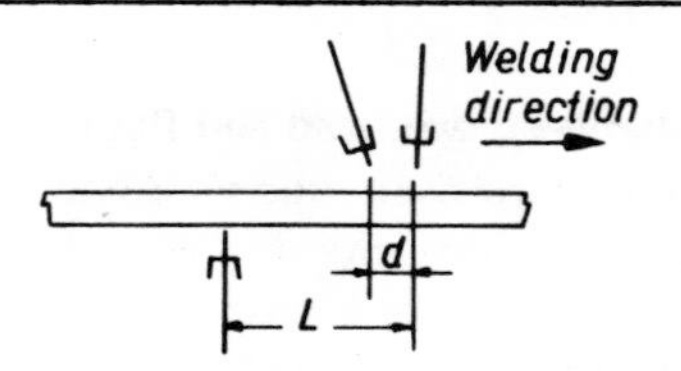

result is obtained for thickness differences of up to 3mm, and also with scarfed joints, as indicated by the macrosection, Fig.7.

If there is no mismatch in the weld joint on the reverse side of the plates, Simul arc welding can achieve a satisfactory result, even with a 3mm root gap. However, just as in conventional double-sided welding, there is a tendency for burnthrough by the leading flat welding process when the joint is mismatched, especially when the root gap is greater than 2mm. To overcome this problem where the root gap is over 2mm a sealing bead is made on the top surface of the groove, for which CO_2 gas-shielded arc welding was found to be more effective and give deeper penetration than MMA welding. Thus, even with a 3mm root gap, the operation can be carried out under the same conditions as for no root gap. Figure 8 shows the results of the test for those situations in which a root gap and mismatch are both present. From this it is concluded that, concerning the tolerance limit to the root gap and mismatch, the general criterion is:

The amount of the root gap + the amount of the mismatch
$\leqslant$ 3.0mm

Table 2 Chemical composition of weld metal

Plate thickness, mm		Chemical components, %				
		C	Mn	Si	P	S
12	Flat	0.14	0.74	0.30	0.020	0.018
	Overhead	0.16	1.11	0.36	0.022	0.015
25	Flat	0.11	1.08	0.27	0.014	0.009
	Overhead	0.11	1.24	0.35	0.020	0.006

Distortion in thin plate welding

In welding thin plates the degree of thermal distortion must be taken into account together with bead appearance and properties of weld metal. Since this new welding method operates on both sides of the steel plates almost simultaneously, and with reduced heat input compared with one-sided welding, it is confidently expected that the distortion produced by welding is also very small. To confirm this aspect a test was performed on a plate, 12mm thick, 3000mm wide, and 6000mm long, and the distortion measured at the positions indicated in Fig.9. As shown, the degree of weld distortion is small, the largest deviation being 5mm compared with the measurements taken

Table 3 Mechanical properties of welded joints

Plate thickness, mm		Results of tensile test		Results of impact test, J/cm^2			Results of bend test	
		Tensile strength, N/mm^2	Elongation, %	-10°C	0°C	+20°C	Face, reverse bend	Side bend
12	Flat	535	33	44	50	78	Good	Good
	Overhead	502	32					
25	Flat	490	34	47	67	105	Good	Good
	Overhead	537	35					

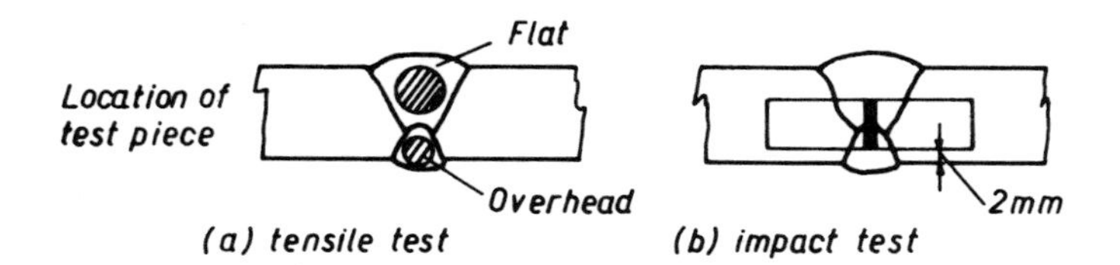

(a) tensile test (b) impact test

prior to the test weld. Thus, this process is considered to be very effective in thin plate welding.

CONCLUSION

An automatic, synchronised, two-sided butt joint welding method has been developed to replace the previous double-sided, as well as one-sided, welding techniques. This new process has many advantages, including reduction of thermal distortion stemming from the decrease in heat input, and widening the range of utilisation of the I-shaped groove joint. It is expected that this new method will be successfully applied to thin plate joints as well as to seam and butt joints in structural tubular members.

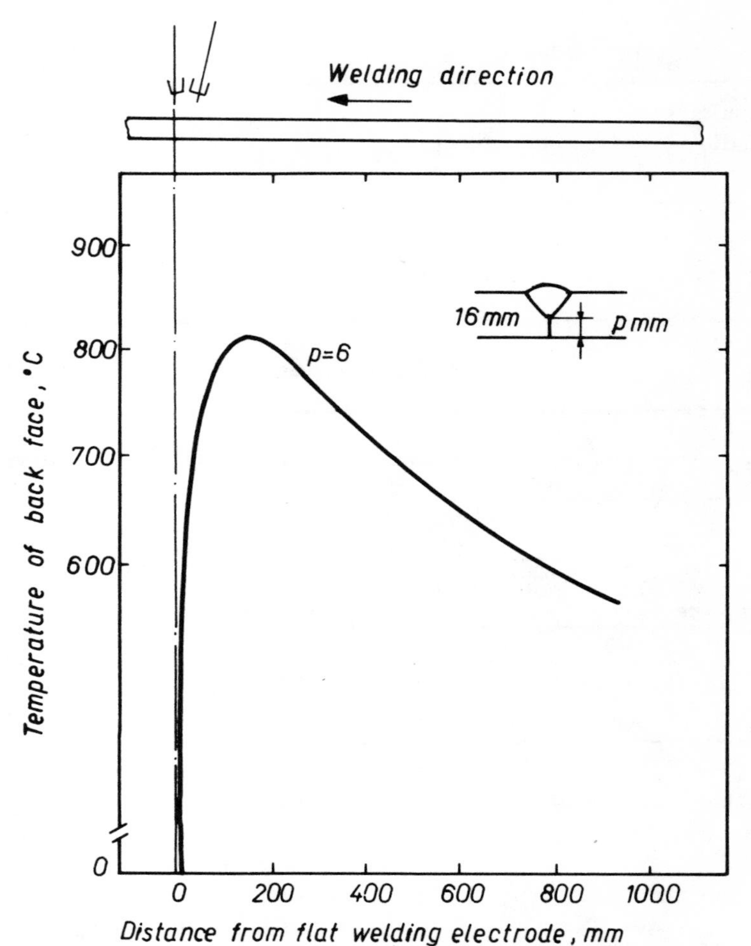

1 Relationship between backface temperature and distance from flat welding electrode

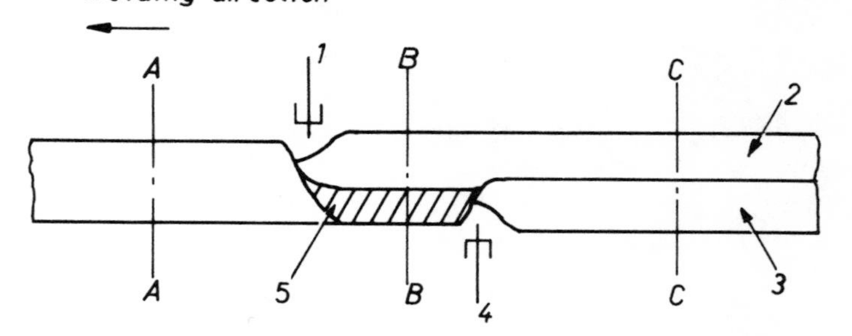

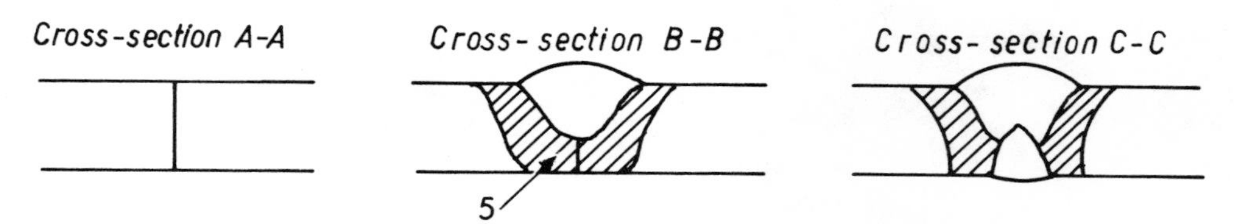

2 Basic principle of Simul arc welding process. 1 — flat welding electrode; 2 — flat weld bead; 3 — overhead weld bead; 4 — overhead welding electrode; 5 — zones heated by flat welding

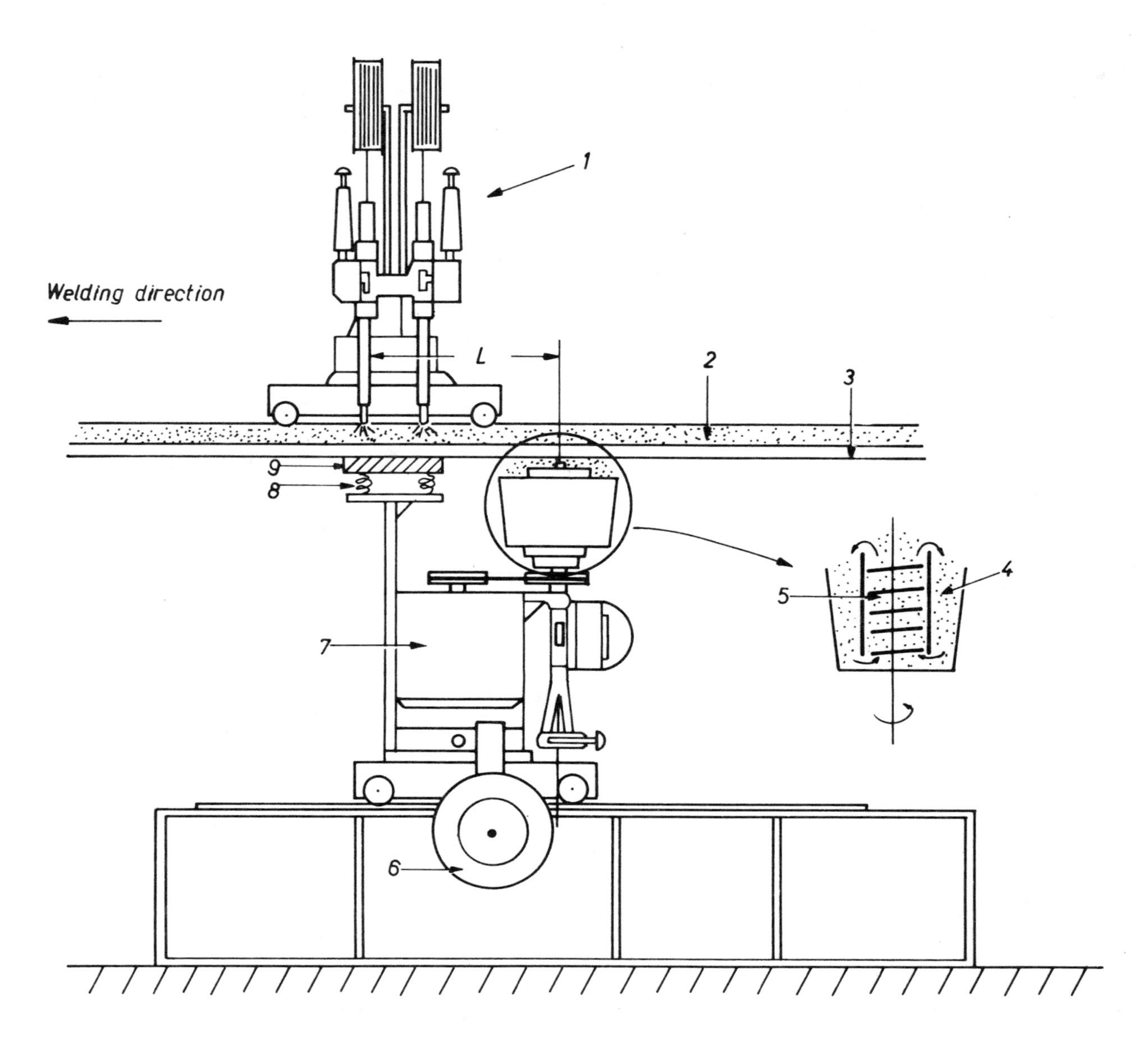

3 Schematic diagram of complete welding installation. 1 — submerged-arc equipment for flat welding; 2 — flat welding flux; 3 — plate; 4 — overhead welding flux; 5 — screw; 6 — wire; 7 — motor; 8 — spring; 9 — copper shoe

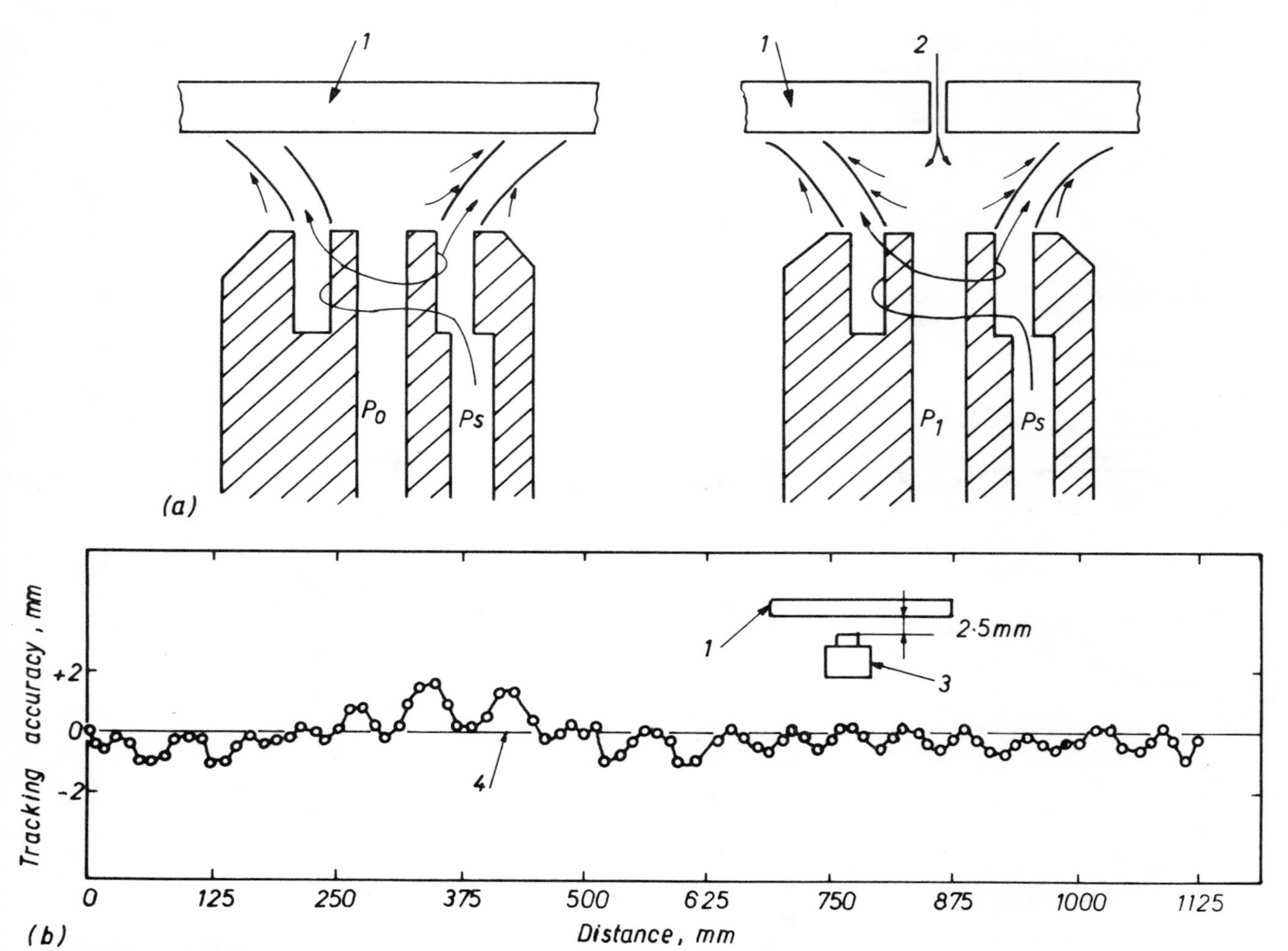

4 Seam tracking by pneumatic detector: (a) principle of detector (left) under solid plate (right) under weld line; Ps — supply pressure; P_o, P_1 — detector pressure; (b) tracking accuracy in I groove. 1 — plate; 2 — outside air; 3 — detector; 4 — weld line

5 Simul arc welding equipment

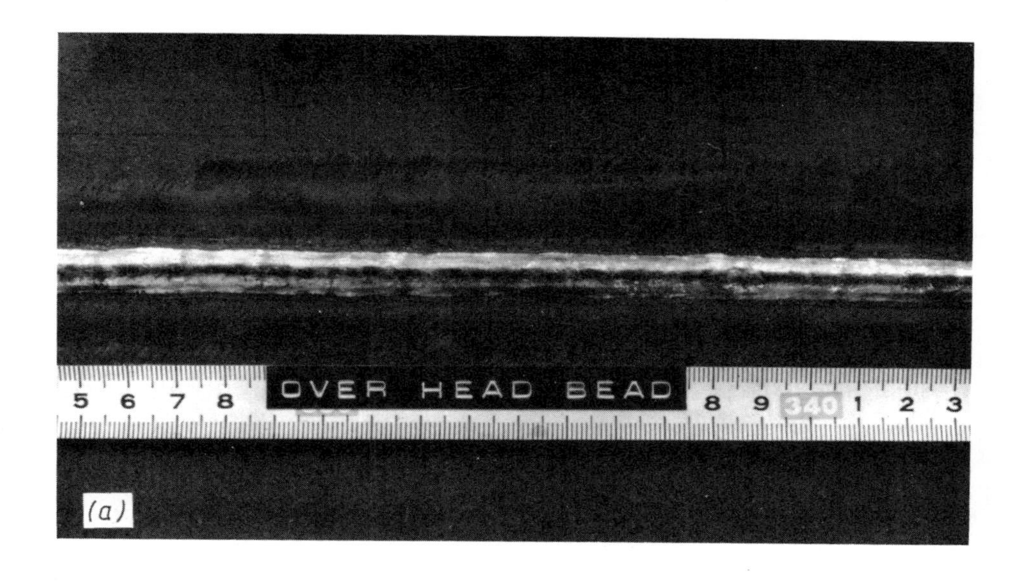

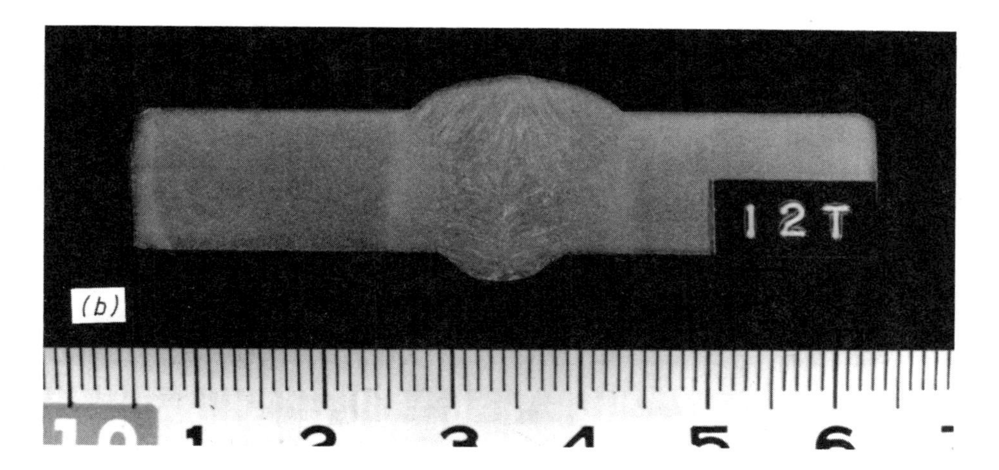

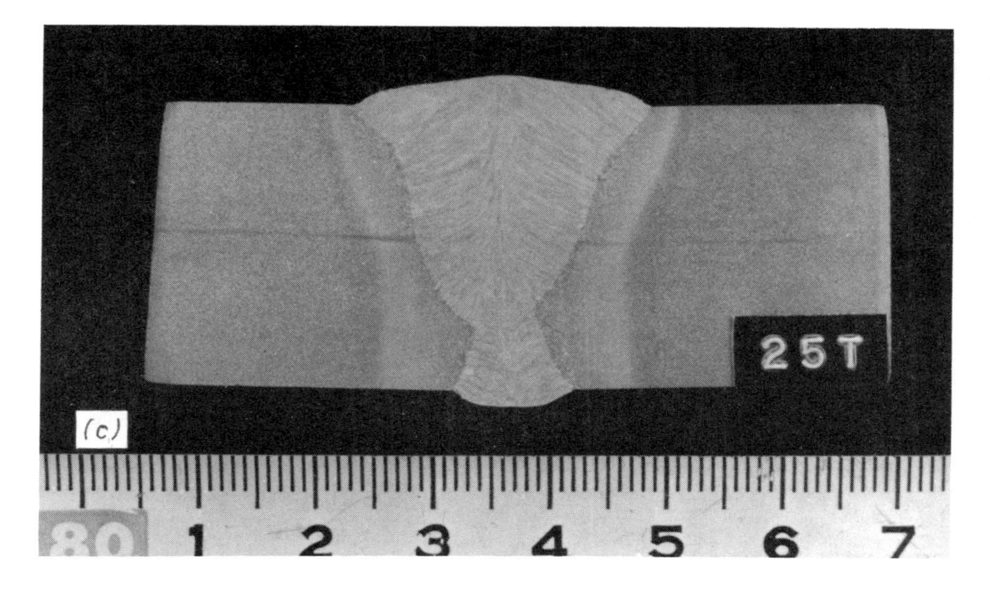

6 Weld bead profiles: (a) underbead appearance (overhead submerged-arc weld), (b) macrosection of weld in 12mm thick plate, and (c) macrosection of welding in 25mm thick plate

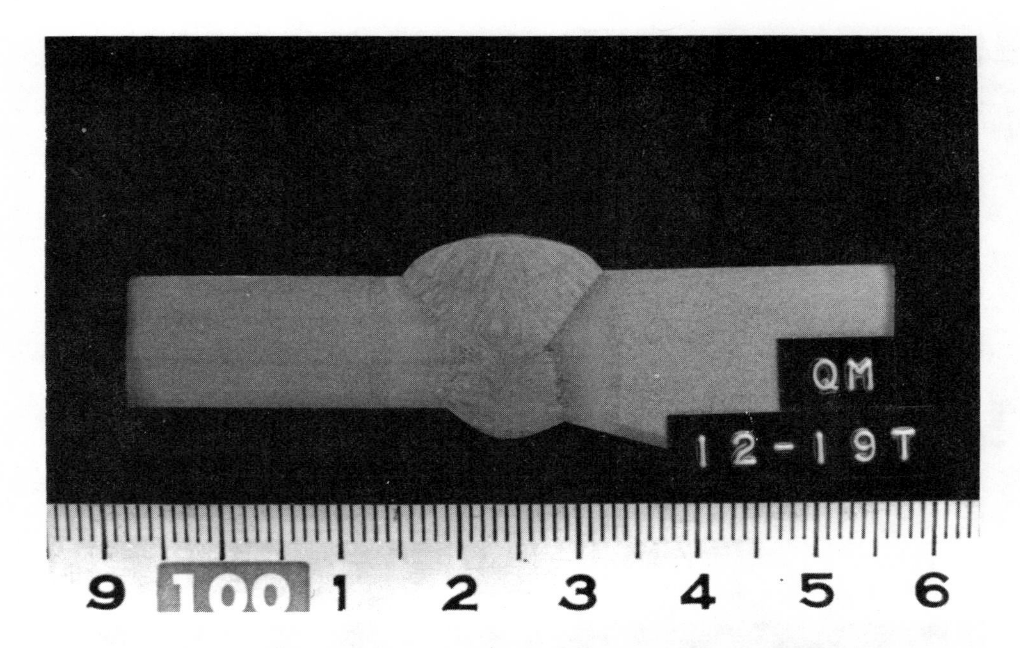

7 Macrosection for scarfed joint between 12 and 19mm thick plates

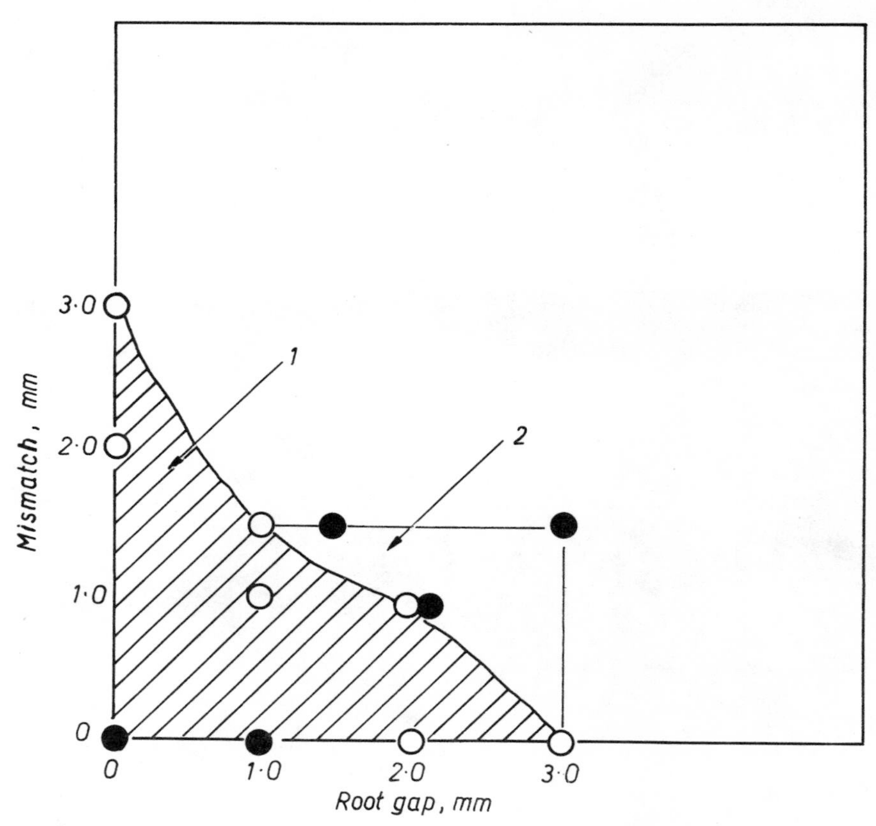

8 Tolerance limit for mismatch and root gap. Domains for good welding results without (1) and with (2) sealing. Mismatch and root gap values which gave good results without (○) and with (●) sealing

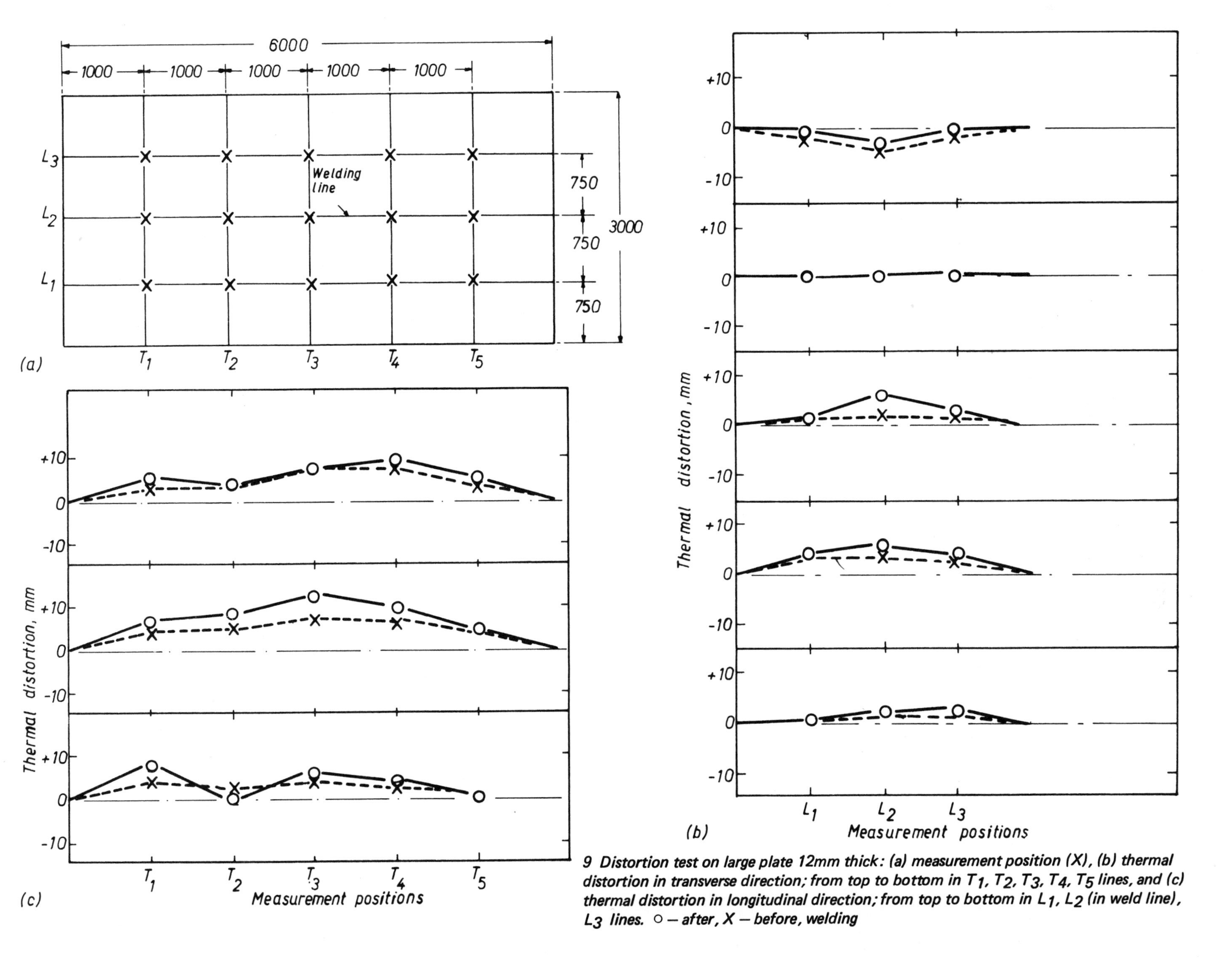

9 Distortion test on large plate 12mm thick: (a) measurement position (X), (b) thermal distortion in transverse direction; from top to bottom in T_1, T_2, T_3, T_4, T_5 lines, and (c) thermal distortion in longitudinal direction; from top to bottom in L_1, L_2 (in weld line), L_3 lines. ○ – after, X – before, welding

PAPER 18

Control of manual metal-arc weld quality by deposition sequence

P.J.Alberry, BMet, PhD, T.Rowley, BSc(Met), and D.Yapp, BSc, MMet

Variability in metallurgical quality can be a major problem in the manufacture and performance of welds in power plant. It is shown that an increased understanding of the interaction between the welding process and a material can lead to the development of weld procedures to control weld quality. Two applications are described: firstly, weld bead deposition sequences are used to improve the ultrasonic inspectability of austenitic welds in a nuclear power plant, and, secondly, control of weld bead overlap is used to improve the resistance of ½CrMoV steam pipe compositions to heat-affected zone cracking during stress-relief heat treatment.

INTRODUCTION

One of the objectives of welding research at Marchwood Engineering Laboratories is to ensure that high average levels of weld quality can be maintained in combination with a low level of variability in behaviour. In this Paper two examples are described which show how an increased understanding of the structure and properties of manual metal-arc (MMA) welds has led to the ability to control weld quality. In the first example the grain structure of austenitic weld metals is controlled by specifying the weld bead deposition sequence, details of which would be normally left to the welder. The specific weld bead sequence produces highly aligned columnar grains with a low scatter of grain orientation, and, as a result, the weld metal properties in particular directions are reliably reproduced. This technique has been applied to give considerable improvement in the ultrasonic inspectability of austenitic welds in a nuclear power plant.

The second example describes the control of cracking in heat-affected zones (HAZs) of welds during stress-relief heat treatment (SRHT) by control of the HAZ microstructure.

The authors are all Research Officers at the Marchwood Engineering Laboratories of the Central Electricity Generating Board.

Experience with power plant fabrication in ½CrMoV compositions has shown that the chance of cracking is increased if the HAZ contains a high proportion of coarse-grained material. However, it is possible to reduce the incidence of cracking by careful control of the welding position and weld bead sequence to produce a high degree of grain refinement in the HAZ. A detailed physical model has been developed which allows the amount of refinement to be predicted in a given welding situation. Welding techniques which have resulted from this work have been used in power plant construction to improve the weld quality, so that repair rates from HAZ cracking during both fabrication and service have been substantially reduced.

Both these examples illustrate that careful control of the detailed deposition sequence of the MMA process can be applied to produce welds of a defined quality with low variability.

CONTROL OF GRAIN ORIENTATION IN AUSTENITIC WELDS

Production of welds with controlled grain alignment

It is well established that columnar grains can grow with the same crystallographic orientation from weld bead to weld bead in multipass austenitic welds. Figure 1 shows a conventional butt weld in which this epitaxial growth is clearly evident. However, it is also apparent that the orientation of the columnar grains changes with position in the weld, and in some areas new grains are nucleated and block the growth of existing grains. It will be shown below how the weld bead deposition sequence is critical in the determination of both the degree of grain alignment and the orientation of grains within a weld.

The initial indications of those factors which are important in the determination of grain orientation were obtained from observations on a small weld pad made for test purposes. The pad was produced by depositing planar layers of weld beads using a 3.25mm diameter electrode on to a vertical base plate (horizontal-vertical welding), with the weld beads placed one above another. The bead sequence can be seen in Fig. 2a and the resultant grain orientation is shown in Fig. 2b. A high degree of grain alignment is present in this pad and the orientation of the grains is the same over the whole of the pad area. Both these effects are a direct result of the sequence of welds beads, deposited in planar layers. It was subsequently found that the particular orientation of grains in this pad is a characteristic of the horizontal-vertical (H-V) welding position on a vertical plate. The MMA process produces a relatively shallow weld pool, and the geometrical form of the pool is not greatly influenced by welding parameters. For H-V welding with beads placed one above another, the grains grow at an angle of about 30° relative to the normal to the base plate. This mean grain orientation varies by only about $\pm$ 5° even for relatively large changes in welding variables, including electrode diameter, travel speed, current, and angle of electrode.

By making a series of weld pads it became clear that a characteristic grain orientation is produced for each welding position. For instance, in the flat position the grains make an angle of about 15° relative to the base plate normal, compared to 30° for H-V welding. Similar observations were made for other possible welding positions, including more complex situations such as welding vertically upwards with weaving. Hence if welds are deposited with planar layers of weld beads in a defined sequence, it is possible to predict the grain orientation which will result.

As a result of the observations made on these small pads it was concluded that:

1 A high degree of grain alignment is produced by making welds with planar layers of weld beads

2 The mean orientation of the grains is a characteristic of the MMA welding process and is only slightly affected by changes in welding parameters

Physical model for grain orientation

Figure 2a and b reveals that in some regions of each weld bead the columnar grains are not perpendicular to the fusion boundary. Epitaxial growth has still occurred, however, in spite of the changes in fusion boundary orientation from one bead to the next. It appears that an 'averaging' process is taking place: the grains are growing perpendicular to the average fusion boundary. and are hardly affected by local deviations of the fusion boundary from the average.

A simple heat flow model of the factors determining the average fusion boundary orientation is shown in Fig. 3. For the H-V welding position, Fig. 3a, the effect of gravity produces relatively thick layers of weld beads, so that when a new bead is added in sequence a significant component of heat flow is into the previous bead, compared with that into the underlying layer of beads. As a consequence, the resultant heat flow vector deviates by about 30° from the base plate normal. However, when welding in the flat position the effect of gravity produces relatively thin layers of weld beads, and the resultant heat flow vector deviates by only 15° from the base plate normal, Fig. 3b. Thus the grain orientations will be 30° relative to the base plate normal for H-V welding, and 15° from the base plate normal for flat position welding, since the resultant heat flow vector and the average fusion boundary orientation are mutually perpendicular.

This simple model for grain orientation has been extended to enable predictions of grain orientation to be made for all welding positions. Where a weld is multipositional, e.g. a butt weld on a horizontal pipe, account can be taken of the effect of gravity in intermediate positions, so that the grain orientation can be predicted at each point round the weld.

It is possible to produce a desired grain orientation in a multipass weld by specifying the orientation of the planar layers of weld beads and the sequence in which the weld beads are to be deposited. Figure 4 shows a butt weld in which planar layers of weld beads were deposited at a specified angle, θ, relative to the weld preparation to produce a grain orientation perpendicular to the weld surface. A range of grain orientations can be produced in butt welds, although there are limits to the angles at which layers of weld beads can be deposited. The deposition of welds with planar layers of weld beads is controlled in practice by using a series of templates. The welder then has to ensure that the layers match the templates as welding proceeds. Welders involved in the development work have found little difficulty in achieving the required deposition sequences, and the techniques have now been applied in power plant construction.

The degree of grain alignment depends to a large extent on the welding process as well as on the bead sequence. Extensive epitaxial growth is necessary to produce a high degree of alignment, and this can occur only when the changes in fusion boundary orientation from bead to bead are relatively small. Those welding processes which, unlike MMA, produce a deeply penetrating weld pool cause large changes in fusion boundary orientation which limit epitaxial growth and increase the scatter about the mean grain orientation.

Effect of grain structure on weld quality

Aligned grain structures occur to some extent in all multipass austenitic welds. The techniques described above can be used either to increase the extent of the alignment and to control its orientation, or similar techniques can be used to increase the scatter and to produce a less well oriented weld. The grain structure of austenitic welds is important because of the effect it has on their mechanical and physical properties. Some of these are strongly affected by the orientation of the grains but the effect is slight in others. For instance, for a specimen with grains parallel to its axis, the secondary creep rate is lower by factor of 2 (538°C, 300MN/m^2) and the ultimate tensile strength is higher by only 10%, compared with a specimen with grains perpendicular to its axis. One of the most important effects, which has already been put to practical use,[1] is the influence of grain orientation on ultrasonic attenuation. For pulse-echo measurements with a 2MHz compression beam on a 50mm thick weld, the apparent material attenuation can be reduced to 5dB for the best orientation (compression wave at 45° to grains) compared to 40dB for the worst (compression wave perpendicular or parallel to the grain orientation). This observation has been applied in construction to improve the inspectability of heavy section welds in a nuclear power plant, where the grain orientation techniques described above were used to produce a weld with grains at 45° to the ultrasonic beam direction.

A further use for the controlled grain orientation techniques has been to produce specimens in which the effect of grain structure on properties can be studied. It is now possible to investigate the effect of grain orientation on properties for the aligned regions in conventional welds, as well as for specially aligned welds.

In a conventional weld the grain alignment and orientation will vary from point to point within it, and also from weld to weld, depending on the sequence in which the welder happens to deposit the weld beads. In a controlled grain orientation weld the variation of properties in the weld can be predicted in advance and will be reduced from weld to weld, and it is possible to manipulate the structure to produce specified properties.

CONTROL OF HAZ STRUCTURE IN FERRITIC WELDS

HAZ cracking in ½CrMoV welds

A large proportion of welds in recent CEGB generating plant involves thick section ½CrMoV steels fabricated using MMA techniques and preheat temperatures of 250°C. Under these conditions the HAZ will be predominantly bainitic. The bainite present in the HAZ exhibits a range of prior austenitic grain sizes which, for the purpose of this Paper are divided into coarse- and fine-grained regions according to whether the grain size is greater or less than 35μm. At a level of 35μm the hot tensile ductility is expected to be approximately 5% in this material composition.[2] Practical experience suggests that those regions of the HAZ with a ductility $\leqslant$ 5% will be 'at risk' during SRHT. Circumferential HAZ cracking has caused extensive problems in both fabrication and services over the past few years.[3] Cracking occurs in the coarse-grained HAZ during SRHT or during service and can lead to full circumferential failure of the component. Service experience and recent research[4-6] have indicated that cracking is influenced by material composition and SRHT and is more likely in those welds which contain a high proportion (> 80%) of coarse-grained bainite in the HAZ. Furthermore, cracking will be more serious in such welds since crack linkage between extensive regions of coarse grains can easily occur across the dividing ligaments of fine-grained bainite. Six examples of the proportion and distribution of coarse-grained bainite from service components are shown in Fig.5. In the two instances where stress-relief cracking was observed, Fig.5e and f, it can be seen that a high proportion of coarse-grained bainite is present throughout the weld thickness, with a very high proportion towards the weld outer surface. Once cracking has initiated at the outer surface, where the residual stress is expected to be higher,[7] it can propagate easily through the crack-sensitive, coarse-grained HAZ microstructure.

Production of welds with refined HAZ structure

To demonstrate the effects of welding procedure on HAZ structure, four laboratory welds were made, Fig. 6, on a thick section test block of the following composition:

	C	Cr	Mo	V	Mn	Si
wt%	0.13	0.7	0.69	0.37	0.55	0.29
	Ni	Cu	S	P	Al	Sn
	0.21	0.15	0.14	0.009	0.007	0.02

Co, Pb, B, Nb, Ti, W, Sb, As — all <0.01 wt%

Welding was carried out using procedures in accordance with CEGB standard 23584. Trials 1 and 2 were welded in the flat position with a low overlap between weld beads (10-20%). Trials 3 and 4 were made by buttering on a near vertical face to simulate welding in the flat position into a double J preparation with a 30° included angle. This weld geometry automatically produces a high degree of bead overlap in the range 50-80%.

The effects of the overlap of weld beads in producing low levels of coarse-grained structure is shown in Fig. 7. Welds 1 and 2, with 10-20% overlap, have 70-80% coarse-grained bainite in pockets 8-9mm long. In comparison, welds 3 and 4 with 50-80% overlap of welds beads have only 15-25% coarse-grained bainite in pockets 3-4mm long. It is clear from these trials that the degree of overlap is a critical factor in the determination of the proportion of coarse-grained bainite in the HAZ. Note that the degree of overlap and hence refinement is approximately constant in the laboratory welds in contrast to the service welds, Fig. 5.

The way in which a high degree of overlap produces a low proportion of coarse-grained bainite in the HAZ is indicated in Fig. 8. Coarse- and fine-grained structures exist in the HAZ of a single-pass run as shown in Fig. 8a. In this particular steel composition the hot tensile ductility has been used as the criterion to divide 'fine' grains from 'coarse', although in reality there is a gradation of structure throughout the HAZ. Those grains of size >35μm have been designated 'coarse' and those <35μm 'fine'. Physically the dividing line between these two regions occurs in the vicinity of the 1100°-1200°C isotherm, the exact position being determined by the heat input and the steel composition. Thus the region between the 1100°-1200°C and the 950°C isotherms act as a 'grain-refining zone'.

Figure 8b shows that for welds on a horizontal plate with low overlap, the grain-refining zone can produce a fine-grained structure in only a small proportion of the HAZ adjacent to the fusion boundary. The HAZ therefore contains large areas of coarse-grained crack-prone structure separated by small pockets of refined material. However, if a high degree of overlap is produced, Fig. 8c, when the weld beads are deposited on a vertical face, a far larger proportion of the HAZ is refined and it is the coarse-grained structure which exists only as small pockets. In effect this will limit the maximum defect size possible in the coarse-grained region which has to be taken account of in fracture mechanics assessments.

In practice, flat position welding on to a vertically faced weld preparation automatically produces a high degree of overlap. It is more difficult to obtain the same degree of overlap when welding on a horizontal plate because of the shape of the weld beads. It will be shown later that the practical application of these results to produce crack-free welds generally involves the use of techniques which inherently produce a high degree of overlap.

Physical model for HAZ structure

A physical model has been developed which allows quantitative predictions of the HAZ structure in multipass welds to be made.[2] The principles of the model will be outlined here and its predictions for the four trial welds described above will be compared with measurements of the HAZ structure.

Given the basic information on the degree of overlap, heat input, and material composition, the model predicts the:

1. Grain size distribution in the HAZ
2. Proportions of martensite, bainite, and ferrite in the HAZ
3. HAZ width and depth
4. Proportion of 'coarse'-grained bainite in the HAZ
5. Proportion of 'fine'-grained bainite in the HAZ

The detailed equations and empirical correlations used in the model will be discussed elsewhere.[2]

The model has been applied to the

four laboratory welds in the following stages:

(a) austenite grain growth and transformation kinetics (which vary with composition[8-10]) were measured[2,11] for the weld trial base plate

(b) three-dimensional heat flow theory[12,13] was used to calculate temperature profiles and thermal cycles in the HAZ for each particular heat input. The heat flow data were combined with the metallurgical data from the base plate to predict the HAZ grain size distribution and microstructures

(c) the heat flow data were combined with the empirical correlations of heat input and weld bead height to calculate the mean bead height, mean bead width, and mean weld bead overlap. This procedure was used because direct measurement of weld bead dimension for a particular weld is not always possible

(d) the geometrical construction, Fig. 8, was used to produce the detailed spatial distribution of HAZ grain size and microstructure in the multipass weld for the mean weld bead overlap and the respective heat input of a particular weld. It should be noted that there is some variability of overlap in the four trial welds between the ranges shown in Fig 7

The agreement between the model predictions and measurements of the four welds, Fig. 7, is encouraging particularly in view of the limitations of the model, e.g. the use of an average value of weld bead overlap and an idealised weld bead geometry. It should be noted that the model predictions of 'coarse'- and 'fine'-grained areas relies on an arbitrary 5% hot tensile ductility criterion. However, for this material composition, 'fine' grains have a grain size of <35μm and 'coarse' grains of 35-136μm. These grains sizes correspond closely with the average measured values of 'fine' grains (ASTM 8-9, 25-35μm) and the 'coarse' grains (ASTM 7-4, 49-139μm) for the four trial welds.

In view of the agreement between theory and measurements the model of HAZ structure can be used to make quantitative predictions in ½CrMoV steels given information on the heat input, expected degree of overlap, material composition, and metallurgical properties. Since the mean degree of overlap can be calculated from details of the welding procedure and weld geometry, it is now possible to quantify the spatial distribution of areas 'at risk' in a weld HAZ during SRHT so that, if necessary, the welding procedure can be modified to reduce the incidence of cracking.

Although the present example of HAZ grain refinement by control of the welding process applies strictly to MMA welding, the analysis is, in principle, applicable to a wide range of welding processes, weld geometries, and material compositions.

Application of techniques to refine HAZ structures

From the limited survey of the range of commercial welds it is clear that within the framework of a weld specification, e.g. CEGB standard 23584, it is possible to produce ½CrMoV welds with a wide range of relative properties and spatial distributions of the component HAZ microstructures. The laboratory weld HAZs demonstrate that overlap of weld beads is a critical parameter in limiting the absolute size and proportion of coarse-grained bainite in a weld HAZ This parameter is not normally specified in welding procedures but will clearly have an important bearing on the overall weld HAZ performance both during SRHT and in service.

In practice the easiest way to achieve guaranteed high levels of weld bead overlap is to choose those welding parameters and procedures which automatically produce a high overlap. For example, flat position welding a rotating pipe with a small included angle double J preparation produces a situation similar to trials 3 and 4 with a high degree of weld bead overlap. Techniques can be devised for many welding situations which consistently produce high levels of refined structures. There has also been an improvement in HAZ refinement in all-positional repair welding situations with restricted welder access. The welding technique here involves two-layer HAZ refinement.[14] The first layer is deposited using a small diameter electrode (3.25mm) at approximately 50% overlap by welding into the toe formed by each previous bead. The second layer is then deposited on top of the first using a higher heat input, larger diameter electrode (4mm diameter). With the correct balance of heat input between the first and second layer the 'refining zone' of the second layer penetrates the first and can produce total refinement.[14] Techniques of this type are being used increasingly in the fabrication and repair of CEGB plant.

CONCLUSIONS

Manual metal-arc welding processes can be

devised to produce welds of defined and improved metallurgical quality. An understanding of the way in which the welding process interacts with a given material is of great benefit when specifying welding procedures for the production of consistent quality welds.

ACKNOWLEDGEMENTS

The contribution to this work of the authors' colleagues at Marchwood Engineering Laboratories, Central Electricity Research Laboratories and North West Region Scientific Services Division is gratefully acknowledged. This Paper is published by permission of the Central Electricity Generating Board.

REFERENCES

1 BAIKIE, B.L. et al. 'Ultrasonic inspection of austenitic welds'. J.British Nucl.Energy Soc., 15 (3), July 1976, 257-61.

2 ALBERRY, P.J. and JONES, W.K.C. To be published.

3 TOFT, L. and YELDHAM, D. 'Weld performance in high pressure steam generating plant in the Midlands Region CEGB'. Conference 'Welding Research Related to Power Plant', CEGB, Southampton, 17-21 September 1972, 5-19.

4 CHEW, B. et al. 'Control of microstructure by welding procedure to prevent cracking in CrMoV weldments'. CEGB Report no WRM53/76, 1975.

5 GLOVER, A.G., JONES, W.K.C., and PRICE, A.T. 'Assessment of resistance of low alloy steels to reheat cracking using the Vinckier test'. Metals Technology, 4 (6), 1977, 326-32.

6 MYERS, J. 'Effect of deoxidants and impurities on simulated stress relief cracking of ½CrMoV steel'. Conference 'Welding Research Related to Power Plant', CEGB, Southampton 17-21 September 1972, 356-68.

7 CHUBB, E., FIDLER, R., and WALLACE, D. 'Development and relaxation of welding stresses'. Conference 'Welding Research Related to Power Plant', CEGB, Southampton, 17-21 September 1972, 143-64.

8 ALBERRY, P.J., CHEW, B., and JONES, W.K.C. 'Proir austenite grain growth in the heat-affected zone of a 0.5CrMoV steel'. Metals Technology, 4 (6), 1977, 317-26.

9 JONES, W.K.C. and ALBERRY, P.J. 'The role of phase transformations in the development of residual stresses during the welding of some fast reactor steels'. Conference 'Ferritic Steels for Fast Reactor Steam Generators', British Nuclear Energy Society, 1977, Paper 78.

10 MYERS, J. 'Grain growth in vacuum-melted steels during HAZ simulation'. Metals Technology, 4 (8), 1977, 411-12.

11 ALBERRY, P.J. and JONES, W.K.C. To be published.

12 ROSENTHAL, D. 'The theory of moving sources of heat and its application to metal treatments'. Trans ASME, 68, 1946, 849-66.

13 CHRISTENSEN, N. et al. 'Distribution of temperatures in arc welding'. British Welding J., 12 (2), 1965, 54-75.

14 ALBERRY, P.J., MYERS, J., and CHEW, B. 'An improved welding technique for HAZ refinement'. Metals Technology, 4 (12), 1977, 557-66.

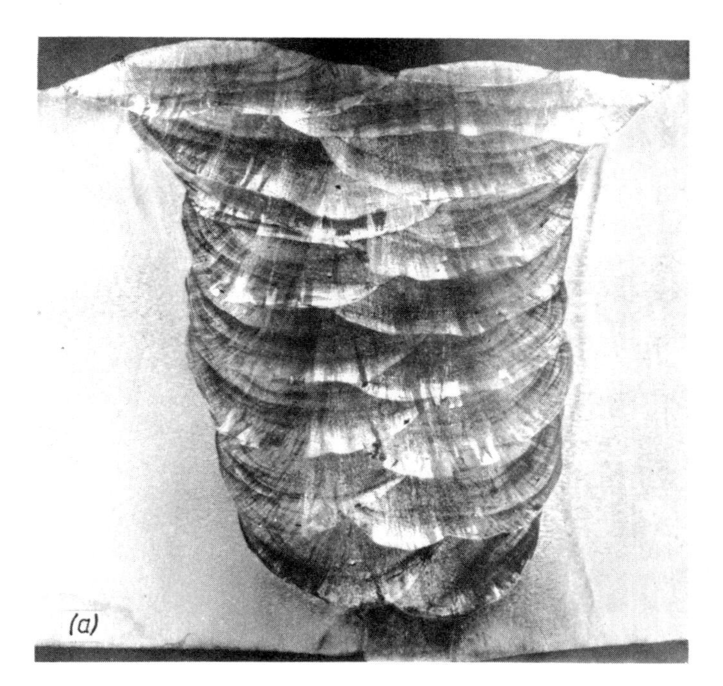

1 Conventional butt weld in 25mm plate. x3

2 Horizontal-vertical weld pad: (a) bead structure, (b) grain structure. x5

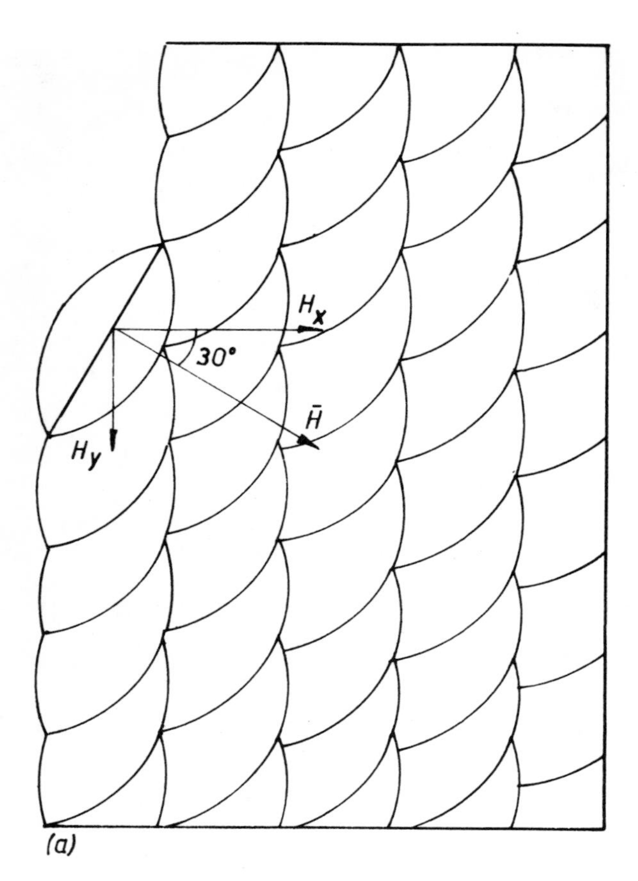

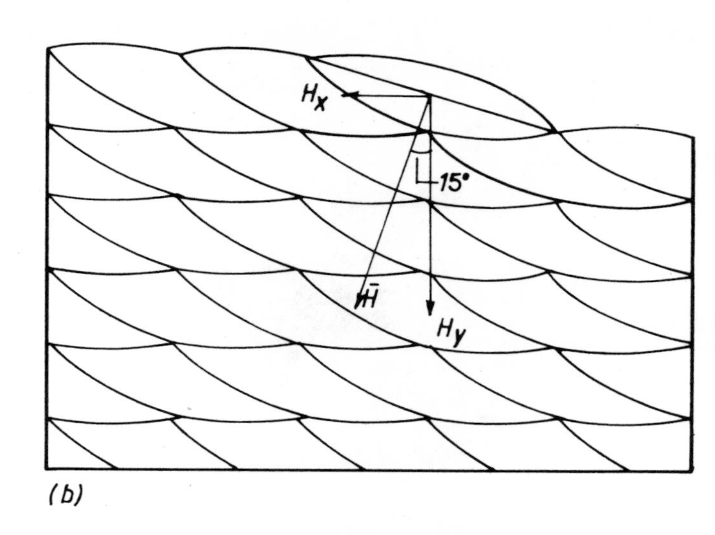

3 Schematic diagram of heat flow for: (a) H-V weld, (b) weld in (b) weld in flat position

4 Aligned grain structure produced by controlled grain orientation technique in 75mm thick weld

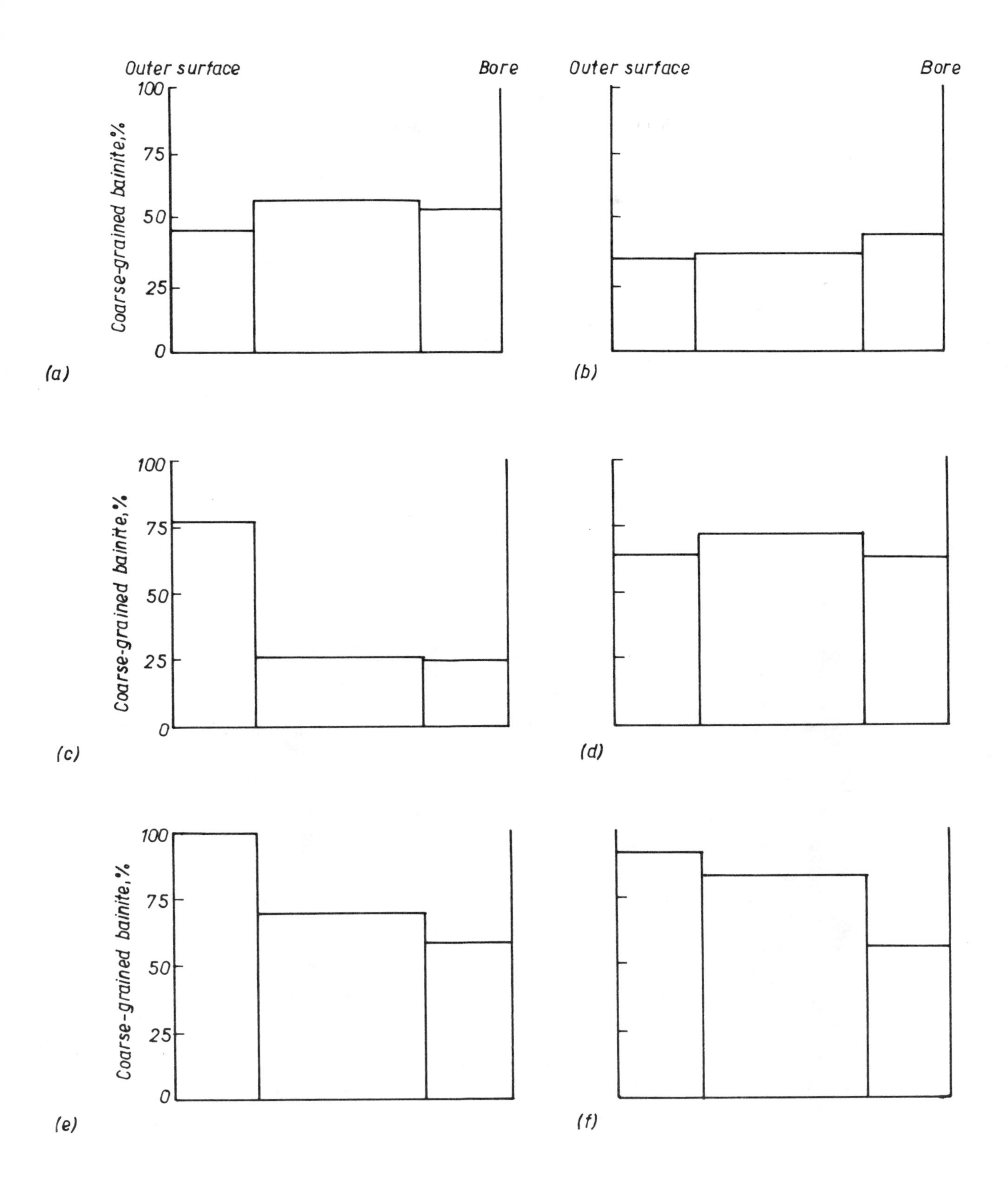

5 Distribution of coarse-grained bainite in ½CrMoV components from service: (a) main steam, pipe 1, (b) main steam, pipe 2, (c) main steam, pipe 3, (d) flange, (e) valve strainer (cracked), and (f) valve (cracked)

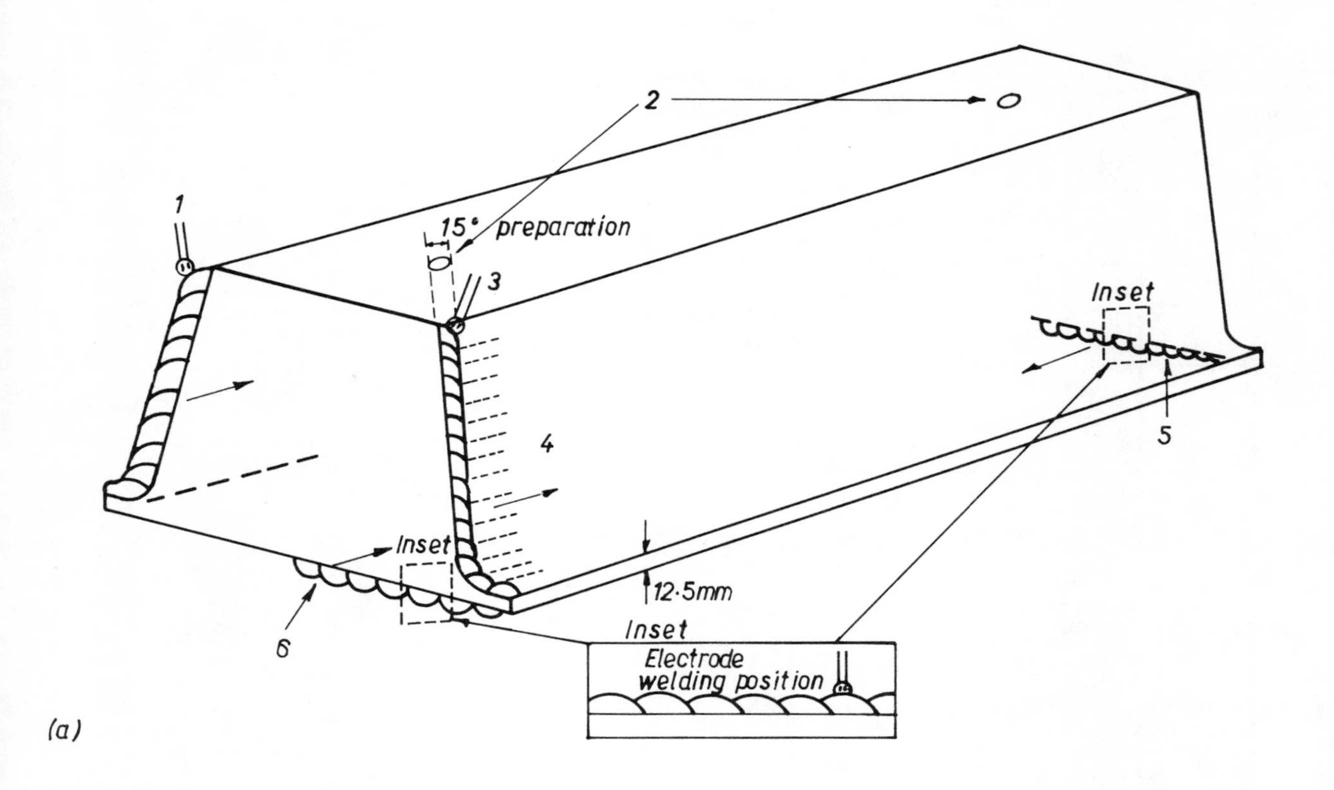

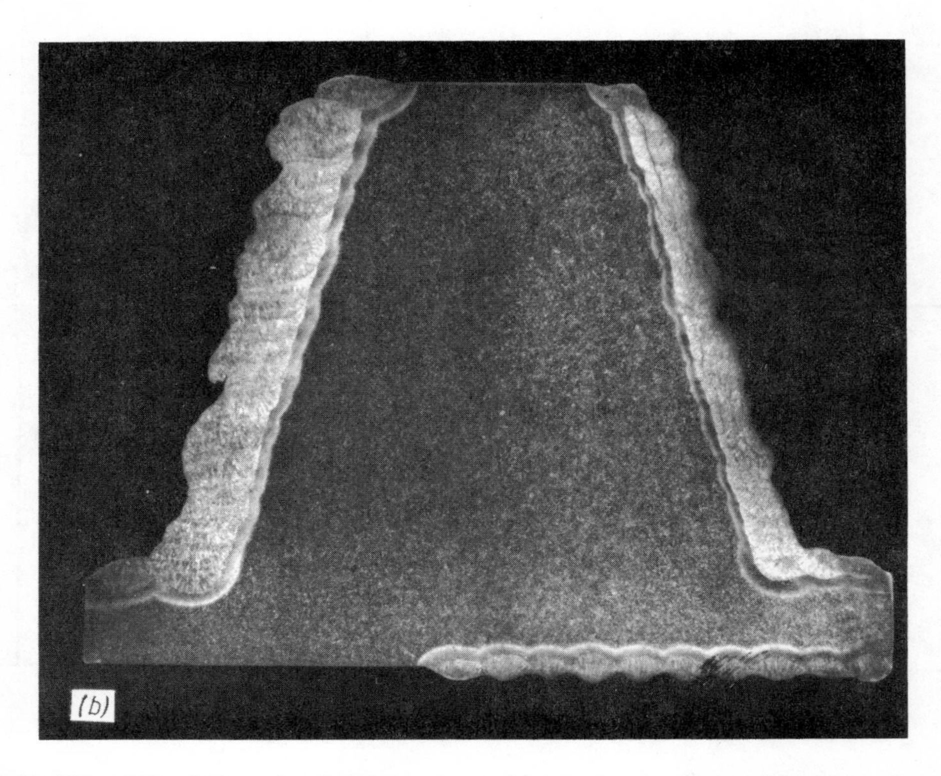

6 Test block (½CrMoV, 100 x 100 x 200mm long): (a) showing weld deposits; electrodes: 2CrMo, 4 and 5mm, (b) macro-section. 1 — 5mm electrodes, H-V butter; 2 — thermocouple holes 25mm deep; 3 — electrode welding position; 4 — 4mm electrodes, H-V butter; 5 — 4mm electrodes, bead on plate; 6 — 5mm electrodes, bead on plate

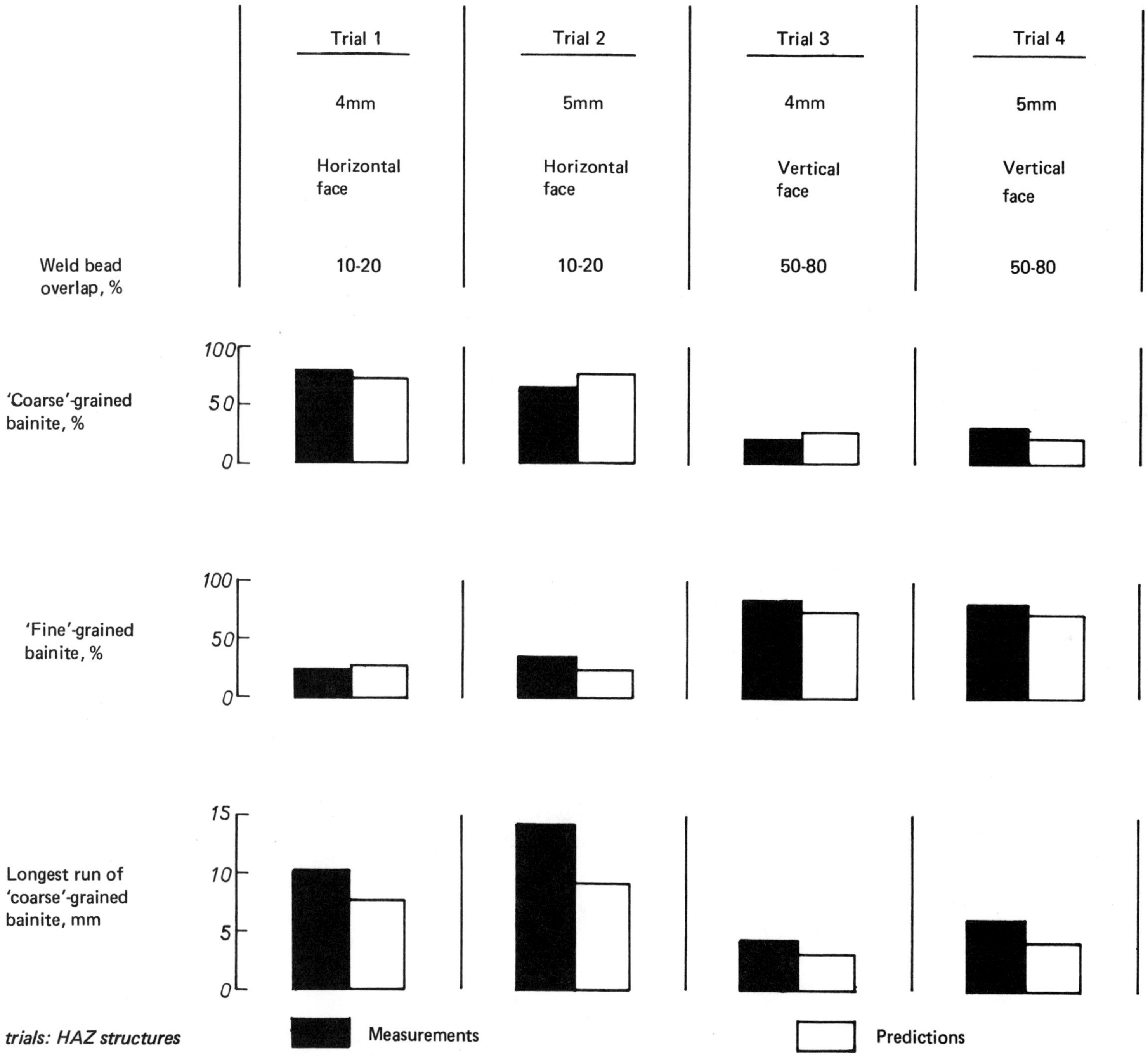

7 *Weld deposition trials: HAZ structures*

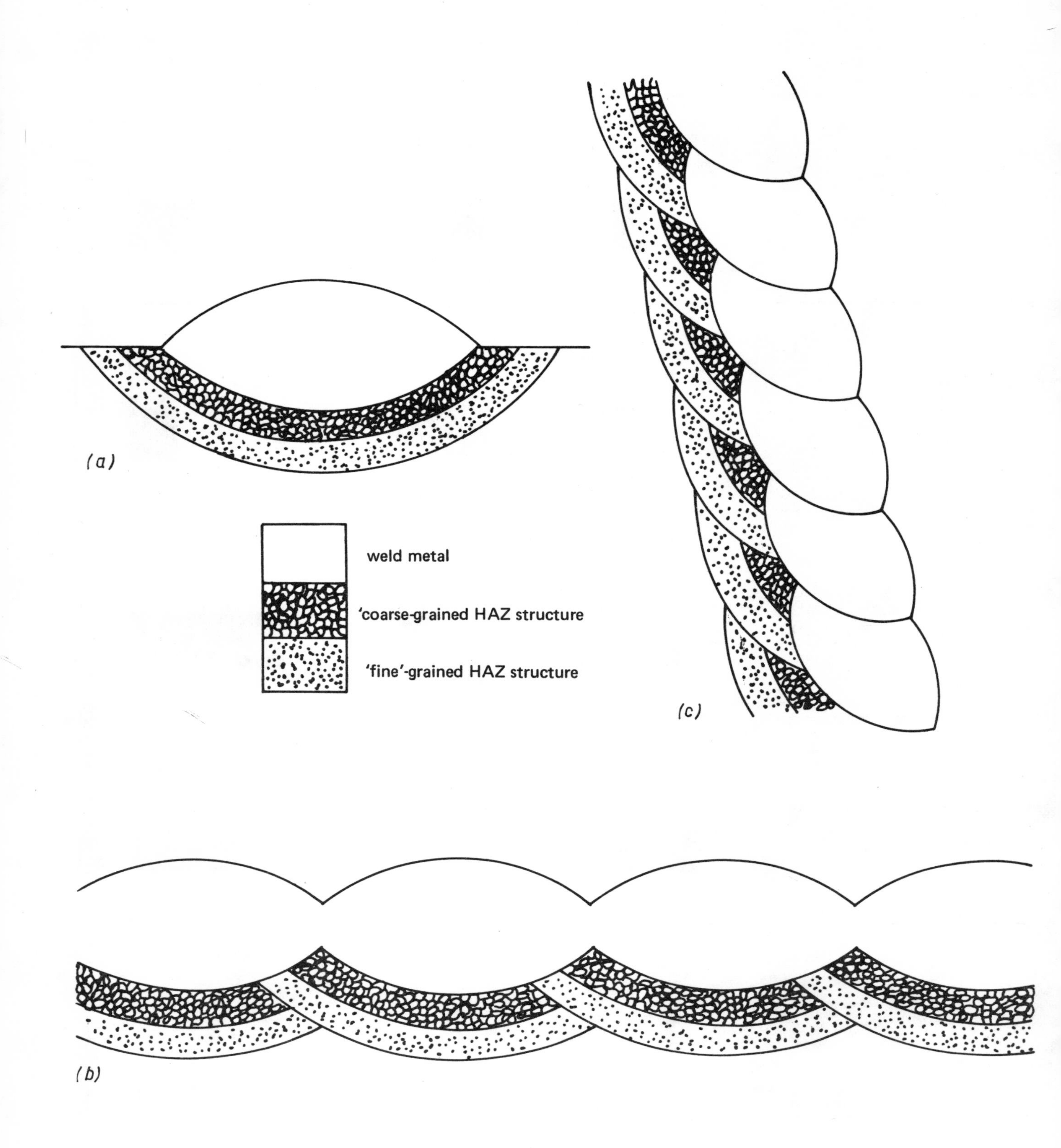

8 Idealised HAZ structures: (a) single weld bead, (b) low overlap weld in flat position, and (c) high overlap on vertical face

PAPER 44

Alternating current MIG - welding

W.Lucas, PhD, MIM, MWeldI

The fundamental requirements for stabilising the AC MIG-welding arc are described within the context of the subsequent development of an arc re-ignition system. In an evaluation of AC MIG-welding as a practical process, the essential operating characteristics are compared with DC MIG-welding, with regard to stability of the arc, metal transfer characteristics, wire feed/welding current relationship, and weld bead penetration. Typical welds are also presented, using conventional joints not only to demonstrate the versatility of the process but also to exemplify specific features which could be exploited in welding other materials.

INTRODUCTION

Despite the extensive use of alternating current (AC) for manual metal-arc (MMA), submerged-arc, and TIG-welding processes, MIG-welding with solid wire is restricted to direct current (DC) and almost exclusively to electrode positive polarity. The inability to apply AC to MIG-welding is understandable since, even for flux-shielded processes, the requirements for reliable re-ignition entail the use of a relatively high open circuit voltage (OCV). For instance, despite the presence of arc stabilising elements in the electrode covering, the OCV in conventional MMA welding can be as high as 100V, but in TIG-welding, even with the thermionic properties of the tungsten electrode, the arc is stabilised only at an OCV of over 150V.[1] To reduce the OCV of the latter to more reasonable levels — less than 100V — an ancillary system (surge injection) is employed to assist arc re-ignition.[2] The MIG arc is essentially similar to that in TIG-welding for electrode positive half-cycles in that, as the arc cathode must form on a bare nonthermionic surface, a high OCV is likewise required for arc stability. Although the use of a transformer with an OCV as high as 250V would ensure arc stability this would be considered too dangerous, particularly for manual welding. However, by using a modified TIG re-ignition system, in which a voltage surge is supplied at each change of polarity, the arc can be operated at an OCV as low as 70V.

Dr Lucas is Group Leader in the Process Operation and Control Department of The Welding Institute.

Stabilisation of AC MIG-welding arc

Close examination of the re-ignition behaviour has shown that the AC MIG arc is more allied to that found for the TIG than for the MMA welding arc.[3] Thus, when the arc extinguishes at current zero, which is associated with cessation of the arc root mechanism, the high re-ignition voltage is required to re-establish the cathode arc roots for the next current half-cycle. A typical oscillogram of arc re-ignition (at a high OCV of some 250V), is shown in Fig.1a, for the negative to positive polarity changeover. Here the arc extinguishes at current zero and re-ignition occurs only when the voltage across the gap rises to approximately 220V. The level of the re-ignition voltages for the positive re-ignition varies over a range and can be as high as 300V.

Thus, to reduce the OCV requirements to a more reasonable level, the design of a suitable arc re-ignition system can be specified: a voltage surge is required to re-form the arc, the magnitude of the surge voltage should be capable of accommodating the most difficult re-ignition, that is, arc extinction followed by a re-ignition voltage of at least 300V, and both positive and negative re-ignitions must be assisted at each change of polarity. The operation of the surge during re-ignition is shown in the oscillogram of the positive re-ignition, Fig.1b. The exact timing of the pulses is fairly critical and, for its greatest effect, the pulse must be synchronised with current zero or applied immediately after zero. The effectiveness of the surge system for assisting re-ignition can be judged from its ability to reduce the OCV from approximately 250V to less than 75V.

The re-ignition system, Fig.2, injects a high voltage/short duration surge within 30msec of the instant of current zero. The surges are supplied by two precharged capacitors, and an electronic switch, comprising a thyristor and detector/firing circuit, ensures that the pulses are injected synchronously with current zero.

ARC AND METAL TRANSFER

There are two unique features in the AC MIG-welding arc which are directly associated with the alternate periods of electrode positive and electrode negative polarity: the arc climbing up the electrode and a pulsed-type metal transfer. With regard to the former, two types of cathode are observed with electrode negative polarity: single spot and diffused arc roots. In the single spot cathode the arc roots form on the molten wire tip, whereas with the diffused type the arc roots are situated on the wire above the droplet. It is significant that arc instabilities, associated in DC electrode negative welding with the diffused type of arc root,[4,5] are caused by the arc roots climbing up the wire, and, after having consumed the available oxide, transferring suddenly to the electrode tip (single spot). This problem does not occur on AC which is attributed to the short time interval of the negative half-cycle (10msec) before the electrode assumes positive polarity.

The metal transfer characteristics of the AC MIG-welding arc are quite unique in that metal transfers predominantly when the electrode has a positive polarity, Fig.3. The effect is twofold in that it extends the operating range for a given wire diameter and reduces the jetting action associated with high current DC arcs, as described hereunder. The operating range for spray transfer DC MIG arcs is limited by the deterioration to globular metal transfer at low currents and by excessive jetting at high currents. In AC MIG-welding the sinusoidal current gives pulsed metal transfer and process stability at welding current levels below the normal spray range. For example, as shown in Fig.4, the lower limits for 1.2mm diameter mild steel wire can be reduced from nearly 300A to 200A without any deterioration in arc and metal transfer stability.

The AC has an equally beneficial effect at high current levels. The jetting action of the high current DC positive arc does not occur because of the continuous interruption in the arc forces. As shown in Fig.3, although jetting (note the excessive tapering of the electrode tip) occurs during electrode positive polarity, the arc becomes more diffuse (electrode tip assumes a rounded profile) when the electrode polarity changes to negative. Thus, the periodic interruption of the jetting forces in the arc permits much higher welding currents than would otherwise be possible with DC welding. In the above example of 1.2mm diameter mild steel wire, Fig.4, a current of 450A could be operated quite satisfactorily and even this current level was limited only by the capacity of the wire feed system and not by the arc characteristics.

The effect of AC on the arc and metal transfer behaviour applies also to other materials. For aluminium, Fig.5, the burnoff rate is about 50% greater than with DC positive welding and, since the maximum current is also at least 50% greater, the deposition rate with AC is about double that for DC, e.g. for 1.6mm diameter wire a welding current of at least 450A can be safely operated with 15m/min feed rate.

WELDING CHARACTERISTICS

Alternating current welding is similar to DC MIG-welding in that conventional materials can be readily welded over a wide range of current levels.[6] Shielding gas is selected as in DC welding according to material type, e.g. argon for aluminium and a gas mixture from the argon/oxygen/CO_2 system for steels. It is noteworthy that, as AC is practised only in the spray transfer mode, the CO_2 content must be limited to not more than 20% to avoid excessive spatter formation.

Mild steel

The weld bead penetration profiles obtained with the AC arc are intermediate between those of DC positive and DC negative welding. For instance, at welding currents up to 300A, using 1.2mm diameter wire, the weld beads have deeper penetration than DC negative but not as great as DC positive welding, Fig.6. At higher welding currents the penetration profile is more similar to DC positive welding but with a distinct advantage: it is more bowl-shaped and does not contain a pronounced 'finger', Fig.7. The finger in DC welding is particularly undesirable in that it is narrow at the most critical position, i.e. on the joint line. Hence, in practice, in terms of useful penetration and its capacity to be tracked along the joint line, it must be ignored. Thus, considering only the bowl-shaped penetration, AC compares quite favourably with DC, having uniform penetration with an excellent surface profile at the toes, Fig.7.

The AC MIG process has been applied to the welding of a relatively wide range of plate thicknesses and joint configurations; typical operating parameters are given in Table 1 with macrosections through the joints shown in Fig.8. It can be seen that the process is equally applicable to welding thin sheet and plate with both butt and fillet-type joint configurations. A particularly noteworthy example is welding 12.5mm plate where the above-mentioned capacity for high rate deposition was fully exploited in the role of joint filling; the absence of jetting permitted a welding current of 393A to be used with a 1.2mm diameter wire without any tendency for jetting or for the formation of a finger-type penetration profile.

Aluminium

As a similar range of weld bead penetration profiles can be obtained for aluminium, AC MIG can be used for welding both thin sheet and plate. Typical operating parameters are given in Table 2 and macrosections through the welded joints are shown in Fig.9. An important feature in welding aluminium is that the weld pool is much smaller than the corresponding pool size for DC. This facilitates greater control of the behaviour of the weld pool and enables positional welding to be carried out at high currents.

DISCUSSION

Research at The Welding Institute has clearly established that, by using an arc re-ignition system and a suitable shielding gas, MIG-welding can be readily practised with AC. Although this in itself is of academic significance, since it now dispels previous doubts on the use of AC for MIG-welding, the questions must now be posed, 'What of the future of AC MIG: will it remain a research curiosity or does it have a much wider role to play in production technology?'. To answer this question it must be emphasised immediately that on economic grounds alone there would be no advantage, because of the present low cost of welding rectifiers, to be gained in the use of transformers, particularly as they must now include the cost of an arc stabilising system.

Table 1 AC MIG-welding parameters for mild steel

Plate thickness, mm	Welding technique	Root face, mm	Root gap, mm	Welding procedure	Shielding gas	Wire diameter, mm	Welding current, A	Wire feed speed, m/min	Arc voltage, V	Welding speed, m/min
2	SECB	-	-	Single pass	Ar-5%O_2	1.0	155	4.0	21	0.33
6	SEOB	-	3	Single pass	Ar-2%O_2-10%CO_2	1.2	345	10.5	23	0.33
6	90° HV fillet	-	-	Single pass	Ar-2%O_2-10%CO_2	1.2	255	8.9	25	0.33
13	Butt 60^oV	1	-	1st pass	Ar-2%O_2-10%CO_2	1.2	325	9.8	22	0.25
				2nd pass			395	12.5	28	0.25

Welding wire - mild steel (Si-Mn deoxidised, BS 2901)

SECB - square edge close butt

SEOB - square edge open butt

Table 2 AC MIG-welding parameters for aluminium

Plate thickness, mm	Welding technique	Root face, mm	Root gap, mm	Welding procedure	Shielding gas	Wire diameter, mm	Welding current, A	Wire feed speed, m/min	Arc voltage, V	Welding speed, m/min
3	SECB	-	-	Single pass	Ar-1%O_2	1.6	245	9.3	21	1.4
6	SECB	4	-	Single pass	Ar-1%O_2	1.6	285	11.0	17	0.5
6	90° HV fillet	-	-	Single pass	Ar-1%O_2	1.6	295	10.1	22	0.4
13	Butt 60^oV	6	-	Single pass	Ar-1%O_2	1.6	380	12.5	21	0.4

Welding wire- aluminium (99%Al, G1B)

SECB - square edge close butt

Rather the incentive to change to AC will arise from a desire to exploit the benefits of improved penetration profiles at high welding currents. In particular, the reduced finger-type penetration and the reduced weld pool disturbance could be used to advantage when filling joints at high deposition rates. In this context the AC MIG process could well fulfil an important role in welding nonferrous material, and may even become competitive with the higher power processes such as submerged-arc welding where the absence of a flux may prove advantageous both in controlling the weld pool and in seam tracking.

Although economy is usually the major factor, other benefits can often influence the choice of a welding process. The special features of AC MIG are tolerance to magnetic fields, low porosity, and low parent metal dilution. The process has proved to be especially tolerant to the effects of magnetic fields in that the AC arc can be operated in fields as high as 80 Gauss with only a minor disturbance of the arc. Even at this extreme level the arc merely becomes broader through the alternate deflections on change in the electrode polarity. Thus, in practical welding of ferritic steels, where arc blow problems are currently experienced such as arise from stray magnetic fields and at the beginning and end of the joint seam, AC MIG-welding should greatly reduce if not eliminate problems of arc instability. The alternative solution of demagnetising the plates is wasteful of both time and effort and sometimes proves abortive.

The low porosity and low parent metal dilution characteristics are associated with the reduction in the finger of the weld bead. For example, porosity can be a recurring problem in welding stainless steel using DC and the spray arc welding technique.[7] Careful experimentation has shown that the porosity is invariably associated with the finger, as shown in Fig.10. On the other hand, AC with its more bowl-shaped penetration profile produces almost porosity-free welds, Fig.11. The improved penetration profile may also be exploited in surfacing where the absence of a finger would limit the degree of parent metal dilution in the deposit.

CONCLUSIONS

It is not considered that AC will replace DC for spray metal welding. It is believed, however, that it will exist in its own right having several unique and desirable features, i.e. more stable arc and metal transfer characteristics at very high current levels, higher burnoff rate, better shaped weld bead penetration profile in the spray transfer mode (at high current levels), tolerance to magnetic fields, and elimination of porosity associated with the finger. Therefore, it should be considered as a useful alternative to the various existing arc welding processes so that it may be adopted where it offers the particular advantages of quality and economy.

REFERENCES

1 ORTON, C.H. and NEEDHAM, J.C. 'Some electrical aspects of inert gas-shielded arc welding'. British Welding J., 2 (10), 1955, 419-26.

2 NEEDHAM, J.C. and ORTON, C.H. 'AC argon arc welding at less than 50V rms open circuit'. Trans Inst.of Welding, 15 (6), December 1952, 161-5.

3 LUCAS, W., STREET, J.A., and WATKINS, P.V.C. 'Solid wire AC MIG-welding'. Welding Inst.Members Report P65/75, January 1975.

4 AMIN, M. 'Electrode negative MIG, mild steel arc studies'. Welding Inst. Confidential Report to BOC, C303/3/1970.

5 NORRISH, J. 'High deposition MIG with electrode negative polarity'. Welding Inst.Conference 'Advances in Welding Processes', Harrogate, 1974. Paper 16, 121-8.

6 LUCAS, W. and NEEDHAM, J,C. 'Why not AC MIG-welding?'. Welding Inst. Research Bull., 16 (3), 1975, 63-7.

7 MILLINGTON, D. 'Porosity formation in in the MIG-welding of stainless steel'. Welding Inst.Research Bull., 13 (10), 1972, 293-300.

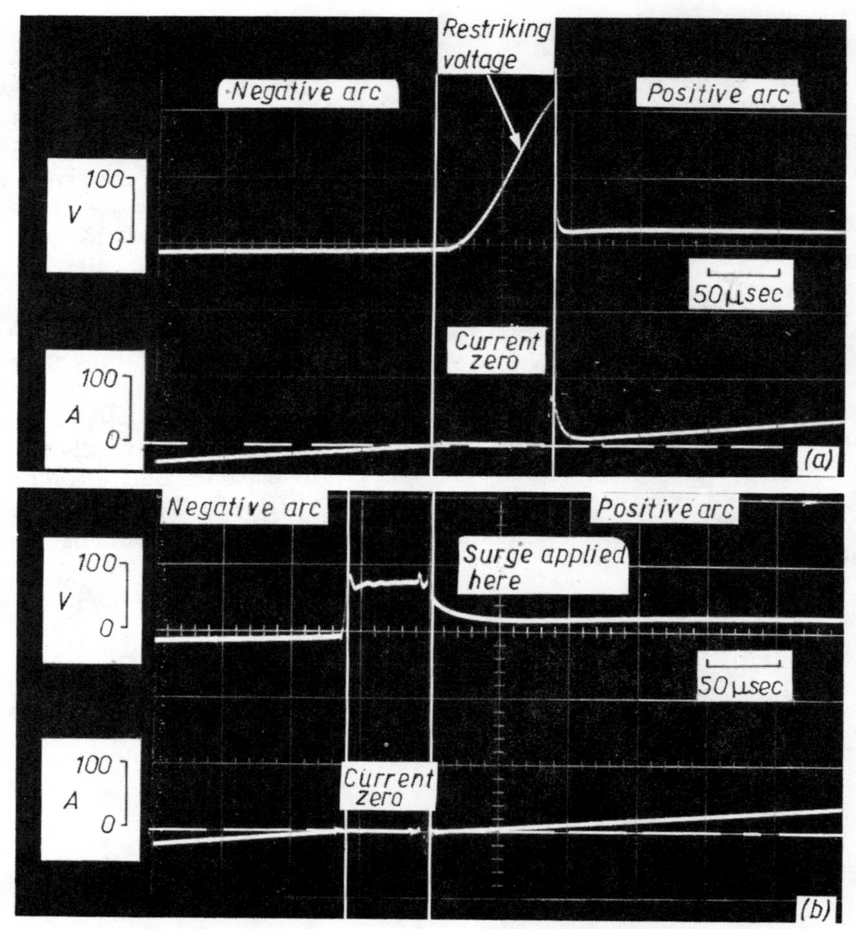

1 Negative to positive arc re-ignition, 1.2mm diameter mild steel at 350A: (a) self re-ignition from high OCV; note momentary current zero pause, (b) surge-injected re-ignition with 75V OCV

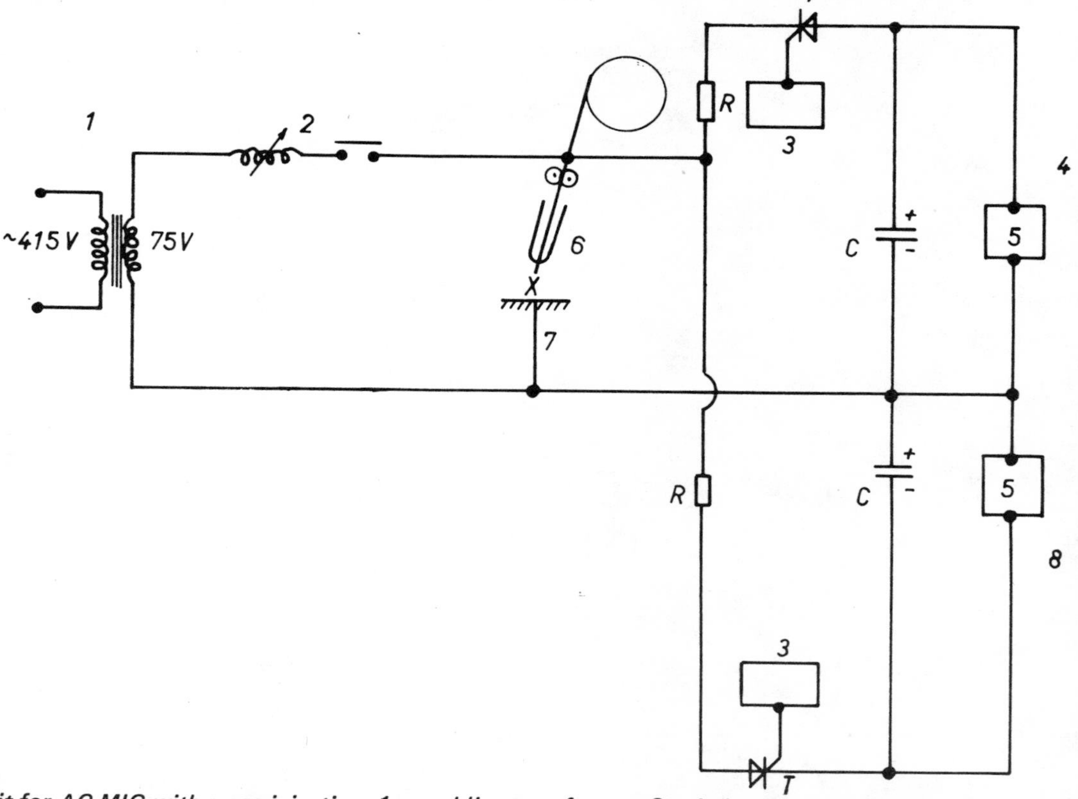

2 Welding circuit for AC MIG with surge injection. 1 — welding transformer; 2 — inductor contactor; 3 — detector firing circuits; 4 — positive re-ignition; 5 — charge circuits; 6 — welding gun; 7 — workpiece; 8 — negative re-ignition; T — thyristor; C — capacitor; R — resistor

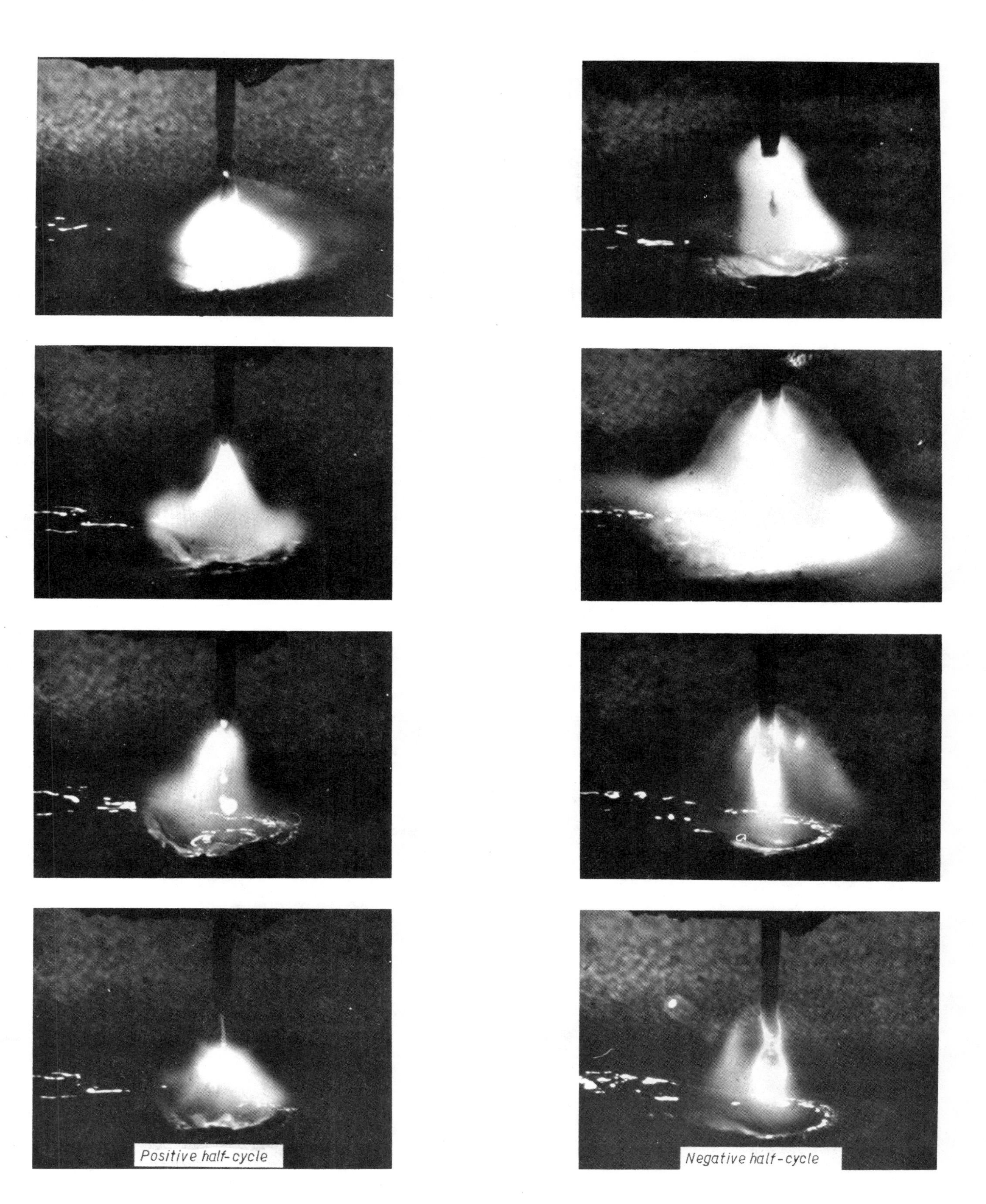

3 High speed photographs of metal transfer in AC MIG-welding (8000 frames/sec). Note: streaming transfer mode during electrode positive half-cycle and no transfer during negative half-cycle (Ar-5%O_2 shielding gas; 10m/min wire feed speed)

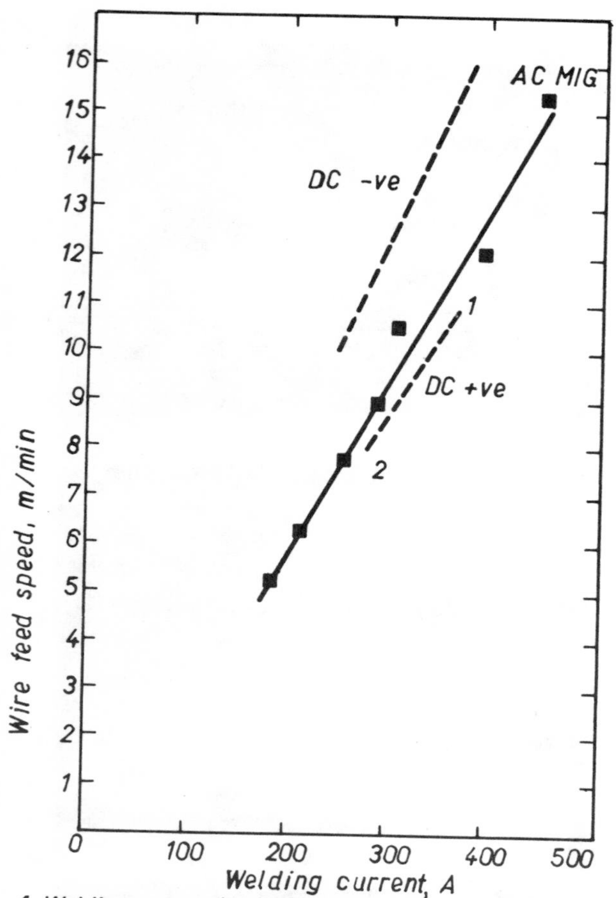

4 Welding current/wire feed speed relationships for mild steel. Note: extended working range of AC MIG compared with DC. 1 – jetting and 2 – globular transfer regions (1.2mm wire)

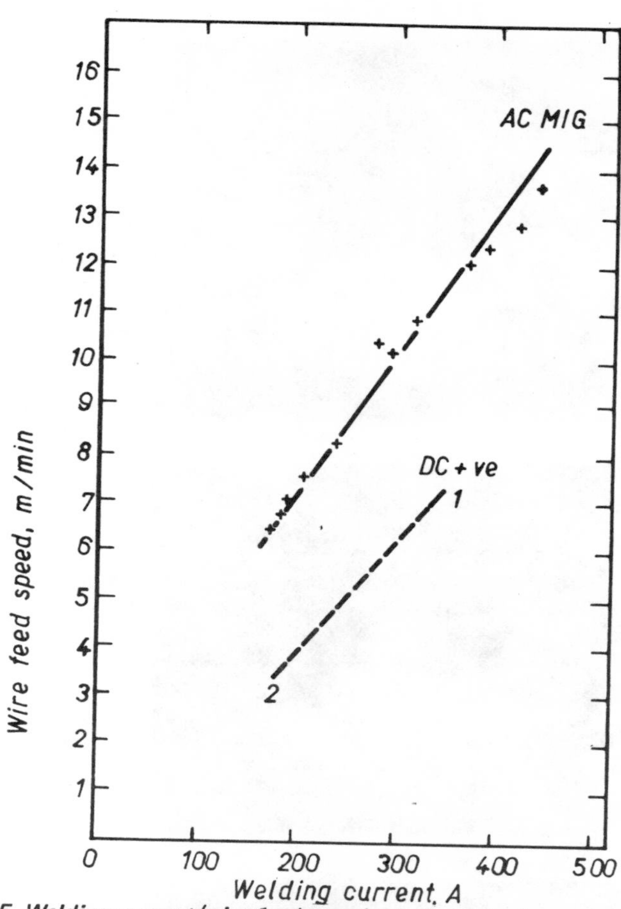

5 Welding current/wire feed speed relationships for aluminium. 1 – jetting and 2 – globular transfer regions (1.6mm wire)

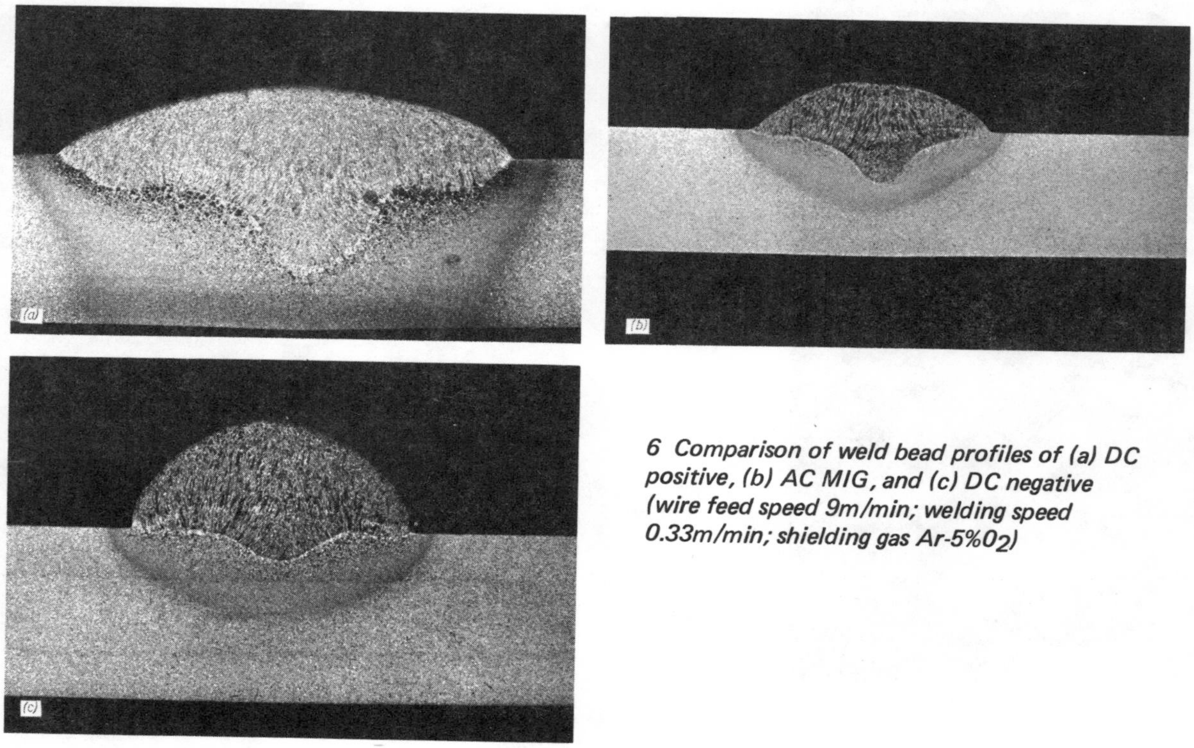

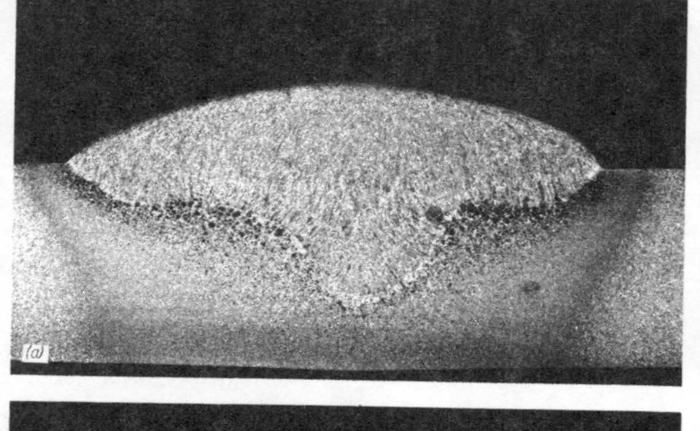

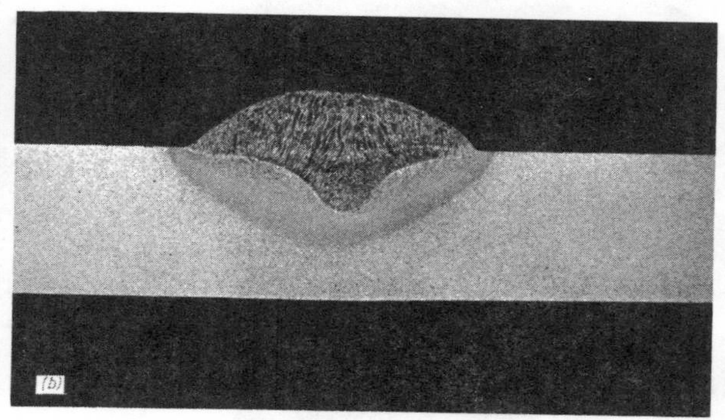

6 Comparison of weld bead profiles of (a) DC positive, (b) AC MIG, and (c) DC negative (wire feed speed 9m/min; welding speed 0.33m/min; shielding gas Ar-5%O_2)

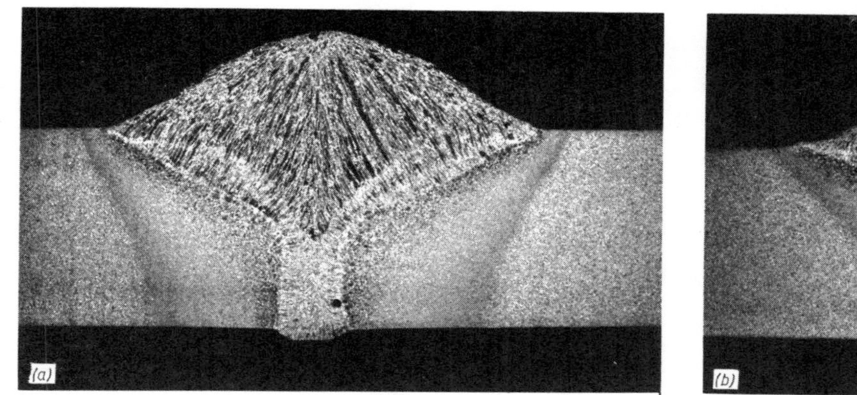

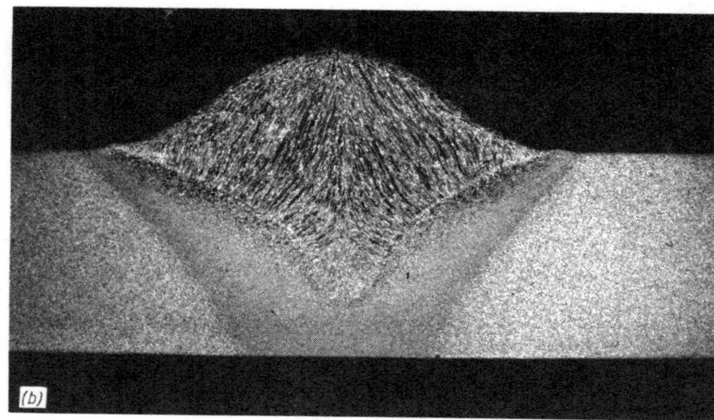

7 Comparison of weld bead profiles of (a) DC positive, (b) AC MIG (wire feed speed 15m/min; welding speed 0.5m/min; shielding gas Ar-5%O_2)

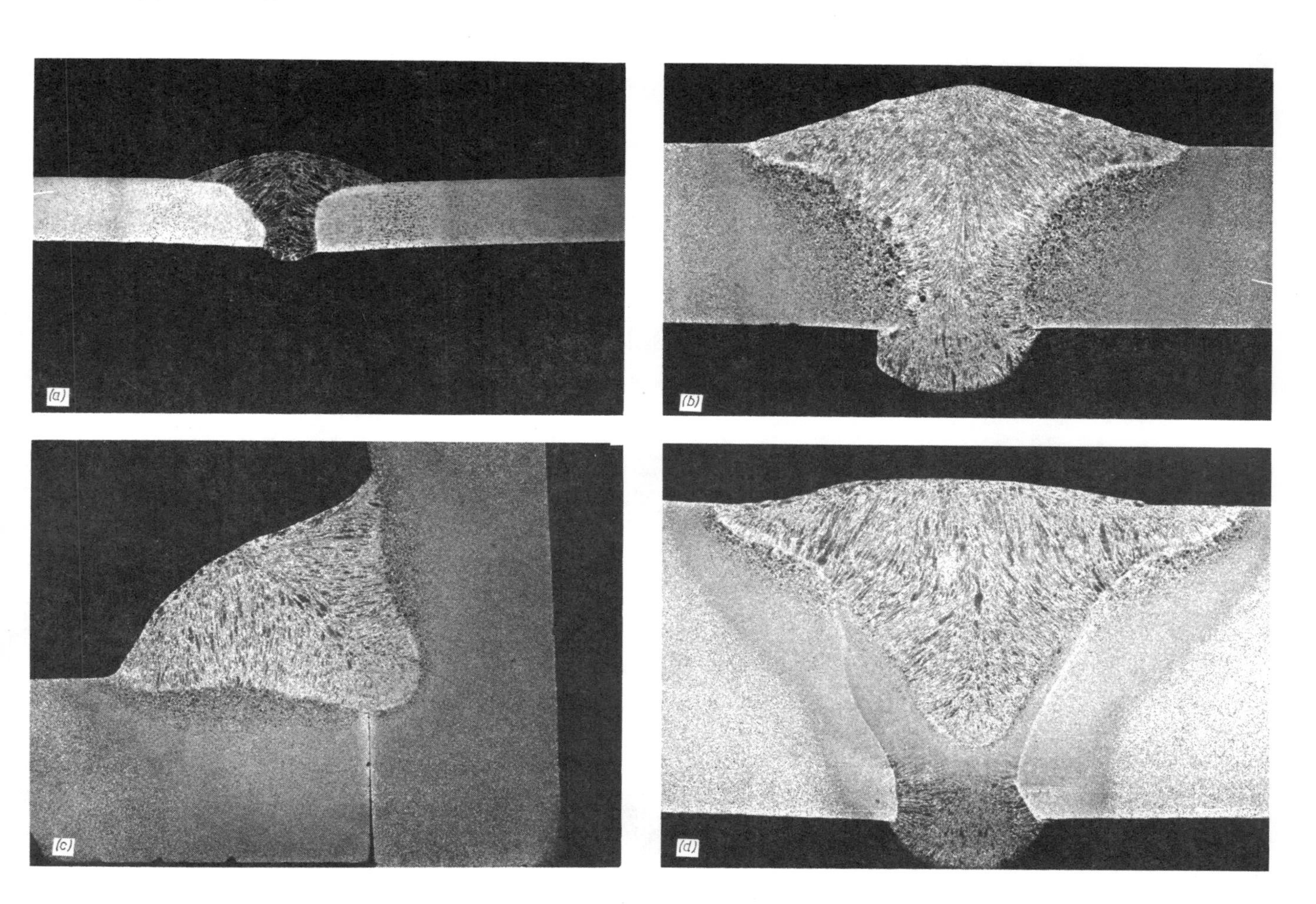

8 Typical welded sections for mild steel with joint preparations, mm: (a) 3, (b and c) 6, and (d) 12

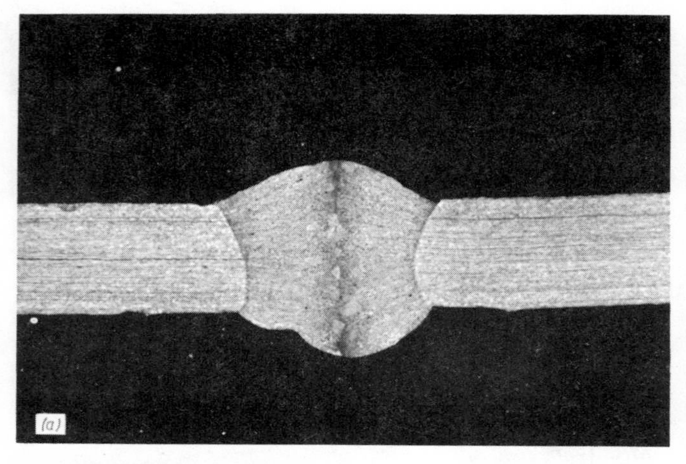

9 *Typical welded sections for aluminium with joint preparations, mm: (a) 3, (b) 6, and (c) 12*

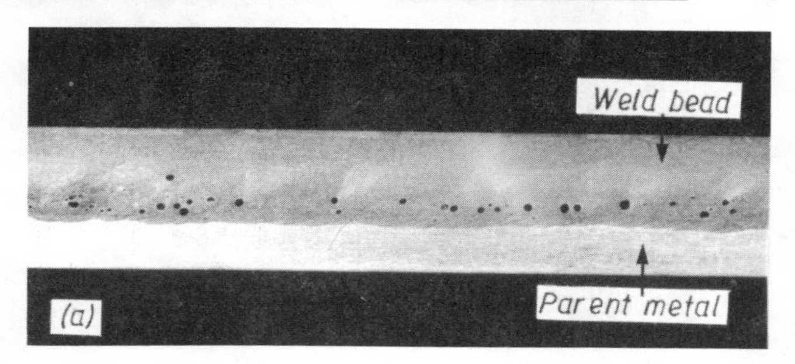

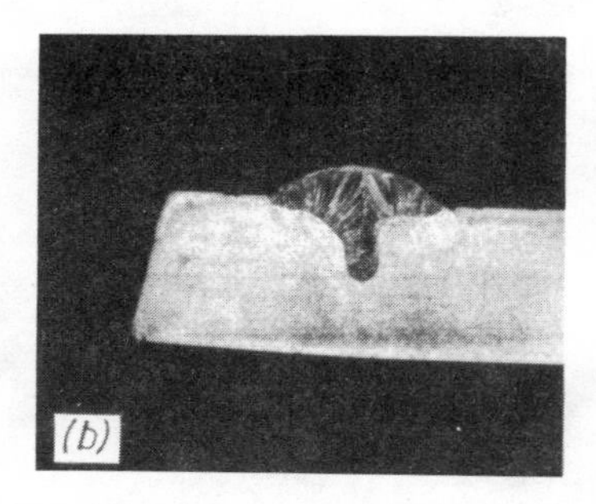

10 *Bead-on-plate stainless steel welds with DC MIG showing porosity associated with finger penetration. Welding parameters: welding current 260A; shielding gas Ar/1%O_2; welding wire 19/10/1Nb, 1.2mm diameter. (a) longitudinal section, (b) transverse section*

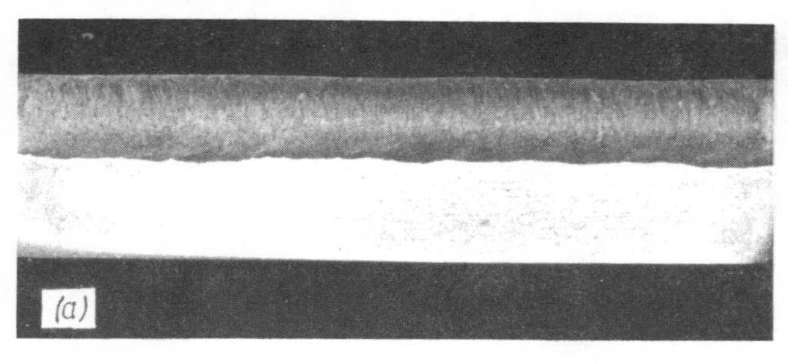

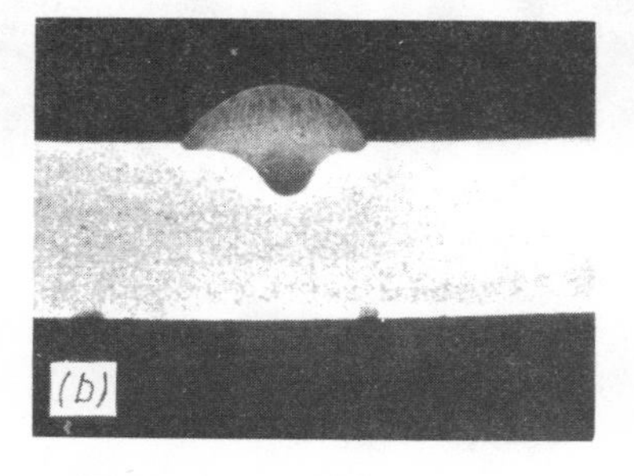

11 *Bead-on-plate stainless steel welds with AC MIG showing bowl-shaped penetration profile and low porosity. Welding parameters: welding current 245A; shield gas Ar/2%Ar; welding wire 19/10/1Nb, 1.2mm diameter. (a) longitudinal section, (b) transverse section*

The twist arc welding process

S.Kimura, Y.Nagai, and T.Kashimura

Twist arc welding is a flat position narrow gap gas metal-arc process in which two intertwined wires are used for the welding electrode. The rotational movement of the welding arc, which is a feature of this process, prevents such defects as lack of sidewall fusion, which are the cause of unreliable quality in the conventional narrow gap technique. Experiments established that, if the groove width is held within the range 12 to 18mm, weld joints of stable quality can be obtained simply by adjusting the travel speed according to the groove width without changing either the welding current or arc voltage.

Moreover, the process employs an automatic control system using a TV camera, which has the function of adjusting the electrode position so that it is always on the weld line, controlling the travel speed if the groove width in the workpiece fluctuates, and remotely controlling the welding parameters. This automatic control system makes it possible to hold the electrode position within 0.5mm of the centre of the groove gap in real time, and frees the welding operator from the severity of the working environment by virtue of its remote-control function.

INTRODUCTION

With the recent increase in the size of structures, more high quality plates have come into use than ever before. Welding processes which achieve a stable weld quality as well as reducing the overall welding cost have long been desired.

The high efficiency welding processes currently applied to thick plates include electro-slag, submerged-arc, and narrow gap welding adopting the gas metal-arc process. Of these three methods the electro-slag process is considered to be the most economic, taking into account both the costs of the welding installation and repair of defects. However, this process is limited in the kinds of joint to which it can be applied, as it needs postweld heat treatment to ensure a high quality weld zone. As for the submerged-arc and gas metal-arc narrow gap welding processes, the latter is said to be economically less advantageous because it demands the use of costly specialised equipment and also because it has a limited tolerance range of welding conditions within which sound welds can be assured. Accordingly, the submerged-arc process is most widely adopted for welding thick plates.

Experimental studies aimed at the elimination of the abovementioned uneconomic factors in the gas metal-arc narrow gap welding process have led to the development of the

Mr Kimura, General Manager — Technical Centre, Mr Nagai, Senior Researcher, and Mr Kashimura, Welding Engineer, are all with Kobe Steel Limited, Fujisawa, Japan.

'twist arc welding process . This a flat position narrow gap gas metal-arc welding process in which a twisted wire, made of two intertwined wires, is specifically used as the consumable electrode and combined with the 'TV-monitored automatic welding control system' in which TV images are used to adjust the electrode position so as to be always on the weld line and to control the travel speed according to the groove width.

TWIST ARC WELDING PROCESS

Wire characteristics

The consumable electrode comprises two intertwined wires. When used in arc welding various characteristics not seen with ordinary consumable electrodes are observed. These relate to the rotational movement of the arc, melting rate of the wire, and the influence of the wire on the sectional macrostructure of the deposited metal, as described below.

Arc rotation

The following data have been obtained from high speed ciné films of the arc in gas metal-arc welding using the twist wire.

Where two wires, differing in diameter, are used as the twist wire, the arc deflects in the axial direction of the larger-diameter wire. The smaller-diameter wire supports the arc less frequently and the arc therefrom is absorbed by that from the larger-diameter wire. Accordingly, the arc makes a continuous rotational movement with the melting of the wire, Fig.1. Furthermore, the droplets from the smaller-diameter wire integrate with those from the larger-diameter wire in transferring to the workpiece. Thus, the smaller-diameter wire behaves like a kind of filler wire only.

However, when two wires of the same diameter are intertwined, the arc is generated alternately from the tips of the two wires, so that the rotational movement of the arc is discontinuous. Rotational movement is also observed at low welding currents where globular transfer is present, but the rotation is less smooth than that observed in the spray transfer range.

The radius of gyration of the arc increases with decrease in the pitch of the twist in the wire. For the same pitch the speed of rotation increases with increase in the welding current, and the radius of gyration becomes greater with increase in the arc voltage. However, the rotational movement of the arc is not clearly observed when three wires of the same diameter are intertwined into a twist wire.

The rotational movement, which is characteristic of the twist wire, can be ascribed to the arc stiffness and to the influence of the magnetic field formed by the welding current in the vicinity of the electrode tip.

Deposition characteristics

Figure 2 compares the melting rates of the twist wire with ordinary solid wires. As shown in the Figure, the melting rate for the twist wire is about 10% greater for the same cross-section. This is attributed to the spiral form of the two wires which makes the actual wire extension greater and gives rise to an increase in resistance heating, even for nominally the same overall stickout distance. In addition, it is also presumed that, as one of the twisted wires touches the other from which the arc is generated, it exhibits a behaviour similar to that of a filler wire.

Figure 3 compares the macrosections of welds made with the twist wire and those with ordinary solid wire. As seen in the Figure, the bead formed by the twist wire tends to a decreased penetration depth but an increase in sidewall fusion. This is attributed to the rotation of the off-axis arc which causes some fall-off in the gouging forces beneath the tip of the electrode, as well as to the active convection and agitation of the molten pool resulting from the arc motion.

As illustrated in Fig.4 this is a flat position, narrow gap, gas metal-arc welding process in which the twist wire is used as the consumable electrode. It is intended to eliminate the lack of sidewall fusion (a common problem experienced in narrow gap welding as currently applied) by the rotating arc function of the twist wire without weaving the electrode mechanically or modulating the welding current or arc voltage at regular intervals.

Operating characteristics

High quality joints free from defects like lack

Table 1 Optimum welding conditions for twist arc process

Power source	DC drooping characteristic
Shielding gas	Ar + 20%CO_2
Wire diameter	2.0 x 2.0mm
Pitch of twist	9 ~ 10mm
Groove width	12 ~ 18mm
Welding current	550 ~ 600A
Arc voltage	31 ~ 32V
Welding speed	400 ~ 25mm/min

Table 2 Welding conditions for practical trials

Power source	DC drooping characteristic		
Shielding gas	Ar + 20%CO_2		
Parent material (plate thickness)	ASTM A387, Gr.11 (50mm)	ASTM A387, Gr.22 (50mm)	ASTM A516, Gr.70 (120mm)
Welding wire	AWS, SFA5.18 E70S-GB	AWS, SFA5.18 E70S-GB	AWS, SFA5.18 E70S-GB
Groove width	14mm		
Welding current, voltage	550A, 32V		
Welding speed	300 ~ 350mm/min		

of fusion can be obtained at all times because the fusion to the groove sides is increased by the rotational movement of the arc. Also the welding machine is very simple in construction and can be handled with ease because it dispenses with special regulating devices such as for mechanical weaving.

Table 1 gives the optimum conditions for twist arc welding as obtained from experiments in machined square grooves aimed at maximising the penetration to the grooved sides. The Table implies that this process is capable of producing joints of reliable quality if the deposition only is adjusted according to variations in the groove width without changing the welding current or voltage.

To study the applicability of the twist arc process to pressure vessels, boilers, etc. trials were conducted on welding steel plates as typically used for their construction. The welding conditions adopted are shown in Table 2, together with the welding materials, steel plates, and chemical composition of weld metal in Table 3. The test results are given in Table 4, with side bend testpieces and macrosections of the welded joints shown in Fig.5. All testpieces exhibited excellent performance, indicating that the process is capable of producing high quality welded joints.

Table 3 Chemical composition of test material and deposited metal

Material		Chemical composition, %							
		C	Si	Mn	P	S	Cr	Mo	Total N
ASTM A387, Gr.11	Parent plate	0.13	0.67	0.53	0.004	0.006	1.30	0.50	-
	Wire	0.075	0.47	1.13	0.010	0.011	1.48	0.56	-
	Deposited metal	0.090	0.54	0.91	0.008	0.010	1.32	0.53	0.0068
ASTM A387, Gr.22	Parent plate	0.12	0.35	0.52	0.008	0.007	2.32	0.96	-
	Wire	0.092	0.55	1.06	0.011	0.012	2.24	1.20	-
	Deposited metal	0.097	0.42	0.84	0.010	0.012	2.26	1.06	0.0076
ASTM A516, Gr.70	Parent plate	0.26	0.21	0.81	0.011	0.013	-	-	-
	Wire	0.10	0.82	1.41	0.013	0.007	-	-	-
	Deposited metal	0.11	0.59	1.10	0.013	0.013	-	-	0.0075

Table 4 Mechanical results

Parent material	Postweld heat treatment	Joint tensile strength, (JIS Z3121)	N/mm^2	Charpy impact value, J (JIS Z3112, 2mm V side notch, 0°C) Weld metal		Fusion line		HAZ	
ASTM A387, Gr.11	670°C x 7hr	623 620	(621)	82 69	(75)	158 178	(168)	239 254	(246)
ASTM A387, Gr.22	690°C x 7hr	659 660	(659)	123 137	(130)	155 143	(149)	189 197	(193)
ASTM A516, Gr.70	625°C x 5hr	583 577	(580)	130 124	(127)	167 179	(173)	257 225	(241)

AUTOMATIC CONTROL SYSTEM

Automation of arc welding presupposes the automatic adjustment of the position of the welding wire along the weld line, and automatic control of the welding parameters for inaccuracies in the workpiece groove. In narrow gap welding especially, highly reliable seam tracking and adequate control of welding parameters are indispensable in avoiding weld defects and safeguarding the contact tip and other component parts from damage. Furthermore, in welding pressure vessels with very thick plates, a system capable of remote control of the welding operation is necessary to improve the working environment for the welding operator who is otherwise exposed to high temperature, dazzling arc light, and fume for many hours.

For these reasons, the twist arc welding process employs an automatic control system in which the image of the square groove in the vicinity of the welding arc is registered by a small TV camera. Seam tracking and control of the welding parameters are then performed by analysing and processing the video signal from the camera as outlined below.

As shown in Fig.6, the automatic welding control system comprises a TV camera, video signal processing unit (VSPU), TV monitor, welding parameters control unit (WPCU), and welding machine. Figure 7 shows the experimental twist arc welding setup in the laboratory in which a conventional gas-shielded arc welding machine travelling on guide rails is used.

Video tracking

The system performs the following detection and control functions. The positions of the left and right edges of the square groove in the vicinity of the welding arc are detected, and the position of the welding wire is adjusted to the weld line, i.e. the centre between the left and right sides. From the positions of the left and right edges as detected, the width of the square groove is determined. The measured value is then fed back to the welding speed control so as to adjust it to suit the detected width.

The image of the square groove in the vicinity of the welding arc is registered on a small, high-efficiency TV camera fitted to the welding machine. The camera is fitted with a neutral density filter combined with an infrared interference filter so that it provides a clear image of the square groove in the vicinity of the arc. The extracted video signal shows a satisfactory signal-to-noise ratio, and the groove edge image can be clearly distinguished from that of the welding arc.

The VSPU uses the signal from the TV camera to detect the width of the groove and the weld line, as indicated schematically in Fig.8, where the hatched shape represents the welding arc. The video signal obtained is converted to a two-level signal with a suitable threshold level to convert the image to a digital picture. The TV camera is so designed that the horizontal scanning spot is examined consecutively in the vertical direction from top to bottom in Fig.8. Then the position of the extreme end of the arc image on line A (which is adjusted to the centre of the wire on the TV monitor screen and stored), i.e. the lower intersection of the arc picture and line A, is detected. The reference line B is generated at a set position (higher by a specified amount tha this lower intersection) immediately below the welding wire.

The positions of the left and right sides of the square groove to be detected on the reference line B show good correspondence with the vertical position of the wire tip. When the arc moves up or down, for some reason or other, the reference line B tracks it automatically, and the point at which the digital arc picture changes from dark to bright is detected as position C of the right-hand edge of the square groove. Similarly, the point at which the picture changes from bright to dark is detected as position D of the left-hand side. The right- and left-hand groove edge signals thus detected are gated and those judged to be valid are sampled (number α) and the average of these α values used as the datum position for the width of the square groove. Furthermore, position E, equidistant between edges C and D, is obtained to represent the weld line.

Output data on the deviation between the wire centre position A and the weld line position E, and on the width between C and D, are obtained in digital form from each set of horizontal scan lines taken as the minimum unit. Unless an error signal is involved, the output is updated as frequently as α/60sec. Suitable values for α range from 20 to 50, although this varies according to the reliability and sensitivity of the original data.

Operator console

The TV monitor displays line A (preset to the wire centre position), reference line B, lines C and D showing the positions of the groove edges as detected, line E for the weld line position, detection gate limits (not shown in Fig. 8), and the image of the groove in the vicinity of the welding arc taken by the TV camera. By watching the monitor the operator is aware of the condition of the welding wire, arc, and molten pool and checks the detected data and seam tracking performance. If the tip of the electrode wire curves for some reason during the welding operation, he need only reset line A manually by watching the monitor screen.

The WPCU includes all operational functions, i.e. sequence circuit and voltage control circuits, carriage motor drive, stepping motor drive for seam tracking, and instrumentation for welding current, voltage, and speed.

The digital information on the width of the groove produced by the video signal processing unit is converted to an analog signal to control the carriage motor so that the product of the welding speed and the detected groove width maintains a constant value. If the carriage motor is thus controlled and the welding current and voltage are held constant, the height of the deposited metal per pass is maintained constant even when the width of the groove changes. The carriage motor drive is steady with less than 1% fluctuation and has a frequency response of about 40Hz. Experiments established that the variation in height of the deposited metal per pass was less than $\pm$0.2mm for variation in groove width from 12 to 18mm.

The deviation between the wire centre position and the detected weld line produced by the video signal is used to control a stepping motor which drives the slide base mounted on the carriage in the transverse direction to match the wire centre with the weld line. The tracking accuracy is determined by the smallest digit on the line scan of the TV monitor, and varies according to the image magnification and the scanning pitch. If the groove width is so set that it is displayed within one-third of the width of the TV monitor screen, the tracking accuracy turns out to be about 0.2mm because the number of effective line scans per field is 240. However, experimentally the accuracy obtained was within 0.5mm because of variations in the detected data.

CONCLUSION

It has been established that the twist arc process is a reliable flat position narrow gap gas-shielded arc welding method in which the rotational movement of the arc produced by the twist wire overcomes the unreliable weld quality problem inherent in conventional narrow gap welding.

Also, the TV-monitored automatic control system incorporated in the twist arc process ensures highly reliable seam tracking by real time detection and improves the working environment for the operator by its remote-control facility.

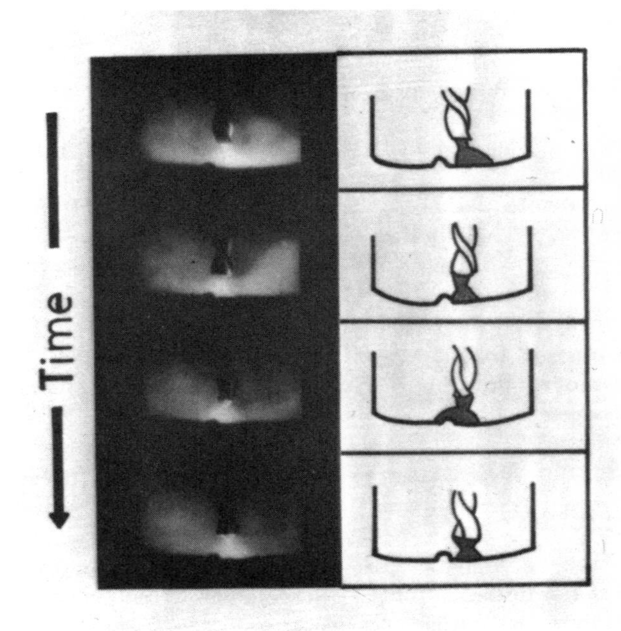

1 Rotational movement of welding arc produced by twist wire

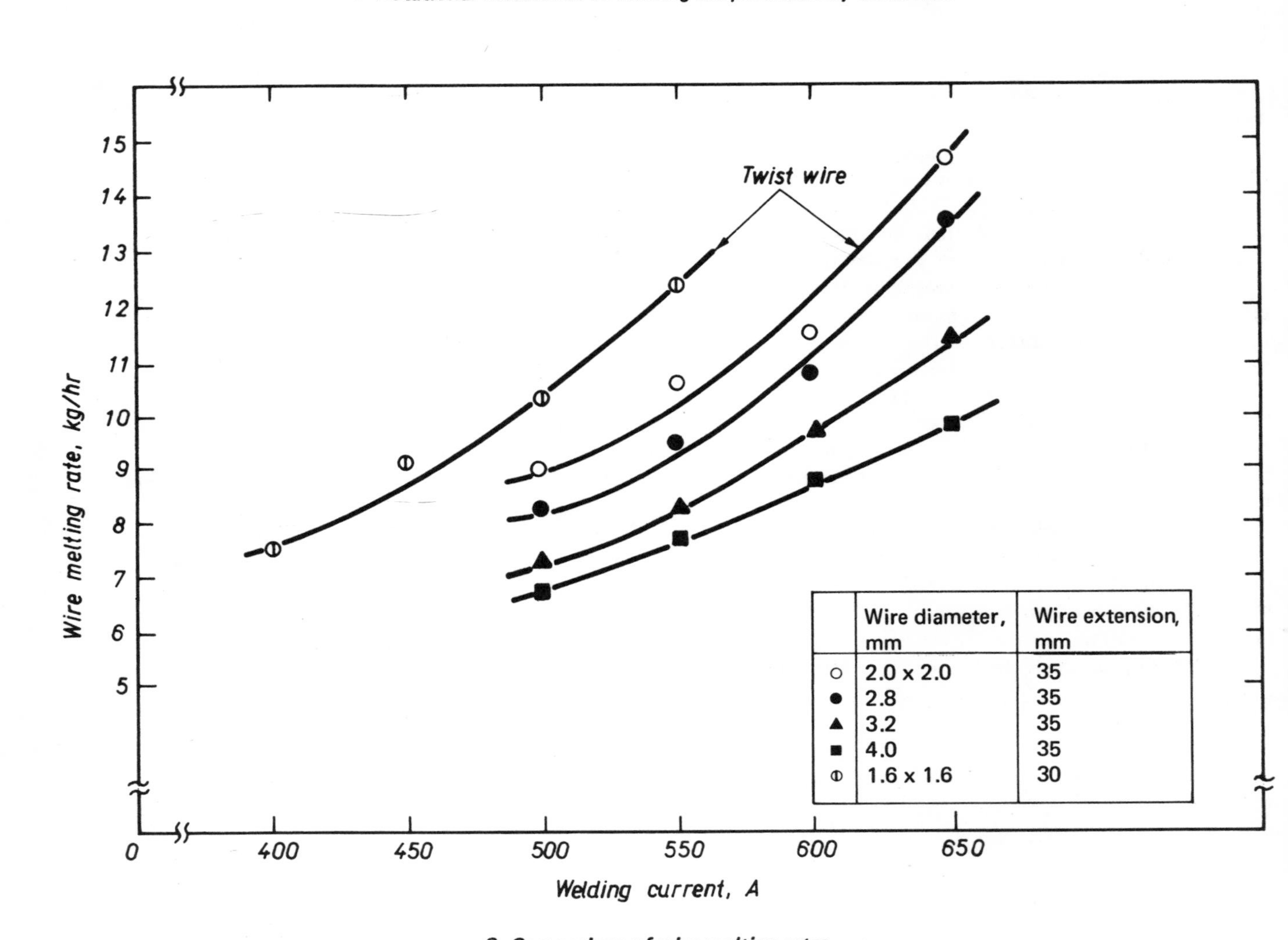

2 Comparison of wire melting rates

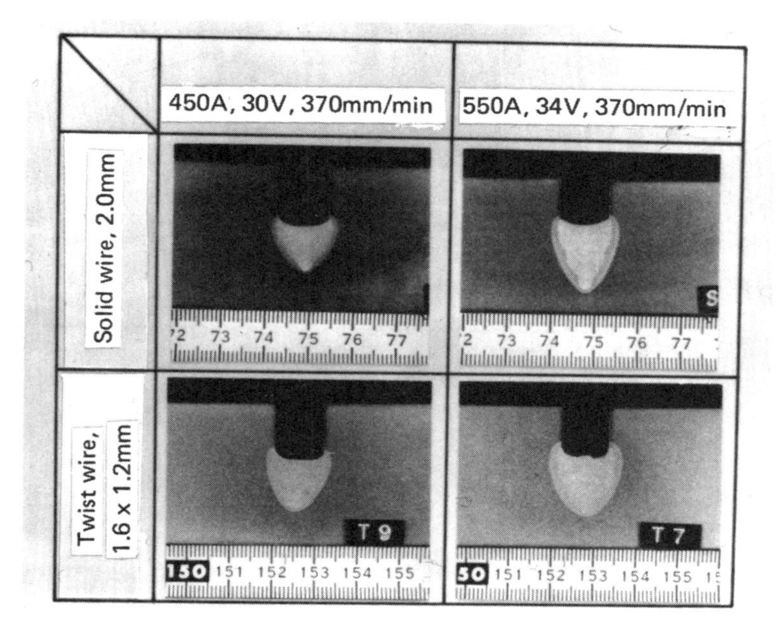

3 Comparison of deposited metal cross-section

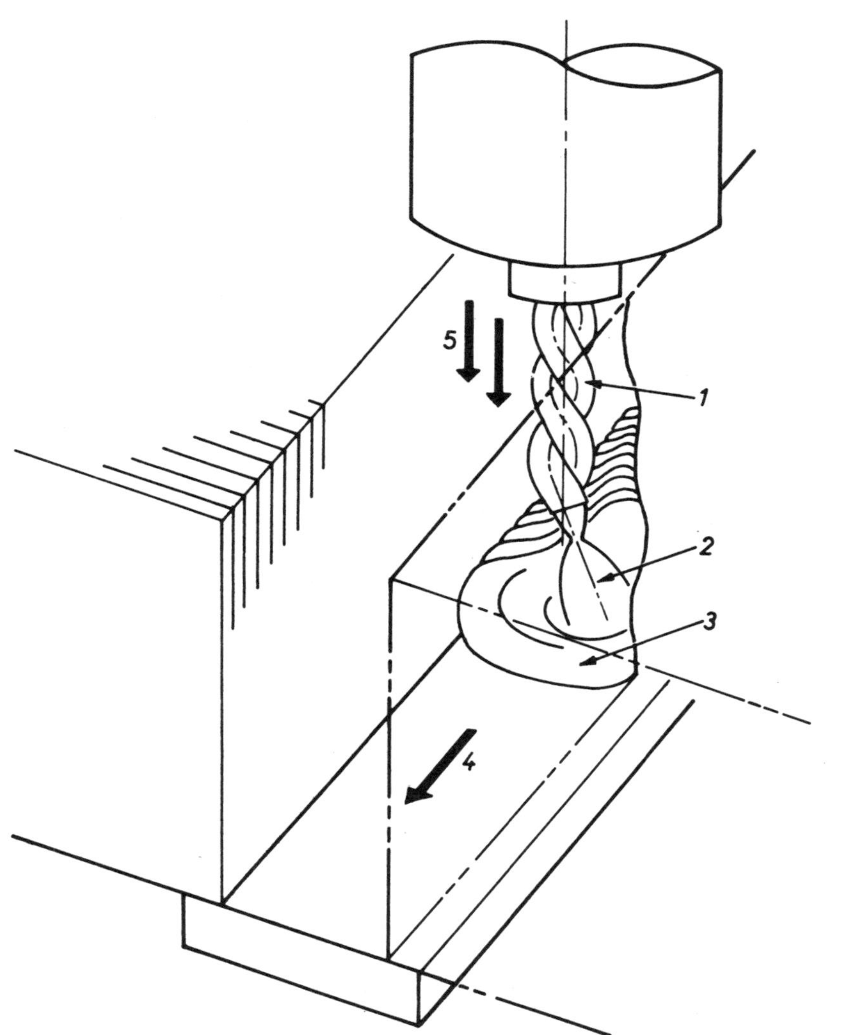

4 General arrangement of twist arc welding process. 1 — twist wire; 2 — arc; 3 — molten pool; 4 — direction of welding; 5 — shielding gas

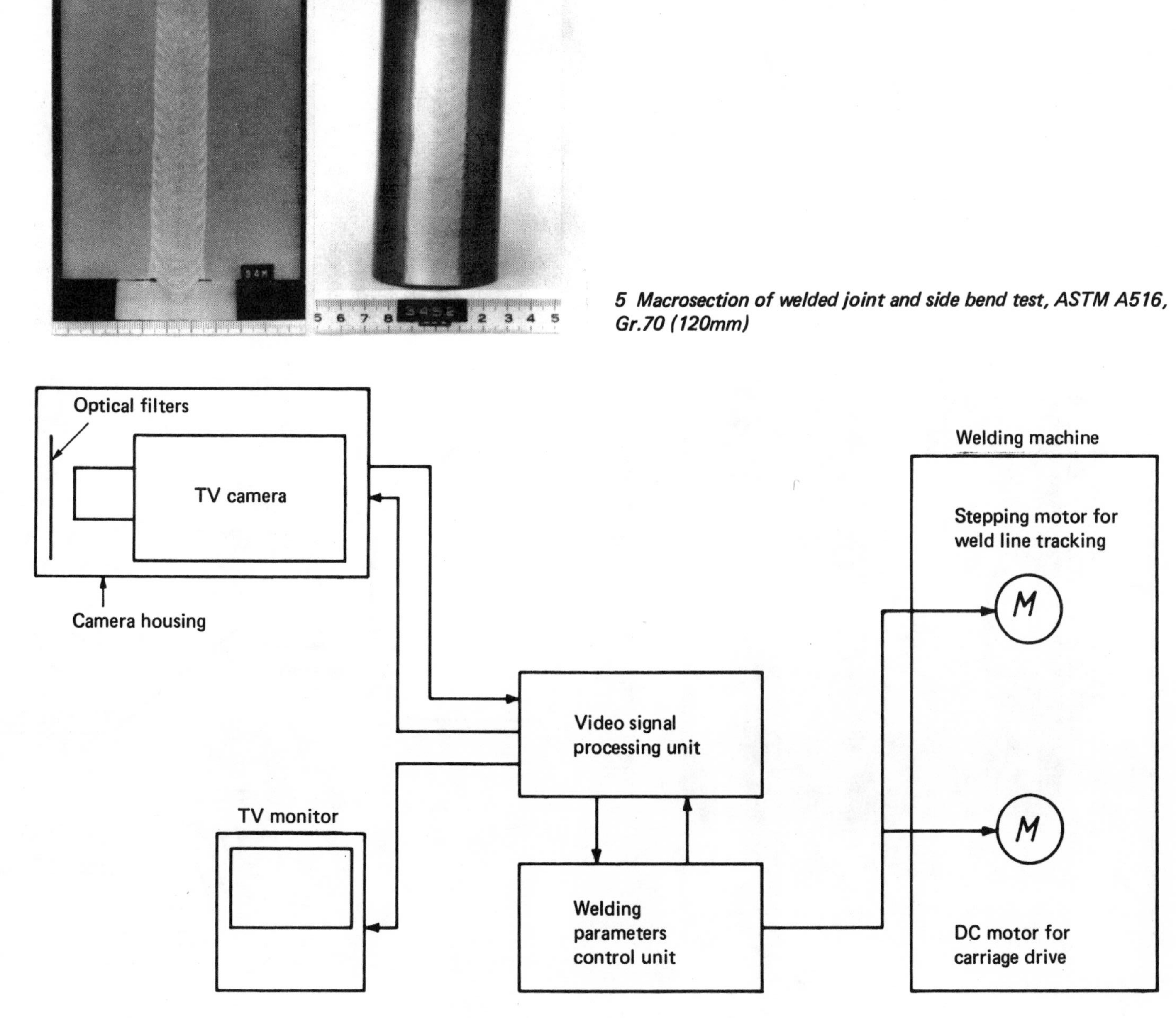

5 Macrosection of welded joint and side bend test, ASTM A516, Gr.70 (120mm)

6 Functional diagram of welding control system

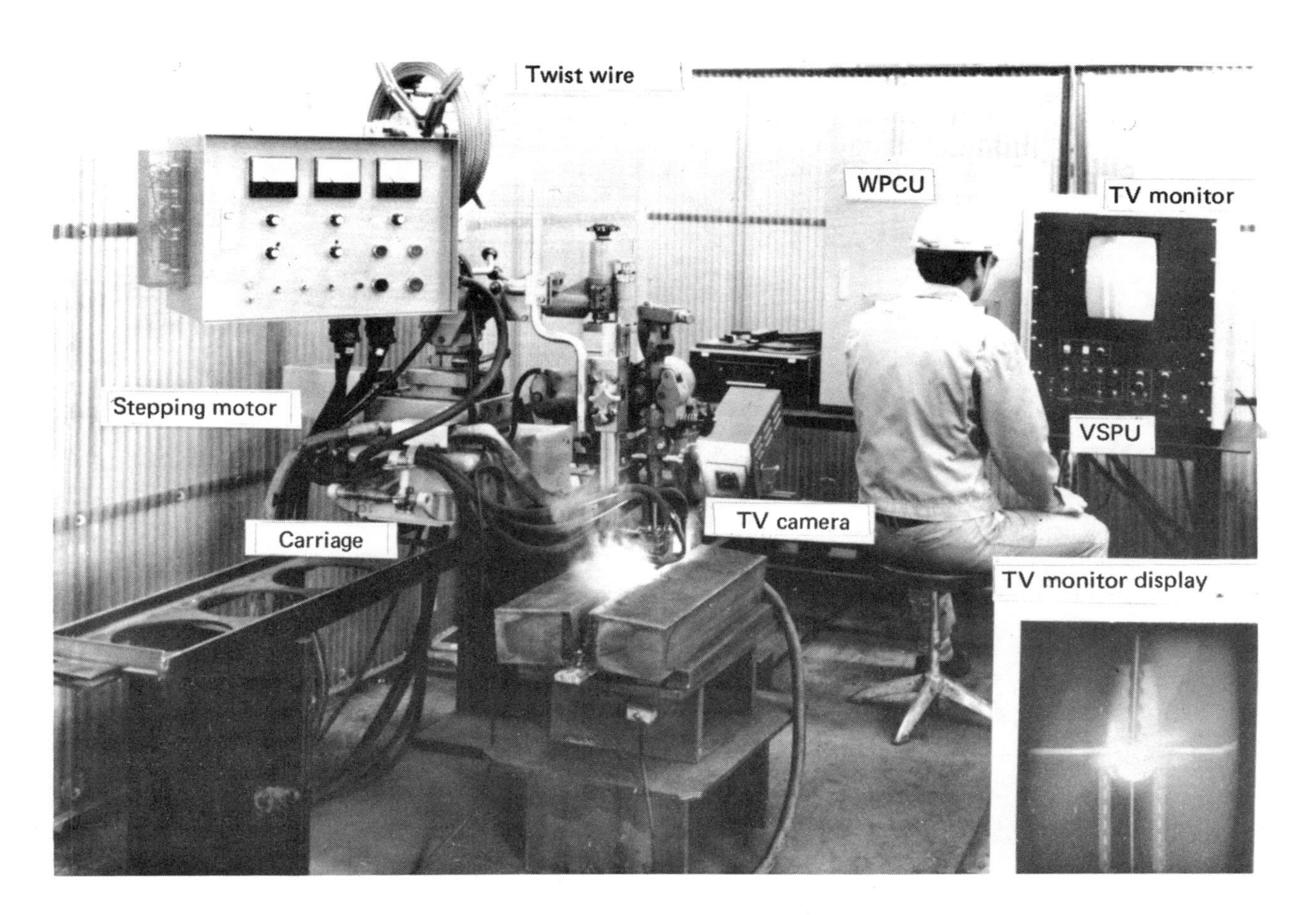

7 Twist arc welding in laboratory

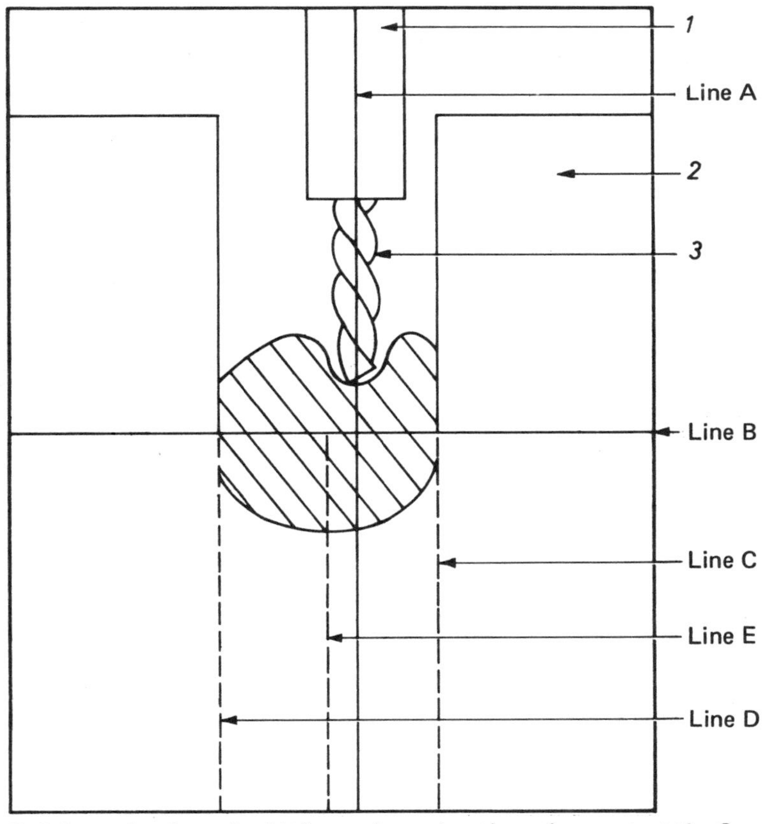

8 Schematic of required information extraction. 1 — contact tip; 2 — workpiece; 3 — twist wire

PAPER 45

Switched arc MIG-welding

K.W.Brown, TD, CEng, MIMechE, FWeldI

INTRODUCTION

Automatic welding processes are already widely used for both component and structural fabrication, using mainly flat and horizontal-vertical positions. Even so, inconsistent penetration and irregular weld metal shape are encountered arising from variations in edge preparation and joint fitup. Despite these shortcomings there is an increasing demand for automatic welding in the vertical and overhead positions to facilitate the construction of large fixed structures. Clearly any process innovation that could provide improved control of the weld pool would be welcomed.

In semi-automatic welding the operator attempts to regulate the weld pool metal by employing a complex pattern of weave movements. By these means a skilled welder exerts a degree of control over weld shape, penetration, and sidewall fusion and compensates for variations in joint fitup and preparation.

Automatic MIG-welding techniques have been developed which simulate the electrode weaving motions used by skilled welders, invariably carrying out these cyclic movements at manual-like frequency.[1] Although arc manipulation by mechanical means predominates, arc sweeping has also been obtained by fluidic (gas jet)[2] and magnetic methods.[3] Furthermore, the mechanical weave is supplemented by a change in welding parameters with the lateral electrode movements, e.g. by using a combination of spray and short-circuit transfer operation in each weave cycle.[4] Although many of these techniques will undoubtedly play an important role in mechanised welding, it is believed that the traditionally low frequency motion, based on manual weaving techniques, may not be expedient for interfacing with the fast response transducers and electronic control systems that will be used in automated MIG-welding applications.

The development of welding as a major tool in industry continues to accelerate, mainly because of the availability of precision hardware and power sources which have both benefited from the rapid developments in low cost integrated circuits and semi-conductor elements. It was against this background (described by Houldcroft[5] in the 1977 AWS Adams Lecture) that alternative means for weld metal shaping were sought, taking advantage of the high speed response capabilities of the arc. Where appropriate, process innovations should be capable of interfacing with sensing transducers for tracking the joint and controlling weld quality while the weld is being made. Such transducers, many of which are still under development, will be able to supply digital information permitting automatic welding systems to operate in a similar manner to present-day advanced numerically controlled (NC) machine tools.

Mr Brown is a Principal Research Engineer in the Process Operation and Control Department of The Welding Institute.

ARC OSCILLATION TECHNIQUES

It should be noted that NC machine tools attain their control and precision by carrying out their function in a coordinated series of very small discrete steps, whereas MIG-welding (with spray or short-circuiting metal transfer) is in general a random process with respect to the detailed metal deposition. However there is one well-established exception, the pulsed-current controlled transfer mode reported first by Needham[6] in 1962. Here the process parameters can be set so that the current pulse just detaches one droplet from the wire tip, with the continuously feeding electrode being melted in the intervening low current period in readiness for the ensuing pulse transfer. Problems of excess weld pool fluidity experienced with conventional spray transfer, and the risks of lack-of-fusion defects common with short-circuiting arcs, are both assuaged with controlled transfer. Thus it would appear that pulsed-current MIG provides an ideal subject for investigation because, as well as being essentially 'digital' in nature, it has in its own right a basic potential for positional welding. Moreover pulsed metal transfer offers scope for new arc oscillation techniques because, with its cyclic nature, it becomes relatively easy to synchronise lateral oscillation of the arc with the periodic metal transfers. Using this characteristic feature two novel high speed arc oscillation techniques have been devised, one based on electrode movement synchronised with current pulses (stepped arc welding)[7] and the other inducing arc translation by current commutation between two or more electrodes (switched arc welding).[8] The former relies on relatively high speed mechanical oscillation in which the arc is cyclically oscillated (lateral to the weld path) during a welding or overlaying operation. Here the phase of the cyclic oscillation is correlated with the phase of the pulse-controlled metal transfer from the electrode to the workpiece, so that each transference of metal takes place at a preselected instant during the cyclic transverse movement of the arc. In this manner the heat and metal transfer input along the weld joint line can be programmed to provide heat and metal where it is most needed, down to time increments of the order of 0.1sec.

In the electrical oscillation method,[9] instead of oscillating a single electrode two or more arcs are operated in turn from two or more electrodes. To the question 'Why not operate two or more electrodes simultaneously?', the answer is that normally arcs operated in close proximity interact with each other and disturb the essential characteristics of each separate arc. To overcome this problem a commutation system has been devised to operate arcs close together by switching the current to each electrode in turn. Thus a single arc is in effect positionally translated by the commutated current.

Although the switched arc principle can be applied to both consumable and nonconsumable electrode processes, this Paper describes the effects obtained with a twin electrode MIG-welding system. Welding current is applied to the electrodes at two levels: a continuous low level background current to both electrodes, and a higher level 100Hz pulsed current applied sequentially so that in turn each electrode experiences a 50Hz modulated current. The background arcs exist simultaneously but at a current level which is insufficient to cause any significant arc interaction problems. If used alone the low current would produce globular metal transfer, and the high level current (if maintained continuously without pulsing) would provide spray type transfer. Thus, in combination, the background current provides a molten droplet on each electrode tip which is completed and detached by the sequential pulses of high level current.

SWITCHED ARC PROCESS

To demonstrate the feasibility of a twin electrode MIG system and to investigate its basic characteristics bead-on-plate tests were made in the flat position on mild steel. A duplex (capstan style) multigrip wire feeder, Fig.1a, supplied two similar electrodes (electrically insulated from each other) to the twin electrode water-cooled MIG-welding head, Fig.1b and c. Welding power was obtained from a prototype sinusoidal pulsed current power source and a commercial constant current TIG unit which supplied the background current via separation resistors, Fig.2.

A series of weld deposits, each some 250mm in length, was made using six different electrode configurations and covering a range of seven wire feed speeds from 4 to 10m/min (total for the two wires) in 1m/min steps. For direct comparison of weld metal shapes the welding speed was set for each test at one-twentieth of the total wire feed speed to provide a constant metal deposition rate. All tests were carried out with 1.2mm diameter mild steel double deoxidised wire, constant contact tip to workpiece height of 12mm, and Ar-2% O_2 shielding gas.

The effects of the three main bead-shaping parameters, i.e. the extent of electrode separation, the total current level, and the orientation of the electrodes with respect to the weld path, are illustrated by the transverse section photomacrographs.

ELECTRODE SEPARATION

Changes in the distance between the twin electrodes were achieved by simply bending the contact tips, Fig.1c, as required. Figure 3 shows the range of weld metal shapes which can be obtained merely by changing electrode separation at a given constant total current, 200A, for three separations 12.7, 9, and 6mm, as measured at the centres of the contact tip extremities.

TOTAL CURRENT

The effect of the total current level at a given separation is illustrated in Fig.4 for a constant deposition rate. Similar control of weld metal shape to that for electrode separation is found, ranging from the separate weld beads (at low current), Fig.4a, to a broad but fairly uniform penetration profile, Fig.4d.

ELECTRODE ORIENTATION

The effect of changing the alignment of the electrodes with respect to the welding direction from in-line to transverse is illustrated in Fig.5 for a fixed current, 200A. These are compared with a single arc deposit, Fig.5a, made at a similar current and deposition rate.

DISCUSSION

Commutation current technique

With multiple nonconsumable electrodes there would be, as far as is known, no limitation as to current magnitude and duration. Any suitable low current could be used for the background or pilot arc, and any higher current at any repetition frequency for the commutated pulse. The system is not limited to simple sinusoidal communication, although this is a particularly convenient approach, using half-wave rectified supplies. In principle a common DC source could be applied to a group of electrodes using, say, transistor switches to divert the current to each arc in turn.

Compared with normal spray transfer MIG-welding in Ar-2%O_2 gas (about 280-340A for 1.2mm steel wires) the operating current range with the commutated twin MIG system is very wide : some 160-400A for the same wire size. Even wider operating ranges should be feasible, down to less that 100A with 1mm wire and up to about 700A with 1.6mm diameter wires. Higher currents should also be possible with triple electrodes compared with the present twin electrode system, and could give operating levels up to, say, 600A with 1.2mm and about 1000A with 1.6mm diameter wires. It will be appreciated that in the switched arc method each electrode functions at a shared total current level, so clearly the average arc force on the individual wires is considerably reduced compared with that for single electrode operation.

Penetration profile

The possibilities for weld metal shaping by the switched arc twin MIG system have clearly been demonstrated in bead-on-plate tests on mild steel. Although the present system used relies on the pulsed-current MIG-welding technique, the latter merely obtains the correct total heat input for the complete weld zone; it does not by itself cause a redistribution of heat input and metal transfer pattern with respect to the weld cross-section. With the switched arc technique it has been shown that pronounced changes in weld metal geometry are readily obtainable, the energy distribution being regulated by varying the spacing of the electrodes or their alignment relative to the weld path, both of which are influenced by the total current level.

It should be noted that, with the design of twin electrode gun used for these tests, the angularity of the electrodes increases as the separation is reduced. In an extension of this work[10] it has been shown that electrode inclination angles of 10-20° produce significant effects on the weld metal shape, resulting for instance in asymmetry and reduced penetration. Furthermore, the bending curvature of the tips determines the emergent angle of the electrode wire and hence the weld pool spacing of the coaxially transferred drops. From drop-on-plate tests at high rotary traverse speeds it can be seen, Fig.6, that the metal transfer remains substantially coaxial with the electrodes whether they are parallel at 12.7mm, Fig.6a, or when the contact tips are bent to 7.5mm separation, providing 3.5mm projected wire spacing at test plate level, Fig.6b. In certain applications inclined electrodes with axially projected transfer might be used with advantage, for example with crossed electrodes working in a V joint preparation.

If necessary, the central fusion in a twin penetration profile, as in Figs 3b and 4b, can be increased by providing a local preheat, as in the series MIG-TIG system. For example, the current from a commutated twin MIG system could be returned via a TIG arc (argon shielded) positioned slightly ahead of, and central with, the MIG electrodes. This gives a central penetration comparable with those at high current as in Figs 4d and 5d. The positioning of the TIG preheat midway between the path of the following twin MIG electrodes, Fig.7, proves remarkably efficacious for weld metal shaping compared with the pulsed MIG-TIG system using a single MIG electrode.[11] This simple expedient causes the two separate beads to merge, Fig.8b, and, Fig.8d, considerably increases the central penetration to give a substantial penetration depth over the major portion of the fusion width.

Deposition characteristics

Apart from the benefits of directing greater heat towards the sidewalls which should be advantageous for multipass welding in V or U joint preparations (with a reduction in the tendency to sidewall lack-of-fusion defects), the switched arc system may also be useful for the root pass. Here the metal transfer is directed to the side edges of the preparation rather than the centre, which, if not fitting perfectly, can lead to excess penetration, and even projection of the electrode through the gap between the nominally abutting weldment components. Similarly the twin MIG system may be beneficial in welding thin sheet, the transfer from the two electrodes being directed at the sheet surface with the overall heat input still sufficient to allow fusion between the sheets. Furthermore, by more accurate control of the heat and metal transfer distribution across a welded joint, it is possible to obtain a broader central penetration and to compensate for slight mistracking of the joint. Thus it may well prove that the twin MIG switched arc system is more tolerant to joint gaps, or other mismatches, than the equivalent single electrode system.

There is also evidence that the tandem molten metal droplets entering the weld pool on either side produce a stirring action which may assist in eliminating porosity. This mixing action can be seen in Fig.5: two individual beads are shown in Fig.5a, but, although the two electrode wires are of the same specification, they were taken from different batches and the etched deposits are distinguishable; in Fig.5b and c the mixing is discernible. Figure 4 also indicates a similar effect. Similarly the spacing of the metal transfers may beneficially change the solidification pattern of the weld pool metal and lead to improvements in the mechanical properties and weld metallurgy.

The twin MIG switched arc system may show advantage as a positional technique. For example, when vertical welding thick plate the total heat input must be limited to prevent the molten pool from dropping away; here the more logical distribution of heat and metal may permit higher deposition rates than normal, with improved sidewall fusion. This may promote mechanised welding in position as there is no need to weave as in a single electrode system, and the greater tolerance to joint geometry may be essential in using simple machines to emulate a skilled manual operator. The high frequency lateral arc oscillation of the twin MIG system permits high speed vertical-down welding without the penetration finger wander frequently experienced with single electrodes oscillated at conventional frequencies.

Practical application

Tests were carried out in the flat position to establish the viability of the switched arc system to make butt welds with improved weld metal shape. Figure 9 shows typical macrosections of single pass welds in 3, 6, and 9mm thick mild steel, at total current levels of 200, 300, and 350A respectively, and Fig.10 illustrates a two pass weld in 12mm plate. The use of the switched arc method for joint making clearly shows indications of improved weld metal shaping; this has been further investigated in a subsequent test programme.[12]

In vertical-up welding the different heat input pattern of the switched arc method did not have sufficient influence to prevent metal runout at lowest current (160A) available from the system operating with twin 1.2mm wires. However it is believed that vertical-up deposits would have been practical with smaller electrodes operating at lower currents (say 120A).

With vertical-down operation bead-on-plate runs were readily accomplished over a wide operating range (170-450A). Unlike a conventional short-circuiting arc a narrow stringer bead is not formed, but rather a wide shallow fusion zone with twin penetrations.

The switched arc system could be applied to multiple electrodes for overlay welding, using a single-phase rectified supply for twin electrodes or three-phase to operate three twin electrode guns in juxtaposition. At present MIG overlaying is employed primarily on components which will later have attachments welded by a similar process. The early techniques used overlapping stringer beads to form a continuous layer of weld metal; sometimes an auxiliary cold filler wire is added to increase the deposition rate and to minimise dilution. However the characteristic deep papillary depression of each fusion line into the parent metal results in excessive dilution. But it was observed during the switched arc tests that beads of somewhat unusual shape, Fig.11 (favouring overlay applications), were produced when arcing on a vertical surface with relative downhand movement of the MIG gun. Exceptionally thin well-wetted deposits permit an overlay of very low dilution to be rapidly built up without consuming a large quantity of the invariably expensive cladding material. The introduction of an electrode lead inclination may also prove beneficial[10] as indicated in the single electrode bead, Fig.11a. Details of further overlaying tests have been reported elsewhere.[12]

CONCLUSIONS

A simple technique has been developed for weld metal shaping in MIG-welding without electromagnetic or mechanical oscillation apparatus. In principle a multiple electrode system is used with the welding current switched to each electrode in turn. Sequential current pulsing mitigates the effects of arc interaction associated with multiple DC arcs operated closely together.

The shape of the weld metal can be controlled by the spacing of the electrodes, their orientation about the joint line, and the welding conditions.

In principle the technique can be applied

with advantage to other arc processes such as TIG-welding, and to hybrid systems such as the series MIG-TIG combination.

Potential advantages of the method include a wide operating range, improved sidewall fusion, a reduced tendency to finger porosity, and easier joint tracking with the wider penetration characteristics. The system concept could find wide application, from surfacing with multiple electrodes to positional welding with the facility for controlling the penetration profile.

ACKNOWLEDGEMENTS

The author wishes to thank the many colleagues who maintained an interest, proffered advice, and provided assistance throughout the work.

REFERENCES

1 NIPPON STEEL CORPORATION. UK Patent 1 276 015.

2 INAGAKI, M. and OKADA, A. UK Patent 3 838 243.

3 DEMINSKII, Yu A and DYATLOV, V.I. 'Magnetic control during gas-shielded arc welding with a consumable electrode'. Automatic Welding, 18 (4), 1963, 82-3.

4 ARIKAWA, M. et al. 'SS-arc process'. IIW Doc.XII-B-108-72, 1972.

5 HOULDCROFT, P.T. 'Developing precision assembly by welding'. Metal Constr., 9 (8), 1977, 337-44.

6 NEEDHAM, J.C. 'Control of transfer in aluminium consumable electrode welding'. Symposium 'Physics of the Welding Arc', London, 29 October-2 November 1962, 114-22. London, Inst.of Welding, 1966.

7 BROWN, K.W. UK Patent 1 332 059 (stepped arc welding).

8 BROWN, K.W. UK Patent 1 450 912 (switched arc welding)

9 BROWN, K.W. 'Commutated current multiple arc welding'. Welding Inst.Report P70/75, 1975.

10 BROWN, K.W. 'Spraying in the right direction . . .?. Welding Inst.Research Bull., 17 (9), 1976, 233-6.

11 WILSON, A. 'Arc and bead characteristics of series MIG-TIG arcs'. Welding Inst. Report P25/68, 1968.

12 CARTER, A.W. 'Switched arc twin MIG-welding of joints'. Welding Inst.Report 30/1976/P, 1976.

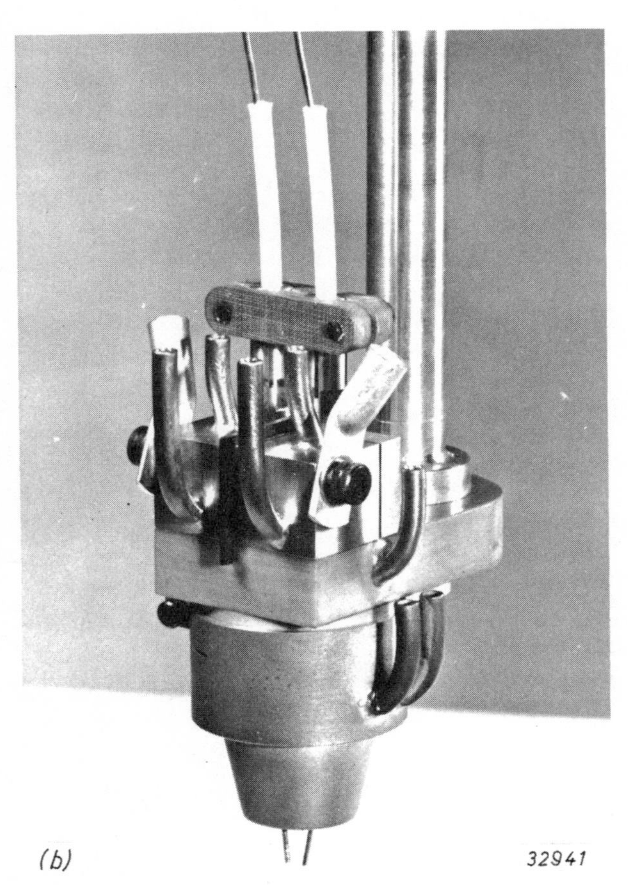

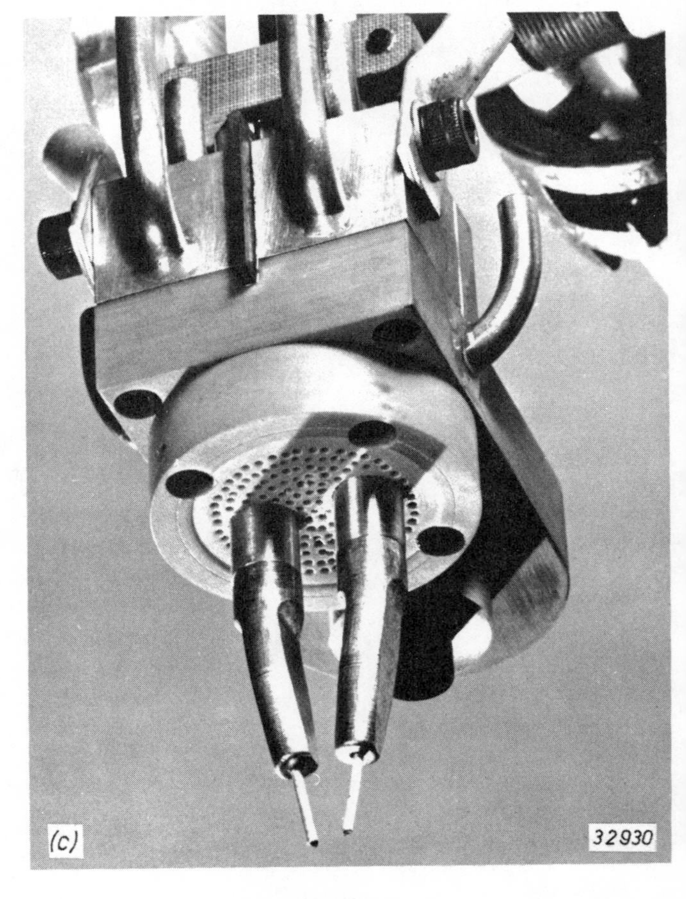

1 Twin wire MIG-welding apparatus: (a) wire feeder with tandem traction wheels ('multigrip' capstan type), (b) and (c) welding gun for mechanised tests: (b) general arrangement, (c) view with gas nozzle removed

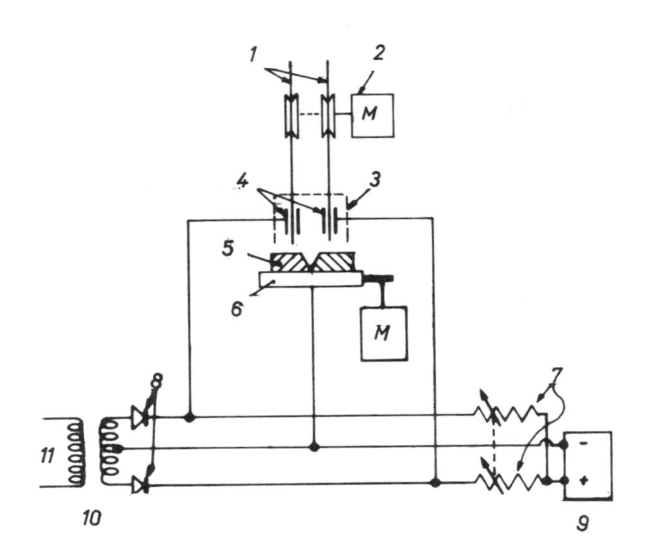

2 Circuit diagram of twin wire system with shared background supply and simple biphase half-wave pulse source for commutated current. 1 – twin electrode wires; 2 – wire feeder unit; 3 – gas nozzle; 4 – contact tips; 5 – work; 6 – traverse; 7 – resistors; 8 – diodes; 9 – background power supply; 10 – pulse power supply; 11 – AC mains

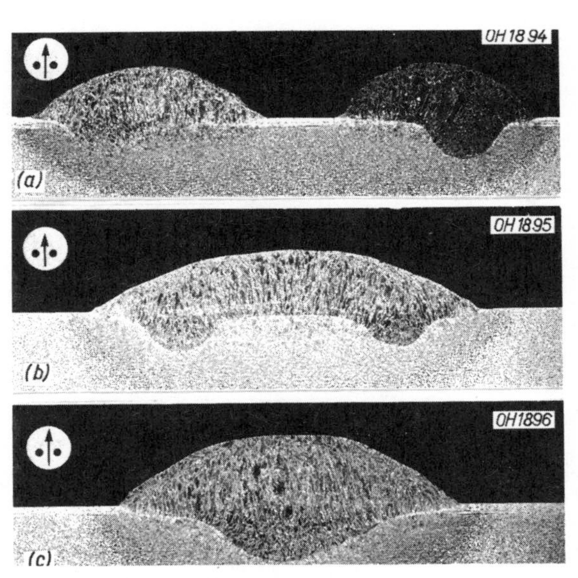

3 Effect of electrode separation on bead and penetration profile at constant wire feed rate and travel speed. Transverse twin electrodes with contact tip spacings, mm: (a) 12.7, (b) 9, and (c) 6. Welding conditions: total wire feed 5m/min (2.5m/min each wire); total current 200A (100A each wire); arc potential 20V; travel speed 250mm/min x 3½

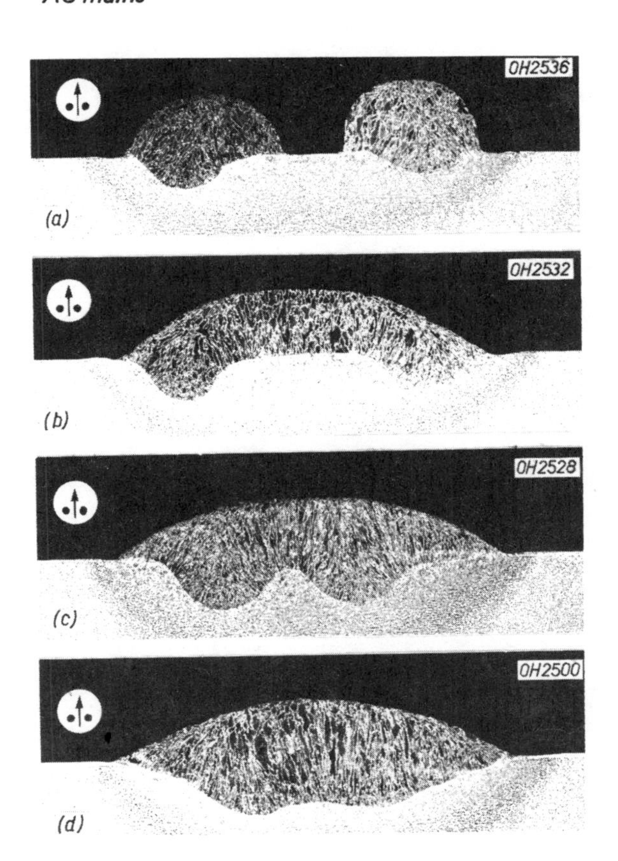

4 Effect of current on weld metal shape with constant ratio of feed rate to travel speed (constant excess weld metal section). Transverse electrodes 9mm apart. Total wire feed, m/min (A): (a) 4 (160), (b) 6 (240), (c) 8 (320), and (d) 10 (400) x 3½

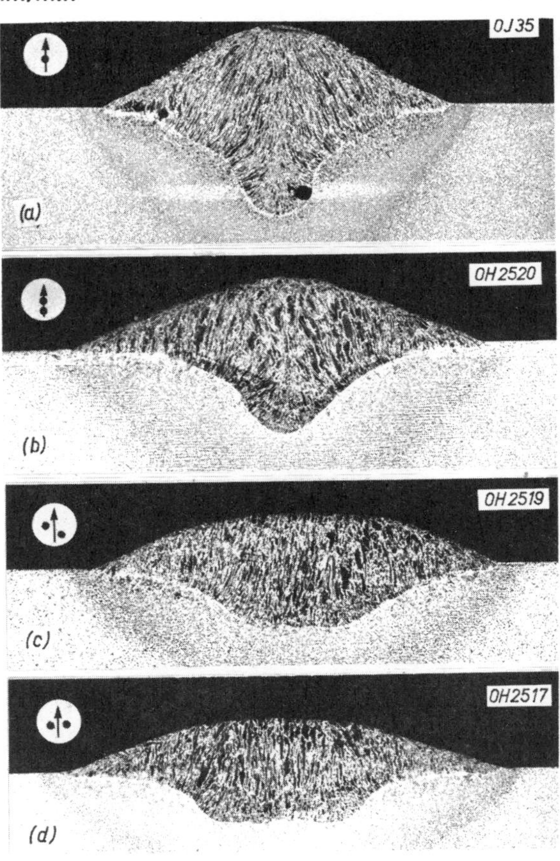

5 Effect of electrode position about joint line at constant wire feed rate (current) and travel speed: (a) single electrode, feed 9m/min (360A approx.), (b), (c), and (d) twin electrodes 6mm spacing, total feed rate 9m/min (360A), with electrodes: (b) in-line, (c) at 45°, and (d) transverse x 3½

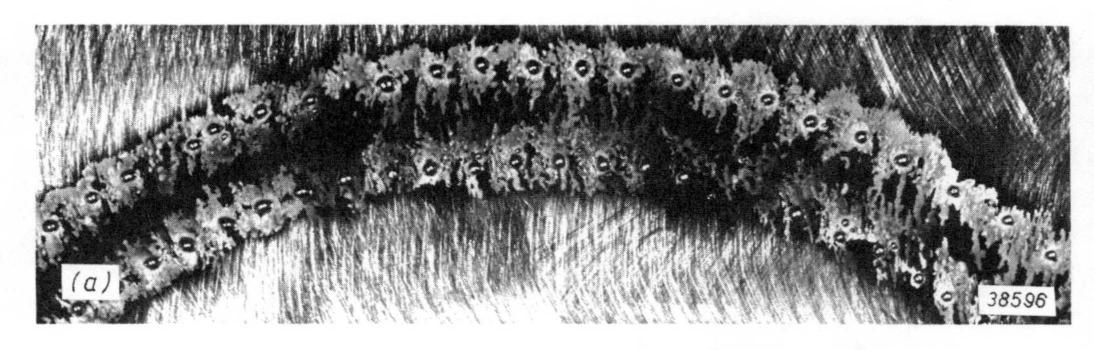

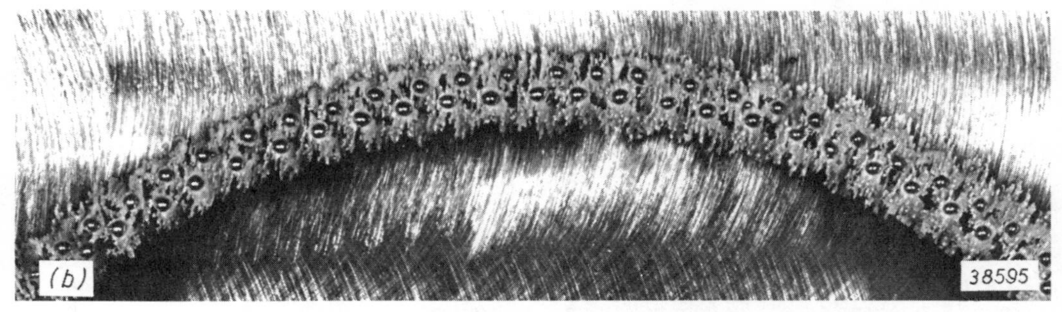

6 Effect of electrode separation on high speed traverse delineation of metal transfer for 50Hz sequential current pulsing at 5m/min total wire feed (200A). Transverse twin electrodes with contact tip spacings, mm: (a) 12.7, (b) 7.5 x 1

7 Arrangement of twin MIG and single TIG electrodes for series pulsed MIG-TIG bead-on-plate tests

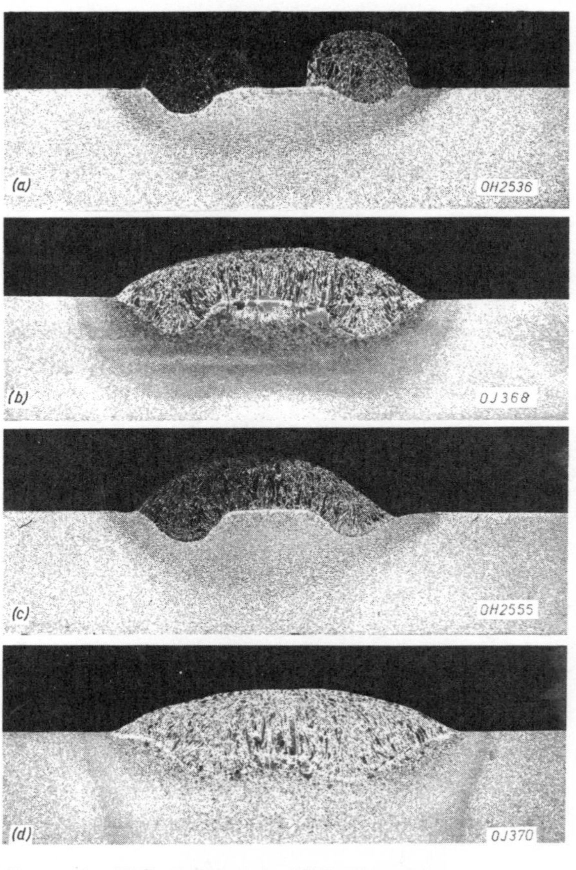

8 Effect of adding series TIG arc to twin MIG electrode system on weld metal shape. Transverse twin MIG electrodes with 9mm spacing: (a) twin MIG, total wire feed 4m/min, (b) as (a) but with leading central series TIG arc, (c) twin MIG, total wire feed 5m/min, and (d) as (c) but also with leading TIG arc x 3½

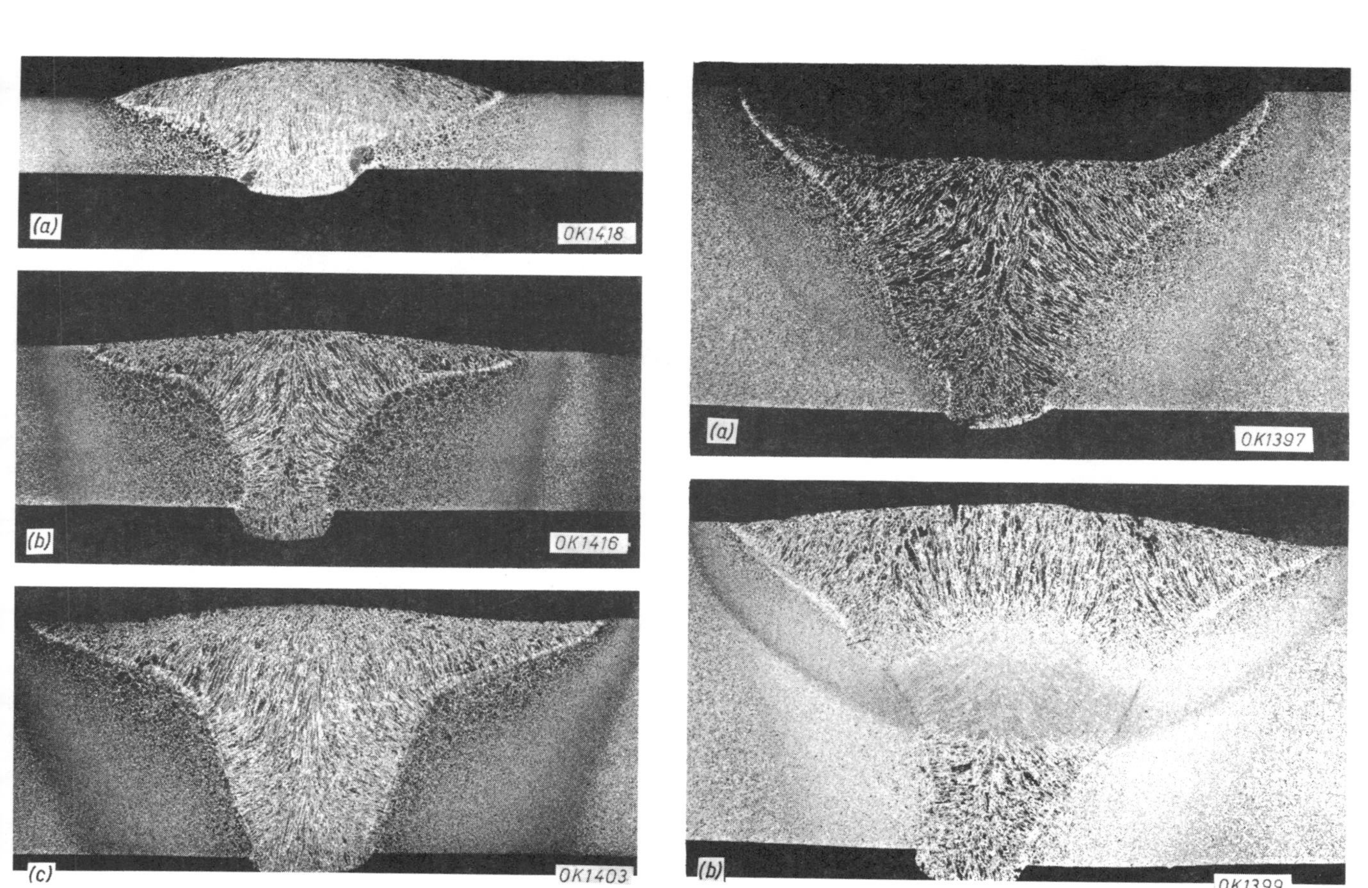

9 Switched arc welds in mild steel with 6mm spaced transverse 1.2mm electrodes and 3mm root gap: (a) 3mm sheet, square butt preparation, 200A total current, 250mm/min travel speed, (b) 6mm plate, 60° preparation, 1mm root face, 300A total current, 250mm/min travel speed, and (c) 9mm plate, 60° preparation, 1mm root face, 350A total current, 200mm/min travel speed x 3½

10 Two-pass switched arc weld in 12mm mild steel plate, 60° preparation; each pass 400A total current at 200mm/min travel speed: (a) first pass, 3mm root gap, 1mm root face, 6mm spaced transverse 1.2mm electrodes, (b) second pass, 10mm spaced electrodes x 3½

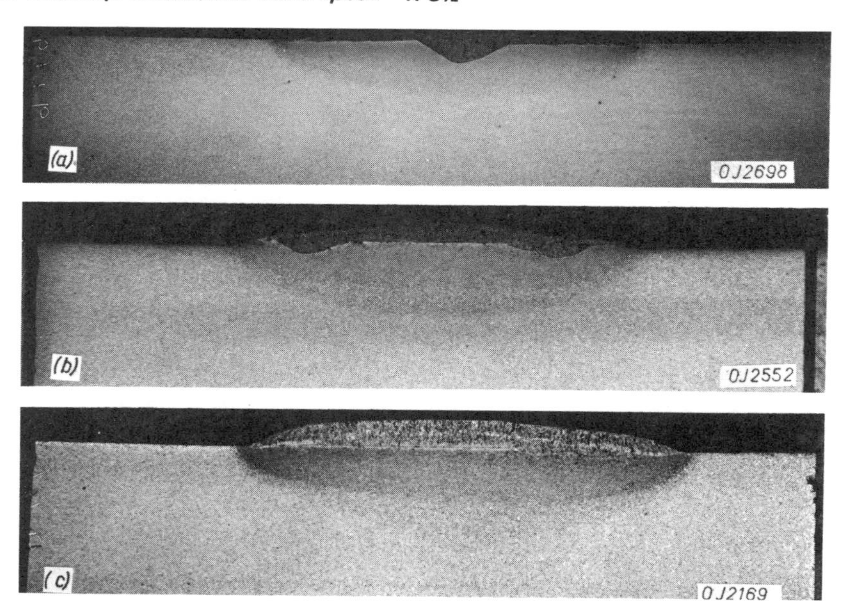

11 Overlaying with mild steel to mild steel in vertical-down position, miscellaneous deposits: (a) single electrode bead (270A), travel 1m/min, (b) twin electrodes (250A), 9mm transverse spacing, travel speed 600mm/min, and (c) twin electrodes (320A), 9mm transverse spacing, travel speed 700mm/min, molten metal runout imminent x 3½

Argon/oxygen and argon/CO_2 gas shields for the deep penetration gas metal-arc welding of aluminium alloys

T.Watanabe, BSc, MSc, DEng, and H.Marumoto, BSc

Studies were carried out to clarify the effects of oxidising gases mixed in argon on the appearance of the bead shape and depth of penetration, wettability of deposited metal, porosity, and microstructure of weld metal in MIG-welding A 5083 alloy. Mixing several volumes per cent of oxygen to argon is effective in increasing the depth of penetration, improving wettability, and decreasing porosity.

INTRODUCTION

In TIG- and MIG-welding aluminium and its alloys high purity argon has been used as the shielding gas. However, there are recent interesting reports[1-3] which state that porosity in weld metal can be considerably reduced when a small amount of active gas is mixed with argon and used as the MIG shielding gas.

In this connection a systematic study has been conducted on the effects of the composition of the shielding gas (such as Ar/O_2 and Ar/CO_2) which were considered to be potentially useful in respect of the bead shape, penetration, and porosity.

The results of these tests are briefly reported below.

TEST PROCEDURE

Materials

In these tests 10mm thick plates of 5083-Q were used as parent metal with electrode filler wire of 5183-WY 1.6mm diameter.

Shielding gas

100%Ar, Ar + 2.9 or 5.63%O_2, and Ar + 5.76%CO_2 were used as shielding gases at 20 litre/min flow rate.

Welding machine

An Air Reduction Company pulse arc welding machine, Model PA-3, and Super Midget welding torches were used throughout for these tests.

Dr Watanabe, Manager — Inspection Department, and Mr Marumoto, Welding Inspector, are both with the JGC Corporation, Yokohama, Japan.

Test method

Bead-on-plate runs were carried out on workpieces 200mm long x 150mm wide, and bead appearances investigated together with shape, penetration, and porosity in the weld metal. The welding conditions for the bead-on-plate runs are detailed in Table 1 for the specimens shown. These tests were conducted for the purpose of selecting the welding conditions necessary for the root pass of a one-sided weld without backing strip, and hence a comparatively low current range was used accordingly.

TEST RESULTS

Bead appearance

Figure 1 shows the appearance of beads deposited under constant conditions, i.e. 180A current and 400mm/min welding speed, in which the gas mixture ratio was varied. Wherein oxygen is added to the argon gas, Fig.1a and b, the bead surface is covered with a white powder which seems to be Al_2O_3. When the powder is brushed off the weld metal appearance looks good with a fine ripple. However, if the oxygen content is large, spatter will increase.

Figure 1c shows the corresponding bead appearance for the carbon dioxide addition.

Figure 2 shows the effect of changing the current on the bead appearance at constant gas

Table 1 Welding conditions for bead-on-plate tests

Specimen / Condition	Average current, A	Average voltage, V	Peak voltage, V	Background voltage, V	Pulse rate, sec	Travel speed, mm/min	Wire diameter, mm	Gas composition, %	Output characteristics	Thickness, mm	Temperature, ^{o}C / Humidity, %
86	180	26	66	28				2.91%O_2			
87	180	26	66	28				5.63%O_2			
88	120	24	66	24							
89	140	23	60	24	100	400	1.6	2.91%O_2	Constant potential	10	26 / 46
90	160	25	66	28							
111	180	26	66	28							
112	120	23	66	24				5.76%CO_2			
113	140	23	60	24							
114	160	26	66	28							

composition, which should be compared with those made at 180A shown in Fig.1.

In the oxygen addition, Fig.2a, as the welding current increases (up to the limit of 180A) the spatter in the neighbourhood of the weld toe decreases.

The bead appearances for carbon dioxide addition to argon gas, regarding the quantity of spatter and crater shapes, are very similar to those for the addition of oxygen. However, the bead surfaces are not as smooth compared with those for oxygen addition, and are especially uneven when deposited at a low current (120A), Fig.2b.

Bead shape and penetration

Effect of gas composition

On the condition that only the kind and degree of gas admixture is varied while the welding current, 180A, and the welding speed, 400mm/min, remain constant, measurements were made to determine the changes in reinforcement height, depth of penetration, bead width, contact angle, and dilution ratio. For convenience the contact angle mentioned here is the angle, θ, indicated in Fig.3, where the angle is measured from bead cross-sections enlarged ten times.

The angle of contact and dilution ratios are indicated in Fig.3a, together with the bead reinforcement height and width and depth of penetration, Fig.3b. The corresponding macrosections for oxygen addition are shown in Fig.3c.

If the oxygen content is increased the contact angle becomes smaller and the dilution ratio increases to become 23% greater than that in the argon gas only. This is because the depth of penetration continues to increase more than the change in reinforcement height, cf Fig.3b. The shape of the bead and penetration is virtually semi-circular, as shown in Fig.3c.

Thus the effect of adding oxygen is, under these conditions, to decrease the reinforcement height by about 1.3mm compared with pure argon gas. If the oxygen content is further decreased the reinforcement height will decrease a little. The depth of penetration also increases with the oxygen content, being deeper than that for pure argon gas only.

The bead width is narrower by nearly

1mm compared with that for argon gas only, but changes little with further increase in the oxygen content.

Similar tests were also conducted for reference on the effects of adding carbon dioxide to argon gas. It was found that the effect was much the same as for oxygen additions. This is believed to be because oxygen is produced by thermal dissociation as $CO_2 \longrightarrow CO+O$.

Effect of current

The changes which occur in the reinforcement height, depth of penetration, bead width, contact angle, and dilution ratio were investigated for variations in current at constant welding speed 400mm/min.

Figure 4a indicates the changes in bead dimensions and Fig.4b summarises those in bead shape, i.e. contact angle and dilution ratio. These are shown in more detail in the macrosections, Figs 5, 6, and 7, for pure argon, argon+2.9% oxygen, and argon+5.8%CO_2 respectively.

With oxygen addition to the shielding gas the reinforcement height does not always increase with current, and it is generally lower than that for argon gas only. The depth of penetration increases as the current is increased and is significantly deeper in comparison with that for pure argon.

The bead width increases substantially linearly with current and generally tends to be a little wider than that for argon only.

The contact angle decreases with current for argon alone, but only slightly with increase in current for the mixed gas. It should be noted that the contact angle at low current, 120A, is 30° less than that for pure argon.

As the current is increased up to 180A the dilution ratio also increases, presumably because the depth of penetration increases more rapidly than the bead height.

It is clear that with the oxygen additions, Fig.6, there is a major difference in the dilution ratio compared with that for argon only, Fig.5. The bead shape and penetration is nearly circular and lack of fusion (at the weld toe) is not found, even with a low current.

With carbon dioxide addition the effects of current on the reinforcement height, penetration, bead width, contact angle, dilution ratio, and bead shape are almost the same as those for oxygen addition to the argon gas, presumably for the same reason as mentioned under 'Effect of gas composition'.

Porosity

The change in porosity produced according to

Table 2 Nominal pore diameter and volume

Diameter of pore		Pore volume at average diameter, mm^3
Range, mm	Average, mm	
0.2-0.6	0.4	0.033
0.6-1.0	0.8	0.268
1.0-1.4	1.2	0.904
⩾1.4	-	-

the kinds of gas used was also investigated. Regarding all porosity as spheres, the number of pores observed in each X-ray film per 150mm length were measured according to their diameter and classified as shown in Table 2. The total pore volume was thus calculated for each film based on the average diameter and number of pores, and the results summed to obtain the gross volume.

Pores less than 0.2mm diameter were neglected and those of more than 1.4mm diameter were measured separately to calculate their respective volumes. Figure 8 shows the incidence of porosity under conditions used for pure argon and the admixtures. This indicates that the addition of oxygen to the argon is remarkably effective in the elimination of porosity.

Microstructure

The microstructures were examined for the various kinds of gas shield when welding at the given current and speed, 180A and 400mm/min respectively. Figure 9 shows the microstructures of weld cross-sections for the three gas shields. Figure 9b is very similar to that for Fig.9a, except that there are no inclusions. Something which looks like an inclusion can be observed in Fig.9c.

DISCUSSION

Bead width

Assuming that the weld reinforcement is a part of a cylinder of weld metal of width, W, mm, and contact angle, θ, radian, with liquid density (aluminium), ρ, kg/mm^3, welding speed, v, mm/min, and specific deposition rate (wire fusing rate), α, kg/min A, then for current, I (A), the bead width, W, can be calculated according to the following equation

$$W = 2\sin\theta\sqrt{\frac{\alpha I}{\rho v\left(\theta - \frac{\sin\theta}{2}\right)}} \qquad [1]$$

For example, bead widths were calculated for the condition 180A current and 400mm/min travel speed for oxygen admixtures of 2.91 and 5.63%. The resulting values for the contact angles shown in Fig.4b were 13.5 and 13.8mm. On the other hand, those actually measured were 13.5 and 14.0mm respectively.

The value of α used for this calculation is 1.59×10^{-4} (kg/min, A) and 2.66×10^{-6} (kg/mm^3) for the value of ρ. From the above, it appears that the bead width can be calculated with tolerable accuracy using eq.1.

Penetration

It is easily understood that the penetration becomes larger as the current is raised because the electrical energy increases and the magnetic pinch effect also increases to enhance the impact velocity (cavitation action) of the melting droplets. However, in spite of the constant current and travel speed the penetrations differ markedly according to the shielding gas composition (see Fig.3b and c).

To explain the above, the arc voltage should be taken into account, e.g. the voltage drop at both electrodes (cathode and anode voltage drop), positive column voltage drop, and/or voltage drop caused by dissociation. If oxygen is added to argon the arc column will decrease owing to the dissociation of oxygen. Thus for a constant overall arc voltage, the positive column voltage will be reduced according to the equation below, and so the arc length is reduced

$$V_T = V_E + V_L + V_D$$

where V_T is total arc voltage
V_E is cathode and anode electrode voltage drop
V_L is positive column voltage
V_D is voltage drop caused by dissociation

Consequently the arc forces work more effectively and the arc concentration increases.

Thus it is believed that the depth of penetration becomes greater than for pure argon shielding gas under nominally constant conditions (current, arc voltage, and welding speed).

Other tests were therefore conducted to establish the relationships between the depth of penetration, arc length, and arc voltage, Fig.10, for a range of oxygen admixture. As this Figure shows, nearly the same depth of penetration is obtained at the same arc length to a first approximation. In the region of the shorter arc lengths the penetration is slightly larger than for argon only, even if the arc length is kept the same. The heat of reaction between aluminium and oxygen is assumed to be a contributing factor.

Porosity

It is presumed that, when oxygen or carbon dioxide is added to argon gas, the number of pores in the weld will be reduced because the partial pressure of hydrogen in the arc column is reduced. However this does not explain the apparent superiority of oxygen in this respect, and this aspect should perhaps be the subject of further investigation. Nevertheless, on the basis of the above general results, it seems that studies of high speed procedures for welding aluminium pipes are worthy of further exploration.

CONCLUSION

The effects of oxidising additions to the shielding gas in MIG-welding aluminium alloys have been investigated and the following conclusions drawn:

1 Added oxygen reduces the contact angle and increases the depth of penetration considerably and is effective in the prevention of porosity. The shortened arc length associated with the addition of oxygen causes the depth of penetration to increase

2 Adding carbon dioxide to the argon gas shield produced largely similar welding results to those gained in the $Ar+O_2$ shielding gas

REFERENCES

1 KOBAYASHI, T., KUWANA, T., and AOSHIMA, I. 'The argon-nitrogen gas metal arc welding of aluminum'. J.Japan Welding Soc., 39 (9), 1970, 44-50.
2 MASUMOTO, I. and SHINODA, T. 'Effect of addition of nitrogen or oxygen in shielded gas on porosity of MIG aluminum weld metal'. J.Japan Welding Soc., 38 (12), 1969, 84-93.
3 NAMBA, K., FUKUI, T., and SUGIYAMA, Y. 'On the addition of nitrogen and oxygen gases to shielding gas of MIG-welding in aluminum and its alloys'. J.Japan Welding Soc., 39 (12), 1970, 30-38.

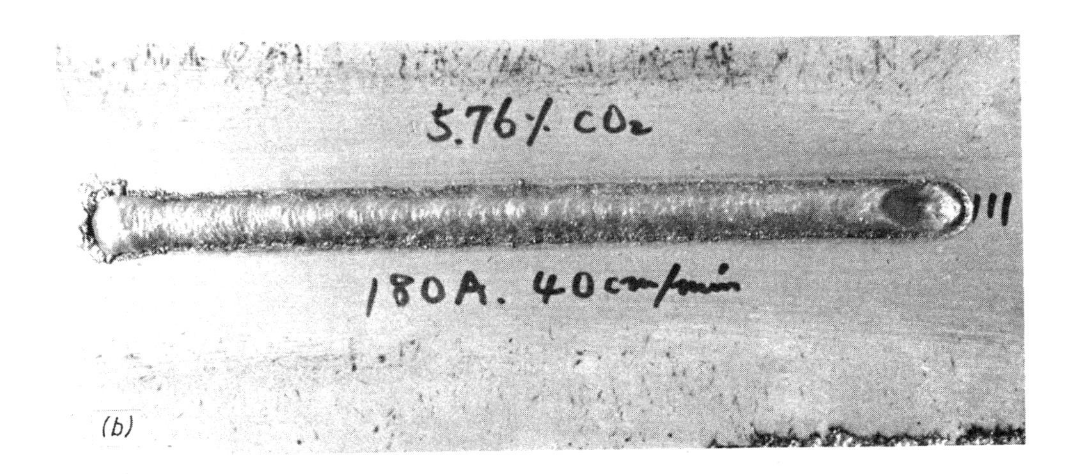

1 Effect of gas composition on appearance of bead at constant welding conditions (180A, 400mm/min)

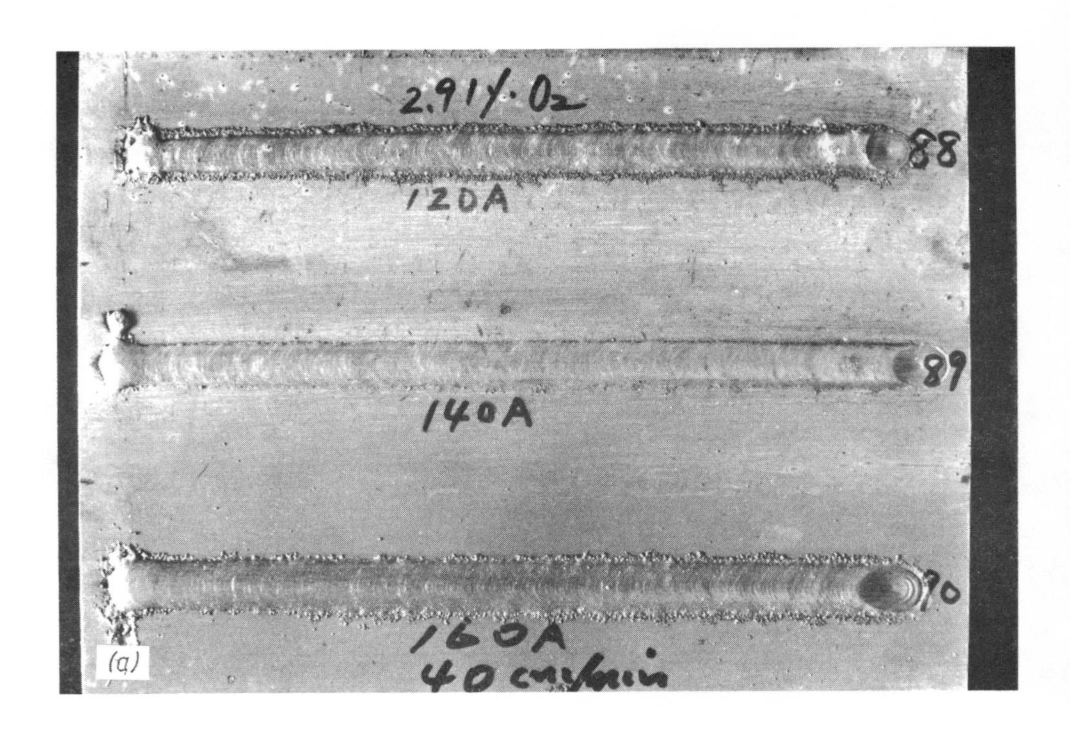

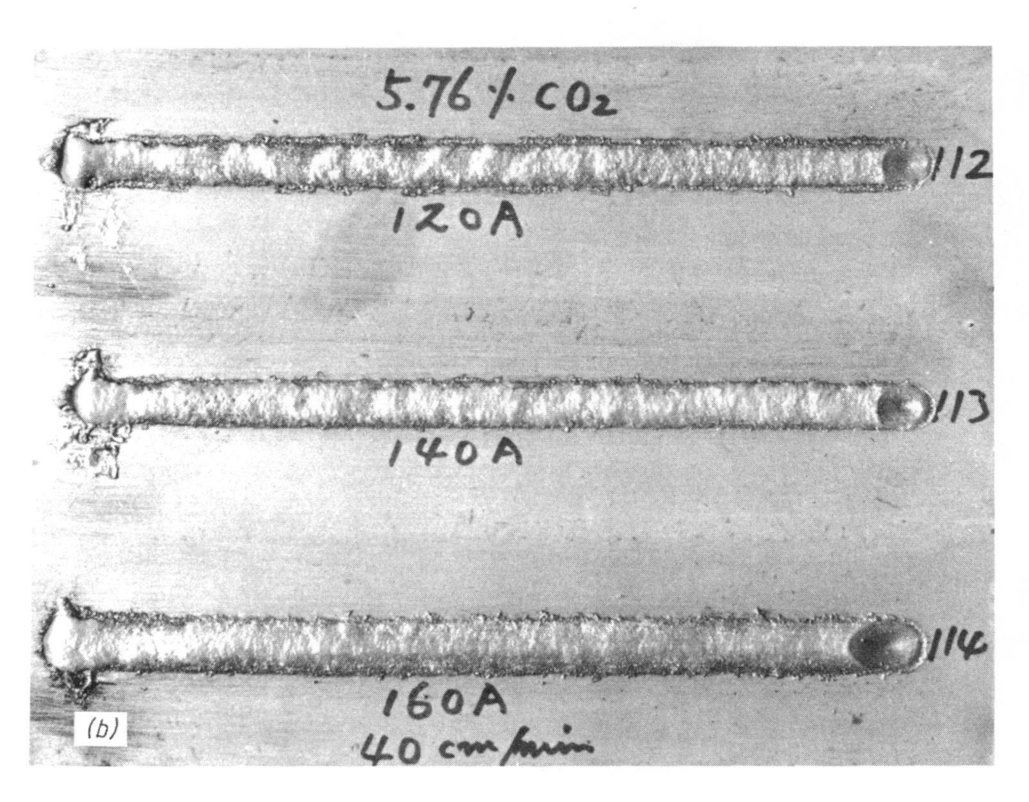

2 Effect of welding current on appearance of bead at constant travel speed (400mm/min) for: (a) Ar+ 2.9%O_2, (b) Ar+ 5.8%CO_2, admixtures

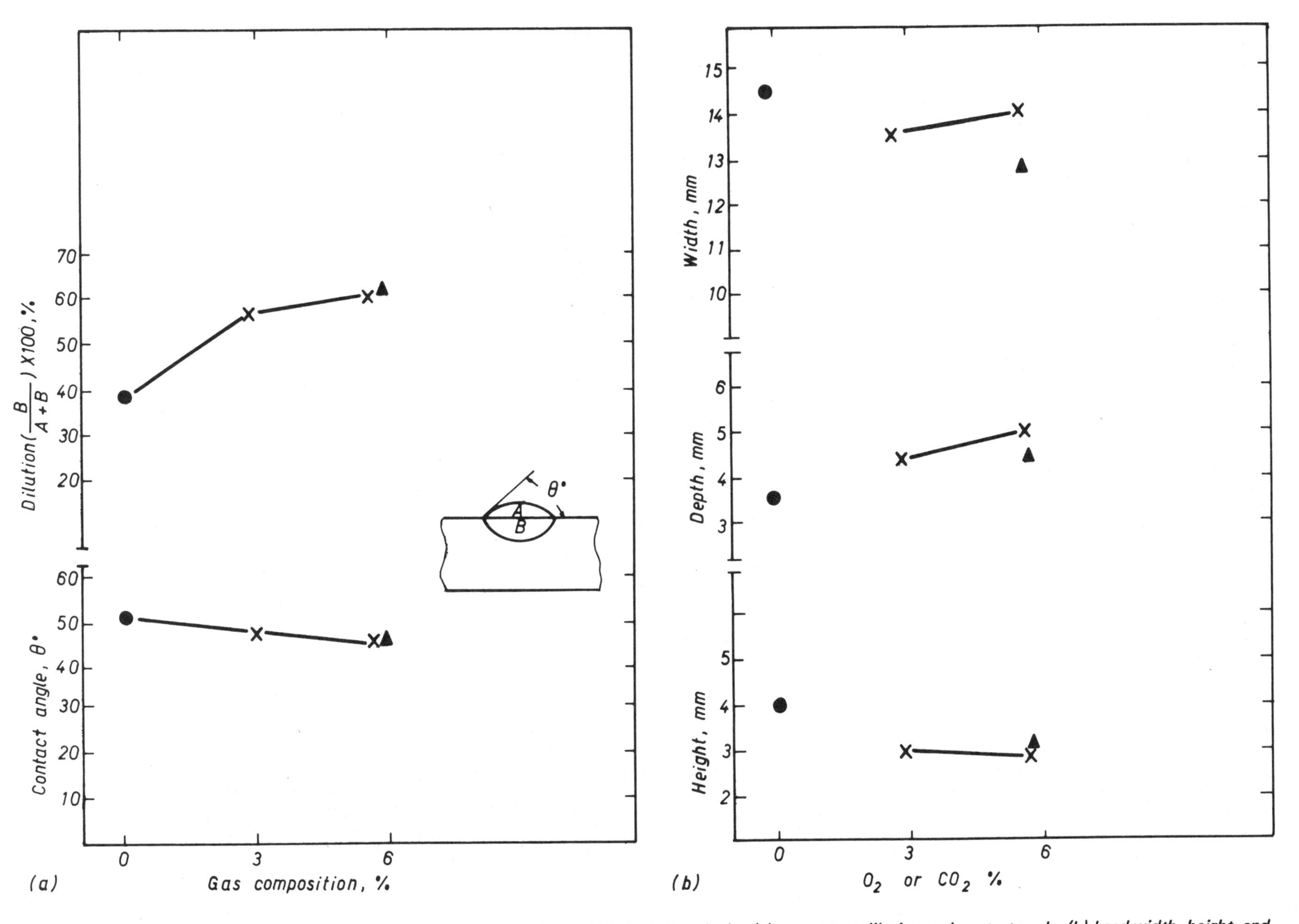

3 Effect of gas composition on weld bead dimensions at constant conditions (180A, 400mm/min: (a) percentage dilution and contact angle, (b) bead width, height, and penetration depth. ● – Ar; X – Ar+ O_2; ▲ – Ar+ CO_2

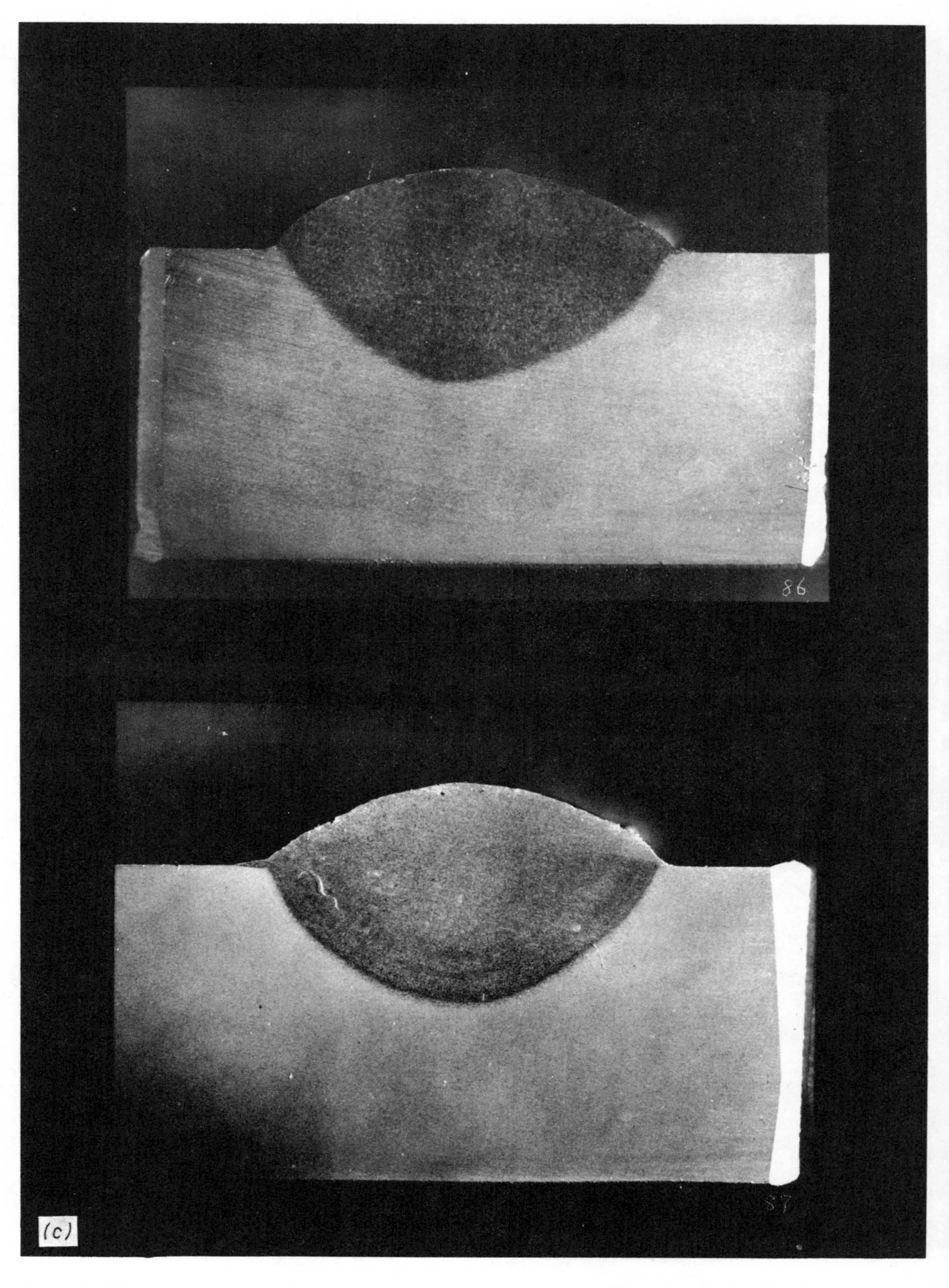

3 contd Effect of gas composition on weld bead dimensions at constant conditions (180A, 400mm/min: (c) macrosections (top) Ar+ 2.9%O_2 (bottom) Ar+ 5.6%O_2 x5

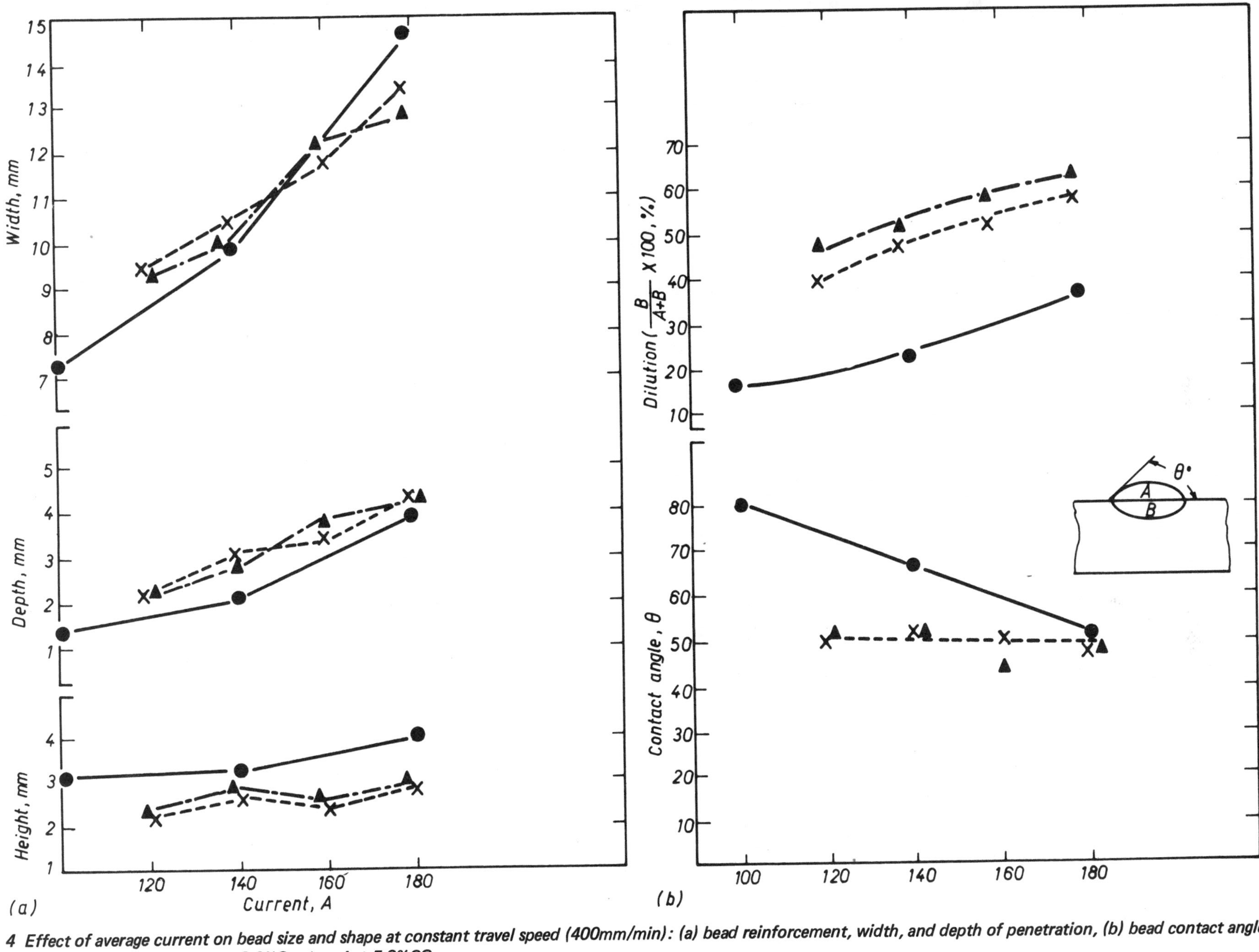

4 Effect of average current on bead size and shape at constant travel speed (400mm/min): (a) bead reinforcement, width, and depth of penetration, (b) bead contact angle and dilution. ● – Ar; X – Ar+ 2.9%O_2; ▲ – Ar+ 5.8%CO_2

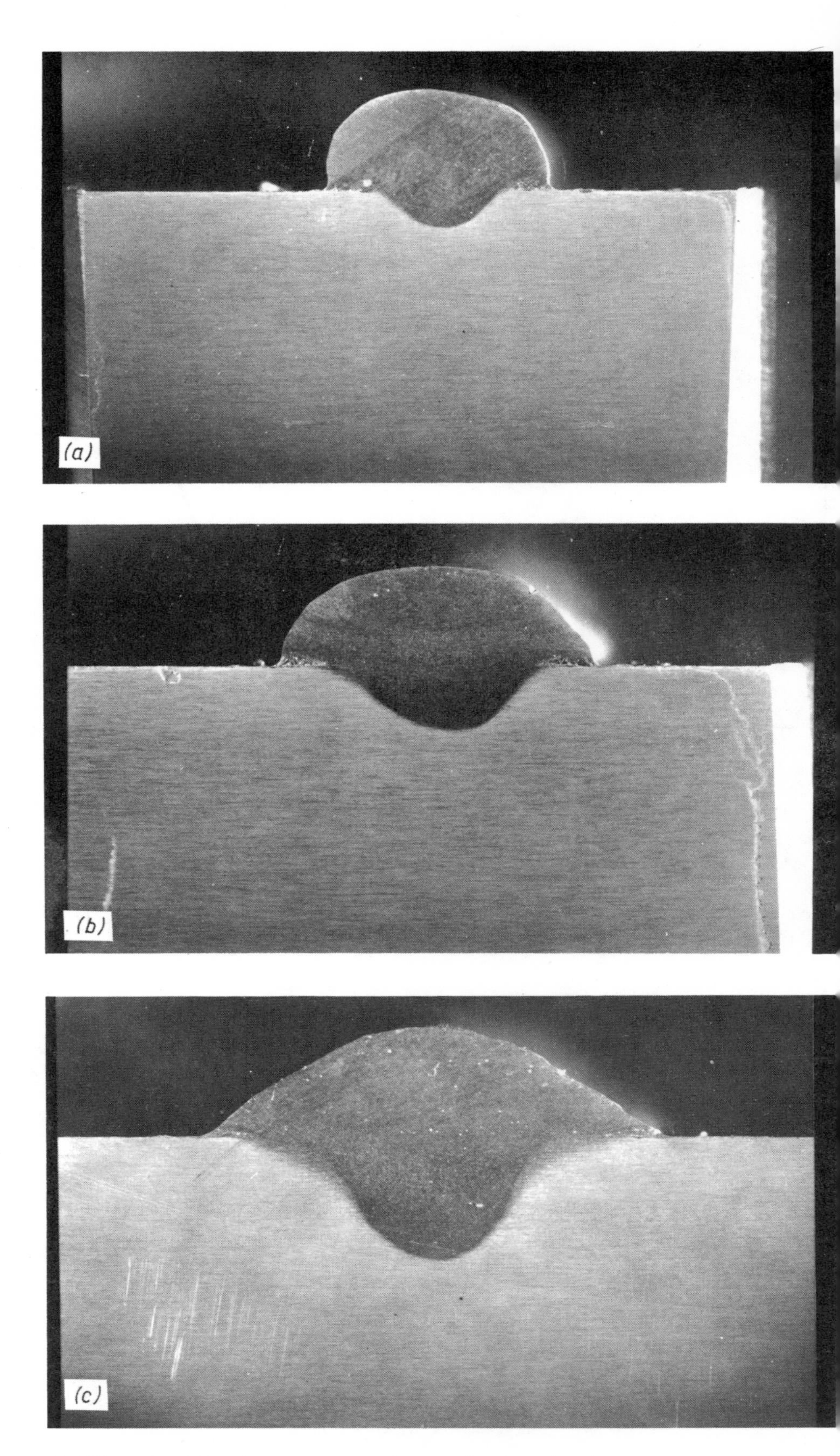

5 Effect of welding current on bead shape and penetration in argon at constant travel speed (400mm/min), A: (a) 100, (b) 140, and (c) 180 x5

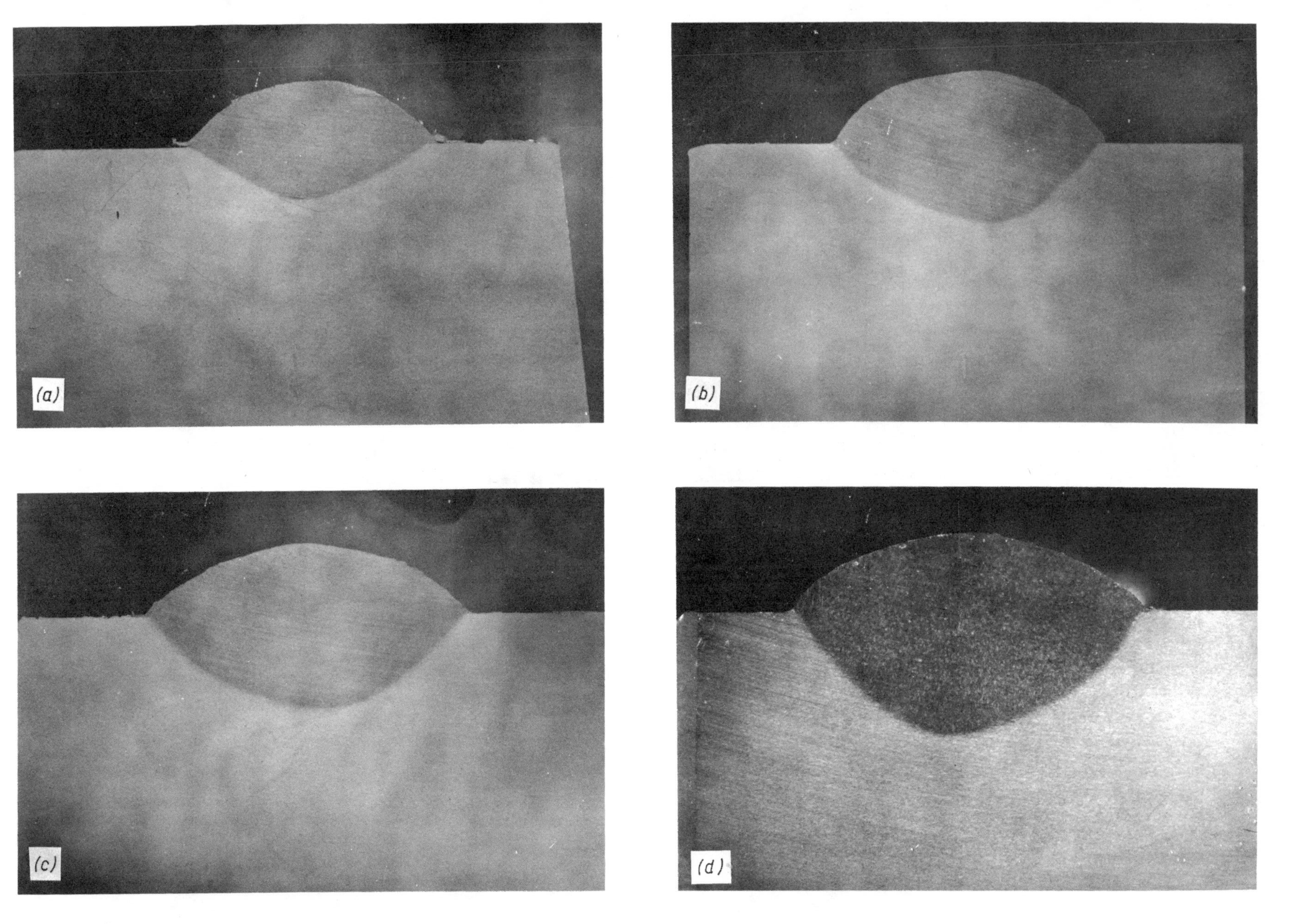

6 Effect of welding current on bead shape and penetration in Ar+ 2.9%O_2 at constant travel speed (400mm/min), A: (a) 120, (b) 140, (c) 160, and (d) 180 x 5

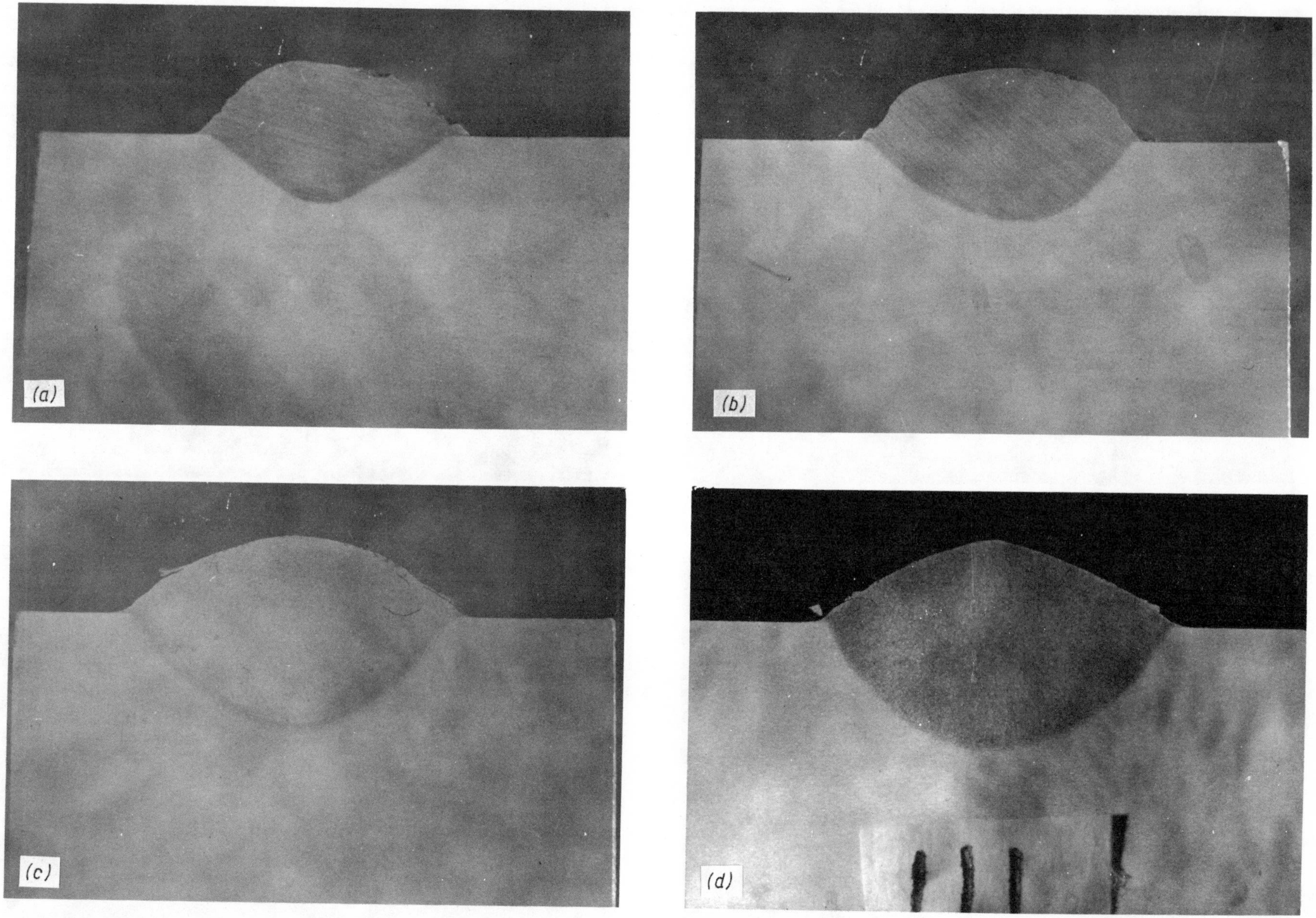

(a)

(b)

(c)

(d)

7 Effect of welding current on bead shape and penetration in Ar+ 5.8%CO_2 at constant travel speed (400mm/min), A: (a) 120, (b) 140, (c) 160, and (d) 180 x 5

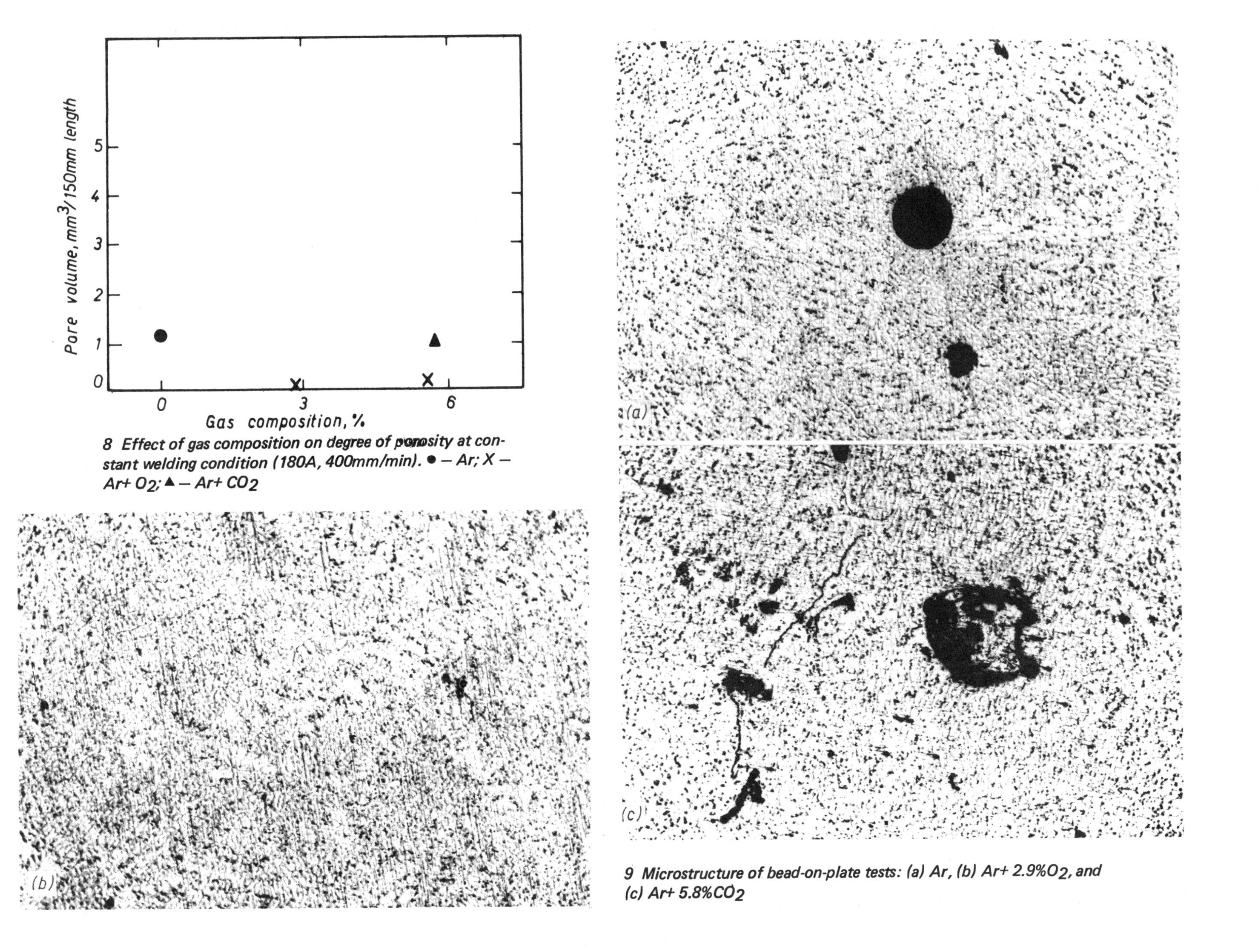

8 Effect of gas composition on degree of porosity at constant welding condition (180A, 400mm/min). ● – Ar; X – Ar+ O_2; ▲ – Ar+ CO_2

9 Microstructure of bead-on-plate tests: (a) Ar, (b) Ar+ 2.9%O_2, and (c) Ar+ 5.8%CO_2

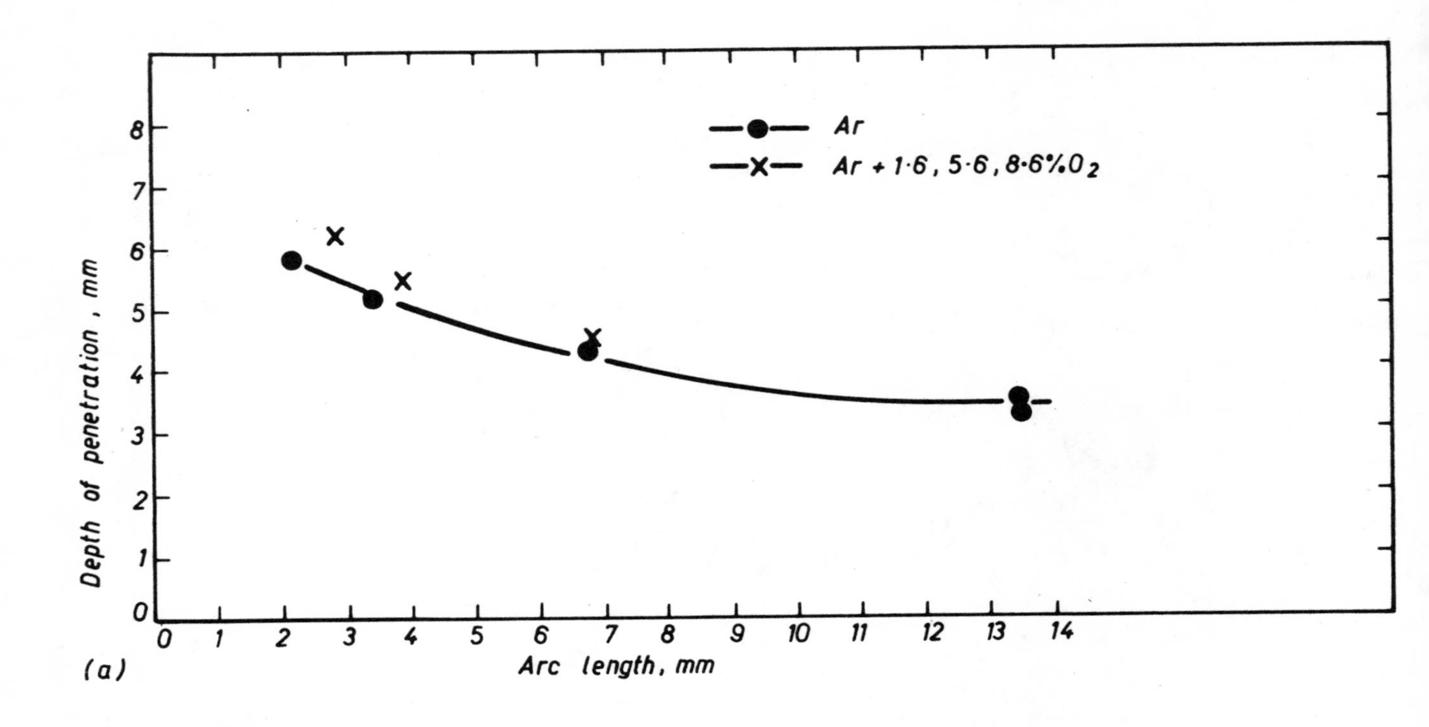

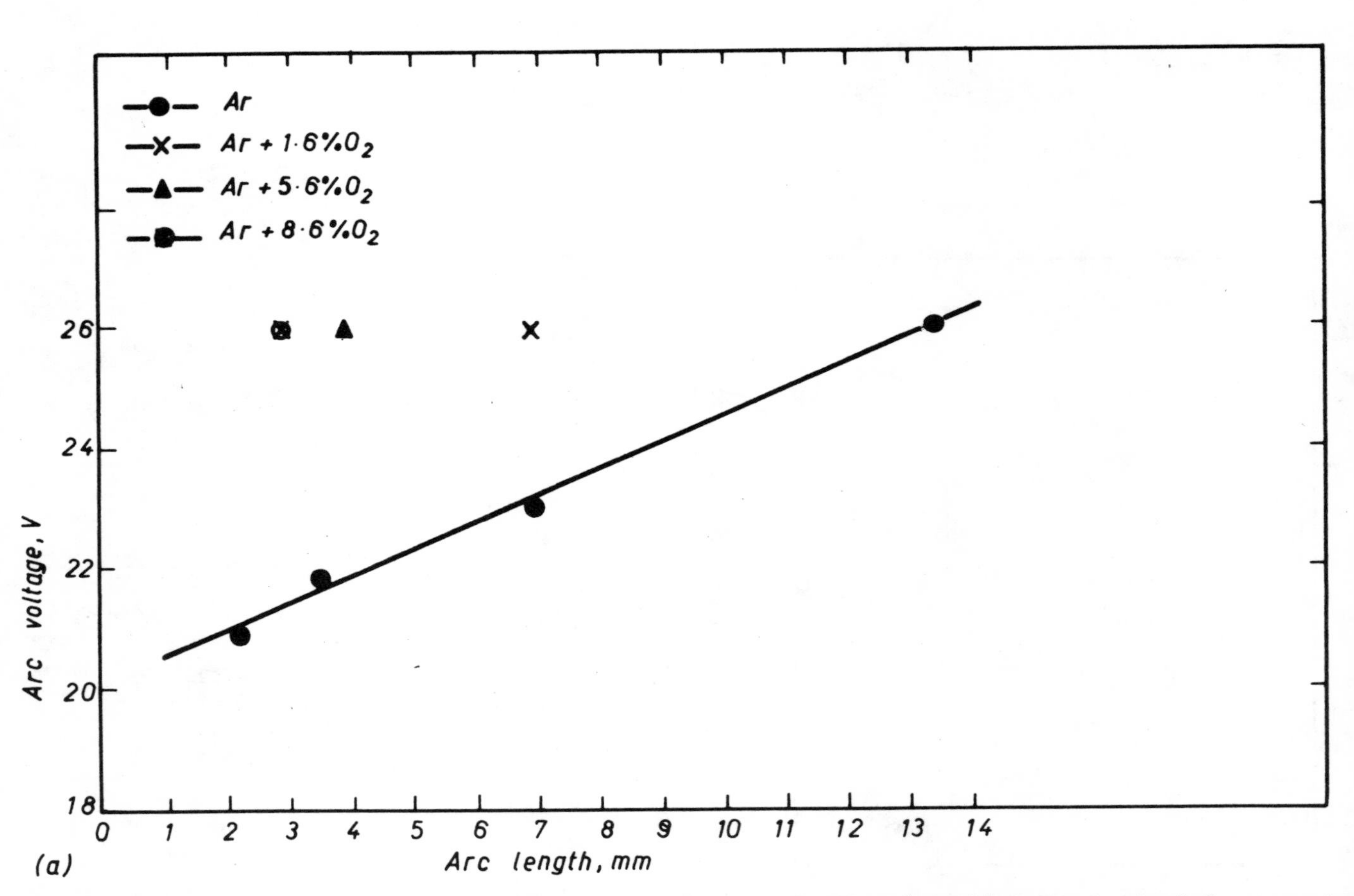

10 Effect of arc length on penetration depth and arc voltage at 180A, 400mm/min: (a) penetration depths for range of oxygen additions, (b) arc voltage/length characteristic and effect of oxygen addition

A new concept for AC/DC power sources for TIG-welding

J.Lowery, BSc(Met)

A new system for welding aluminium has been devised using three-phase rectified current with a subsequent chopper system to give a square wave AC output. The resulting arc properties allow features to be introduced which can result in higher welding speeds than with conventional units and eliminate the need for high frequency stabilisation of the arc.

INTRODUCTION

The conventional AC/DC current power source is based upon a single-phase transformer with an appropriate control system and a rectifier unit. When the AC side is required, as when welding aluminium, the single-phase unrectified current is taken direct from the transformer. When DC is required the rectifier unit is switched into the circuit.

This system, although fairly simple, has disadvantages both in AC and DC welding. For the latter, as the AC/DC power sources use a single-phase transformer, the primary current input is relatively high. For example for a 300A welding unit this would be in the order of 60A, which means thick primary cables and large fuses and the power drawn from the mains is unbalanced.

On the output side the rectified single-phase current gives a large amount of ripple, unless very large inductors are used. The ripple on the output can be regarded as a nuisance with TIG-welding, but as these power sources are also used for manual metal-arc (MMA) welding the use of special electrodes of the high alloy and cellulosic types becomes very difficult if not impossible. Spatter and arc instability are problems.

For AC welding the conventional AC/DC power sources use a single-phase transformer and hence, in principle, have a sinusoidal output. On the primary side the same problems occur as above, but on the secondary side a few new problems become apparent. Alternating current arcs go out each time the current passes through zero and the arc must be reignited. This normally happens naturally when the electrode goes from positive to negative, but the negative to positive changeover for the electrode is more difficult. To stabilise the arc in MMA welding, certain elements can be introduced into the electrode covering, which give easier reignition.

In TIG-welding however the usual solution is to use a high frequency generator running continuously across the arc. In practice this is not fully reliable and it is often noticed that a few periods of instability will occur which give rise to poor cleaning of the weld area. The TIG arc also has an inherent rectification effect as it is easier for the current to go through the arc when the electrode is negative. This gives an imbalance to the current wave shape and subsequently a poorer arc cleaning. This must be compensated for by the addition of a large capacitor bank to the unit.

In general it can be stated that the AC/DC power source in its conventional form is a convenient way to solve the immediate needs of industry for a versatile all round welding machine. However with the advent of new electronic components much better ways can be found allowing the best advantage in both AC and DC techniques.

The ultimate in DC welding is to use three-phase rectified current. This can then be chopped into square wave AC to give new features never before possible. Thus, instead of the conventional AC/DC power sources the new approach can be classified as DC/AC type.

BASIC RECTIFIER FOR DC

The TIG 300 represents the apex of a development programme to produce an integral range

Mr Lowery, Marketing Manager, is with AGA Welding Limited.

of equipment for MMA, TIG, and plasma welding. Each of the components in the system has been developed to be used either on its own or in combination so that the ultimate properties could be obtained. It was therefore necessary to begin with a basic rectifier for the system. It was decided to make this in two sizes: one for normal electrodes and TIG, and the other of higher capacity for high efficiency electrodes and other high current processes. These two are identical, apart from the higher capacity of the current-carrying components, so it will suffice to describe the smaller unit named R 300.

A rectifier for TIG or MMA welding is required to have two different functions. Firstly, it shall provide a stable smooth current at a maximum value of 300A and 32V (according to ISO standard). The voltage need not be higher than this except when abnormally long arc lengths or long secondary cables are used. The second function, however, is arc striking, which requires a voltage of 70-80V at very low currents.

These two conflicting requirements mean that the transformer is usually designed to give 300A at about 70V, with a consequent waste of both construction materials and running kVA demand. The R 300 was purpose-built and contains two circuits, one for the 80V at very low current to give reliable striking, and the other for 42.5V at 300A where the extra 10V allows for extra long cables etc.

To produce a stable system which would permit various control functions, it was decided to use an electronic system incorporating power thyristors. This allowed for such items as continuous current control, remote control, hot start, slope-up, slope-down, spot welding, pulsed current welding, and so forth. It also meant that a feedback control could be used whereby the voltage and current could be monitored. The current could then be stabilised and the reference voltage signal used for various functions.

It is also very important to smooth the welding output from a thyristorised unit to limit the amount of ripple. As the range of the unit extended from 7A up to the maximum, a special design of inductor was necessary so that low current range could also be smoothed. A continuously saturable inductor was designed with a drooping characteristic. This gives maximum inductance at low currents and as the current is increased the inductance value drops. This allowed considerable savings in the iron content of the inductor and drastically reduced its size. The resulting output gave a stable arc even with the more difficult cellulosic electrodes.

The final DC rectifier unit designed is shown in Fig.1.

Other controls could then be build on, with this as the basic unit, by adding a few functions to the basic rectifier for DC TIG-welding. These include stepless slope-up and slope-down controls, spot welding timer, and both control and remote switching of current level. The voltage signal was also used to switch off current if the voltage dropped below 7V, i.e. if the electrode was about to short circuit to the pool. Plug-in accessories were developed for remote control of current, pulsing of current, and instrumentation.

Functions for striking the arc, gas and water controls were incorporated in a separate box. The arc striking was obtained using a single ignition spark to limit interference normally encountered with high frequency to a minimum. The complete DC system for TIG-welding is shown in Fig.2.

THE AC CONVERTER

The AC converter or chopper system is the opposite of a rectifier bridge and the basic circuit is shown in Fig.3. It consists of four power thyristors which switch the current obtained from the power source from positive to negative and vice versa. Thyristors can be opened, i.e. made to conduct, by applying voltage to the gate, but close only when the current is reversed. However, since there is only DC on the primary side it was necessary to superimpose a reverse pulse of current through the thyristors. This was possible with a capacitor across the thyristor, in series with a small secondary thyristor, to trigger the pulse. In this manner they can be opened and closed at will. To produce the AC the thyristors are switched in pairs which results in a square wave output characteristic as shown diagrammatically in Fig.4. The time of the periods for each polarity can be selected at will, though the frequency was fixed at 50Hz for convenience. The percentage time of positive and negative periods on the output terminal could be varied between 10 and 90% in a continuously variable manner. (In practical trials these limits were found to be too wide for the applications used and were reduced on the final machine to 30-70%.)

When connected to the welding arc it was found that the change of current as the arc went through zero was entirely stable without the need for any superimposed high frequency voltage. It was necessary only to strike the arc with the same system used for DC

TIG-welding and the arc would then continue of its own accord.

To simplify arc striking with blunt electrodes of the type used in AC welding the arc was first struck with electrode positive and held in this mode for a short period to heat the electrode, and then the chopper circuit was initiated. This gave safe reliable striking and a very stable arc.

The individual units in the R 300 system were first used separately. However, on the production model all the necessary units were built into one chassis to simplify the unit for the operator. The complete unit was named TIG 300/AC/DC, Fig.5. It was found that both the DC and AC characteristics of the machine were excellent for the process involved, but, as the AC side gave new possibilities to the welding process, the description hereunder will be confined to this.

PERFORMANCE WITH SQUARE WAVE AC

Arc characteristics

As the arc was struck in the DC mode, even for AC welding, the arc striking reliability was found to be much better than with conventional units. The time period in the DC mode with the electrode positive was adjusted so that the electrode was heated sufficiently to sustain the AC arc subsequently.

The arc was found to be completely stable in the range 40-300A, without the need for any extra voltage stabilisation. This was the situation even with very long arcs and even if the workpiece was almost oxide-free. The arc produced a crisp sound particularly if the welding speed was too low. This occurred as the arc tends to initiate on the plate from oxide islands, and if the weld pool is too clean the arc must transfer a larger distance giving this typical ionisation sound. The welders found this sound helpful when manually welding as they could 'tune' their welding speed.

On aluminium plate the arc cleaning could be varied by varying the balance control. A more positive electrode produced a wider sputtered zone, Fig.6. It was also noticed that the speed with which the torch could be moved depended to some extent upon the oxide cleaning and that a manual welder could weld much faster with the electrode more positive.

Penetration and heat input

It is commonly assumed as a rule of thumb that in a DC TIG system 70% of the heat is produced at the anode and 30% at the cathode. With AC welding therefore with a normal sinusoidal wave a 50/50 relationship is produced for the heat input to the electrode and plate. It would be surmised from this that, when using an unbalanced system, the more electrode negative the greater the heat input to the plate. The opposite, however, was found in practice. In Fig.6b it can be seen that, as the positive period on the electrode is increased, so the width increases of both the penetration and the bead. This can also be seen in Fig.7 where, by going from 30-70% positive, the welding speed can be increased by 100% while retaining virtually the same penetration. Thus by using the balance control the penetration and speed can be increased. Alternatively for thin sheet manual welding an opposite balance can be selected to retain control of the bead.

The disadvantage of using high electrode positive unbalance is that the electrode also becomes hotter and a larger electrode is necessary than with a balanced wave.

The reasons for the increase in heat input have not been fully investigated; however, it can be noticed that, when the arc is burning with the electrode positive, the arc voltage is some 4V higher than with the opposite polarity. The increase in arc voltage and consequent increase in arc power could well account for the increase in heat input to the plate. In practice it has been found that increase in welding speed of up to 50% can be obtained using the square wave unbalanced system over a conventional AC system.

Pulsing the welding current

Pulsed current TIG-welding has been discussed in various articles particularly with reference to stainless steel welding. Not as much work has been carried out on aluminium welding mainly as AC TIG pulsed equipment has not been so readily available. As the TIG 300 system is electronically controlled it is fairly simple to incorporate a pulse, and the arc stability ensures a smooth arc even during the low current periods. The same advantages then apply to aluminium welding as for stainless steel: Fig.8 shows a sample of thick to thin welding in aluminium. Control of the weld pool is enhanced with pulsed TIG and the heat input to the workpiece is reduced.

CONCLUSIONS

1. A complete system has been developed for AC/DC TIG-welding using an electronically controlled rectifier and a chopper system to produce square wave AC current

2. The square wave AC current produced

gives a stable arc without the need for high frequency

3 By varying the unbalance of the positive and negative cycles, arc cleaning and penetration as well as heat input can be varied

4 The more positive the electrode becomes, the greater is the heat input and arc cleaning

5 Pulsed AC TIG-welding (on aluminium) gives similar advantages to those reported with DC TIG (on stainless steel).

1 R 300 standard rectifier

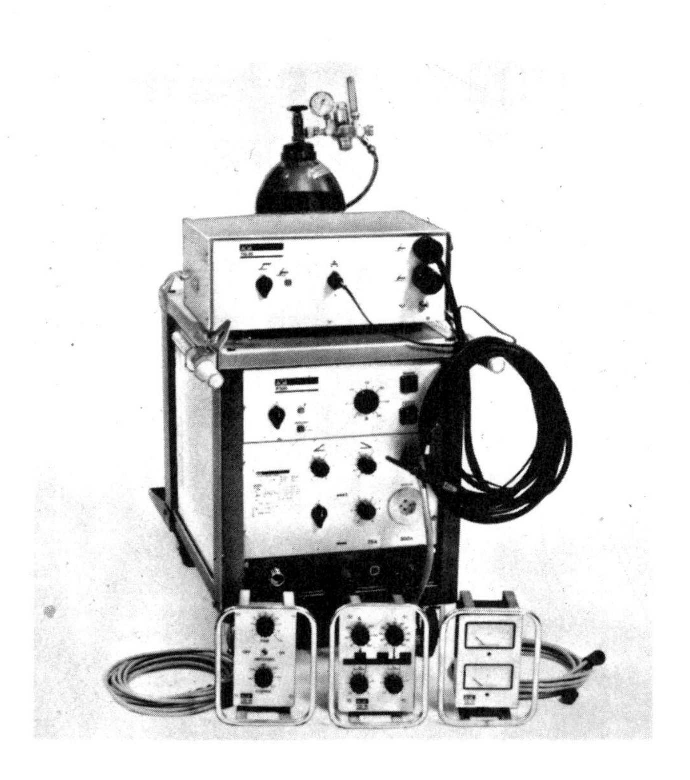

2 R 300 TIG S with accessories for DC TIG-welding

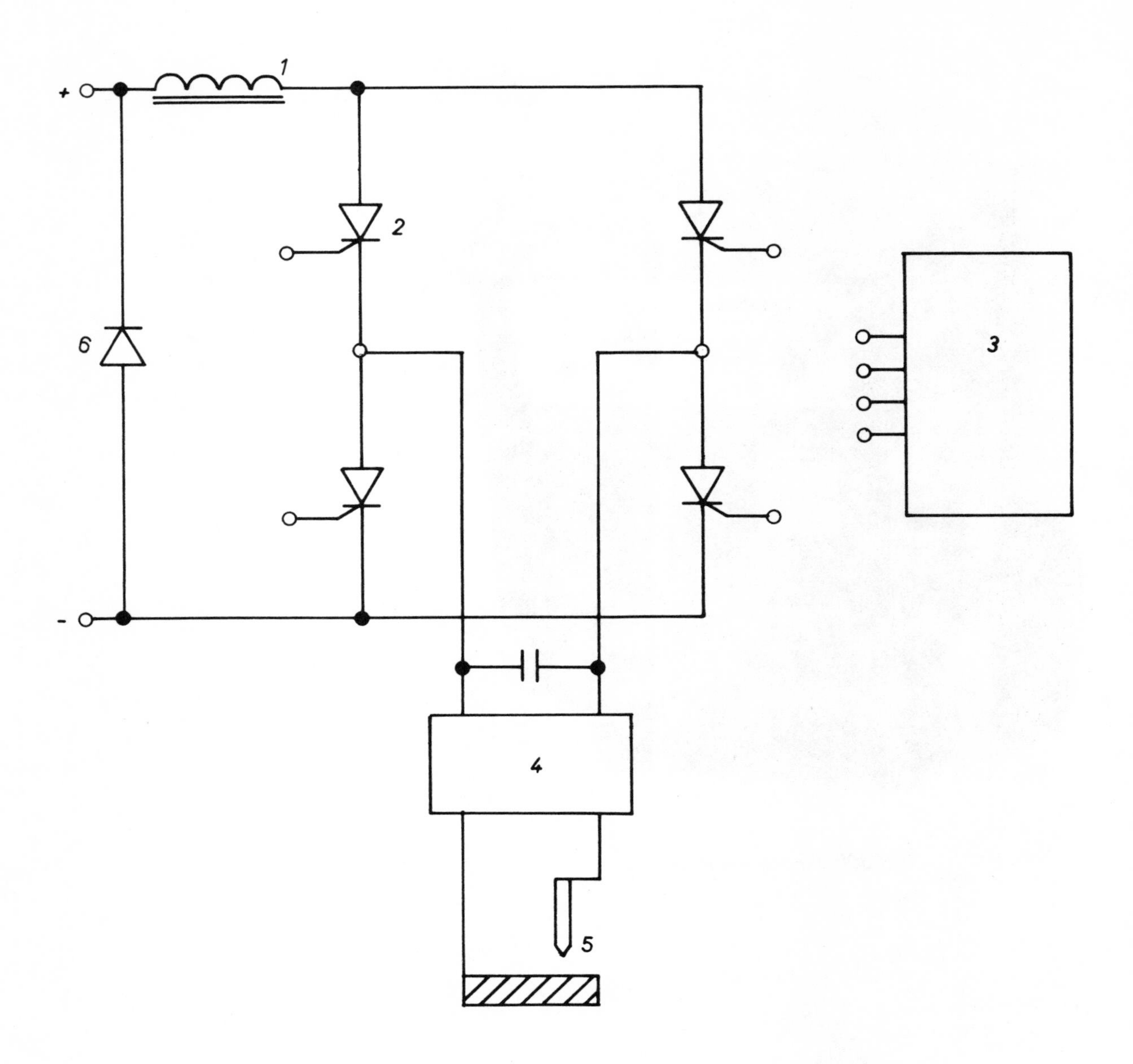

3 Schematic diagram of converter unit for production of square wave AC. 1 — inductor; 2 — power thyristors; 3 — electronic control; 4 — arc striking unit; 5 — AC arc; 6 — free-wheel diode

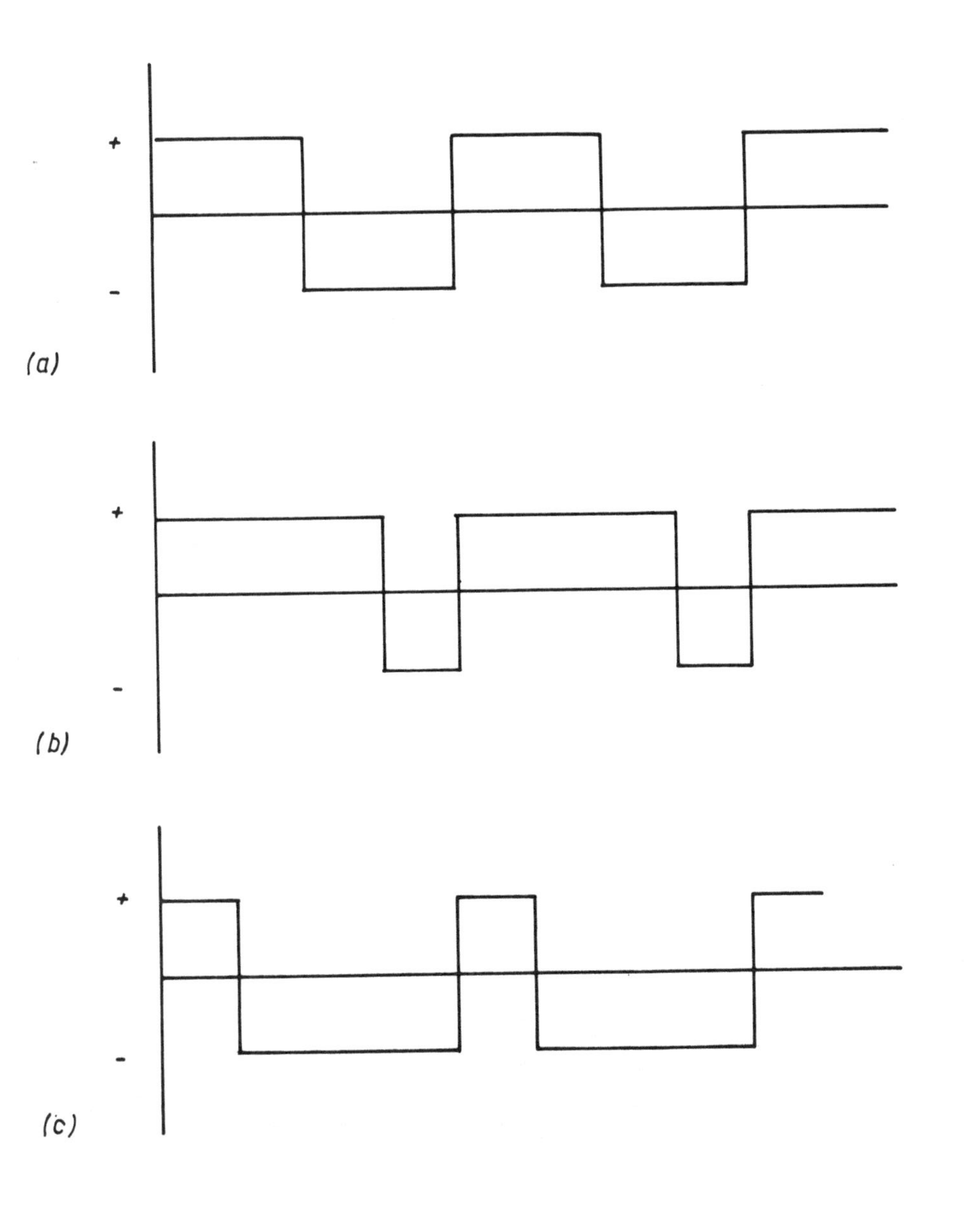

4 Schematic diagram of square wave balance control: (a) 50% balance, (b) 70% electrode positive, and (c) 70% electrode negative

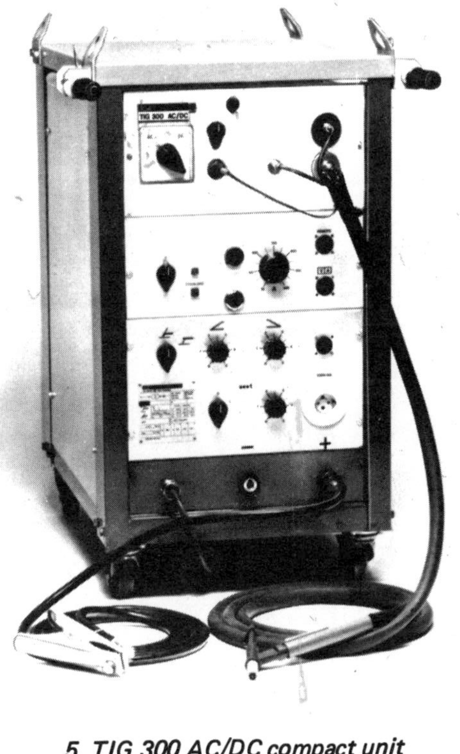

5 TIG 300 AC/DC compact unit

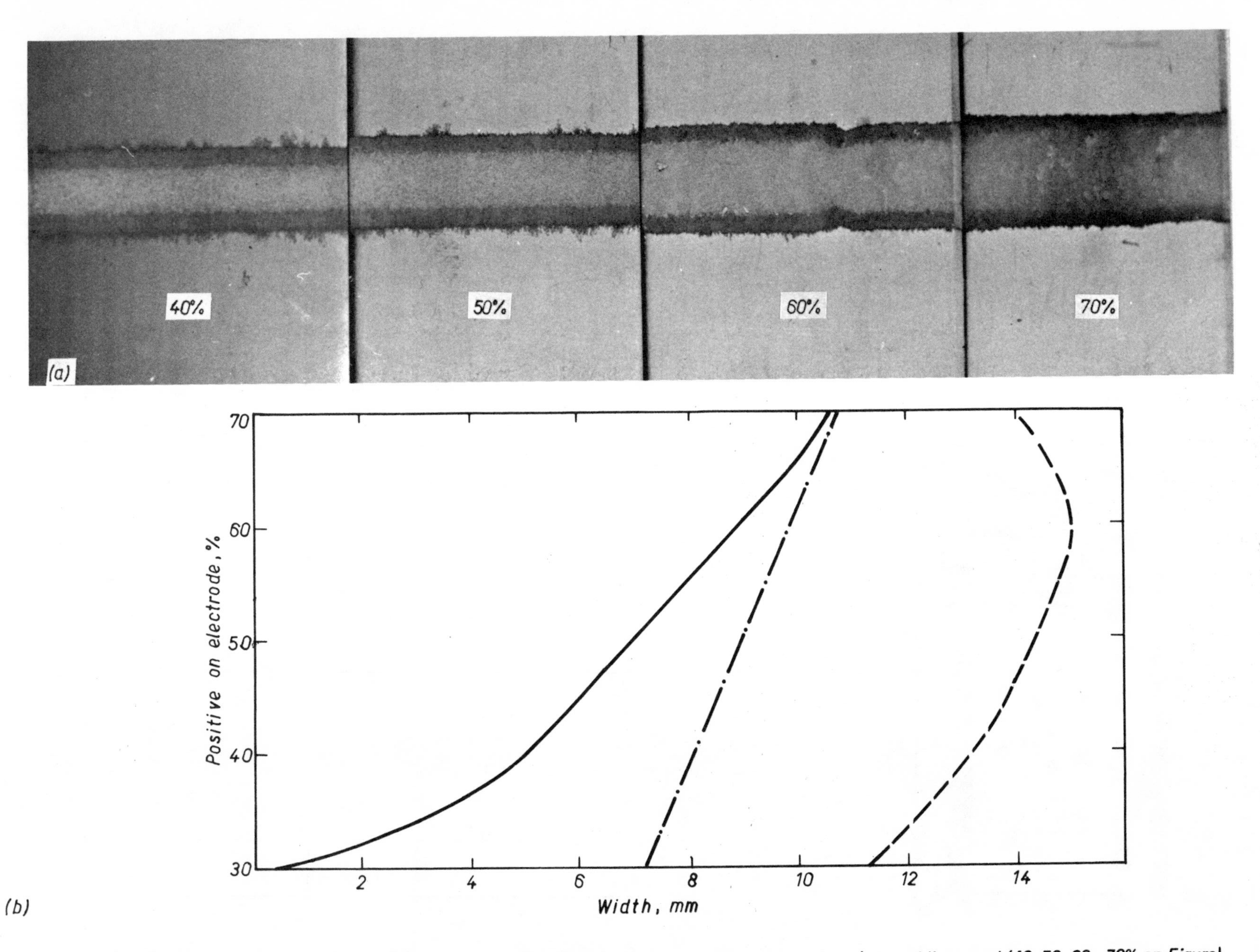

6 *Effect of percentage unbalance on weld pool and sputtered zone width: (a) melt runs at 50A, 150mm/min welding speed (40, 50, 60, 70% on Figure), (b) relationship between unbalance and bead characteristics for 3.2mm aluminium plate. Welding conditions: 150A, 200mm/min travel speed. ——— underbead; —·—·—top bead; — — sputtered zone*

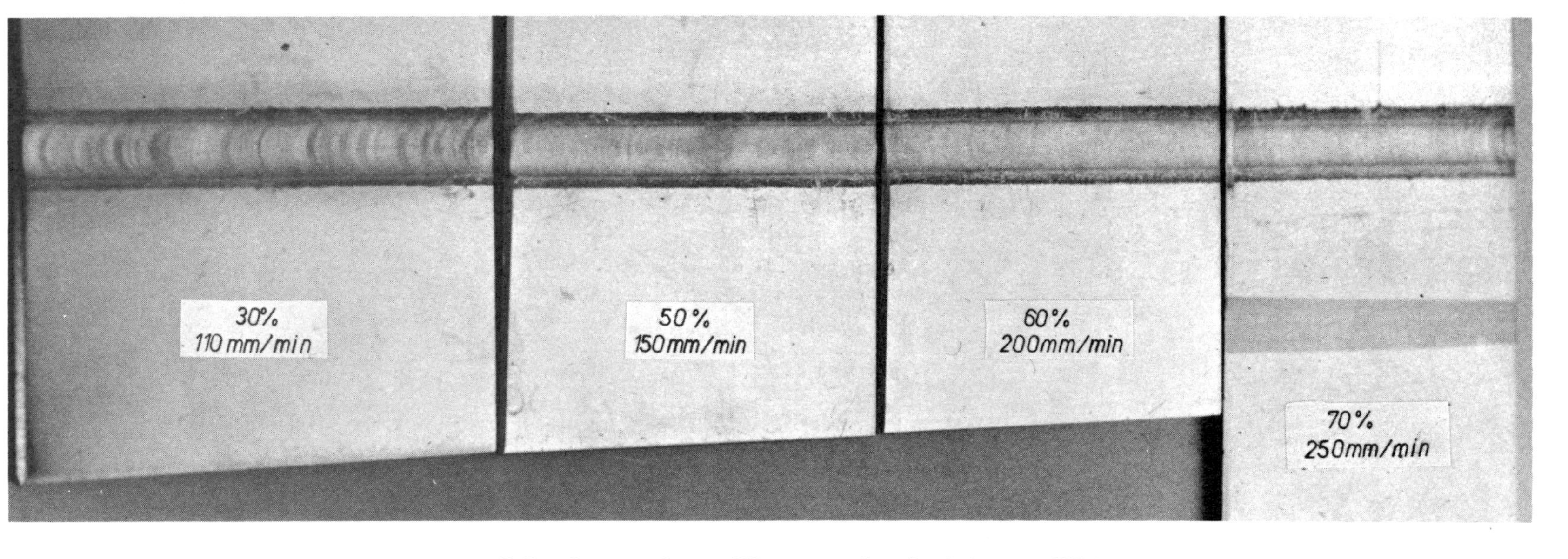

7 Equal penetration at different speeds and unbalance at 150A

8 Example of pulsed TIG-welding in thick to thin plate

A study of DC electrode negative TIG arc welding of aluminium alloys

T.Watanabe, BSc, MSc, DEng, and H.Marumoto, BSc

With a given plate thickness, from 6.0 to 25.4mm, and under given welding conditions, the depth of penetration of a weld made by electrode negative DC TIG arc can be approximately calculated from the equation of interpolation. Moreover the underbead reinforcement size is easily controllable by using the electrode negative TIG arc.

INTRODUCTION

In the welding of aluminium and its alloys, AC TIG and DC electrode positive MIG arc welding have been primarily employed.[1] Particularly to forming a uranami* bead in one-sided welding, without backing in relatively thin aluminium pipes as used for LNG, AC TIG-welding is used exclusively since MIG-welding may cause burnthrough.[2] However, AC TIG-welding has drawbacks in that it requires a high level of skill and is inefficient. For this reason the development of a new welding technique for the root bead has long been desired.

In this context the authors examined the possibility of applying DC electrode negative TIG-welding which has not previously been extensively used for one-sided welding of aluminium pipes. As stated above, in one-sided pipe welding the formation of a smooth, well-fused uranami bead is the prime factor in the determination of weld quality. Therefore, various experiments were conducted regarding the formation of a uranami bead using the DC electrode negative TIG arc. As a result it was found that DC TIG-welding is a very effective technique: it is more efficient than AC TIG-welding and the range for uranami bead formation is wide.

This Paper summarises various studies which the authors have conducted on the formation of a uranami bead in aluminium and its alloys and on the establishment of welding parameters for DC electrode negative TIG-welding.

*Uranami implies a root pass of good underbead profile made from the top side only.

Dr Watanabe, Manager — Inspection Department, and Mr Marumoto, Welding Inspector, are both with the JGC Corporation, Yokohama, Japan.

TEST PROCEDURE

Materials

Parent plate:
A5038P-O, thickness 6.0-25.4mm
5083TD-O, 21mm OD, thickness 7.5mm

Electrode:
Thoriated tungsten, diameter 4.0-6.4mm, filler wire A5356WY, diameter 1.2 and 1.6mm

Shielding gas:
Argon

Welding equipment

TIG power supply	: AIRCO AC/DC Heli welder, capacity 300A
Torch	: Osaka Henatsuki AW12, air cooled
Pulsed arc power supply	: AIRCO Model PA-3
Torch	: Airco super midget

EXPERIMENTAL RESULTS

Penetration with TIG arc

Melt runs were made with the DC electrode negative TIG arc in the 5083-O workpieces (thickness 6.14 and 25.4mm, size 150 x 200mm) without using filler wire. The depth of penetration of the parent metal was determined corresponding to the changes in the heat input. The argon gas flow was set at 15 litre/min.

For comparison, a similar experiment was conducted on the penetration of the AC TIG arc. The preparation of the plate before welding, i.e. degreasing and removal of oxide film, was accomplished by conventional methods.

Figure 1 shows the depth of penetration of the parent metal with variation in welding speed for several current levels and polarities for 6mm plate. With DC, especially electrode positive, the depth of penetration is found to increase abruptly as the welding speed is decreased,[3] although this is to be expected. It is also observed that, under the same nominal current conditions, DC TIG gives far greater depths of penetration than AC TIG. Macrosections of the DC electrode negative TIG melts at 140A are shown in Fig. 2 at corresponding welding speeds to those in Fig. 1. The critical heat input required for fusing through the entire plate thickness is obtained when the welding speed is 100mm/min. When the welding speed is decreased below this limit, i.e. the heat exceeds the critical heat input, a convex surface is produced on the back side of the parent metal because of the weight of the weld metal itself, see, for example, Fig. 2 for welding speeds of 70 or 50mm/min.

Figure 3 shows the corresponding relationships between the depth of penetration and welding speed for 14mm thick plate, with macrosections given in Fig. 4. The dotted line in Fig. 3 is the welding data from Fig. 1 for 6mm thick plate at 140A. These data approximately correspond to the relationship for 14mm thick plate at 180A. A similar relationship between the current/welding speed and the depth of penetration is found for 25.4mm thick plate, Fig. 5. A comparison of the data for plate thicknesses of 6 and 25.4mm at 140A welding current and 100mm/min welding speed shows that, while the 6mm plate is entirely fused, the depth of penetration is as low as 1mm in the 25.4mm thick plate.

From the above results, the relationship between the parameter[4]

$$x = 3\sqrt{\frac{I^4}{SE^2}}$$

and the depth of penetration, y, where I is welding, A, S is welding speed, mm/min, and E is arc voltage, V, is found to be as shown in Fig. 6.

Accordingly, the relationship between the parameter, x, and the depth of penetration, y, in millimetres can be given approximately by a linear equation. Based on the data for penetration depth an empirical formula was found for the two extreme plate thicknesses (6 and 25.4mm) from the method of least squares. Three points were then obtained for 14mm thick plate from the formula by direct interpolation as

$$y = (y_2 - y_1)(t - t_1)/(t_2 - t_1)$$

where, y_2: depth of penetration in mm for parameter x at 6mm plate thickness
y_1: depth of penetration in mm for parameter x at 25.4mm plate thickness
t_1: plate thickness, 25.4mm
t_2: plate thickness, 6.0mm
t : plate thickness satisfying $t_2 \leqq t \leqq t_1$
y : depth of penetration in mm for x of plate thickness t

The above three points give a straight line (shown broken in Fig. 6), which approximately corresponds to that obtained from the empirical data (solid line). Accordingly, the depth of penetration for any given plate thickness can be calculated from the above equation. For example, to calculate the depth of penetration for a given plate thickness t mm ($6 \leqq t \leqq 25.4$) the welding parameters (current, voltage, and welding speed) are first determined and the parameter x calculated. Then the depth of penetration y_1 and y_2 for the plate thicknesses 25.4mm (t_1) and 6mm (t_2) (as found by the least squares method) are calculated for that value of x. Substituting the values thus obtained in the expression for interpolation gives the depth of penetration, y, for the given thickness, t, with the welding condition, x.

Uranami bead

The relationship between the heat input and depth of penetration examined so far is applicable to the case of zero root gap in a butt joint. Figure 7 shows the limiting curves for complete fusion of the root face in 6.0mm thick plate modified with a U-type butt joint. Accordingly, from these data the welding

Table 1 Welding parameters for the root bead

	Root face mm	Welding current, A	Welding speed, mm/min	Plate thickness, mm
(a)	3.0	140	200	6.0
(b)	4.5	180	320	

Table 2 Welding condition for pipe

Weld pass	Welding method	Welding current, A	Arc voltage, V	Welding speed, mm/min	Shielding gas flow, litre/min	Weave	
						Width, mm	No./min
1	DC electrode negative TIG without filler	150	14	150	Ar 18		
2	Pulsed arc MIG-welding	90	20	280	Ar 20	5	60
3	Pulsed arc MIG-welding	100	20	280	Ar 20	8	60

conditions enabling complete penetration in the flat position can be found for a given root face or plate thickness. The shaded portion indicates the tolerance range for a uranami weld bead.

For reference, the bead appearance and cross-section for DC electrode negative TIG-welding in the flat position are shown in Fig. 8 under the conditions (Table 1) for a uranami bead for a root face of 3 and 4.4mm from Fig. 7. The surface resulting from welding under the condition (b) sometimes exhibits a cabbaging appearance. It is stated that such a bead tends to be produced when the tip of the electrode is pointed or in high speed welding.[5]

When the welding is to be performed over such a surface, grinding can be used for preferably a DC electrode positive TIG arc applied to smooth the surface as shown in Fig. 8. It can be seen that the underbead appearance is uniform and the amount of reinforcement on the back side can be easily controlled by adjusting the heat input. Figure 8c shows a micrographic section of the butt welded zone and, as can be seen, the butt joint is completely fused. That complete fusion can be obtained, even with a considerably larger root face, is one of the great advantages of DC electrode negative TIG arc welding. Experiments were also conducted on the root bead produced in the vertical (upward) and the horizontal-vertical welding position. The resultant bead appearances are shown in Fig. 9.

Application to pipes

A pipe was welded, with the pipe rotating in the horizontal position, to identify any differences in the welding conditions for the formation of a smooth uranami bead in one-side welding from those for the preliminary experiments on plates. In fact little or no difference could be detected.

For welding the pipe the root bead was formed by DC electrode negative TIG and the reinforcement subsequently built up to the specified amount using pulsed MIG. The welding parameters used are given in Table 2, together with the bead appearance, Fig. 10. It can be seen that the uranami bead in the pipe is almost as uniform as that obtained in butt welds of plate using a DC electrode negative TIG arc. Bend and tensile tests were conducted just for reference purposes only, and the results were satisfactory.

Thus it has been established that the application of DC electrode negative TIG for the root facilitates uniform uranami bead formation and that the tolerance range thereof is adequate.

It may be reasoned that an oxide film is formed when the backface is fused, since the DC electrode negative TIG arc exhibits little cleaning, and that this helps to prevent burn-through.

CONCLUSIONS

The following points have been established from an investigation of the effect of DC electrode negative TIG-welding of aluminium alloy.

1 The depth of penetration of DC electrode negative TIG arc for any given plate thickness from 6 to 25.4mm can be calculated from the formula given by interpolation if the welding conditions, i.e. current, voltage, and speed, are given

2 The formation of a uranami bead is facilitated by the use of DC electrode

negative TIG and the tolerance range in application is adequate in practice

REFERENCES

1 'Aluminium handbook'. Light Metal Inst., 677-82.

2 'Inert gas arc welding in aluminium and aluminium alloys'. Light Metals Inst., 3-4.

3 ANDO. 'Arc phenomena in welding'. Sanpo, 2nd ed., 447.

4 ANDO. 'Arc phenomena in welding'. Sanpo, 2nd ed., 361.

5 ANDO. 'Arc phenomena in welding'. Sanpo 2nd ed., 448.

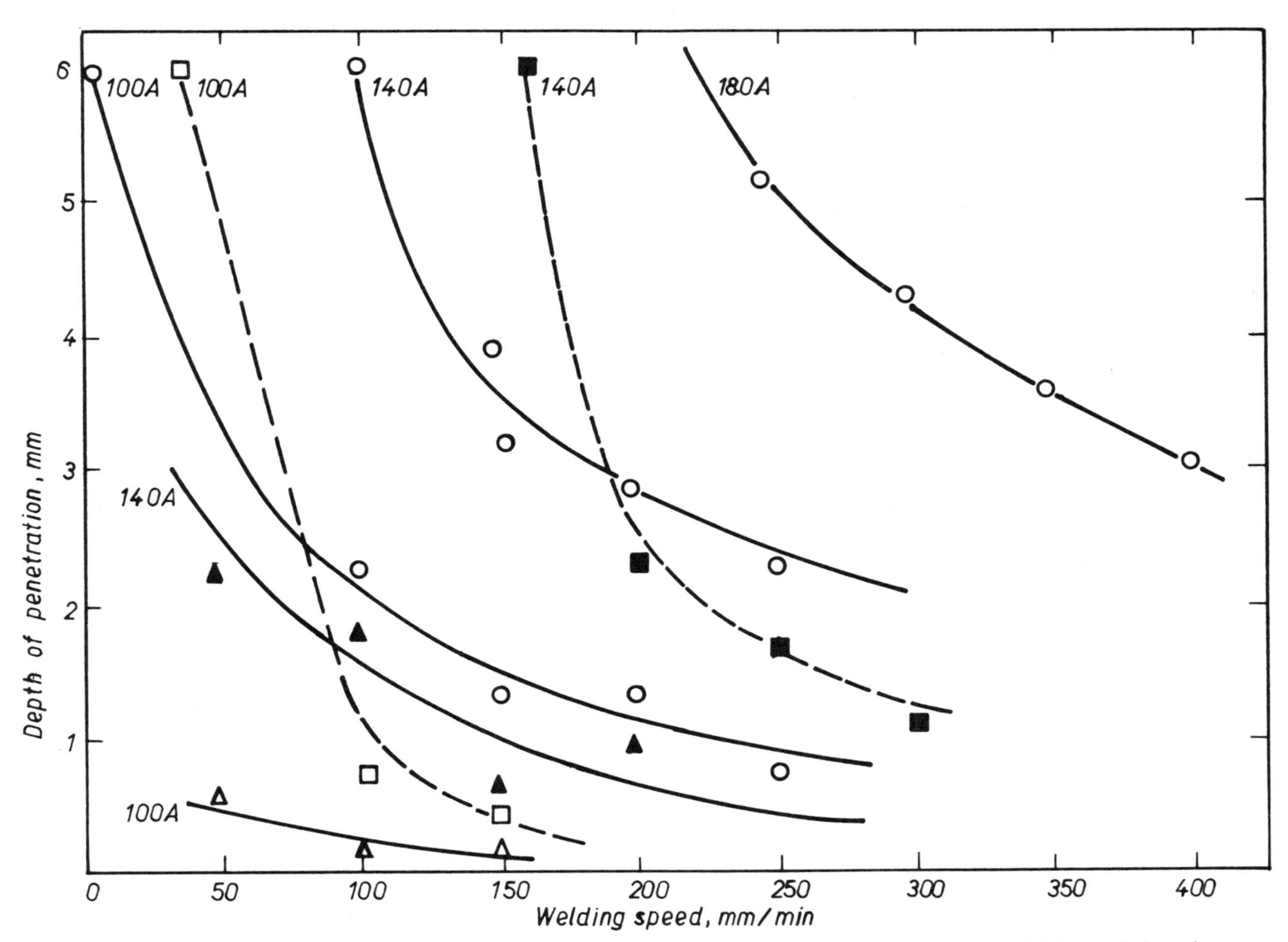

1 Relationship between depth of penetration, welding speed, and current for TIG melt runs (plate thickness 6.0mm). ○ — DC electrode negative; □■ — DC electrode positive; △▲ – AC

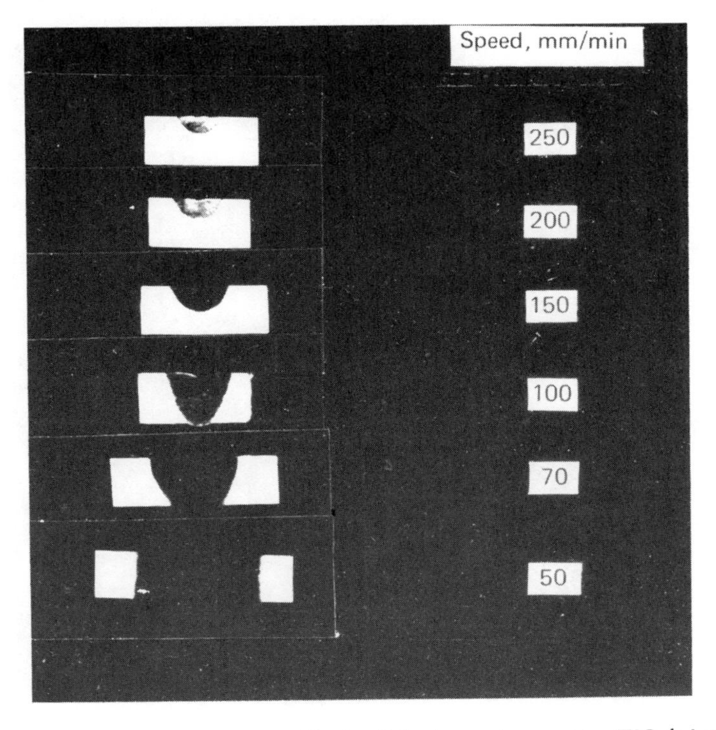

2 Effect of welding speed on penetration profile for DC electrode negative TIG (plate thickness 6.0mm)

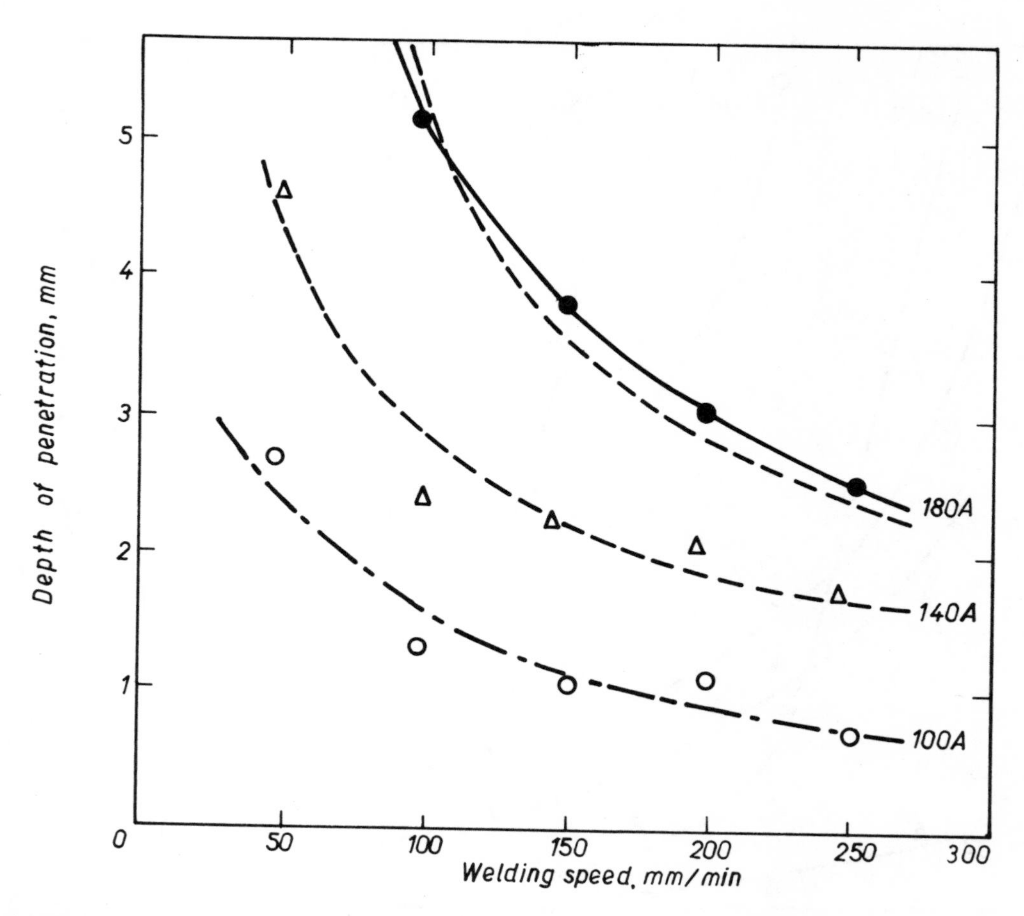

3 Relationship between depth of penetration, welding speed, and current for DC electrode negative TIG (plate thickness 14.0mm). ---- DC electrode negative at 140A, from Fig.1

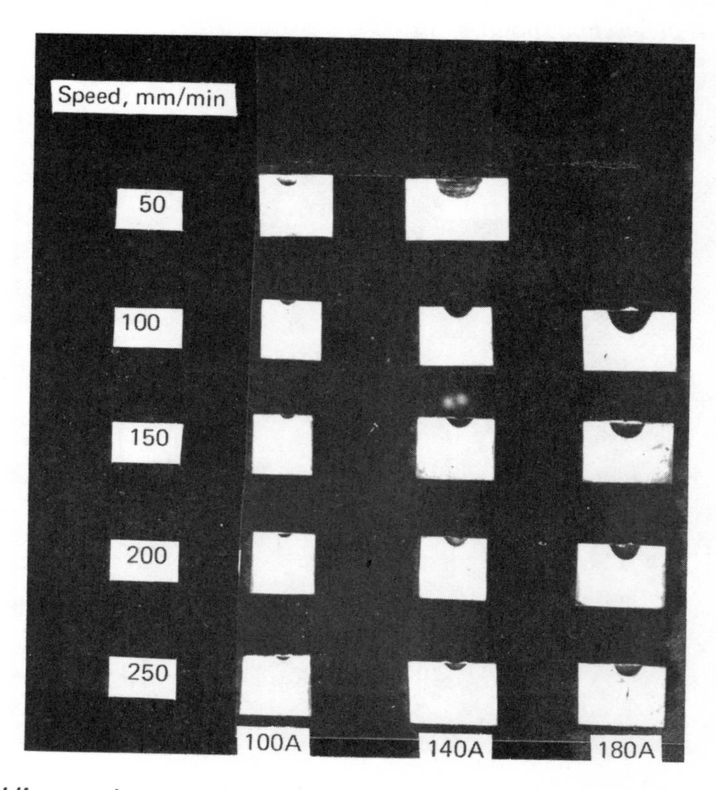

4 Effect of current and welding speed on penetration profile for DC electrode negative TIG (plate thickness 14.0mm)

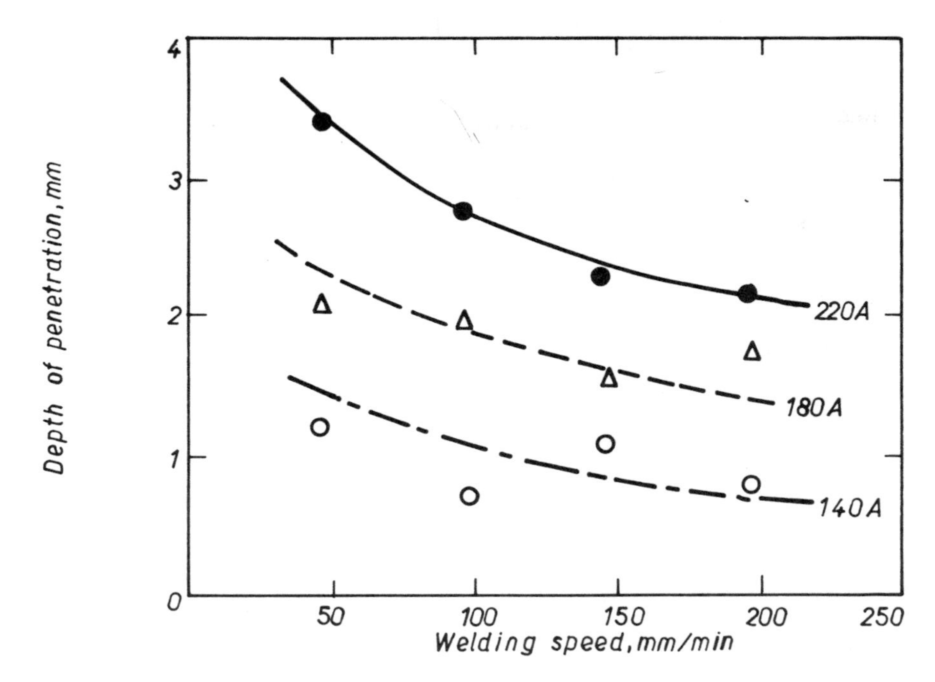

5 Relationship between depth of penetration, welding speed, and current for 25.4mm thick plate

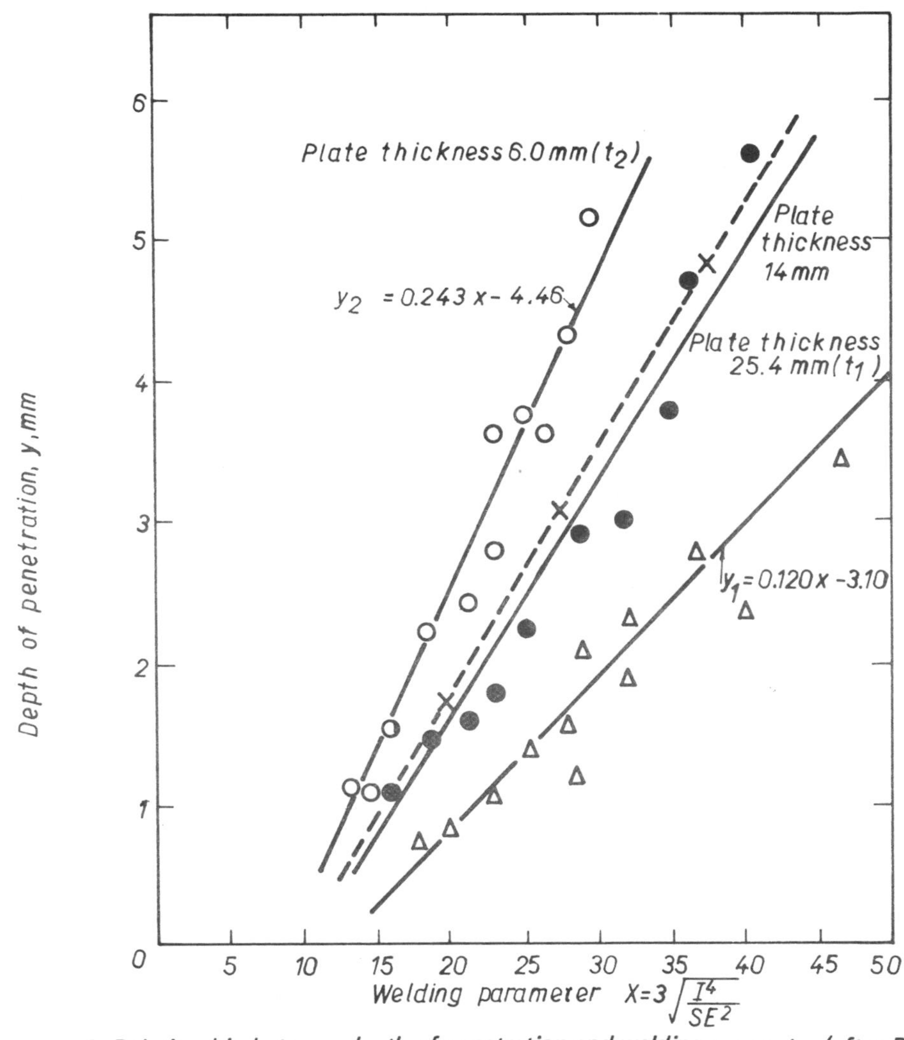

6 Relationship between depth of penetration and welding parameter (after Ref.4)

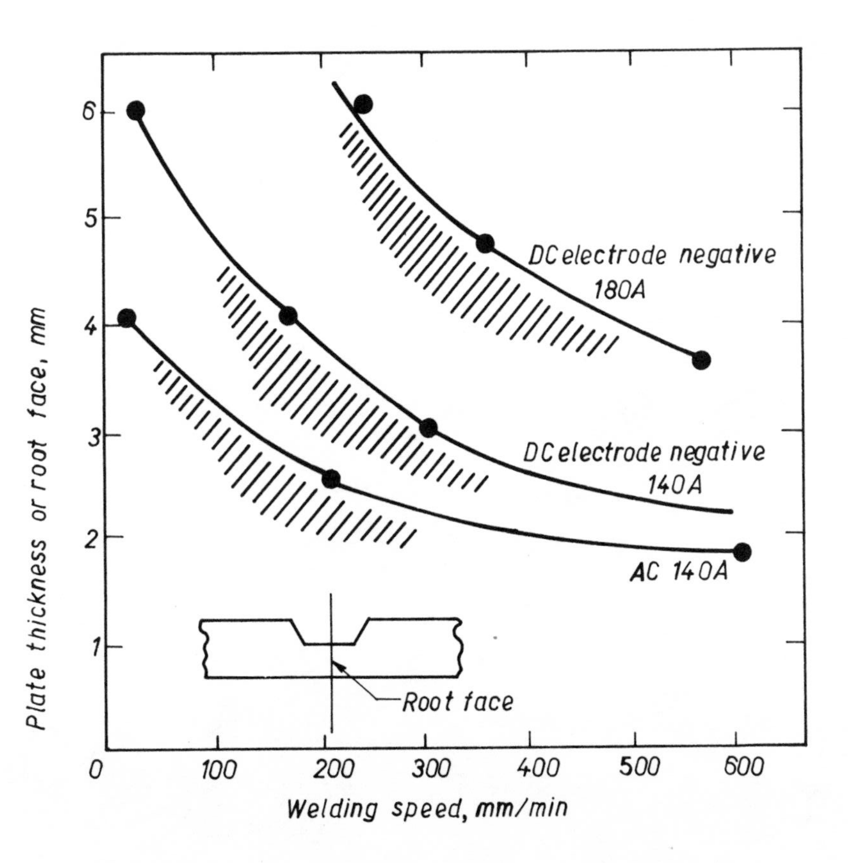

7 Welding condition for full penetration of close butt joint (plate thickness 6.0mm)

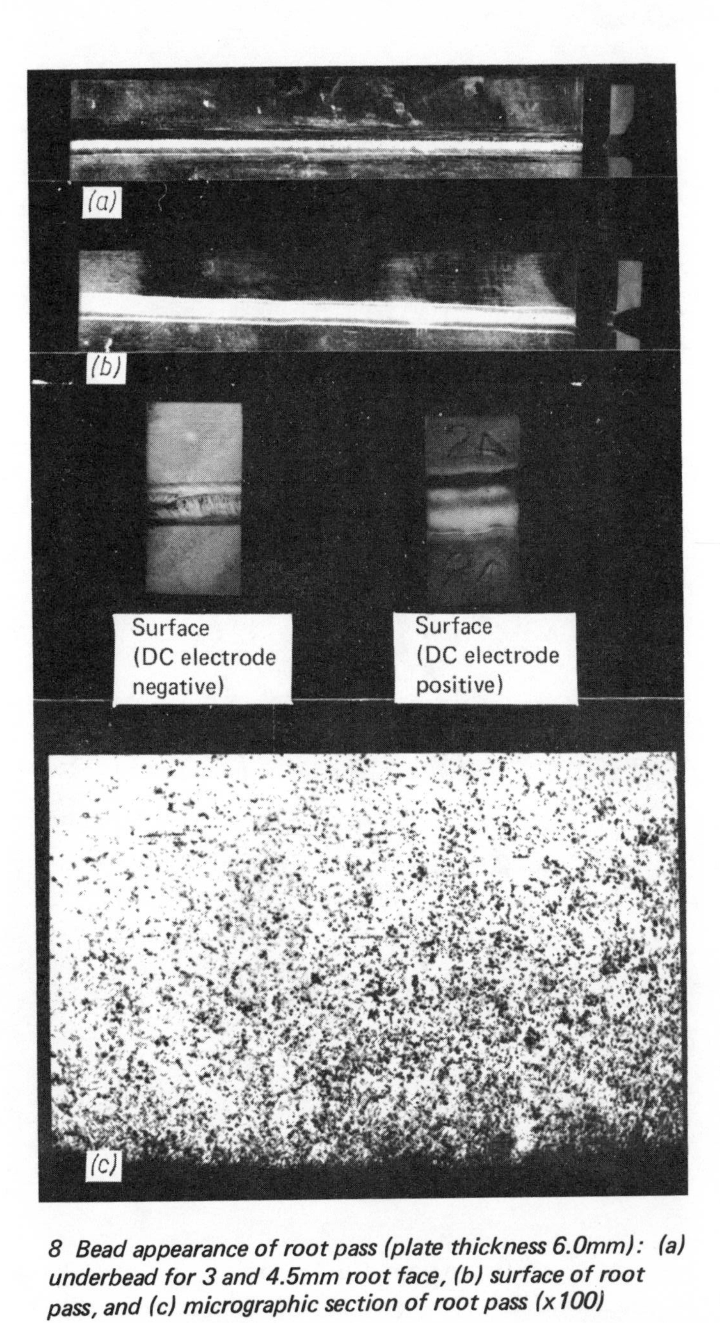

8 Bead appearance of root pass (plate thickness 6.0mm): (a) underbead for 3 and 4.5mm root face, (b) surface of root pass, and (c) micrographic section of root pass (x100)

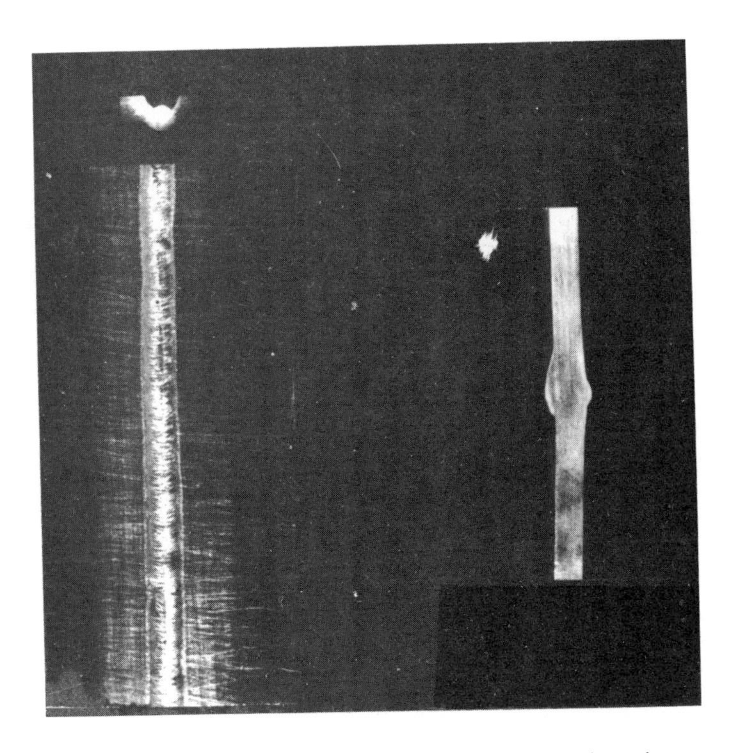

9 Bead appearance in vertical up (left) and H-V (right) positions

10 Example of application to 5083 pipe

Surface heat treatment using a plasma torch with a magnetically traversed arc

D.Goodwin, BSc (Eng), ACGI, and J.E.Harry, BSc (Eng), PhD

INTRODUCTION

Surface treatment is an important industrial manufacturing procedure enabling low cost materials to be used, often with greater ease of manufacture. An example is the heat treatment of the bearing surfaces of components in which the core remains ductile and the surface is hardened. Surface heating processes are inherently more efficient than methods involving the heating of the entire material, enabling energy savings to be made. Only a small amount of heat is required and the duration of the heating cycle is short so that heat losses are small.

Induction heating and flame heating are used for surface treatment but are not always applicable, and in practice, through heating in ovens, furnaces, or salt baths is often used. These methods are preferred where either the duration of the process is governed by surface reactions or the geometry of the component is unsuitable for induction or flame heating. Typical processes which are governed by the rate of reaction at the surface are carburising and nitriding. An example of geometries where only localised areas are required to be heated are cam shafts. On the other hand, large components such as mill rolls would require very high powers.

As well as these areas of application there are others where temperatures and power densities higher than those obtainable by either induction flame heating or ovens, furnaces, and salt baths are required. An example is the fusion of high melting point coatings. For such applications a localised source of uniform high power density, capable of being scanned over the surface, would be useful. The electron beam is capable of this but the necessity to work in a chamber at reduced pressure limits the usefulness, and the capital cost is high. Nevertheless in-line systems for continuous treatment such as in-line wire annealing have been developed. The carbon dioxide laser has also been used for surface heating at power outputs up to 10kW. However, the capital cost is very high: of the order of £250 000 for a 10kW system. The running costs are also very high since the laser efficiency is less than 20%, and the efficiency of conversion of the energy in the beam to useful heat is low because of reflection at the work surface.

Mr Goodwin is with Plasma Equipment Limited and Dr Harry is a Consultant.

The plasma torch is potentially suitable for these types of application. The running costs are relatively low and the capital cost is around £3000 for a 30kW torch. The principal advantages of the plasma torch are the high heat flux and the high temperature obtainable which enable more rapid heating and hence new treatments to be carried out. Surface power densities can be achieved from the plasma jet of the order of 10^5W/mm^2 (compared with 10^3W/mm^2 for an oxyacetylene flame) which is comparable with surface power densities obtainable from lasers.

The temperature and intensity are highly nonuniform in the plasma jet and confined to a very small zone of the order of 1mm diameter. Although surface heating would be possible at high powers it is necessary to move the torch at high velocities with overlapping paths, and even then the high radial temperature gradients are likely to cause nonuniform heating and surface damage.

THE PLASMA TORCH WITH A MAGNETICALLY TRAVERSED ARC

The torch is shown in Fig.1. A magnetic field is applied across the nozzle so that it interacts with the arc current, producing a force perpendicular to the direction of the field and current in the same sense as that for an armature of an electric motor. The field coil is wound on a laminated iron core and supplied with AC producing a field of about 0.05T at a field current of 3A. The deflection of the arc is controlled by the magnetic field strength and is limited by the length of the nozzle aperture. Maximum deflection in excess of 50mm has been obtained.

This torch has been used up to 12kW, limited only by the power available. The operating parameters appear to be relatively tolerant and no difficulty is envisaged in increasing the current or power input, or changing the gas. Virtually no erosion of the electrodes has occurred, the original electrodes still being in use. The efficiency of conversion of electrical input energy to available heat at the outlet of the torch is greater than 70%. The conversion efficiency of the available heat to that used in heating the workpiece is high for operation in the transferred arc mode, since a large part of the available power is dissipated in the anode fall region and at the anode root.

Although the velocity of the arc root on the workpiece in the transferred mode may vary, measurement of the temperature distribution at three positions (2mm below the surface of the workpiece, 10mm apart at the centre and edges of the arc traverse) showed less than 5% variation. This indicates a surprisingly high degree of uniformity within the workpiece, as a result of its thermal diffusion, and no damage to the surface occurred.

The movement of the anode root on the workpiece depends on the material and can vary from continuous movement on mild or stainless steel to an erratic sticking motion on aluminium. This behaviour has also been observed for arcs moving between parallel rail electrodes with both external and self-magnetic fields, and tends to occur over a limited range of arc velocity. Since sticking can cause surface damage some tests were carried out using a 400Hz alternator to supply the field coil. At this frequency the mean velocity for an arc traverse of 25mm was 40m/sec. A considerable improvement in surface finish was obtained.

PRACTICAL APPLICATION

A detailed investigation of the application of the torch to surface hardening cutting blades used in moving machines has been made. A cross-section showing the hardened edge is shown in Fig. 2. Hardening is normally carried out using natural gas burners followed by quenching. A similar result has been obtained with the plasma torch. Although the direct operating costs were comparable, very much higher production rates were possible with the plasma torch enabling more effective use of the work handling equipment.

Many coatings sprayed by flame, arc, or plasma torch can be fused to improve the adhesion and reduce porosity using flames. This is not possible with high melting point coatings, which often have good temperature and abrasion- and corrosion-resistant properties which cannot be fully realised because of the substrate. The uniform and controlled heating obtained with the plasma torch, together with the absence of burnt gases, enables high melting point coatings to be satisfactorily fused. Results obtained for a flame sprayed coating of Ni-Cr alloy fused with the plasma torch, Fig. 3, show no apparent porosity and good adhesion to the substrate.

DISCUSSION

Although a large number of methods of surface hardening are used, a number of problem areas exist. Moreover in some instances direct competition by the plasma system with existing processes is possible. These include surface hardening of localised zones, e.g. cam shafts, flat surfaces, e.g. lathe beds, and large structures, e.g. mill rolls. Ovens, furnaces, and salt baths heat up the entire component with consequent waste of energy and, unless protected, harden all exposed surfaces. Localised heating in small areas of irregular sections and flat surfaces is difficult and inefficient using induction heating, and very high powers are required to harden components of large cross-section.

Carburising and nitriding processes are relatively slow, particularly the latter which may typically take from 12 to 24hr. These two processes involve through heating of components and, because of the long time required, result in a considerable waste of energy. The use of the plasma torch enables the atmosphere to be chosen independently of combustion requirements.

Another problem area in spraying is the preparation of the substrate to obtain good adhesion. The use of a plasma torch to clean the substrate will remove the surface films by a process analogous to ion bombardment and can improve the adherence of coatings.

Hot machining in which hard metals are softened immediately in front of the machine tool tip is another possible area of application. Plasma torches are already used for hot turning; however, the broader and more controlled output of the torch could result in an improvement in performance and also open up the possibility of hot milling.

Continuous in-line wire annealing is a potential application. The elimination of large furnaces and the high efficiency obtainable are attractive advantages.

Table 1 Potential applications

1	Localised surface hardening or hardening of large components including nitriding and carburising
2	Fusing high melting point surface coatings to produce hard or corrosion-resistant surfaces of low porosity, and surface treatment before coating
3	Hot machining
4	Wire annealing
5	Preheating and post-heat treatment of welds and cosmetic welding
6	Edge heating, descaling, scarfing, and flash removal in steel mills
7	Coating using a consumable electrode
8	Tensioning, cleaning, and hardening inside surface of railway lines

The uniformity and controllability of the torch should enable it to be used for preweld heating and postweld heat treatment as well as for improving the appearance of a weld seam by fusing the surface.

A problem in steel rolling is rapid cooling at the edge of the strip: higher surface heating rates are required. Descaling, scarfing, and flash removal after hot sawing are also problem areas in which the torch might be advantageously used.

The torch is not suited to plasma spraying of powders. However, it is possible that it could be used for processes analogous to those used in consumable electrode spraying torches.

Another area of potential application includes tensioning of railway lines, hardening of the inner surfaces on bends, and cleaning the surface to obtain better adhesion, which at present limits the performance of high speed passenger trains on the London to Glasgow route and goods trains.

These potential areas of application, where the plasma torch can be expected to produce substantial advantages in saving energy and materials, are summarised in Table 1.

FUTURE DEVELOPMENTS

The basic system can be modified in a variety of ways. Several cathodes could be installed along a longer orifice: thus several arcs could be made to operate synchronously. These could be positioned so as to produce a line or any desired shape. A variation in which a static magnetic field causes an arc root to move continuously along an endless path has also been devised and in this form might have applications for trepanning.

The ability to operate from a TIG-welding power supply substantially reduces the cost of the process and increases its area of potential application. The feasibility of modifying a TIG torch is being investigated at present and could extend the process still further.

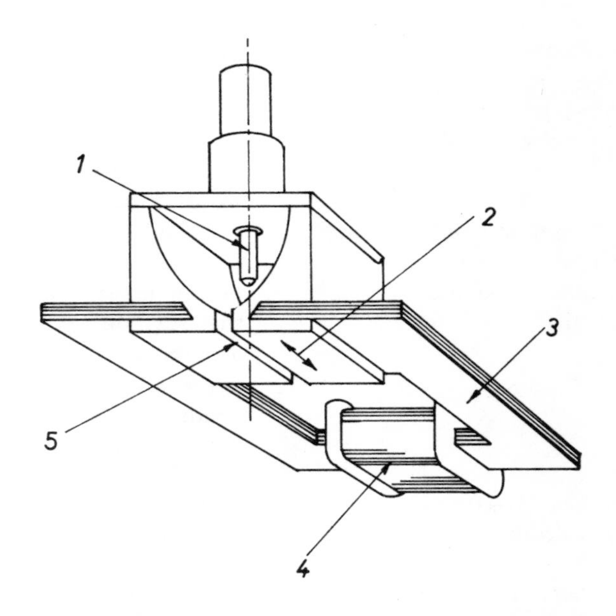

1 Schematic view of plasma torch with magnetically transversed arc. 1 – cathode; 2 – direction of arc motion; 3 – laminated iron core; 4 – field winding; 5 – nozzle

2 Section through blade showing heat treated zone

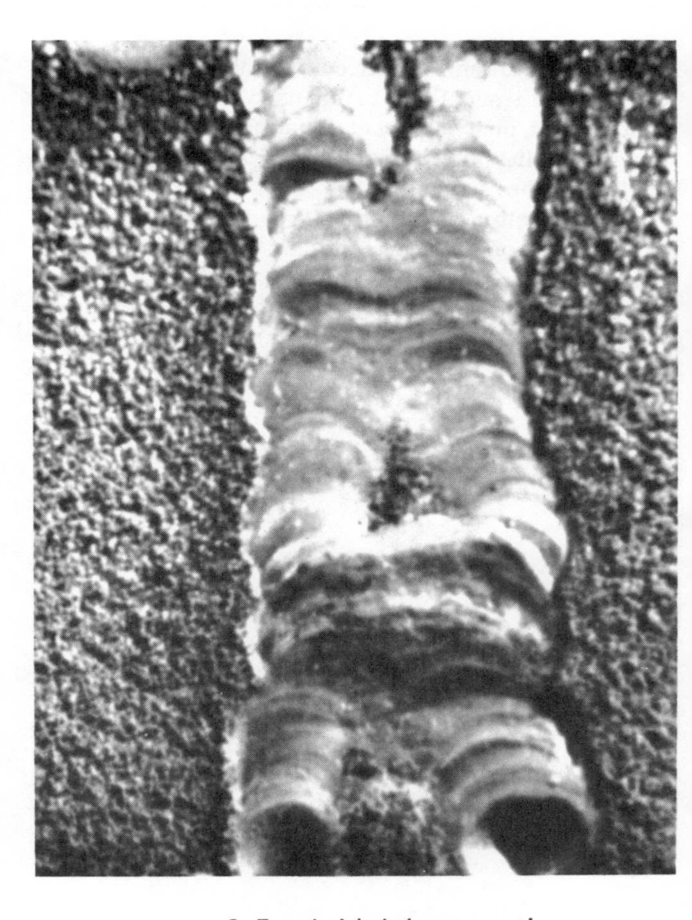

3 Fused nickel chrome coating

Energy optimisation in pulsed TIG-welding

A.P.Bennett, BSc, PhD, FWeldI

This Paper examines how the time to penetrate in pulsed TIG-welding depends upon power level and plate thickness, so as to make pulsed TIG welds with minimum consumption of energy. Good agreement is found between theoretical predictions based upon thermal diffusion from a point source of heat, and experimental measurements made in mild steel, 316 stainless, and Incolloy 800 in the thickness range ≯5mm. The practical applications are felt to be in the prediction of operating conditions which both reduce the time taken to produce welds and minimise energy usage. In thin materials the times taken are longer than point source theory would suggest, and this is tentatively ascribed to the weld pool itself acting as a distributed heat source.

INTRODUCTION

Pulsed current TIG-welding is much used in welding materials of dissimilar thickness, and where minimum energy input is desired to avoid distortion of the workpiece. Pulsed TIG has also been the basis of recent advances in automatic control of weld penetration. Bennett and Smith[1] used the method of backface viewing to control welds made in plate of various thicknesses up to 6mm. The on times were found to lengthen with thickness, as might be expected, but at the greater thicknesses (5 to 6mm) some operating conditions were found in which the on time for a single spot reached over 100sec. There was then a general overheating of the weld region and the weld pool sagged badly. Such welds were therefore slow and of poor quality. The present work examines the conditions which give rise to this combination of bad quality and bad economics so that they might be avoided in future.

The overheating was felt to stem from the excessive times taken, and it was further suspected that this was a matter of the power levels used.

Dr Bennett is a Research Officer at the Marchwood Engineering Laboratories of the Central Electricity Generating Board.

THEORY

Standard diffusion theory[2] describes the temperature field around a continuous steady point source of power, p, switched on at time zero inside an infinite medium of thermal conductivity, k, and thermal diffusivity, $\alpha = k/\rho c$. (Here ρ is density, c is specific heat, and the point source is taken to be at the origin of coordinates.) Converting to a semi-infinite medium, and using the method of images to represent a plate of finite thickness, g, the temperature at a given location in the plate is

$$T(r,\ t) = \frac{p}{2\pi k} \sum_{n=-\infty}^{\infty} \frac{1}{r_n} \operatorname{erfc} \frac{r_n}{2\sqrt{(\alpha t)}} \qquad [1]$$

where $r_n^2 = x^2 + y^2 + (2ng-z)^2$.

For a backface location z=g so the image terms combine in pairs and eq.1 simplifies accordingly. Moreover at the backface centre point (BFCP) r=z=g so eq.1 becomes

$$T = \frac{p}{\pi k} \left\{ \frac{1}{g} \operatorname{erfc} \frac{g}{2\sqrt{(\alpha t)}} + \frac{1}{3g} \operatorname{erfc} \frac{3g}{2\sqrt{(\alpha t)}} + \ldots \right\} \qquad [2]$$

Clearly the time to penetrate for a given power is the value of t which brings the BFCP to melting temperature T_m, which leads to a general relationship between power, thickness, and time to penetration.

From eq.2 the variation of temperature with time for given conditions at a given location, e.g. the BFCP, is generally as sketched in Fig.1. The problem is to find and use an identifiable feature on this curve, such as the maximum of slope, and the time at which it occurs. (This feature incidentally is also of special significance in the context of automatic control systems.) Now the temperature reached at a given time and place is proportional to the power used, so in principle power could be adjusted until the time to maximum slope is also the time to melting. In this way a pair of values t* and p* are defined which can be divided into eq.2 to produce the required general relationship.

Double differentiating eq.2 gives $\hat{t}$, the time to maximum dT/dt and substituting $\hat{t}$ into eq.2 gives the temperature of maximum slope. Finally, power is scaled to make this the temperature of melting. This gives the values $t^* = g^2/6\alpha$ and $p^* = \pi kgT_m/\mathrm{erfc}\sqrt{(1.5)} = 12.02\ \pi kgT_m$ to be substituted back into eq.2 to calculate normalised time, t/t*, as a function of normalised power, p/p*. The relationship is shown in Fig.2, which also shows the effect of power upon pt/p*t*, i.e. normalised energy per pulse. It will be noted that time to penetration falls continuously as power increases, whereas energy to penetration goes through a minimum.

EXPERIMENTAL

A TIG-welding equipment was used with a feedback control device[1] which allowed the moment of penetration to be detected automatically Single weld spots were made over a preset range of welding parameters. After each weld had been made it was examined visually, and the plate moved so that the next weld was made on a fresh portion of the plate. As far as possible all the welds for a given material were made on the same piece of plate. Currents ranged from 80 to 300A, and the on times were up to 200sec. The torch was horizontal with a 3.2 diameter tungsten and set at a 3mm gap from the vertical plate. Weld spots were not allowed to overlap and the minimum time between welds was 5min to allow equalisation of temperature. The local effective plate temperature (preheat) was 20°C. The arc efficiency was taken as 50% and material properties, averaged from 20°C to melting point, as:

	Mild steel	Stainless steel	Incolloy 800
Conductivity, k (w/mK)	24	24	20
Diffusivity, α (mm^2/s)	5	5	5
Melting point, T_m (°C)	1520	1440	1380

In Fig.3 the values of $6\alpha t/g^2$ are plotted against $\rho/12\pi kg(T_m-T_o)$, where T_o = preheat. The agreement between theory and experiment is within 50%, and as it occurs over a fortyfold ordinate range for materials as metallurgically different as mild steel and Incolloy, it is considered to be quite good, particularly in view of the assumptions implicit in the application of simple diffusion theory to welding problems.

In a supplementary series of tests on the same Incolloy plate the preheat temperature was up to 100°C. The results obtained, also shown in Fig.3, are in substantial agreement with the earlier results for plates at 20°C.

In terms of the operation of the equipment (as a control system used upon thick material) no real problems were encountered. At long on times, say >30sec, the recorder traces showed oscillations in the radiant output, and for times >100sec the oscillations grew so large that the precise moment when the control relays tripped appeared to depend as much upon the amplitude of the oscillations as upon the size of the pool itself. However, as these times are so long as to be unrealistic for practical application this is not considered to be a serious problem.

To amplify the results a second set of tests was performed to investigate the effect of thickness. Tapered plates of mild and stainless steels were used, with thickness reducing from 6.3 to 1mm over a distance of 150mm, and isolated weld spots were made as before. The results, Fig.4a, are coded to indicate the actual thickness (three ranges). The superimposed graphs s/g = 1,3,5 refer to a distributed heat source which will be discussed later It will be seen that the results for thicknesses in the range 5 to 6mm are in substantial agreement with those found earlier, but that with thinner material the times to penetration were significantly longer than expected.

In production pulsed TIG-welding a joint seam usually consists of overlapped, rather than isolated, spots, and it could be argued that this is a different case. Accordingly pulsed data for 50% overlapped welds[3] was re-examined, and overlapped welds were also produced on tapered plates prepared as above but welded using the automatic indexing facility. In this work the levels of current were set as main 160A/background 20A, background time was 2sec, and the on time regulated itself automatically. A line of about seventy weld spots was produced on each tapered plate, with a repeat distance of 2.1mm. As the weld spots produced were found to be 10 to 12mm in diameter on the thick material, falling to 6 to 7mm on the thin, this represented a considerable degree of spatial overlap. The results obtained, Fig. 4b, are coded by thickness as before, and in both instances show that for thick material the agreement with theory is good, despite the overlap which would have been expected to cause local preheat and therefore shorter on times. It should be noted that the agreement for thin material is again poor.

DISCUSSION

The present work has established a model for the growth of a weld pool based upon simple diffusion theory for a point heat source. This model is able to describe the variation of time to penetration as a function of power and thickness in materials of thickness greater than 4mm, with a high degree of accuracy. In thinner material the model appears inadequate; the reason for this will be examined later. In the thick material both theory and experiment show that the time to penetration falls sharply as power increases, and that the amount of energy needed per pulse falls over the power range studied in the present experiment. The technological advantages of a tenfold reduction in time per pulse when welding thick material will be obvious, and need not be elaborated upon. However, although the shortening of on time is seen here primarily in terms of assisting productivity and reducing energy usage, there are incidental benefits in the context of feedback control. The shortening of on time restricts the opportunity for lateral spread of heat during the pulse, and would thus appear to minimise the masking signal because of backface radiation around the emergent weld spot, which would otherwise interfere with detection of the weld spot itself.

PULSE ENERGY

Figure 3 is plotted on a log-log basis, so the fact that the slope is steeper than -1 means that over the practical range of currents the product of power and time, i.e. the energy per pulse, falls as power increases up to a level of power about $p^*/4$. (Actual power value is discussed below.) The energy per pulse is then of the order 1.5 p^*t^*, which compares very favourably with the values of 10 p^*t^* represented by points at the low current end of the ranges, say $p^*/20$. The efficiency of use of energy is revealed when the energy per pulse is compared against a minimum estimate for the energy content of the weld pool. The simplest model for the weld pool is a hemispherical volume of radius equal to plate thickness, and all on the point of melting. The energy content for this is $\frac{2\pi g^3}{3} \rho c \, T_m$, and as a simple estimate for the extra effect of latent heat this is multiplied by $\frac{4}{3}$. Dividing energy content into the energy supplied indicates that 28% is used for actual melting in the high power situation. But in the low power more energy is supplied, so that only 5% is used for melting, the remainder going into general heating, leading to widened heat-affected zones and reduced cooling rates. In addition, when welding in confined spaces the general temperature level has a significant effect upon the comfort of the operator, and hence on the quality produced. Moreover it can critically affect the reliability of the mechanised equipment which is now coming into use for bore and orbital welding. Finally, photodiodes used for feedback control of welding are themselves sensitive to temperature, so for all these reasons it is desirable to operate at the smallest possible energy per pulse.

The minimum power required may be read from Fig. 2 as $p = p^*/4 \simeq 9kgT_m$. This corresponds to a net figure of 300W/mm for pulsed welding of mild steel. The old rule of thumb for continuous TIG-welding of 1 ampere per thou (0.025mm) represents a net figure of 200W/mm (for 50% efficiency and 10V), and the agreement is considered to be reasonable in view of the different patterns of heat distribution.

Thin material

Turning now to the discrepancy between theory and experiment for thin material, although the times to penetration are longer than point source theory predicts, the actual times are generally still small compared with other process attributes such as the interpulse spacing (off time) and general handling time of equipment.

From a theoretical viewpoint the concept of a point source of heat when TIG-welding on thin material is open to question on at least two counts. Firstly, Gick et al[4] show that arc roots are typically 1mm across. Secondly, TIG weld pools are often seen to have width:depth ratios >2:1. The ratio can be as much as 5:1 and persist for up to 10sec.[5] If therefore, the argument is extended from a point source to a series of area sources, increasing the size of the heat source for a given total power, p, the power density falls and the time taken to reach T_m on the backface rises. For a disc heat source of diameter 2s the power density $F = p/\pi s^2$ and it is possible to show that the time to penetration falls as total power rises, much as was seen for the point source. The relationship[6] is

$$\frac{p}{p^*} = \frac{(s/g)^2}{2[(t/t^*)-1]}$$

It is thus possible to superimpose the graphs for a series of area sources s = g, 3g, 5g, etc. over the data of Fig.4a. When this is done the graph for s = 3g encloses most of the results for 3-4mm plate, likewise the graph for s = 5g with the 1-2mm plate. This would suggest that on an averaged basis, the effective power density is equivalent to a heat source much more than 1mm across, and possibly >10mm across.

The extended heat source could thus be taken as the flat weld pool itself, which is regarded as a disc of material of enhanced conductivity, supplied with energy from the arc but immediately spreading it to give the average power density mentioned above (cf Friedman's semi-Gaussian distribution). A difficulty with that explanation is that, unless deliberately weaved, TIG weld pools in thin plate are rarely even 10mm across, i.e. it is sometimes necessary to postulate a heat source larger than the pool which is observed. As this represents an overcorrection from a simple point source model it is clear that further work on the growth of weld pools, especially stirring and energy transfer within them, will be needed to complete the picture.

CONCLUSIONS

A relationship has been found between power, material thickness, and time to penetration in pulsed TIG-welding thick (5-6mm) steel, which is believed to have practical application in reducing the time taken to produce welds and in minimising energy usage.

The energy per pulse falls with increase of power; for steel the minimum is reached at a power level of 300W/mm. This relatively high value of power is recommended for use when it is important to minimise general heating of the weld region.

ACKNOWLEDGEMENT

This work is published by permission of the Central Electricity Generating Board.

REFERENCES

1 BENNETT, A.P. and SMITH, C.J. Welding Institute Conference 'Fabrication and Reliability of Welded Process Plant', London, 16-18 November 1976, 13-19.

2 CARSLAW, H.S. and JAEGER, J.C. 'Conduction of heat in solids', 2nd ed. Oxford, Clarendon Press, 1959.

3 BOUGHTON, P. and MALES, B.O. Welding Inst. Report P43/70; also in Welding Research Int'l, 3 (1), 1973, 47-71.

4 GICK, A.E.F., QUIGLEY, M.B.C., and RICHARDS, P.H. J.Phys.D.Applied Physics, 6 (16), October 1973, 1941-9.

5 FRIEDMAN, E. and GLICKSTEIN, S.S. Welding J., 55 (12), 1976, 408s-20s.

6 BENNETT, A.P. CEGB Laboratory Report (to be published).

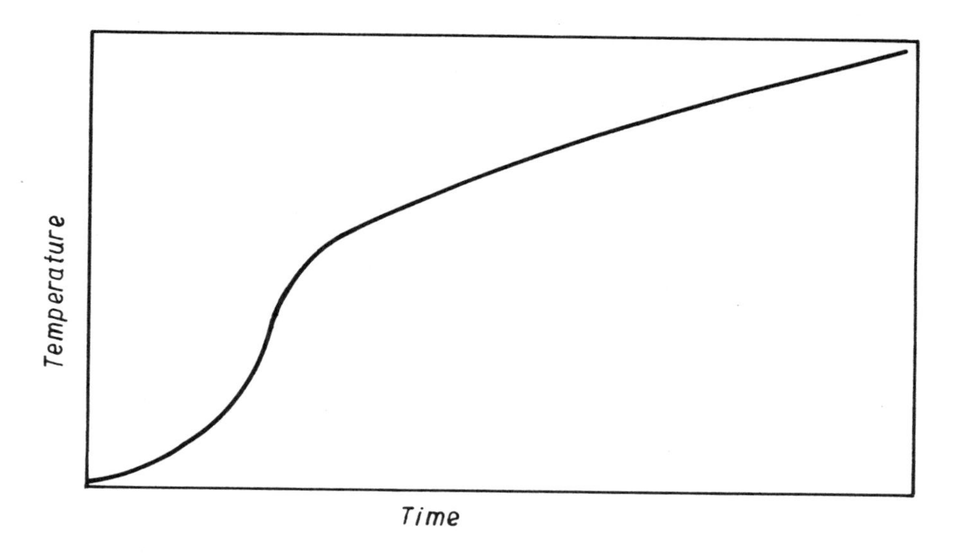

1 Time-temperature variation at BFCP (diagrammatic)

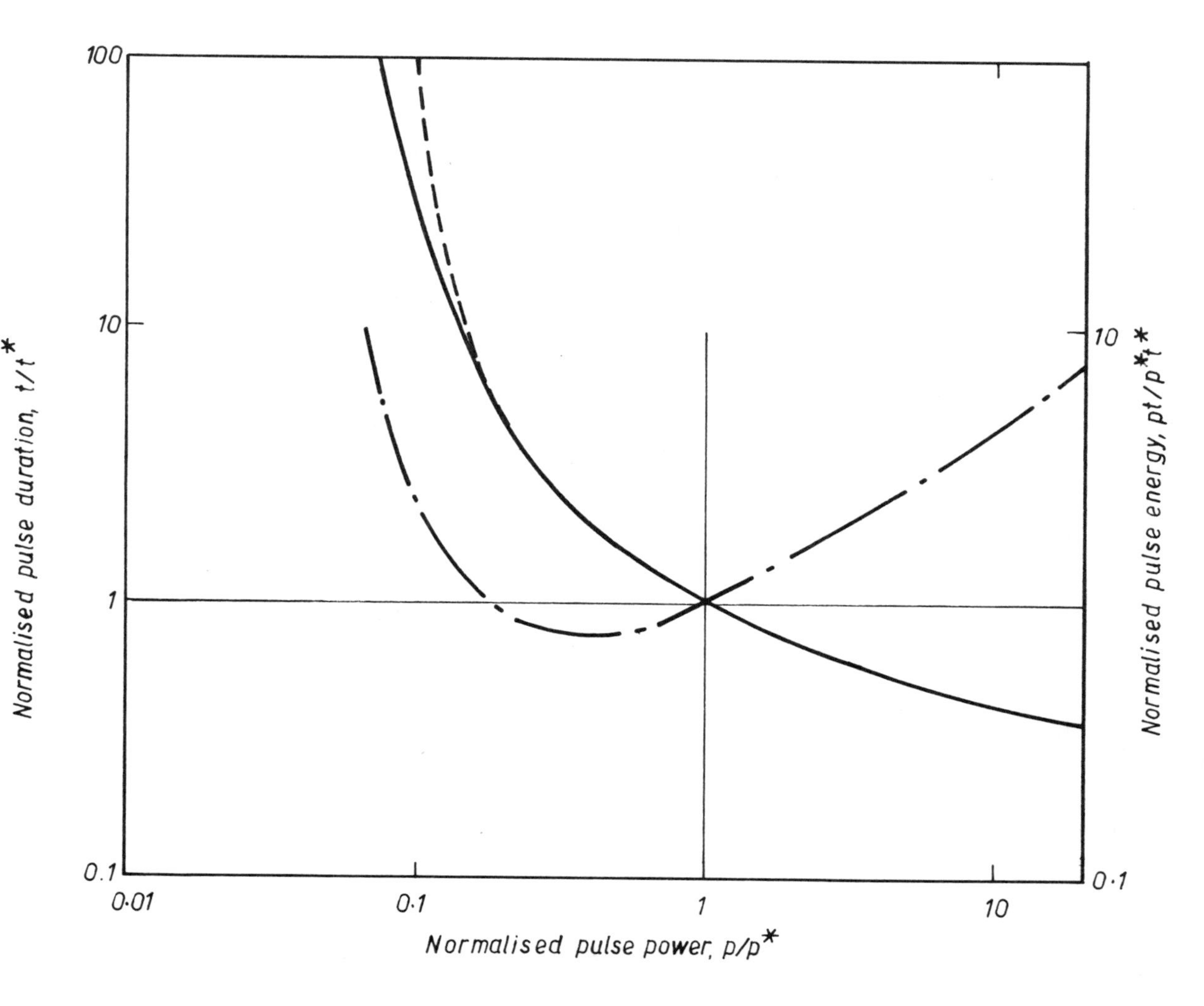

Predicted effect of power level upon time to penetration and energy per pulse (values of time etc. expressed as multiples of responding reference values). —— time, for plate thickness, g; ---- time, for semi-infinite block; —·— energy, for plate ckness, g

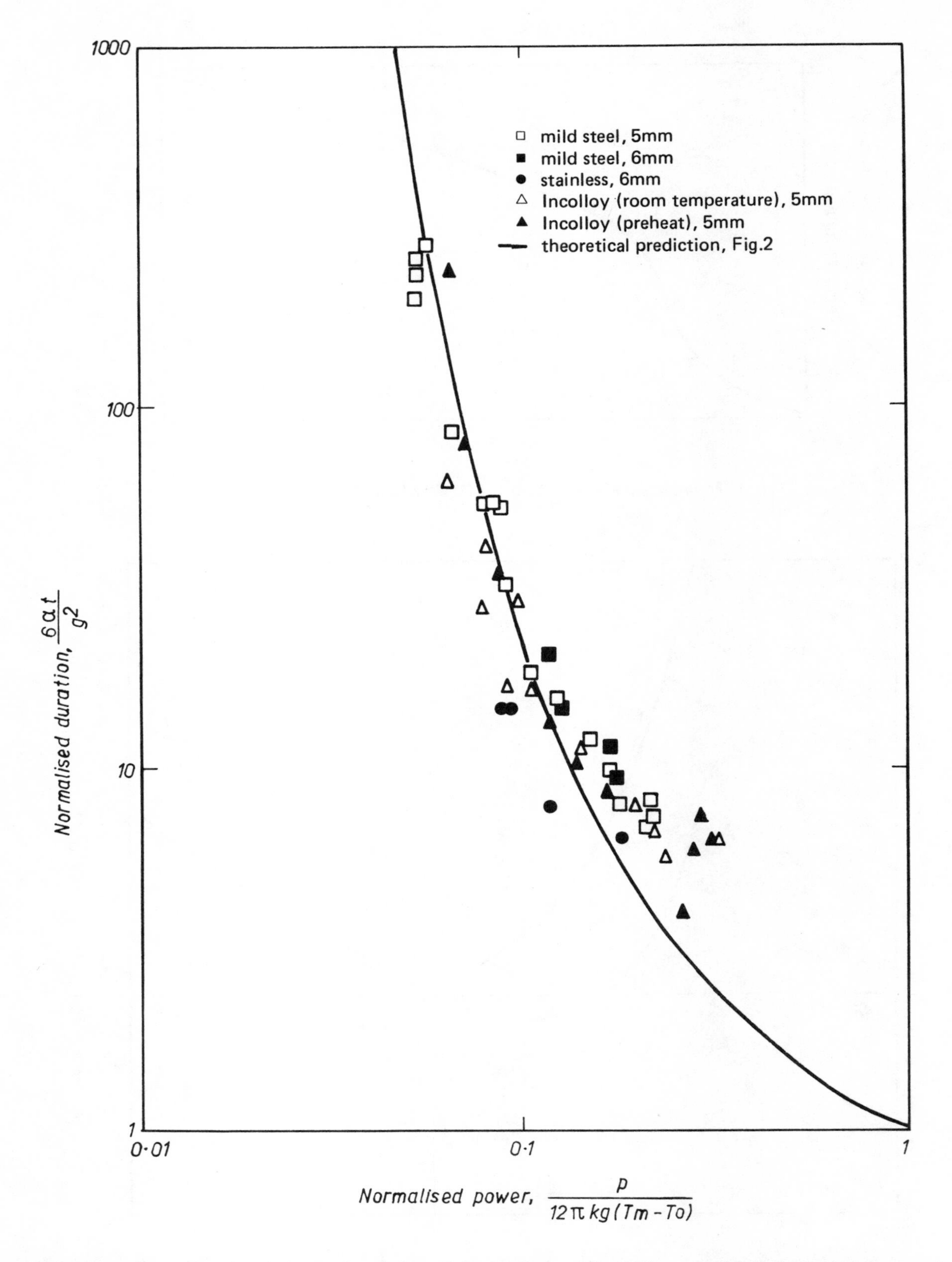

3 Comparison between predicted and observed variation of time to penetration with power for isolated weld spots and plates of material and thickness shown

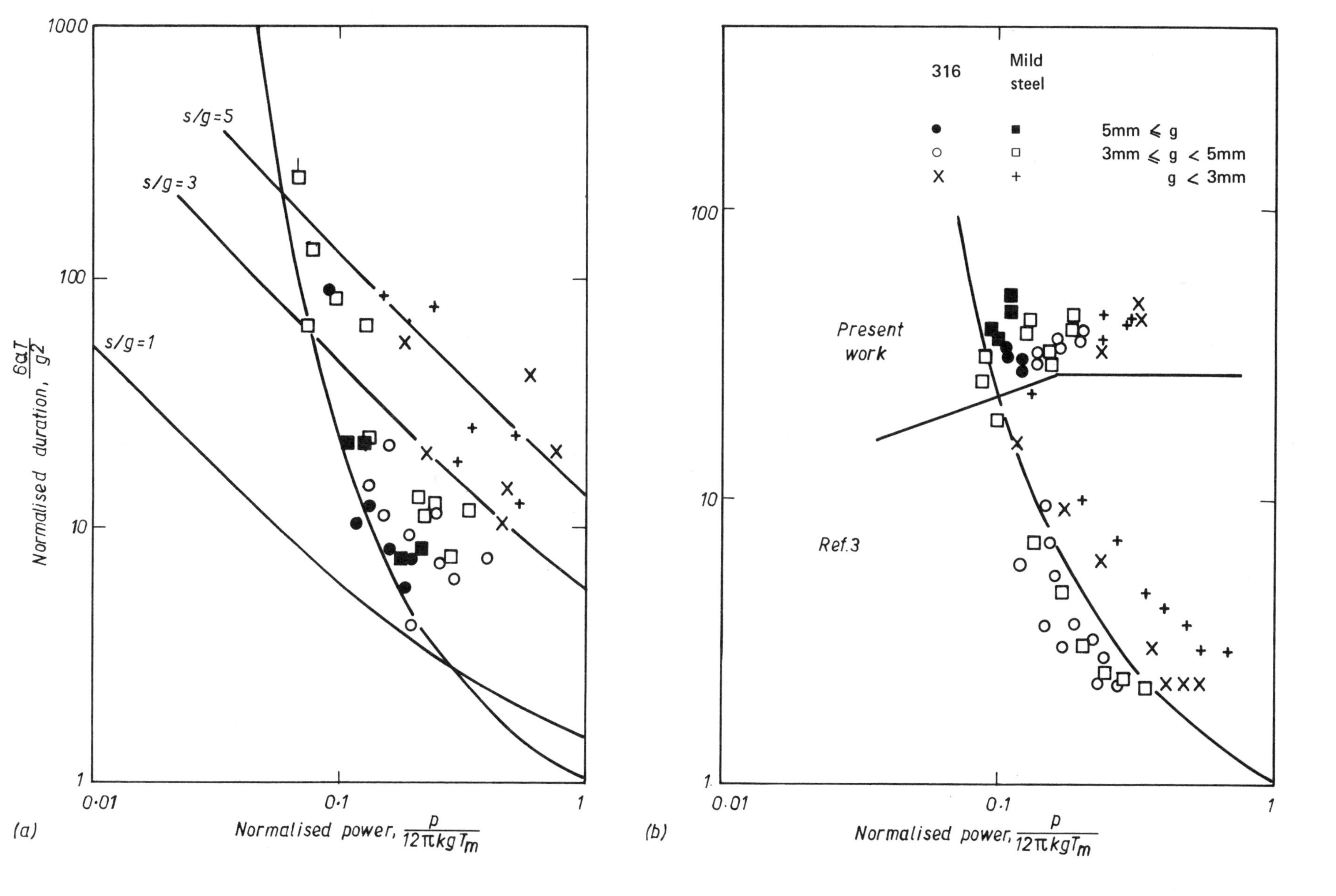

4 Comparisons between predicted and observed variation of time to penetration with power for: (a) isolated weld spots on tapered plate, (b) overlapped weld spots

Seam tracking systems with the arc as sensor

F-J.King, Dr Ing, DiplWirtschIng, and P.Hirsch, Dr Ing

For economic reasons it is continuously necessary to increase the degree of mechanisation in arc welding. Thereby automatic guidance of the heat source, along the joint, is the basic prerequisite for automation of the whole production process. Starting with the demands of industrial practice, an additional leading feeler has disadvantages. These lie in the fact that reliable functioning of the feeler can be upset by the welding process itself and that a complicated extension to the welding unit is necessary.

Using the arc itself as a sensor, it is possible (by studying the shift of the working point on the characteristics) to derive information regarding the distance between contact tube and workpiece. The vertical tracking system based on this technique, which in principle comprises a constant current regulation system, can be extended to a transverse tracking system by using two electrodes which are insulated from one another. The advantages of the twin-electrode system (which is equally suitable for both shielded arc and submerged-arc welding methods) lie in the automatic tracking of three-dimensional curved grooves and in the improved gap-bridging in root and single-pass welds made possible by the special electrode arrangement.

A tracking system based on the same principle can also be used for oscillated torches.

INTRODUCTION

Rationalisation of welding processes, with the aim of increasing their productivity, can be achieved through various technical and procedural measures. The demand for lower cost with constant effort and satisfactory quality can best be fulfilled by mechanisation and/or automation of the production process. Economic considerations, coupled with the increasing dearth of qualified labour and the disproportional increase in the ratio of labour and labour-oriented costs to the total manufacturing cost, compel this substitution in welding methods, even in such fields where at present manual methods predominate.

Dr King, Research Engineer, and Dr Hirsch, Chief Engineer, are both at the Institut für Schweisstechnische Fertigungsverfahren, Aachen, German Federal Republic.

The principal arc welding processes which profit from this development are gas-shielded welding, both CO_2 and with gas mixtures, and submerged-arc welding. The process-dependent economics of these methods, together with the possibility of recording the main welding parameters with relatively simple means, lead to a high degree of mechanisation. This is supported by the fact that fully mechanised processes, because of the possibility of developing special process variants, e.g. the multielectrode technique, guarantee substantial increases in productivity.

The quality of the weld bead, however, must not be neglected in the effort to increase productivity. Substantial economic savings can be achieved through the supervision of welding and prophylactic maintenance which can help to avoid damage. In addition, manufacturing errors which disturb the production process can cost a great deal more. The relief and replacement of supervisory staff through automatic systems thus leads to savings: directly through lower labour costs and indirectly through the assurance of constant quality by the elimination of human error.

The problems in the automation of arc welding processes begin with the transfer of the operator's functions to the machine. The manual welder, functioning as a feedback controller in the partly mechanised gas-shielded arc process, registers the process visually and can adapt to the changing kerf conditions by skilful manipulation of the torch. Even when fully mechanised equipment is used the operator is integrated in the welding process as supervisor and controller.

One of his tasks is the supervision of the correct position of the torch relative to the weld groove. There can be many reasons for the welding torch not being in the centre of the groove. These are, for instance, the geometry of the workpiece, inconsistent weld preparation and poor fitup, and distortion arising from heating, especially in long seams. These can result in defects and make weld dressing necessary.

CLASSIFICATION OF SEAM TRACKING SYSTEM

For these reasons there is a pressing demand to relieve the operator from the task of position correction and to substitute an automatic tracking system.

The difficulties in designing a torch-guiding system are only to a lesser extent concerned with the positioning units. More problematical is obtaining time- and position-based information which can be easily processed to enable exact positioning of the torch. In most situations numerical control is unsuitable since it works only within mathematically defined groove geometries, and the demands on the quality and reproducibility of groove preparation, welding setup, and clamping jigs are very high. This means that the position signals must be obtained either from suitable sensors or derived directly from the parameters of the welding arc.

The development of such sensors is thus the chief problem in the design of seam tracking systems, and intensive work has been carried out in this field in many institutions. A guide to the possible measuring principles suitable for registering the centre of the groove is given in Fig.1. A sensor system which fulfills all demands regarding simplicity, reliability, robustness, nonwearing, and precision equally well for all applications is almost impossible to realise. This is also one of the reasons why there is such a discrepancy between the demand for widespread application and the actual use of such systems.

Special problems resulting from the specific conditions around the arc zone, Fig.2, when indirect position determination is used necessitate placing sensors at some distance from the arc. A seam tracking system with additional sensors for the determination of its momentary position is always subject to difficulties. These problems concern the disturbance of the system through both the joint preparation and the process itself. Thus, for a bead with a small radius of curvature, it is necessary to store the positioning signals. This leads to a relatively complicated design for the welding head, involving separate slides for the sensor unit and welding torch. For this reason it is only logical to use the welding arc itself as a sensor to determine its position relative to bead centre for the purpose of exercising direct control.

Using the static and dynamic characteristic of the arc and its performance characteristic, information can be derived regarding its position relative to the centre of the groove.

MOVEMENT OF THE WORKING POINT IN ARC WELDING

In arc welding processes with consumable electrodes the arc parameters are subject to fluctuation caused either by the process itself or by external disturbances. Thereby in high efficiency welding processes the maintenance of a constant distance between electrode and workpiece is of particular importance. Because of this the power sources used generally have either a slightly drooping or a slowly rising characteristic. These have a typical self-regulating effect as regards arc length. With a constant electrode feed rate a state of balance is maintained through the acceleration or deceleration of the burnoff rate. The dynamic characteristic of the power source is thus of particular importance to the degree and quality of self-regulation. With suitably adapted power source and welding arc characteristics an arc length stability with transient time constants of a few milliseconds are achieved.[1]

Investigations regarding the behaviour of the gas-shielded arc, caused by changes in distance between workpiece, weld pool, and torch, showed that, after the compensation process via the burnoff rate, the original operating point on the static characteristic is not recovered. The remnant displacement of the static operating point stems from the change in the resistance of the section between contact tube and weld pool comprising the electrode extension and arc length.

When traversing a rectangular step on the workpiece the arc length changes suddenly and the operating point shifts from A_0 to A', Fig. 3. This increase in arc length causes a reduction in current. Owing to the effect of the self-regulating process a new operating point, A_0', is taken up. At the end of the step slot the process is reversed, but similar, and the original operating point, A_0, is recovered. Because of the deep penetration and the partly leading position of the weld pool the change in arc length at the edges of the step is not sudden but delayed, so that the dynamic excursions of the arc and those of the power source are only partly effective.[2] The gradual changes in torch to workpiece distance, as they occur under practical conditions, cause slow continuous changes in arc length. The relatively fast self-regulating process is then almost coincident and only the remnant current deviation, $\Delta I_{t\to\infty}$, can be determined with reasonable effort. This deviation, $\Delta I_{t\to\infty}$, is caused by the changes in length of the electrode extension and that of the arc.

The remnant change in the arc length is, compared to that of the electrode extension, very small and constitutes only about 10% of the total change in distance. The actual remnant current deviation per millimetre change in workpiece distance is indicated in Fig. 4. In the range investigated, depending on the shielding gas used, the remnant current deviation, with reference to the nominal current setting, lies between approximately 1.4% (for Corgon) and approximately 2.4% (for CO_2) per millimetre change in distance.

It is thus logical to use the welding current instead of the arc length as the controlled variable with respect to disturbances which cause current variations, in spite of the self-regulating process. Changes in the distance between torch and workpiece have a particularly relevant effect on the welding current and thus on the penetration depth which can lead to defective welds, especially when making root passes.

CONSTANT CURRENT CONTROL BY VERTICAL TORCH POSITIONING

When welding, for instance three-dimensionally curved profiles, as encountered in shipbuilding and tank construction, holding a constant distance between torch and workpiece is thus a prerequisite for keeping the current constant. This can be realised by an automatic vertical torch-positioning system which uses the welding current as a reference value, as shown schematically in Fig. 5. The welding current is measured, using a shunt, and the signal amplified by an operational amplifier. This actual value is compared with an adjustable desired value and the difference used to trigger a solid state regulator. When, because of a change in the distance between torch and workpiece, a remnant current deviation occurs, a difference between the long-term and instantaneous values is registered. This is used to control the torch-adjusting motor so that the set current value is attained.

Since, with the vertical regulation of the torch, holding a constant preset current is of decisive importance to weld quality, it is logical to express the follow-up precision in terms of current deviation. Figure 6 shows the follow-up precision which can be achieved on a plate inclined at 15°. According to the sensitivity of the three-point automatic regulator for this weld setting, with threshold values of ± 4A, the mean value of the current changes is the same, independent of the shielding gas used.

Taking the current deviation of 1.4 to 2.4% into consideration, as determined with different shielding gases, this means a follow-up precision of the torch of ±0.8mm at 200A welding current.

To avoid misinterpretation of process-dependent current fluctuations, e.g. stemming from droplet detachment and transfer and weld pool movement, which would cause a false reaction of the control circuit, the signal is passed through a 20Hz low-pass filter. However, slow disturbances, even when not caused by distance changes between torch and workpiece, are registered by the control unit when the signal is processed in this manner.

Applying this technique to multielectrode processes it is also possible, by evaluating the signals from the two arcs, to gain information regarding the position of the torch relative to the centre of the groove.

TWIN-ELECTRODE WELDING

When welding with two electrodes in a tandem arrangement the main advantage is the higher welding speed which is made possible by the higher deposition rate. The use of two transverse electrodes offers further advantages. Because of the position of the two electrodes (next to each other in the groove) the arcs burn on the groove faces. Thus the pressure of the welding arc and the greatest heat concentration at the weakest point of the groove — its centre — are avoided. On the other hand, it is ensured that the groove faces are fused. This results in improved gap bridging and, at the same time, a reliable fusion of the groove faces. The liquid weld metal runs together in the centre of the groove, behind the arc, and forms a consistently good root bead.[3] Since here even larger gaps together with a suitable backing — within certain limits — are not critical, it is possible with mechanised gas-shielded processes to reliably weld the root pass, which is normally problematical.

A prerequisite for this is, however, that the twin-electrode system precisely follows the centre of the groove. Since, when welding with twin electrodes, a quasi-stable condition is achieved through the two arcs burning on the groove faces, it is logical to use their data for continuously determining the torch position.

The knowledge and experience regarding the shift of the operating point, as a result of changes in distance between torch and workpiece, can be directly applied to the lateral displacement of a twin-electrode torch with transverse wire arrangement. When the torch is moved from the centre of the groove, the distance between contact tube and workpiece increases for one electrode and decreases for the other. As shown in Fig. 7 this causes the operating points to shift on their respective characteristics. The reason for this lies in the changed distances between contact tube and workpiece.

Starting with state A, in which both arcs have the common operating point $A_{1,2}$ with the torch in the groove centre, both arcs experience a sudden change in length when the torch is moved from its central position. The changes in length are, however, in opposite directions and cause, in one instance, the operating point to shift from A to A . This leads to a current increase for arc 1, which has become shorter, and to a current decrease for arc 2, which becomes longer. Because of the effect of the self-regulating process the new stable operating point is A''. As already mentioned this process requires only milliseconds, and, since changes in the position of the torch normally occur gradually, the changes in arc length are further slowed down by melting and fusion of material and are thus relatively slow. Thus, for practical purposes, only the remnant current deviation, $\Delta I_{t\to\infty}$, can be evaluated.

The vertical regulation process, with the twin-electrode torch, is in principle the same as that described earlier for the single electrode process with constant current regulation.

For test welding, a twin-electrode torch was designed in which the electrodes were insulated from one another, but with synchronous feed rates and a common gas shielding cup. As shown in Fig. 8a, the potential difference, u_1 and u_2, corresponding to the individual component currents, I_1 and I_2, are registered with suitable shunts and used for the continuous torch-guiding system. To determine the transverse position of the torch the component currents are subtracted from one another, and for the vertical position they are added. Through comparison with a given current, corresponding to that of operating point A_{tot}, the torch is guided at a constant distance from the workpiece surface, which means that, even with large changes in this distance, the total welding current is constant.

Figure 8b indicates the principles of the regulating unit. The sensitivities of the three-point controls for Right/Left and Up/Down can be set separately on the threshold levels for each of the four directions. The indicating instruments can be switched over from I to I_0 or from $I_{tot.}$ to $I_{tot.\,set}$. For certain applications in which for technological reasons the component current of each electrode is different, the set value for groove centre can be adjusted with a zero offset, so that, despite this, the torch remains in the centre of the groove. The height-regulating motor is triggered via the potentiometer I_{set} for the total current.

Both arcs can be regarded as movable, current-carrying, electrical conductors and as such their magnetic fields must have an effect on each other. When both electrodes have the same polarity the arcs are attracted to one another because of the resulting force. For constant arc parameters this force is a function of the distance between the electrodes. Tests with varied inter-electrode distances showed that, with distances of less than 5 or 6mm, the arcs have a common root. This does not affect the functioning of the seam tracking system but is detrimental to both gap-bridging and tracking precision. Furthermore, when welding with CO_2 the detaching droplets come into short-circuiting contact so that, in some instances, no difference signal can be determined.[4]

The tests for the gas-shielded and

submerged-arc processes were carried out using respectively one or two power sources with a slightly drooping characteristic. This allowed the use of either a twin-electrode torch with a twin-roll electrode feed system or two separate torches, each with its own electrode feed unit. Welding could then be carried out with either one or two power sources.

The optimum distance between the electrodes, for electrode diameters from 0.8-1.2mm, lies between 8 and 10mm. The maximum separation is 12mm where for non-penetrating welds at high speeds two separate beads are obtained. However, when welding under the same conditions in a groove, a common weld pool is obtained. Nevertheless, a reliable root pass is possible only with gaps of less than 2mm.

To obtain perfect root beads in narrower grooves with optimum electrode separation, as required for arc formation and metal transfer, the torch was rotated so that one electrode lags the other and reduces the effective separation. This is of special importance in multi-pass welding, since the electrode separation can be adapted to the increasing width of the groove as welding progresses.

Figure 9 shows a macrograph of a multi-pass weld preparation: 60°-V; zero gap; no root face; sheet thickness 18mm. The root pass and filler layers were welded under Corgon and the final layer with the submerged-arc process. The results of the most recent tests carried out in industry showed that, for short arc welding with CO_2 with 70A/electrode, diameter 1mm, the use of two power sources has a beneficial influence on the process. The presence of a reactance coil in each circuit has a similar effect.

Twin-electrode welding is also of advantage for the submerged-arc process in that it can expand its field of application as shown by the development of the Tiny-Twin method, where two thin laterally displaced electrodes are fed through a common torch. Investigations on the performance of the twin-electrode tracking system, using different fluxes, showed that the precision of the vertical and lateral tracking systems was similar to that determined with gas-shielded arc welding.

In fillet welding with two electrodes these are normally in tandem arrangement because of the greater productivity. In the studies carried out they were laterally displaced, and the maximum angle of rotation, δ, that may be used without impairing the functioning of the twin-electrode tracking system was investigated. Figure 10 shows a horizontal fillet weld with a twin-electrode torch. The electrode diameter is 1.6mm and the distance between the electrode 12mm. To demonstrate the performance of the lateral and vertical tracking systems the web plate was welded in an upward sloping position. With this distance between the electrodes the transverse tracking system functioned perfectly up to an angle $\delta = 75^{o}$ to 80° between the electrode axes and the welding direction.

The tracking precision, which possibly depends on the shielding gas or flux used and the groove angle, lies at ± 0.4mm. To integrate this seam tracking system without major modification into the multiwire welding methods generally used in the production of large diameter pipe, tanks, and shipbuilding, it is necessary to prove its ability to function with these various technologies as depicted in Fig.11. The variant in Fig.11a is that mainly dealt with up to now and has been successfully tested with both the shielded arc and submerged-arc processes. Providing that (with a suitably designed feed unit) the same feed rate can be guaranteed for both electrodes, it is in principle possible to use either a twin-roll or two separate rolls for the wire feed system. In the variant in Fig.11b, welding was carried out with thicker electrodes up to 3.0mm diameter and with a distance of approximately 30mm between electrodes. The variant in Fig.11c was used in conjunction with two DC power sources with synchronised output characteristics. No difficulties regarding the function of the tracking system were encountered, both with like and unlike electrode polarities. An alternative to twin-electrode AC welding is offered by the variant in Fig.11d in which the leading DC torch is replaced by a twin-electrode torch, which takes over its technological functions and in addition permits both transverse and vertical guidance of the complete torch system.

TWO AXES SEAM TRACKING IN TORCH WEAVING PROCESSES

A further possible application of the 'sensing' arc is shown in Fig.12. In certain shielded arc welding applications, e.g. vertical-up welding, the torch is oscillated. The welding units used can be equipped with a two-coordinate tracking system without much expense. Usually the movements made in manual welding are simulated by a triangular movement of the oscillating units used. For this purpose, all necessary parameters such as oscillation amplitude, frequency, and the separate dwell times in the groove centre and on the groove faces, can be set.

Because of the slow welding speed and the low oscillation frequency it can be assumed that the self-regulating process is in sympathy with the oscillating movement. Thus with the torch deviating from its ideal position the remnant current deviation, $\Delta I_{t \to \infty}$, can be used for taking corrective action. This signal is processed further in the following manner: two ideal current values are selected, one for each groove face, and a lower value for the centre of the groove, and compared with the actual value synchronously with the oscillation. If the oscillation is slow (and dwell times correspondingly long) adjustments can be carried out before the next comparison is made. If this is not the situation, two consecutive groove face signals must be compared with one another, and, as in twin-electrode welding, the value obtained by subtraction can be evaluated as a measure of the transverse position of the torch.

The application of this measuring technique to gas-shielded arc welding is of special interest since, as with multielectrode processes, there are enough practical applications where the torch is oscillated for technological reasons alone. In both twin-electrode welding and mechanical oscillation no additional interference with the process is necessary to obtain the values required for vertical and transverse seam tracking.

The object of the current research work is to study the suitability of these systems in short arc gas-shielded welding and submerged-arc welding with AC.

REFERENCES

1 MUNSKE, H. 'Kritische Betrachtung der Eignung von Stromquellen und Steuergeräten für das Lichtbogenschutzgasschweissen unter argon, CO_2, und anderen Schutzgasen mit selbstabschmelzender Electrode'. Fachbuchreihe Schweisstechnik, 17, 1962, 108-14. Düsseldorf, DVS.

2 METZKE, E. and WENDLER, H-D. 'Das Verhalten von Schutzgasschweisslichtbögen bei plötzlichen Längenänderungen'. Schweisstechnik (Berlin), 19 (4), 1969, 154-9.

3 DANE, K. 'CO_2-Zweidrahtschweissung'. ZIS Mitt., 15 (5), 1973, 563-7.

4 SPITSYN, V.V. 'Metal transfer and arcing in the twin arc CO_2 process'. Welding Production, 16 (4), 1969, 7-11.

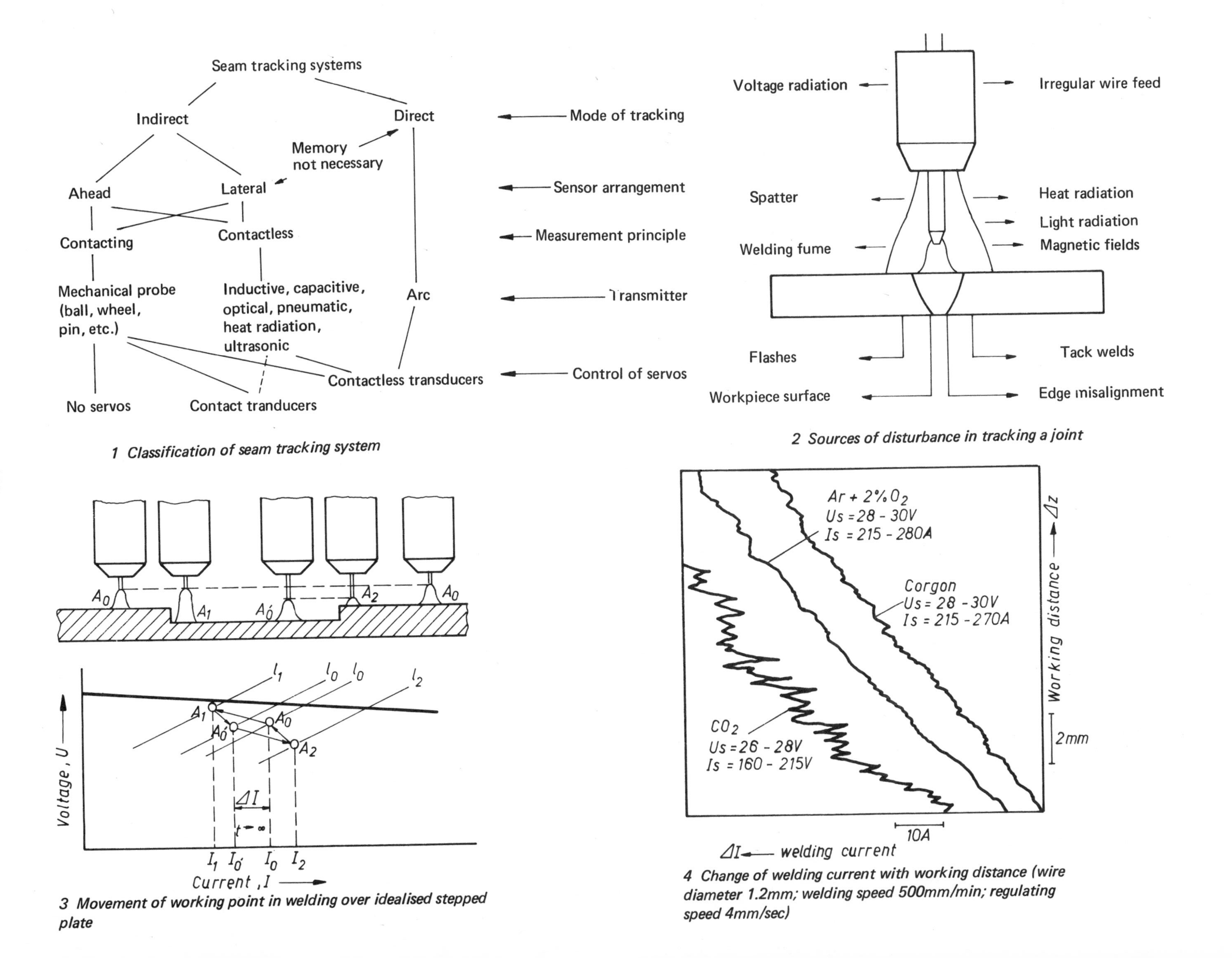

1 *Classification of seam tracking system*

2 *Sources of disturbance in tracking a joint*

3 *Movement of working point in welding over idealised stepped plate*

4 *Change of welding current with working distance (wire diameter 1.2mm; welding speed 500mm/min; regulating speed 4mm/sec)*

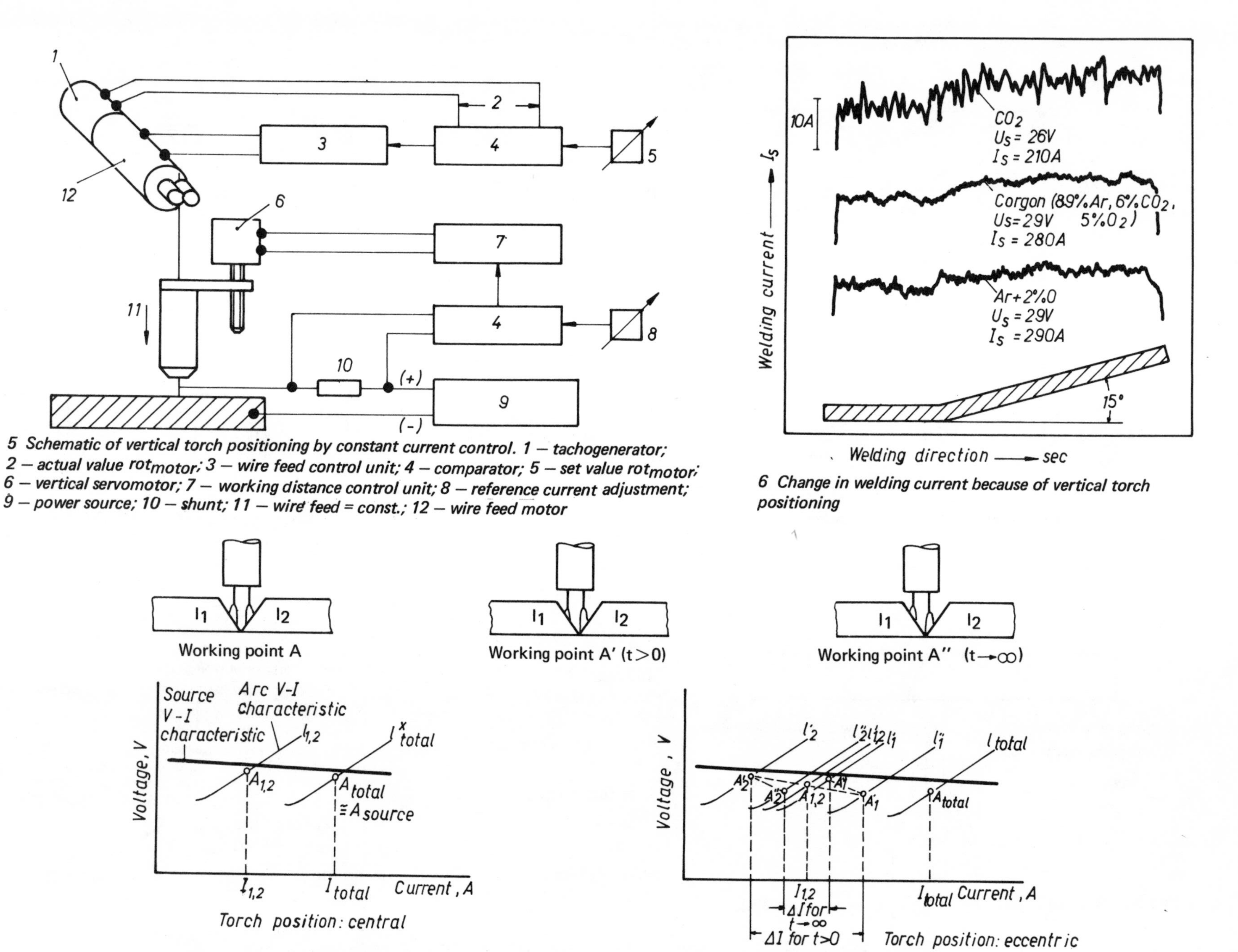

5 Schematic of vertical torch positioning by constant current control. 1 – tachogenerator; 2 – actual value rot_{motor}; 3 – wire feed control unit; 4 – comparator; 5 – set value rot_{motor}; 6 – vertical servomotor; 7 – working distance control unit; 8 – reference current adjustment; 9 – power source; 10 – shunt; 11 – wire feed = const.; 12 – wire feed motor

6 Change in welding current because of vertical torch positioning

7 Movement of working point as result of displacement of twin-electrode torch with insulated wire guides

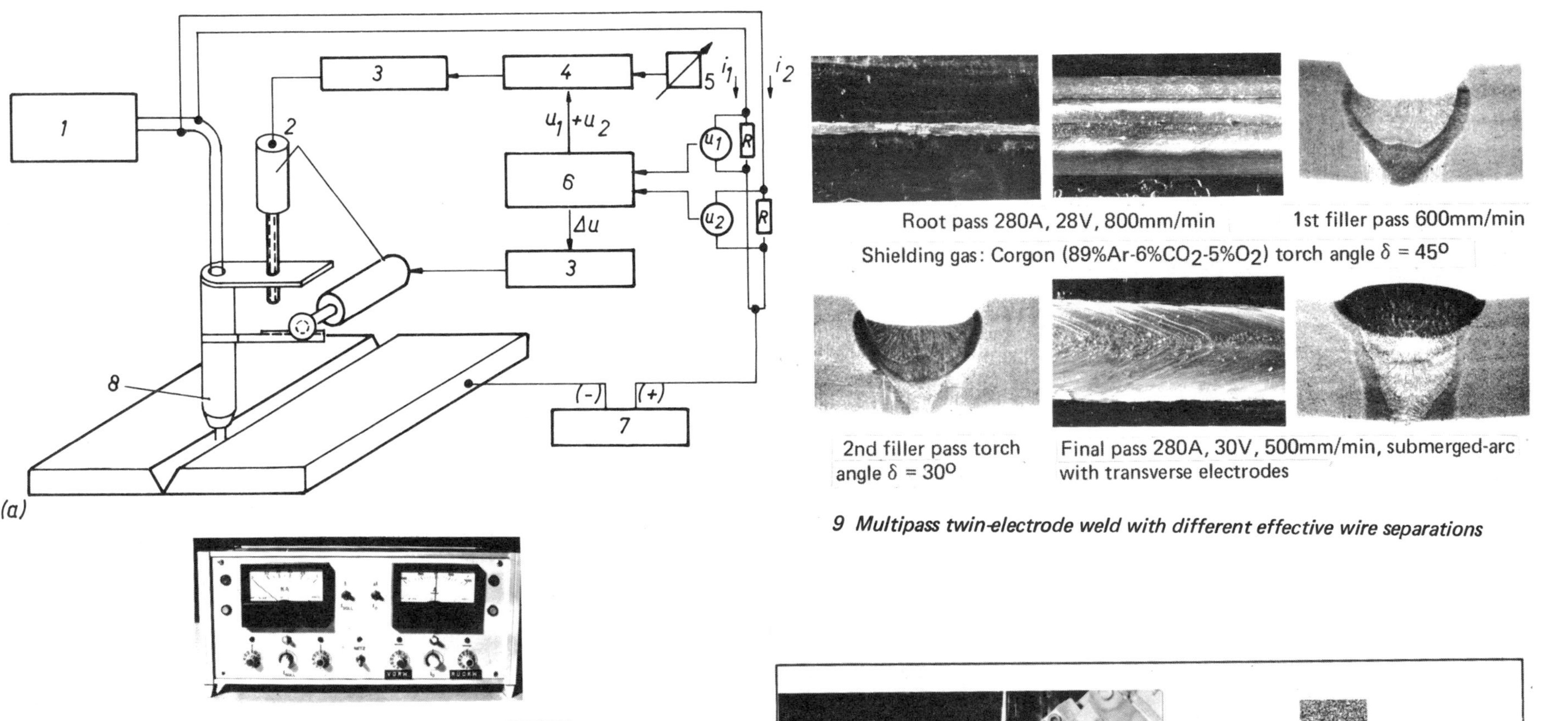

9 *Multipass twin-electrode weld with different effective wire separations*

8 *Twin-electrode welding with adaptive tracking in two axes for control of root penetration: (a) schematic arrangement, (b) basic control circuit. 1 – two-wire feed unit; 2 – servomotors; 3 – servo-amplifiers; 4 – comparators; 5 – reference current adjustment; 6 – controlling unit; 7 – power source; 8 – welding head; 9 – differential amplifier; 10 – reference value for centre position; 11 – horizontal servomotor; 12 – vertical servomotor; 13 – threshold control; 14 – reference value for total current (working distance); 15 – summation amplifier; 16 – shunt*

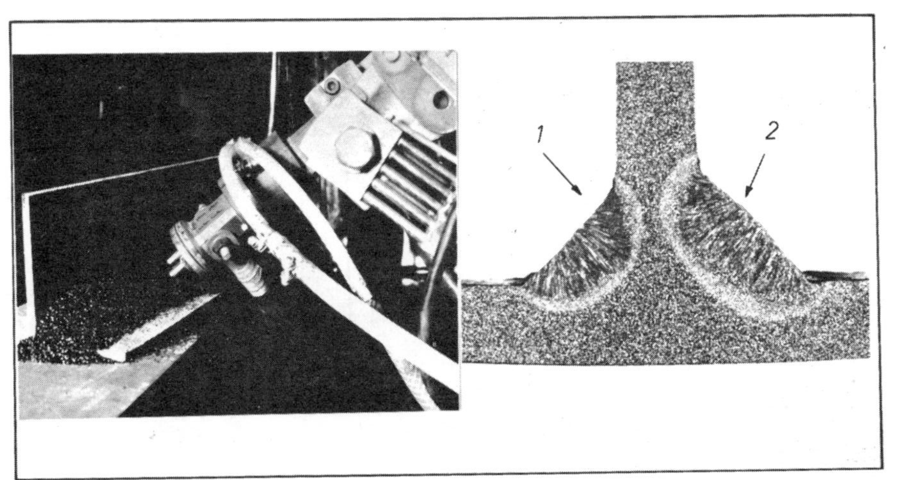

10 *Submerged-arc fillet welding with twin electrode (I_{total} = 480A; U = 31V). Welding speeds: 1 – 800mm/min; 2 – 600mm/min*

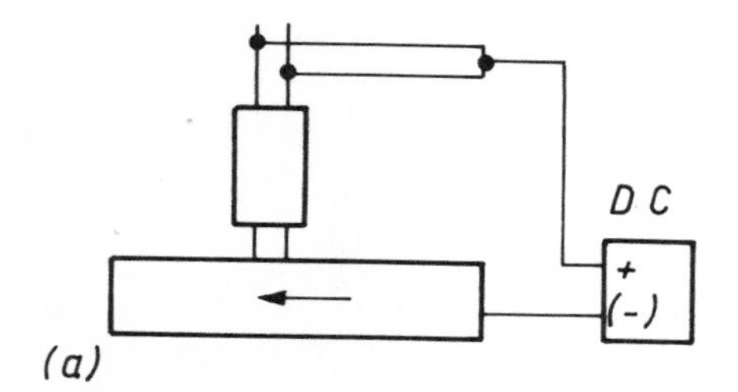

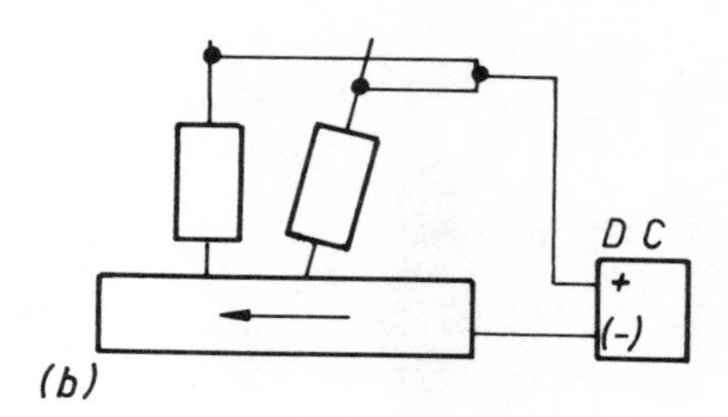

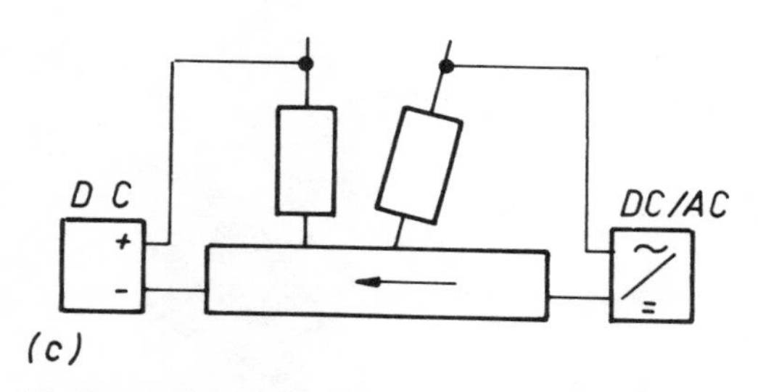

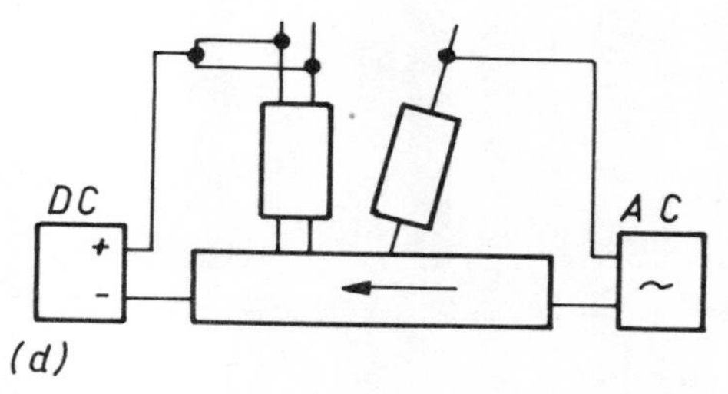

11 Requirements in multi-arc welding with arc sensors for seam tracking: (a) twin-electrode welding, single power source, (b) two-head welding, single power source, (c) two-head welding, two power sources, and (d) twin-electrode plus single-electrode welding, two power sources

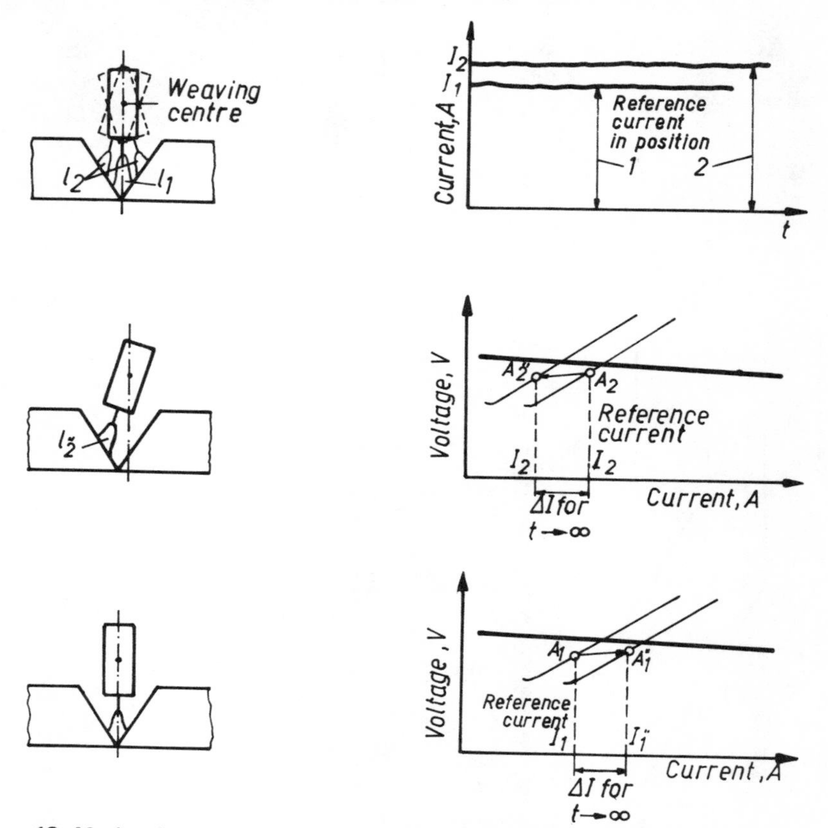

12 Mode of operation and movement of working point with torch weaving for two axes seam tracking. Torch positions: (top) central, working point A; (centre) eccentric, working point A''; (bottom) too deep, working point A''

Feedback control of weld penetration in 1978

P.Boughton, CEng, MIMechE, AMIEE, G.Rider, BSc, AMIEE, and C.J.Smith

Weld quality is controlled by using sampling and reject techniques but, if it were possible, real-time quality control would be preferred on technical and economic grounds.

This Paper describes progress made in research aimed at establishing practical systems for real-time quality control. The work has led to the development of feedback control systems in which measurements are taken from the weld, as it is being made, and the information used to regulate the power and energy input to the weld. One system measures the radiation at the back face of the weld and uses this as a basis for regulating penetration in pulsed TIG-welding. A modification, which enables welding current and travel speed to be regulated in unison, extends this method of penetration control to more conventional welding processes. Another system measures the radiation at the front face of the weld and, by using a mathematical model of the welding situation, generates a feedback signal which is an estimate of the actual penetration. Criteria for acceptable performance are given, and performance data from laboratory test welds are also shown.

INTRODUCTION

Probably the first genuine feedback (FB) control system was James Watt's governor (1775) which was used to regulate the speed of steam engines. The next 150 years saw the development of very many FB systems, each one being intended for some specific application, but it was the 1930s before a general theory was established following the work of Hazen and Nyquist.

The first deliberate use of FB in welding occurred during the 1940s when 'automatic' welding was introduced. These early control systems regulated the welding traverse speed and the arc length (voltage), and are still in use today except that the present versions use modern electronic components in place of the original electromechanical devices.

During the next decade FB control found further application in welding. For example the MIG process is critically dependent upon the good regulation of wire feed rate, and consequently wire feed units incorporating FB have always been a standard item. Also during the 1950s FB control was incorporated into some power sources as a means of regulating and facilitating the control of output current or voltage. Another FB application, introduced about that time, was arc voltage control of the arc length in TIG-welding. More recently about the only other new application of FB has been in the development of seam-following devices. In fact the commercial developments which have characterised the last ten years or so have been limited to sequence control

Mr Boughton, Section Head — Manufacturing Technology, Mr Rider, and Mr Smith, Research Officers, are all at the Marchwood Engineering Laboratories, Central Electricity Generating Board.

or so have been limited to sequence control (programmable welding) and the use of multi-axis manipulators (robots).

Throughout this recent period many workers involved in welding production, research, and development have recognised that in principle FB control should not be limited to the subsidiary role of stabilising drives etc. Much more benefit would be derived if FB systems could be employed to control the 'quality' of the weld directly. The basic difficulty in realising this ideal is in determining what constitutes quality, and even when this is achieved it is usually very difficult to measure the quality parameter directly.

FEEDBACK CONTROL PRINCIPLES

The steady state analysis of a linear FB control loop is simple and well known, however, it is repeated here because it is fundamental to the rest of the Paper.

Referring to Fig.1 (generalised linear FB loop) the equation relating the steady state values of input and output is

$$\theta_o = \theta_i \left(\frac{AF}{1+AF}\right) - \sigma_m \left(\frac{AF}{1+AF}\right) + \sigma_F \left(\frac{F}{1+AF}\right) \quad [1]$$

where θ_o = output
θ_i = input or reference
σ_m = disturbance or inaccuracy before the error amplifier
σ_F = disturbance or inaccuracy after the error amplifier
A = gain of error amplifier
F = transfer function of the welding equipment, weld, etc., e.g. a relationship between power input and weld size

If $AF >> 1$ then

$$\simeq \theta_i - \sigma_m + \sigma_F \left(\frac{1}{A}\right) \quad [2]$$

and if $A >> 1$ then

$$\theta_o \simeq \theta_i - \sigma_m \quad [3]$$

This analysis is the classical way of showing why a FB system will suppress some system disturbances, or noise, σ_F, and why it will not suppress others, σ_m. The performance limitations of a typical household HI-FI system is a well-known practical demonstration of this theory.

In welding, disturbances of the σ_F type stem from variations in: joint thickness, fitup, preparation angle, electrode parameters, shielding efficiency, consumable composition, joint surface contamination, arc length, arc efficiency, mains voltage, travel speed, filler feed rate and position, heat sink, various electromagnetic effects, various electrode effects, etc. In production, where quality matters, great care has to be taken to minimise these σ_F type variations otherwise the weld reject rate on inspection is unacceptably high, or, if inspection is inadequate, the weld failure rate in service becomes unacceptable.

If all these variations are independent and random in occurrence it is valid to define the gross σ_F as a root mean square value, or standard deviation, in which event it will be related to the individual disturbances as

$$\sigma_F^{\ 2} = \Sigma \left(\sigma_1^{\ 2} + \ldots\ldots \sigma_n^{\ 2}\right) \quad [4]$$

where $\sigma_1, \ldots. \sigma_n$ represents the individual disturbances. If none of these disturbances dominate then

$$\sigma_1 \simeq \ldots. \simeq \sigma_n \simeq \sigma$$

and so

$$\sigma_F \simeq \sigma \sqrt{n} \quad [5]$$

where n is the number of sources of disturbance.

For example in mechanised TIG-welding it is estimated that $\sqrt{n}$ is about 4, and the individual σ value seen as a variation in the width of the penetration bead is about 0.2mm. Hence the gross standard deviation, σ_F, on the penetration width should be about 0.8mm.

The significance of eq.5 is that it implies that σ_F will have some minimum value which will be solely dependent on the physics of the welding process. Also the $\sqrt{n}$ term is a law of diminishing returns. So once 'reasonable' care has been taken to select the 'best' welding process for the job and to 'control' the quality of the work, the σ_F which occurs (along with the reject rate) will be near the minimum and will be characteristic.

Thus welding without FB control means accepting the consequences of variable quality, σ_F, of an order given by eq.5.

If FB is used to control the traverse speed, welding current, and arc voltage (still using TIG-welding as the example) the sigmas for these three variables will be reduced to virtually zero and so eq.5 will be modified

$$\sigma_F \simeq \sigma \sqrt{(n-3)} \text{ for three variables with FB control} \quad [6]$$

Thus for the TIG example in which $\sqrt{n} = 4$ the

value of $\sqrt{(n-3)}$ will therefore be 3.6 and σ_F will be reduced from 0.8 to 0.72mm. This is only a marginal improvement. Clearly using FB to reduce the effective number of variables is not very efficient.

If, however, a FB loop could be arranged to encompass the whole welding system, and not just some parts, the effect of σ_F in eq. 5 would tend towards zero. This would be a much more efficient use of FB. But there is a limitation, even with this system, and it is highlighted in eq. 3 which shows that there is always some residual error, σ_m. The main source of this residual error will be in obtaining a measure of the 'output' of the welding system, θ_o. If it is possible to make σ_m small it is worth having a 'grand-FB' scheme; if σ_m cannot be kept small the grand scheme is not viable. To be more precise the criterion for using a grand-FB scheme is

$$\sigma_m \ll \sigma_F. \qquad [7]$$

In practice this criterion has proved to be very difficult to satisfy with welding systems and the problem has been the subject of much research for some years.

QUALITY PARAMETERS

Examples of traditional weld quality parameters are: porosity, cracks, inclusions, lack of penetration, overpenetration, lack of sidewall fusion, etc. This list shows that welds are judged on a scale of defects. Nevertheless defect-free welds are made, and so these traditional quality parameters are capable of quantifying the quality of only a proportion of all welds, i.e. those which are defective.

In any sensible welding procedure the average weld should be free from defects. Marginal and rejectable welds should have a low probability of occurrence. Hence, if a grand-FB system were to be used, it would be required to operate in the defect-free 'region' and the operating point would have to be defined in terms of some quantifiable parameters. Clearly the traditional 'defect type' quality parameters do not meet this requirement.

If the failure mechanisms of welds were fully understood and if the service conditions were predictable, it might be possible to identify the fundamental parameters of weld quality. However, this ideal is still some way from being realised and so in the meantime it is reasonable to rely on common sense and intuition. On this basis a minimum of four parameters seems to be required:

1 The position of the weld relative to the position of the joint
2 The sizes of the weld relative to the size of the joint
3 The solidification rate of the liquid weld metal relative to its composition
4 The cooling rate of the solid hot metal relative to composition

In a grand-FB system one or more of these parameters would need to be measured, or estimated from related measurements, and controlled.

STATE OF THE ART

Feedback control of weld quality has been shown to be possible for two of the parameters (position and size).

Commercial systems are readily available to control the electrode position. The earliest of these 'seam followers' used a mechanical feeler to locate a joint face. More sophisticated techniques have been reported recently. These use noncontact methods such as optical or magnetic transducers,[1-3] TV cameras,[4-5] etc.

Feedback control of weld pool size is relatively new as a concept. A system has been reported to regulate the penetration in submerged-arc welding with variations in fitup of the joint.[5] Also two different systems for the regulation of the penetration of TIG welds have been published.[6-7] There are some industrial applications of at least one of the systems and possibly of the other two as well.

The idea that the solidification rate should be controlled is well established,[8] but as yet no tangible scheme has emerged. Similarly there are as yet no FB control systems specifically designed to control the cooling rate of the solidified weld and its surroundings. However, for many materials it is essential to regulate this parameter to achieve reasonable weld metal and HAZ mechanical properties. Therefore, it is possible that FB might eventually be developed to control these aspects of weld quality too.

MEASUREMENT FOR CONTROL

The prime technical problem in creating any FB control system is to minimise the error in measurement, σ_m. It is axiomatic that the best result is obtained with the method which makes the most direct measurement of the variable which is to be controlled. As the remainder of this Paper is mainly concerned with the control of the penetration in welding, the following discussion will be limited to the measurement of penetration.

If there is access to the rear of a weld, in principle, it is possible to measure penetration size directly. In practice the penetration 'spot' is a hot liquid metal region and so a non-contact method is essential. Hence some form of optical radiation pyrometry ought to be the preferred method for making the measurement. Bennett and Smith[7] describe a specific form of radiation pyrometry. This makes use of optical filters to cut out the longer wavelengths radiating from the HAZ and so minimise the unwanted background signal. Although the above authors show that the optical filters are effective, it can be deduced from their work that $\sigma_m > 0$.

From a typical layout of the photosensitive cells used it is possible to see some of the reasons why this control system is imperfect. The geometrical relationships governing the intensity of radiation (R) falling on the sensors (effective area $100mm^2$) situated at some position defined by r, θ, and ϕ, Fig.2, relative to the weld radiating p watts is given by

$$R = p\,\frac{\cos^2\theta\,\sin(\theta + \phi)}{r^2} \quad \text{watts/mm}^2 \qquad [8]$$

Typical values for a FB control system designed for butt welding tubes are: $\theta = 75^o \pm 1^o$: $\phi = 0^o \pm 5^o$: $r = 25mm \pm 2mm$ (the tolerances represent 3σ). Equation 8 can be used to calculate ΔR and, since the width of the back face spot, W, is proportional to the square root of the level of monochromatic radiation which is being measured, ΔW can be obtained. By this method, and using the above figures, $\Delta W \simeq \pm 0.6mm$ and so these errors contribute 0.2mm to σ_m. In addition to these there are some other small errors arising from inherent variations in emittance, weld pool superheat, etc., but these are more difficult to isolate. However, they are apparently no more significant than those stemming from geometrical effects, and so FB control using back face sensors will probably have $\sigma_m \gtrsim 0.3mm$.

If the aim is to control welds which only partially penetrate the workpiece, some method of estimating the depth of penetration from front face measurements is required. One such method described by Rider[9] uses a semi-empirical relationship[10] between depth, z, front face width, 2y, voltage, V, current, I, and speed, u. A relationship of this kind which is suitable for some TIG-welding is

$$z = \frac{VI}{22yu + 212} - 0.12u - 0.35mm \qquad [9]$$

An independent check of whether the above equation is sufficiently accurate can be obtained by using it to predict the results obtained by Spiller and McGregor[11] in a study of the effect of electrode vertex angle in TIG-welding. This is pertinent because of the wide range of conditions covered. Figure 3 is a histogram showing the errors which occur between the eq.9 predictions and experiment. The standard deviation is $\sigma_e = 0.227$, which is fairly satisfactory when compared with typical values for z of up to 4mm.

Typical values for the variables in eq.9 for a partially penetrated TIG weld on 'thick' stainless steel plate are: $V = 12V \pm 0.5$; $I = 200A \pm 4$; $u = 2mm.sec^{-1} \pm 0.1$; $y = 5mm \pm 0.15 \pm 0.15 \pm 0.15$. (The three tolerances on the pool half width, y, indicate that there are three independent sources of variance, i.e. resolution distance between object and sensor, and alignment of sensor with object axes.) Using eq.9 and the above data the expected value of $z = 5mm \pm 0.8mm$ from which $\sigma_z = 0.8/3$. On the basis that $\sigma_m^2 = \sigma_z^2 + \sigma_e^2$ the errors inherent in this particular method of controlling penetration are such that $\sigma_m \simeq 0.3mm$.

Hence in principle the control of penetration by back face viewing and front face viewing are equally accurate. Also, FB welds should show less variation in penetration than those made without FB. Therefore, although imperfect, either of these two methods of FB is capable of giving a significantly more consistent result than conventional open loop TIG-welding.

PULSE-CONTROLLED FULL PENETRATION WELDS

Any FB loop may be either quasi-stable or astable and this depends upon the gain frequency characteristic of the loop. The overall loop gain/phase angle of the system is given by

$AB\,\underline{|\theta + \psi}$

where $A\,\underline{|\theta}$ = gain/phase angle of FB hardware
and $B\,\underline{|\psi}$ = gain/phase angle of the object being controlled (in this instance the weld pool)

The system will be astable if the following inequality is satisfied

$$|AB| > 1 \text{ when } (\theta + \psi) = \pm\pi \qquad [10]$$

The main components of the FB hardware are the sensor, the error amplifier, and the welding power source. Typical gain/phase angle/ frequency characteristics of these components and the weld pool are given in Table 1.

For the example given in Table 1 the FB

Table 1 Typical gain/phase angle/frequency characteristics for the components of a FB loop based on back face radiation measurement

Frequency	1Hz		3Hz		10Hz		Units
Error amplifier and welding power source	$\frac{200}{10^{-3}}$	$\angle 0^o$	$\frac{50}{10^{-3}}$	$\angle -60^o$	$\frac{10}{10^{-3}}$	$\angle -90^o$	$A.V^{-1}$
Silicon diode sensor	$\frac{10^{-1}}{10^{-3}}$	$\angle 0^o$	$\frac{10^{-1}}{10^{-3}}$	$\angle 0^o$	$\frac{10^{-1}}{10^{-3}}$	$\angle 0^o$	$V.W^{-1}$
Weld pool	$\frac{10^{-3}}{200}$	$\angle 0^o$	$\frac{10^{-4}}{200}$	$\angle -120^o$	$\frac{10^{-5}}{200}$	$\angle -200^o$	$W.A^{-1}$
Loop gain/phase	10^2	$\angle 0^o$	2.5	$\angle -180^o$	5.10^{-2}	$\angle -290^o$	-

system would be astable and would oscillate at around 3Hz. The total phase shift in the loop is mainly because of the weld pool response, and so, changing for a higher grade of power source, for instance, will not make very much difference to the performance of the whole system.

If the maximum and minimum output current excursions of an astable system, such as that described here, are sensibly limited the system will oscillate between the two levels at its characteristic frequency. Also, because the weld pool dominates the response of the system, the excursions in weld pool size will be much less than those in the welding current. Hence the system is naturally a good on/off controller.

A speculative explanation of the physical basis for the phase shift in the weld pool is that it is mainly a manifestation of the time taken for superheated metal to be convected down through the pool from the arc side to the back face. Heat diffusion into the solid material, and the latent heat requirement for melting, will also have some influence on the measured weld pool response.

Figure 4 shows the typical variation of welding current and weld pool radiation with time with FB control operating on a continuous seam. It should be noted that the variation in weld pool radiation seems to be caused more by a variation in the liquid temperature than in the size of the back face spot.

The maximum level of welding current affects the frequency, Fig. 5: an increase in peak current leads to an increase in frequency. The reason for this positive correlation can only be surmised until more research is done. Similarly the magnitude of the peak-to-peak amplitude of the oscillatory radiation output of the weld also remains a matter for speculation. Another peculiarity is that the mark/space ratio of the system will vary with variations in plate thickness etc. It is observed that the mean level of the radiation from the weld varies so as to shift the oscillatory component in relation to the reference level; consequently the on and off times of the welding current vary. This self-regulating ability needs further study if it is ever to be fully understood.

Figure 6 shows the response of a MEL penetration control system to a 'step plate test'. It is obvious that its behaviour is perfectly adequate. Also the accuracy of the system in controlling welds over long runs is consistently measured at $\sigma < 0.4$mm. In a particular experiment involving TIG-welding 3.2mm stainless steel plate, the mean width of penetration was 5.3mm and $\sigma = 0.182$mm. Hence there is no doubt that this pulse control is a valid method.

CONTINUOUSLY CONTROLLED FULL PENETRATION WELDS

TIG arcs are convenient for pulsing but most other arc systems are not. If the inequality in eq.10 is not satisfied, i.e. if $AB < 1$ when $(\theta + \psi) = +\pi$, the FB loop will probably be quasi-stable and continuous control is possible.

Using the previous example, Table 1, the gain at around 3Hz could be reduced from 2.5 to less than 1 by inserting a suitable low pass filter in series with the error amplifier. Without any further modification the system would then operate as a continuous controller: the welding current would automatically vary so as to maintain the size of the weld pool approximately constant.

In practice, the simple system using

welding current as the sole control parameter has a limited performance. There are essentially two limitations: one is that the response to transient changes in the heat sink is restricted by a system time constant of about 0.5sec; the other is that a welding condition cannot be optimised on the current, I, alone. However, if the welding speed, u, was used as a control variable as well as current, in theory the transient response would be significantly improved, and at the same time welding conditions could be more nearly optimised. There is one proviso, and that is that $\frac{du}{dI}$ must be negative so that u and I vary harmoniously. In practice it has been found most convenient to design controllers in which I and u are related as

$$u = - aI^n + b \qquad [11]$$

The typical value of the constants (eq.11) are

$a = 33.10^{-3}$ to 67.10^{-3} $mm.A^{-1}.sec^{-1}$
$b = 5$ to $10 mm.sec^{-1}$
$n = 0.75$ to 1.2

Figure 7 shows a FB loop which incorporates one method of simultaneous control of I and u, and Fig.8 shows how the system performs when used with the TIG process including a cold start and a step plate test. Also in consistency tests using TIG on 3.2mm stainless steel plate, the mean penetration width was 6.99mm and σ = 0.204mm. This provides convincing proof of the efficacy of the continuously controlled full penetration system.

CONTINUOUSLY CONTROLLED PARTIAL PENETRATION WELDS

Rider's technique for estimating the depth of a weld from front face measurements has already been referred to earlier. In this scheme current, voltage, and speed are measured, which present no difficulty. However, a measurement of the front face width of the weld pool is also needed, and this is technologically much more difficult. The solution employed is to use a form of TV camera which incorporates a linear array of tiny (0.1 x 0.1mm) light-sensitive diodes. The scanning of the array is synchronised with periodic (100Hz), brief (10^{-3}sec), arc extinction during which time the pool can be viewed without the overbright arc present.

This FB system is no different in principle from the type discussed in the previous Section of this Paper. However, the method used to obtain the weld pool measurement introduces a phase shift into the FB loop. For the numbers quoted above the phase shift will be 180° at 50Hz, but the frequency response of the top face width of the weld pool is relatively slow, about 10Hz. Thus the system characteristics will not be significantly different from the other control systems. Stable operation requires that the system gain be less than unity at 10Hz, and when this is achieved by putting suitable low pass filters in circuit the response time of the system will be about one second.

Figure 9 is a block diagram of the control system, and Fig.10 illustrates its performance when used to control TIG-welding on 6mm thick stainless steel plate. It should be noted that Fig.10 shows that the system takes one or two seconds to settle down after a step change in the reference level; this is very much as expected.

Steady state tests made on five separate 30mm long x 6mm thick stainless steel plates gave the following results:

Mean top face width	= 5.9mm
Mean penetration depth	= 2.9mm
Number of measurements	= 33
Standard deviation, width	= 0.6mm
Standard deviation, depth	= 0.3mm
Correlation coefficient between width and depth	= 0.3

Clearly the control system works fairly well under these particular test conditions.

The above experiments were done with current as the sole control variable and, by analogy with the work on full penetration welds, an even better performance should be obtained by using the two control variable (I and u) technique. It is hoped to confirm this in due course.

DISCUSSION AND CONCLUSIONS

The idea that weld quality might be controlled by some form of FB is becoming a technological reality. The work reported by the present authors is intended to help promote a better basic understanding of the principles involved. It is also a fact that the research has led to the development of systems which are measurably better than traditional welding methods, i.e. $\sigma_m << \sigma_F$. Some of these new FB techniques have already been transferred from the laboratory to the workshop.

This topic is still very new in welding and there is still a great deal to achieve. There is the need for patient research aimed at extracting the transfer functions from elements of the system. In this context, characterising the behaviour of various types of weld pool and arc systems should be a fascinating task. There are also significant challenges to engineering ingenuity in the development of systems of sensors.

In the longer term it is reasonably probable that someone will discover a practical way of controlling all the aspects of weld quality at once. There is a strong commercial incentive for succeeding in this.

ACKNOWLEDGEMENTS

This work is published by permission of the Central Electricity Generating Board.

REFERENCES

1 HIRSCHMANN, F. 'Seam following by optical means'. Welding and Metal Fab., 40 (12), 1972, 413-15.

2 ARAYA, Y and INOUE, K. 'Automatic control of arc/welding/optical sensing of joint configuration'. IIW XII-V-41-73.

3 ARAYA, T., FUJIWARA, O., and UDAQAWA, T. 'Development of automatic seam follower using noncontact sensor'. IIW XII-K-70-76.

4 WALL, W.A. and STEPHENS, D.L. 'Automatic closed circuit television electrode guidance for welding'. Welding J., 48 (9), 1969, 713-20.

5 WESTBY, O. 'A solution for butt welding automation'. IIW XII-K-74-76.

6 LUCAS, W. and MALLETT, R.S. 'Automatic control of penetration in pulsed TIG-welding'. Welding Inst.Report R/RB/72/75, 1975.

7 BENNETT, A.P. and SMITH, C.J. 'Improving the consistency of weld penetration by back face control'. Welding Inst.Conference 'Fabrication and Reliability of Welded Process Plant', London, 16-18 November 1976. Abington, Welding Inst., 1977, 13-19.

8 GARLAND, J.G. 'Control of weld pool solidification'. Metal Constr., 6 (4), 1974, 121-7.

9 RIDER, G. 'Measurement of weld pool surface size using self-scanned photodiode arrays'. IEE Conference 'Low Light and Thermal Imaging Techniques', 1975.

10 BOUGHTON, P. 'The principle of real-time control of weld quality.' IIW XII-K-46-74.

11 SPILLER, K.R. and McGREGOR, G. Welding Inst.Conference 'Advances in Welding Processes', Harrogate, 14-16 April 1970, 82-8.

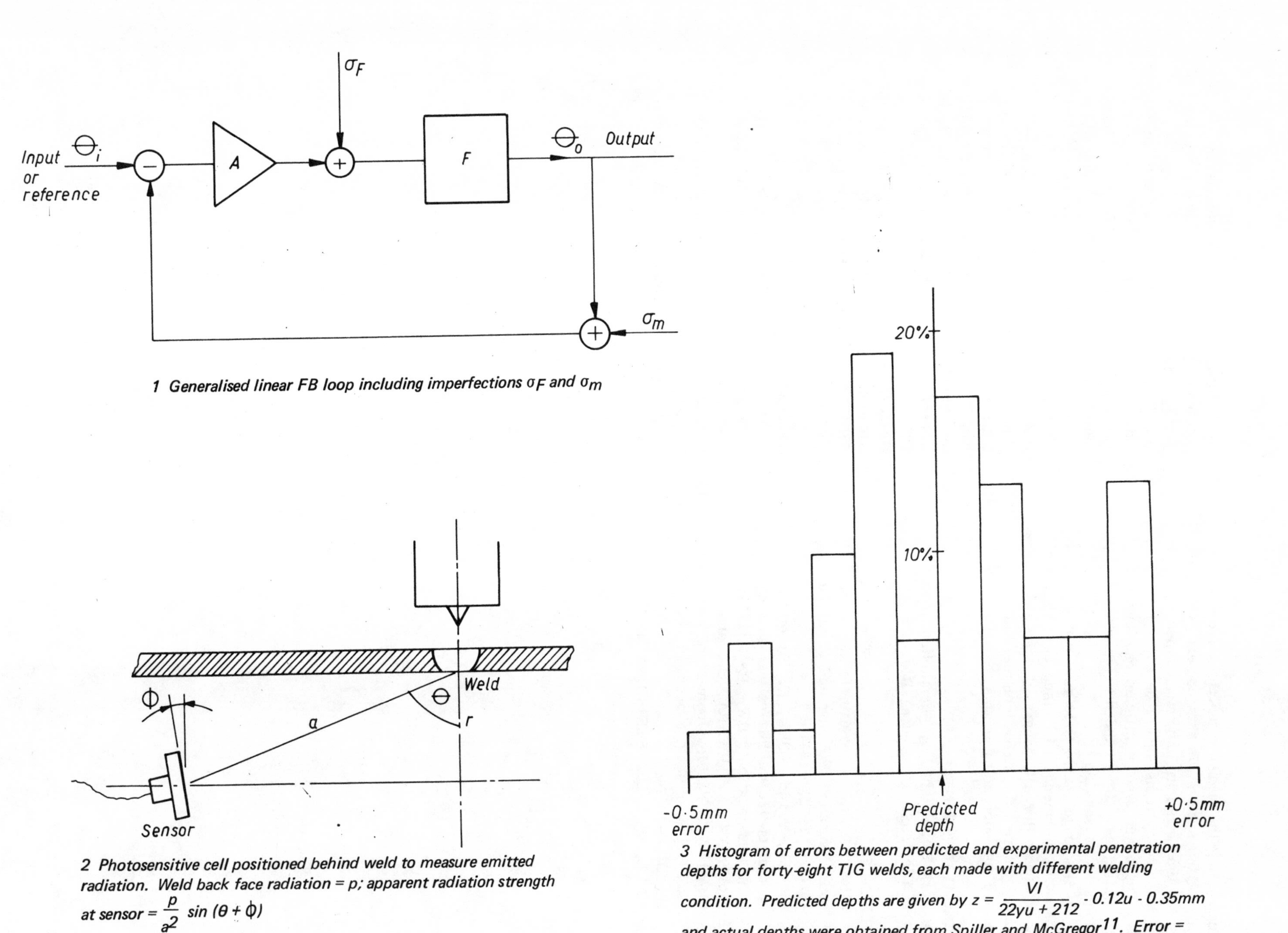

1 Generalised linear FB loop including imperfections σ_F and σ_m

2 Photosensitive cell positioned behind weld to measure emitted radiation. Weld back face radiation = p; apparent radiation strength at sensor = $\frac{p}{a^2} \sin(\theta + \phi)$

3 Histogram of errors between predicted and experimental penetration depths for forty-eight TIG welds, each made with different welding condition. Predicted depths are given by $z = \frac{VI}{22yu + 212} - 0.12u - 0.35\text{mm}$ and actual depths were obtained from Spiller and McGregor[11]. Error = (actual depth) - (predicted depth).

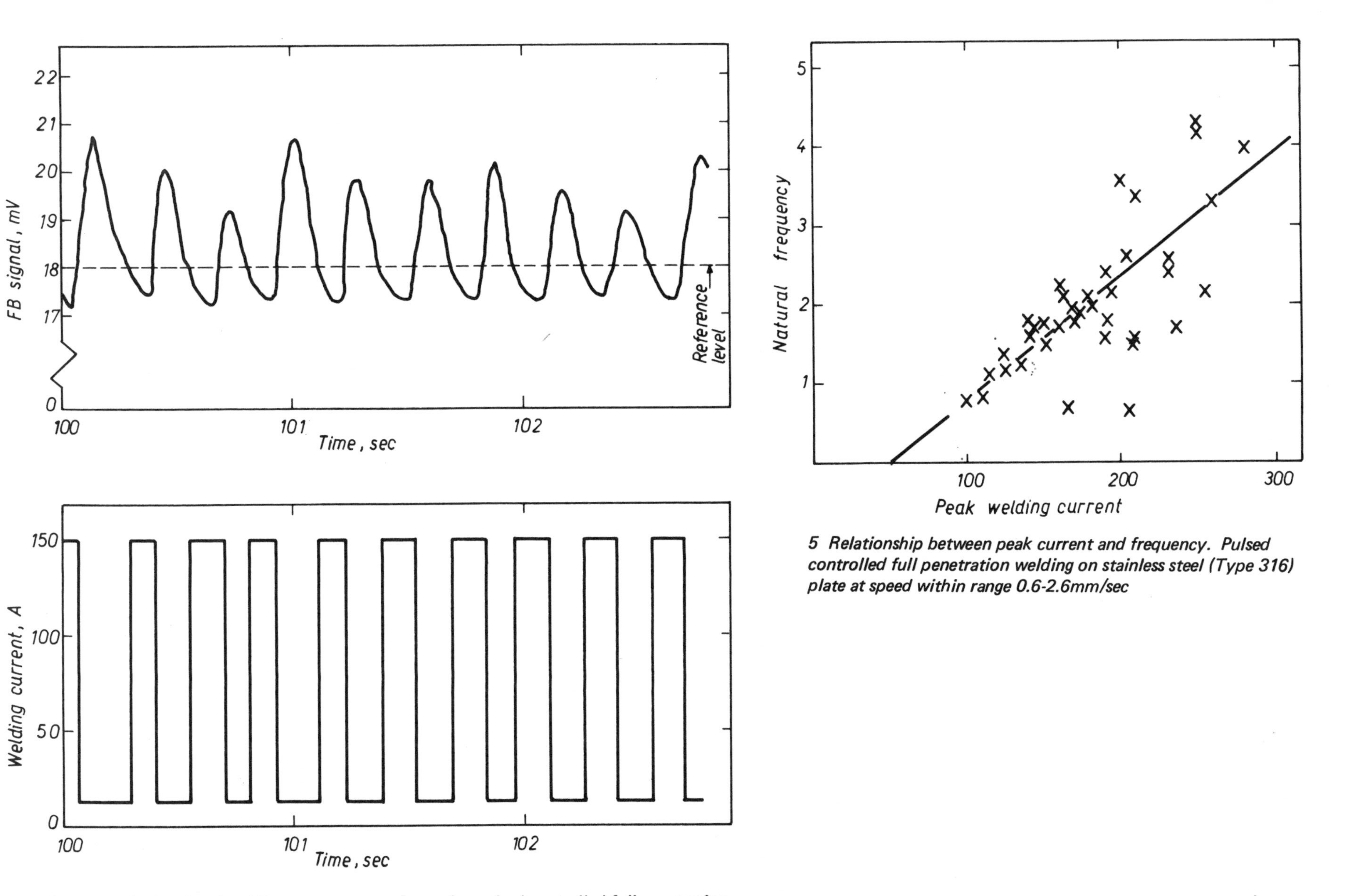

4 Feedback signal and welding current wave shapes for pulsed controlled full penetration welding

5 Relationship between peak current and frequency. Pulsed controlled full penetration welding on stainless steel (Type 316) plate at speed within range 0.6-2.6mm/sec

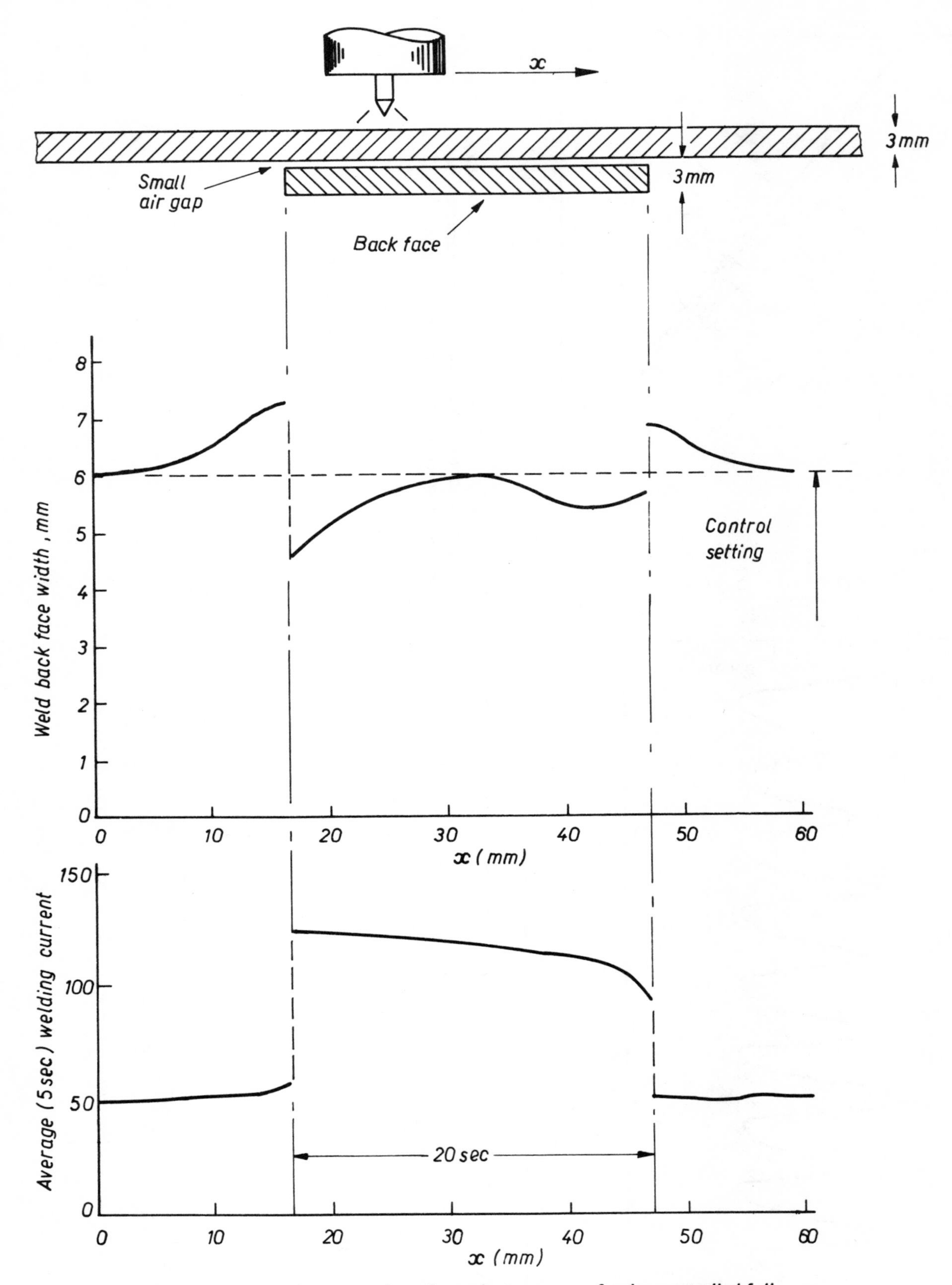

6 Step plate test: demonstration of transient respone of pulse controlled full penetration welding

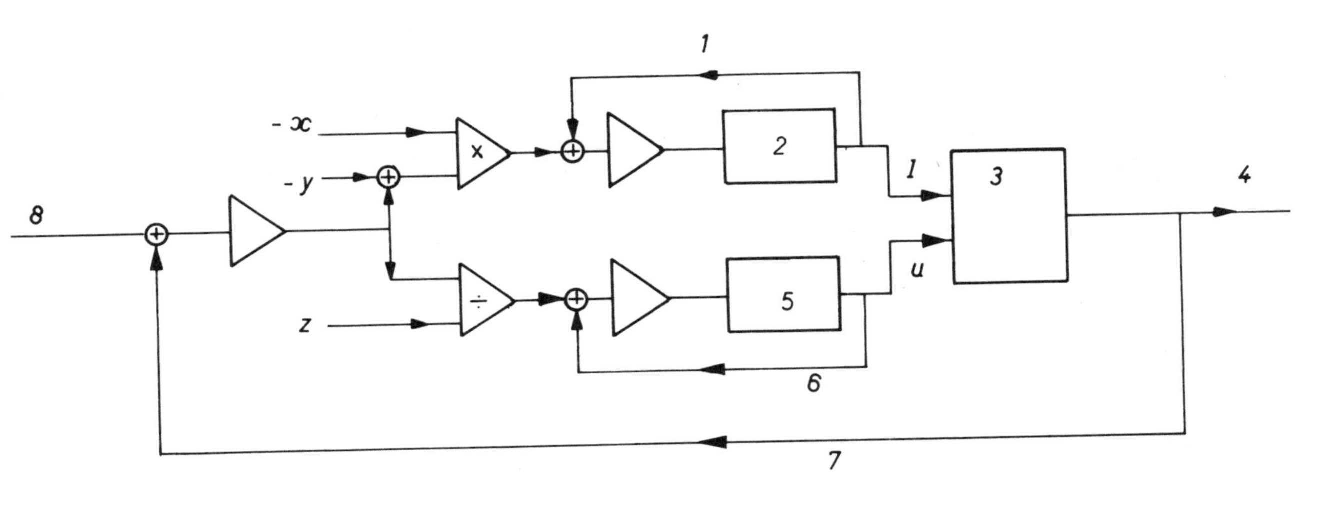

$$u = \frac{1}{z}\left(-\frac{I}{x} + y\right)$$

7 Combined current, I, and speed, u, regulation for continuously controlling full penetration welds. 1 — welding current FB; 2 — power source; 3 — weld pool; 4 — weld penetration; 5 — drive motor; 6 — welding speed FB; 7 — weld penetration width FB; 8 — penetration reference

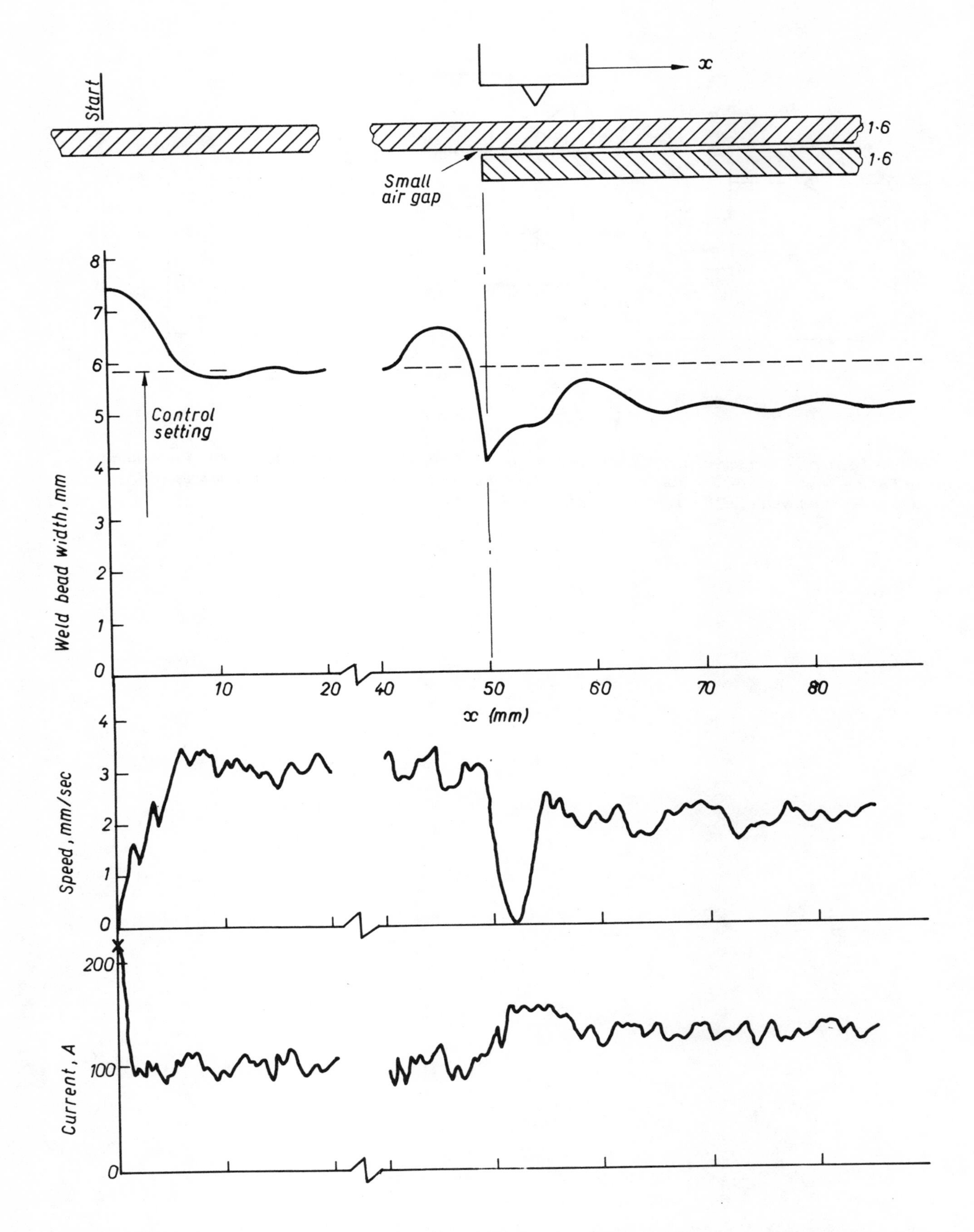

8 Cold start and step plate test: demonstration of transient response of continuously full penetration welding

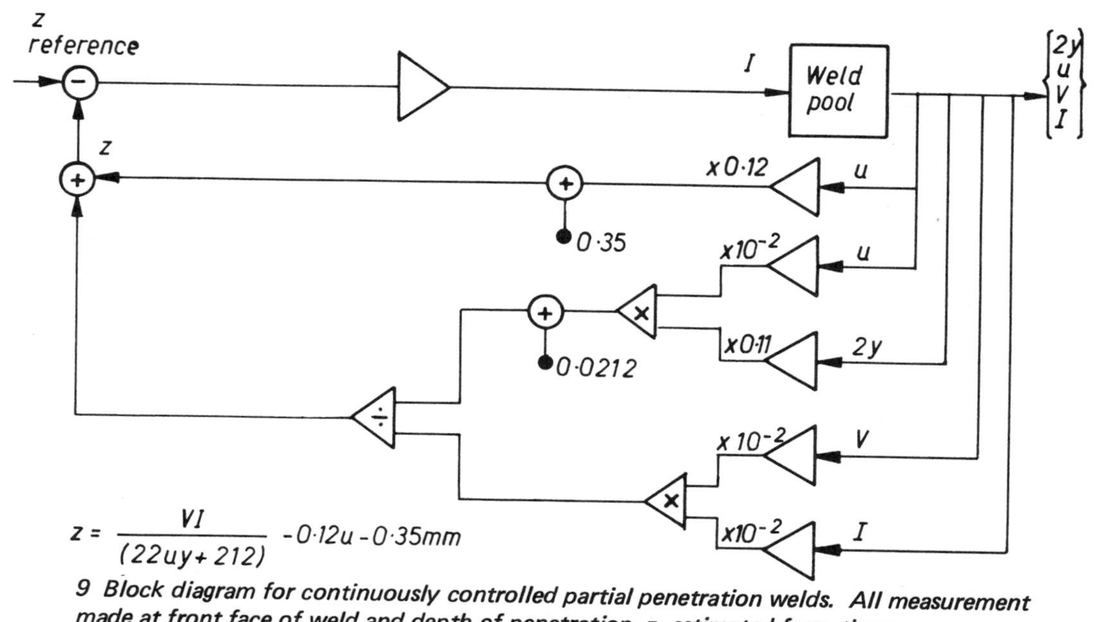

$$z = \frac{VI}{(22uy + 212)} - 0{\cdot}12u - 0{\cdot}35\,mm$$

9 Block diagram for continuously controlled partial penetration welds. All measurement made at front face of weld and depth of penetration, z, estimated from these

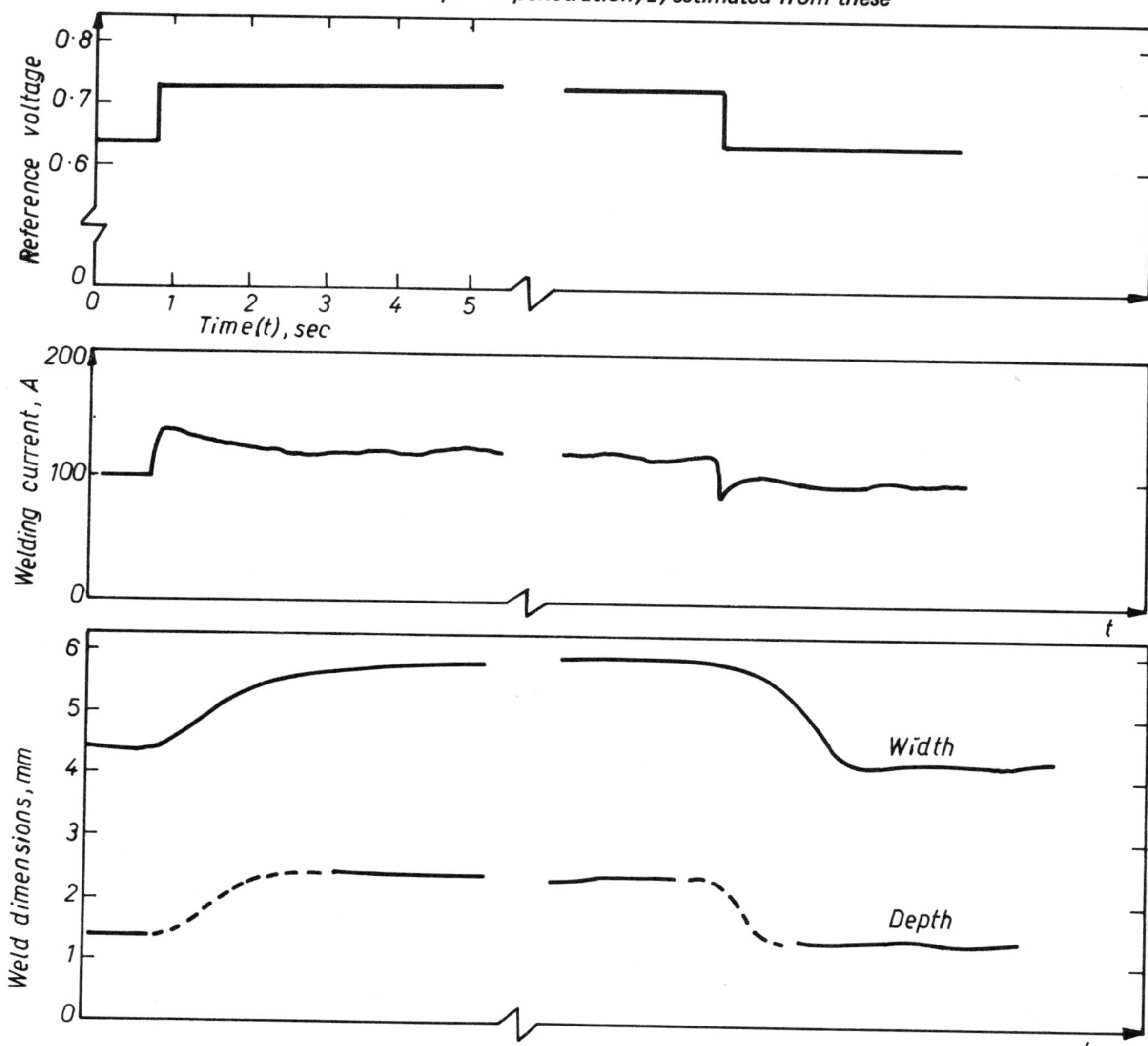

10 Response of continuously controlled partial penetration weld to step changes in FB reference level

Developments in three-phase DC equipment for resistance welding applications

R.Cazes, Engineer ESE, B.Receveur, Engineer EEMI, and G.Sayegh, DSc, DEng

A new range of three-phase rectifier type equipment designed to weld light alloys, heat-resistant materials, and mild steel in the aircraft and motor vehicle industries has been developed.

The Paper first indicates the solutions decided on and gives an insight into the method for calculation of the different currents by computer, by development into a Fourier series. This method takes account of the actual form of the secondary voltage in relation to the phase shift and impedance of the machine.

Secondly, three examples of the industrial applications are described, referring in particular to the welding of light alloys in the motor vehicle industry and the projection welding assembly of extra thick steel carried out on a machine rated at 160kA short-circuit. The three-phase DC rectifiers are compared with alternative systems: single-phase AC, three-phase frequency converter.

INTRODUCTION

Industrial needs in the field of resistance welding call more and more for high power machines, whether these be to weld light alloys in the aircraft and motor vehicle industries, the assembly of extra thick steel plates by spot or projection welding, etc. The electrical power supply to these machines poses serious problems which can be resolved by using the well-known principles of low frequency three-phase converters (Sciaky patent) or three-phase DC rectifiers, which enable the load to be distributed over the three phases, reduce the welding power, and improve the power factor ($\cos \varphi$) of the machine (which approaches about 0.95).

The low frequency converter solution has enabled high performance machines to be applied extensively in the aeronautical industry to weld light alloys and heat-resistant steels.

M.Cazes, Research and Development Manager, M.Receveur, Resistance Welding Laboratory, and Dr Sayegh, Scientific Department Manager, are all with Sciaky SA, Vitry-sur-Seine, France.

Such equipment supplies one or several alternate impulses of a duration which is inversely proportional to the welding current, but requires a large transformer to accommodate the low frequency.

Three-phase DC rectifiers supply a single impulse, the duration of which is no longer linked to the transformer size but is limited only to the power rating of the rectifier unit. This enables both high current and long weld times to be obtained, which is especially interesting in welding extra thick steel. The relatively small volume enables such equipment to be used either as a suspended gun station or as a stationary machine. The field of application is therefore potentially more extensive than that of low frequency converters.

DESIGN AND TECHNOLOGY OF RECTIFIER SYSTEM

Six-phase rectifier assemblies are used with an interphase transformer, which is equivalent to two three-phase rectifiers in parallel and hence is quite suitable for the high currents and low secondary voltage of resistance welding machines. The rectified output voltage shows a low amplitude (4%) six-phase ripple before filtering. To facilitate manufacture and standardisation, three transformers each containing two secondary windings are used, connected to the rectifier according to Fig.1 and coupled to the primary supply in delta. Each one of the transformers is controlled by a conventional inverse parallel thyristor contactor.

The diodes used are made up of n-p junction discs mounted directly on to copper plates forming the secondary conductors. The characteristics of such a diode are shown in Fig.2. Cooling is achieved directly by the copper support, which in turn carries circulating water.

The utilisation of such diode units in welding machines, for which they are specially constructed, is readily accomplished and does not give rise to any superfluous inductive or resistive voltage drop. They are, in fact, designed to be assembled directly on the secondary outputs of the transformers.

AIDS TO DESIGN

Rectifier calculations

Characteristic curves have been compiled for the rectifier assemblies used, giving the average rectified current in relation to the duty cycle of the machine, the welding time, and the temperature of the cooling fluid of the rectifier units, so that the junction temperature remains within the permitted limits.

These curves, Fig.3, show that, for a given rectifier unit, the maximum time for each current impulse is related to the current level and duty cycle. They have been determined on the basis of transient thermal impedance values of the rectifier assembly used, as a function of the duration of the current impulses. These transient thermal impedance values were measured in the laboratory and introduced into a theoretical model to establish data such as are shown in Fig.3.

Calculation of secondary currents

The inductance of the secondary circuit of the welding machine considerably influences the form of the rectified current: on the one hand by increasing the buildup time at the beginning of the current impulse, and on the other by reducing the ripple by filtering. The initial up-slope or buildup effect is particularly sought after for welding light alloys. The leakage reactance of the transformer, which intervenes in the switching of diodes, also modifies the theoretical current form; this all the more so as the machine operates under conditions close to short-circuit.

To determine accurately the form and actual values of the secondary current the development by a Fourier series was used and the problem processed mathematically on a computer. This enables the exact form of the current to be established at the time the machine is being designed, together with the average and rms current of the complete welding impulse and the rms primary current, in terms of the following parameters: welding time, conduction angle of the primary thyristors, inductance of the secondary circuit, leakage reactance of the transfer. All these results are presented in the form of dimensionless curves, an example of which is given in Fig.4.

EXAMPLES OF EQUIPMENTS

Three different systems are described comprising six-phase DC machines equipped with power supplies of similar design but used for quite different purposes.

Suspended gun station with six-phase rectifier

This station has been designed and adapted to weld light alloy doors in the motor vehicle industry. It feeds one or two welding guns by means of very low inductance cables. With a cable 2m long, $250mm^2$ in cross-section, and a gun with 10mm throat depth, it delivers 40 000A. Its welding capacity on light alloys of industrial quality is 2 + 2mm thickness with an average production rate of ten spots/min.

It is constructed in the form of a compact hooded assembly, Fig.5, comprising the three transformers and the rectifier-interphase transformer unit mounted at the bottom. The interphase transformer is made up of a C-core, inside which pass the copper bars of the neutral points of the two three-phase rectifiers. The secondary voltage adjustment switches are housed in the side of the station.

Double protection is provided for the diodes. On the one hand thermostats control the temperature of the cooled conductors, on the other a special device measures the I_t product and cuts off the primary circuit breaker if the critical value is exceeded.

The oil and air assembly to control the force of the guns is arranged outside the station together with the hydraulic assembly to cool the station, cables, and guns. The power thyristors are in a separate cabinet including the programming sequence for the welding cycle and to set the conduction angle of the thyristors.

The main characteristics of this design

Table 1 Principal features of the six-phase gun station

Associated equipment	Gun, throat depth 100mm, maximum force 4000N (5 bar) 'Sciakyflex' cable, 2m long, 250mm^2 in section		
Short-circuit secondary current	10 to 40kA		
Continuous secondary current, 100% duty cycle	7.5kA		
Secondary transformer no-load voltage	7.45-10-15-20V		
Theoretical secondary DC voltage	8.7-11.7-17.5-23.4V		
Maximum short-circuit rating	975kVA		
Conventional rating at 50% duty cycle	258kVA		
Maximum primary line current (380V mains)	1480A		
Maximum time of one impulse for :			
Secondary current I_s	40kA	30kA	20kA
Duration t_{cycles}	10	50	150
Duty cycle	3.5%	6%	14%
Number of diodes per arm	6		
Dimensions (without oil and air panel)	Width 865mm, depth 500mm, height 1125mm		
Weight	930kg		

are summarised in Table 1. Also, the power demand for this six-phase station and an equivalent single-phase AC station are compared in Table 2.

The instantaneous gain in power is negligible because the inductances of the gun and secondary cable are low, but the advantage of the six-phase station lies in balancing the power called for over the three phases, the quite substantial reduction in cross-section of the power supply cables, and in the very low ripple of the secondary current.

Multispot welding machine

The problem posed consisted of assembling plates, thicknesses ranging from 0.65 to 1.2mm, in mild steel and stainless steel on similar metal sections with various dimensions ranging from: height 60mm, width 12mm to height 6mm, width 2mm, welding several spots (two to ten) simultaneously with a C-frame type machine, Fig.6.

In view of the variable number and position of the spots it was not possible to design a traditional multispot machine with several transformers arranged suitably on top. Here it was necessary to use a single transformer.

To resolve the problem of balancing the currents between the different electrodes a three-phase DC supply was adopted. This approach completely eliminates the influence of the inductance of the secondary circuit. With a sensible arrangement of the secondary conductors, Fig.6a, the resistance of the circuit through each spot weld is identical, and current balancing ensured. In addition, the welding power demand was reduced from 640kVA for single-phase AC to 300kVA for the DC system.

P 1200 TCC machine

This high power machine delivers 160 000A short-circuit current and has a 120kN

Table 2 Comparison between six-phase DC and single-phase AC stations

	Six-phase DC station	Single-phase AC station
Associated equipment	Gun, throat depth 100mm, maximum force 4000N (5 bar) 'Sciakyflex' cable, 2m long, 250mm^2 in section	
Secondary short-circuit current	40kA	40kA
Continuous secondary current	7.5kA	7.5kA
Secondary transformer no-load voltage	20V	26V
Conventional rating at 50% duty cycle	258kV	275kVA
Maximum short-circuit rating	975kVA	1040kVA
Primary line current (380V mains)	1480A	2750A

pneumatic pressure head. It is designed to weld connecting straps 8mm thick to the wheel rims of tractors 5mm thick in mild steel by means of four projections each. A wheel having eight straps is welded in 75sec, Fig.7.

The welding head is fed by two six-phase power units coupled in parallel, each comprising three transformers connected in delta on the primary and two three-phase rectifiers connected in parallel via an interphase transformer. Each rectifier element comprises fourteen diodes per arm, of the type described earlier, Fig.2, i.e. 168 diodes total. The diodes are all accessible to facilitate inspection and maintenance. The secondary circuit conductors, rectifiers, and transformers are all water-cooled.

The control cabinet, separate from the machine, contains the three contactors with thyristors and the synchronous control sequence. The operational cycles of this sequence comprise ten functions, including preheating, pulsation welding, quenching and annealing, variable force cycle, etc.

Table 3 indicates the main features of this machine and compares the six-phase system with an equivalent low frequency converter.

The ratings are comparable, although the current balance between the phases is better with the six-phase rectifier approach. The latter clearly shows to best advantage in regard to volume, and above all to the impulse times: a very important point in this situation.

Certain toolings, located inside the secondary circuit of the machine, have been made from nonmagnetic metal. This is because of the repeated electromagnetic forces to which parts in magnetic steel would have been subjected, which could cause a high rate of fatigue and accelerated wear. The choice of nonmagnetic metal has been made, even though this is a DC machine and magnetic masses do not have any influence on the impedance of the circuit, nor do the parts forming the tooling heat up.

CONCLUSION

The three examples described show that the range of application of these six-phase DC sources is very wide.

In the aeronautical industry, to weld light alloys the six-phase DC sources are equivalent to the three-phase frequency converter machines: the welding parameters are identical. The electrical demand on the mains is also the same for the two systems: the power loss in the diodes of the six-phase source is made up for by a higher overall efficiency. Also, concerning the welding of refractory steels (which require much lower currents than light alloys), the same results are found for the two systems. Relatively long weld times are possible on the three-phase frequency converter machines by correctly designing the primary coupling of the transformer. From an economic point of view the low frequency sources cost about 10% less than the DC machine.

On the other hand, suspended welding gun stations, for the assembly of automobile chassis

Table 3 Machine P1200 TCC features

	Six-phase generator	Low frequency converter
Secondary short-circuit current	160kA	160kA
Secondary welding current	120kA	120kA
Continuous current, 100% duty cycle	58kA	58kA
Secondary transformer no-load voltage	3.8-5.1-7.6-10.2V	
Theoretical secondary DC voltage	4.45-6-8.9-11.9V	10.5V
Maximum short-circuit rating	2000kVA	2047kVA
Welding rating	1500kVA	1535kVA
Conventional rating at 50% duty cycle	1027kVA	1050kVA
Primary line current (380V mains):		
Short-circuit	3060A	3125A
Welding	2290A	2345A
Maximum time per impulse for :		
I_s = 160kA, 13% duty cycle	15 cycles	6 cycles
I_s = 120kA, 23% duty cycle	50 cycles	6 cycles
I_s = 100kA, 33% duty cycle	200 cycles	6 cycles
Overall power supply dimensions	1120 x 1000 x 1000mm	1500 x 1000 x 1580mm
Overall machine dimensions	3150 x 1300 x 3065mm	

made from light alloys, would be more difficult to accomplish with low frequency converters because of their large size and weight. As described earlier, the DC rectifier is also preferred for projection welding heavy thicknesses in steel because of the combined need for high currents and long weld times.

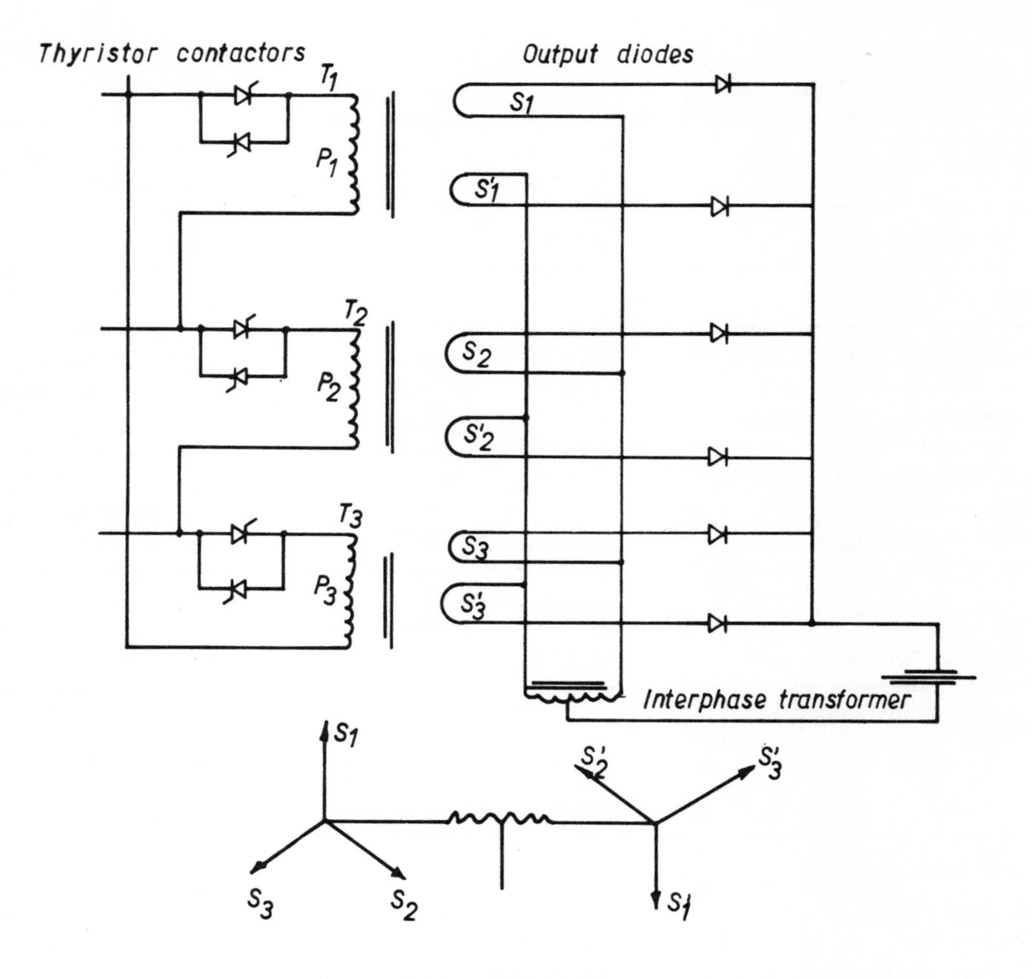

1 Six-phase rectifier with interphase transformer

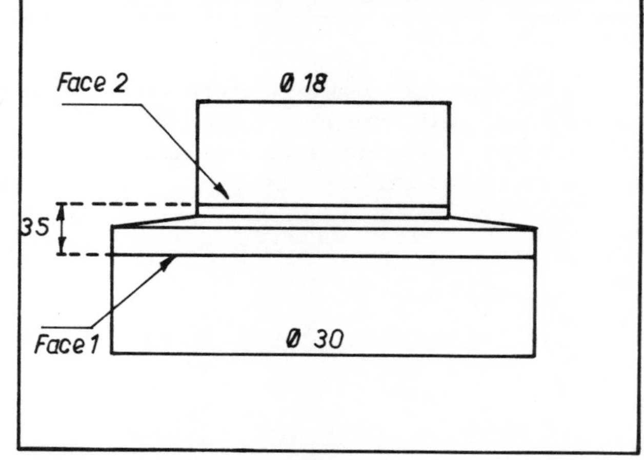

GENERAL CHARACTERISTICS

Technology : diffused silicon
Cooling : by conduction
Appliance : without case
Contact force
on heat sing : 1000 to 1200daN

LIMIT VALUES

Junction -40° to +150°C
Thermal convection (max. value with face 1 pressed on heat sink) : Rth = 0.08°C/W

ELECTRICAL CHARACTERISTICS

With face 1 pressed on heat sink and the temperature, t max., at the heat sink = 100°C
Direct continuous current : IF = 600A
Average rectified current : I_o = 500A
Nonrepetitive peak current (10msec) : IFSM = 10 000A
Direct peak voltage drop at IFM = 1550A and t (j) = 25°C 1.25V

2 Diode construction and ratings

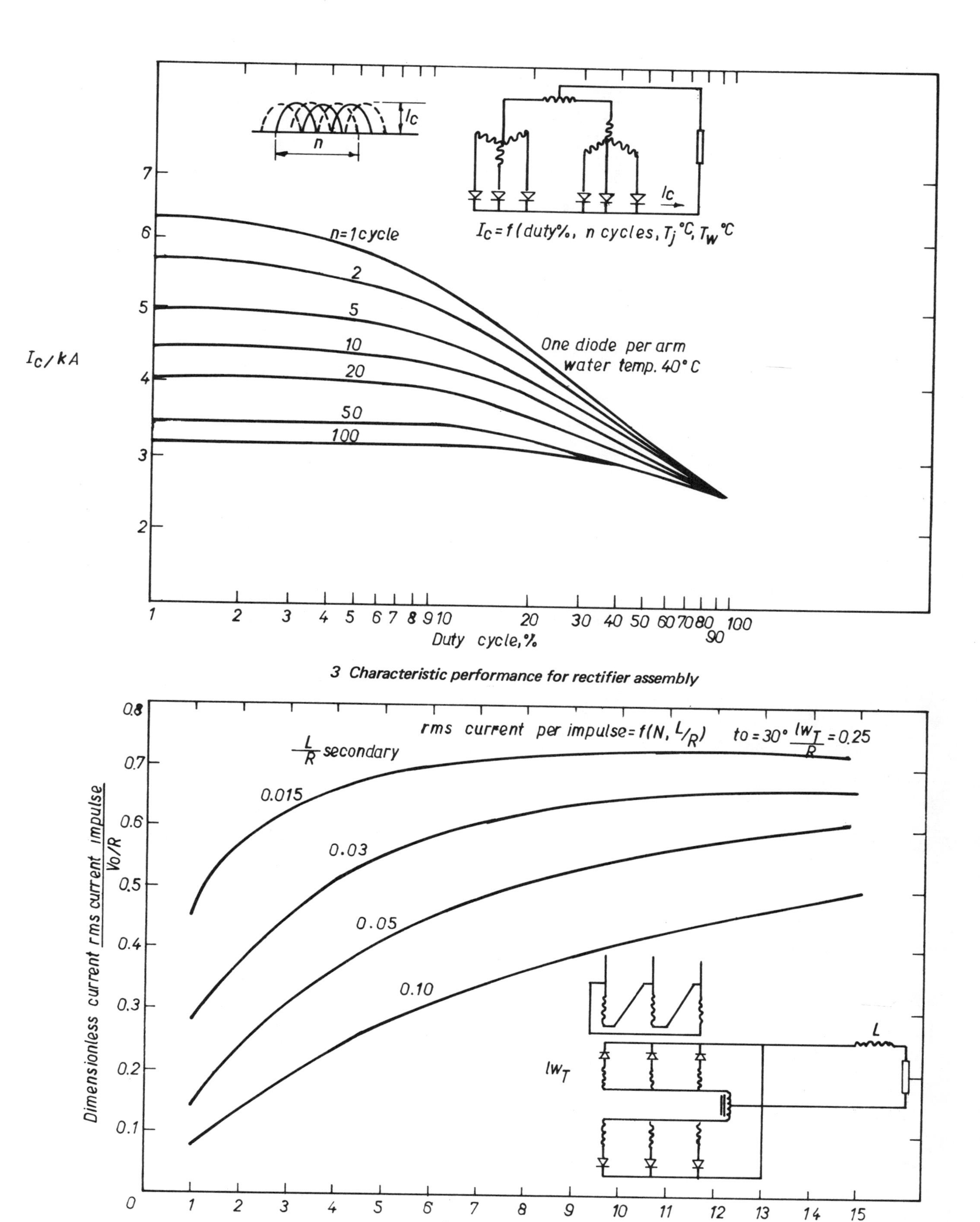

3 Characteristic performance for rectifier assembly

4 rms current per welding impulse. V_o — peak secondary voltage; R — total resistance, including transformer resistance; T_o — phase angle delay; lw_T — transformer reactance; L/R — secondary

5 Suspended gun station with six-phase rectifier

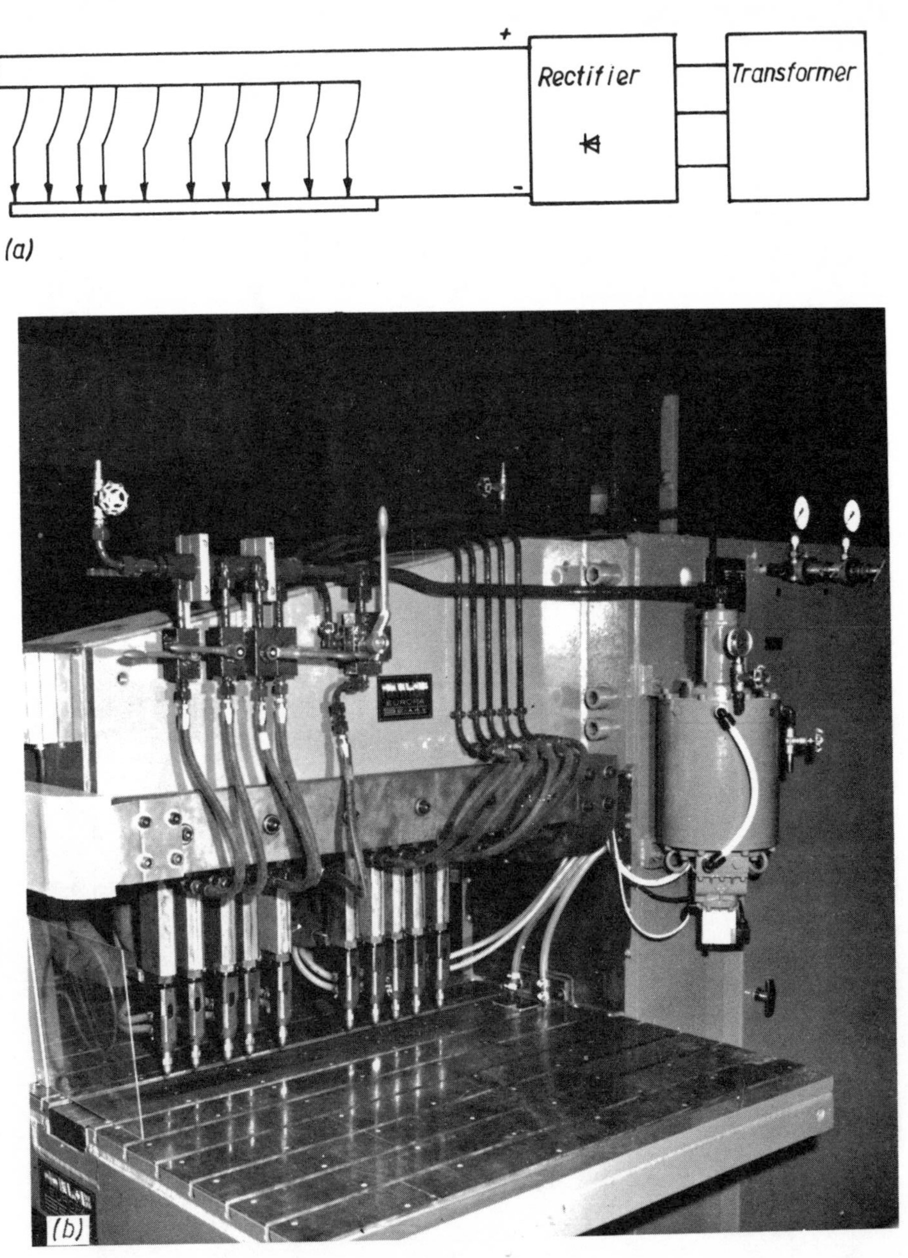

6 Multispot welding machine: (a) arrangement of secondary conductors balancing currents between different electrodes, (b) view of machine

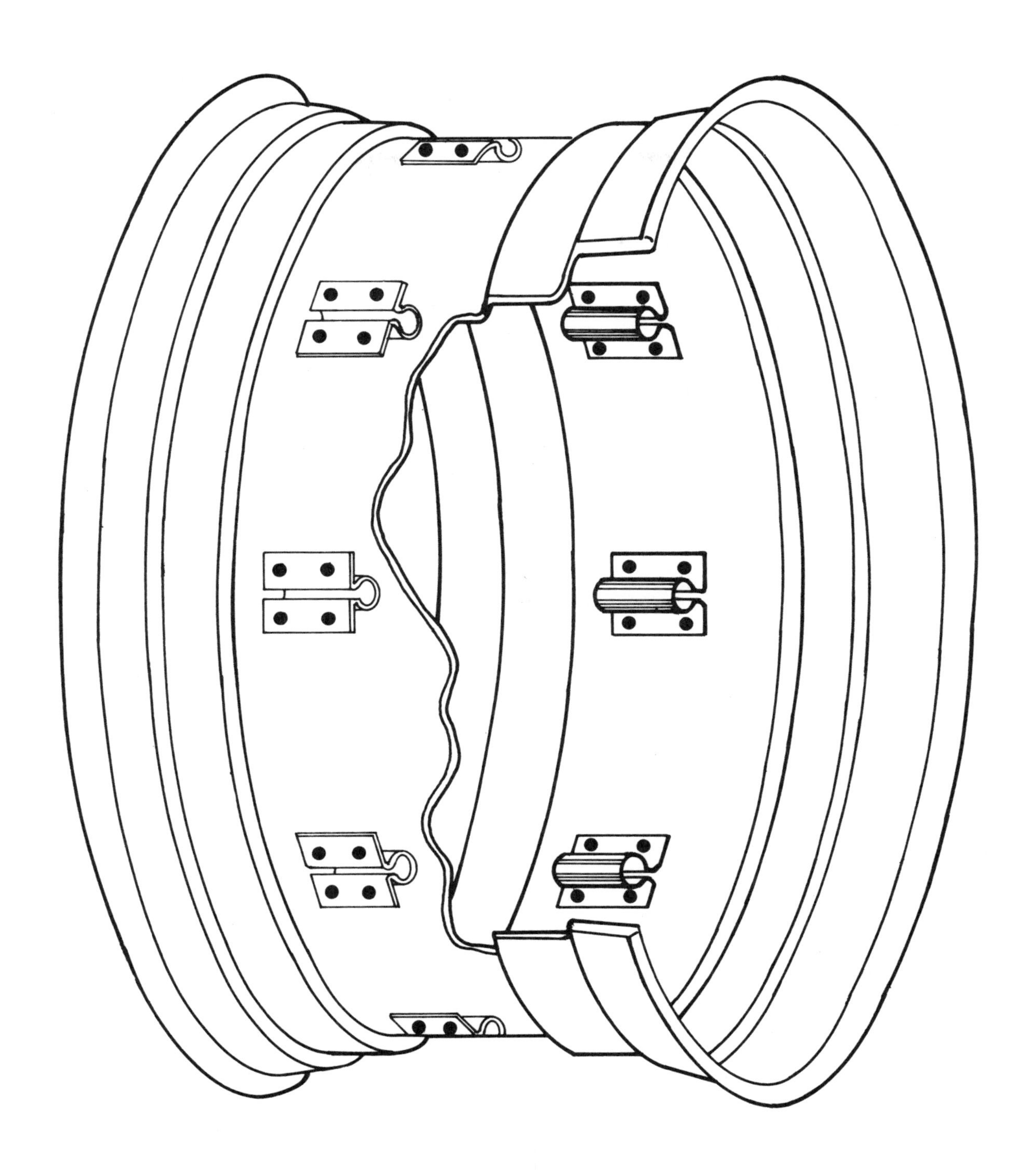

7 View of tractor wheel rim showing eight welded straps. Straps – 8mm thick mild steel; wheel – 5mm thick mild steel

PAPER 31

Control of spot weld quality in coated steel by preprogramming

N.T.Williams, BSc, PhD, MWeldI

The control of weld quality by using a feedback system based on voltage integration (Normalpunkt monitor) has been investigated for coated steels. Choice of initial welding conditions should be based on mid positions on a conventional weldability lobe rather than near the splash limit as is normal practice. Selection of the lower limit integral count markedly influences electrode life, and this should be as high as possible to give best results. Under these conditions electrode life was increased by a factor of five.

Trends obtained from the Normalpunkt were used to calibrate a conventional stepper control unit. This resulted in a greater increase in electrode life as compared with an arbitrary programme. Further increases are feasible by a more precise preprogramming and taking into account the need to step down welding current or to change the calibration of the Normalpunkt because of changes in welding electrode conditions.

It is suggested that a more widespread use of stepper controls be adopted to control both weld quality and electrode wear. This approach would minimise the capital costs of equipment.

INTRODUCTION

The need to apply quality control principles to resistance welding arises for a number of reasons. The quality of welds located in highly critical stress areas has been highlighted particularly in auto-body applications. In addition, economic factors have resulted in a reduction in the number of welds incorporated into a particular component. This sometimes means that welds are required to withstand higher applied stresses in service, and consequently it is imperative in such circumstances that welds are sound and of a minimum acceptable size. Developments in production techniques often create problems in ensuring that such requirements are met, because of the high overall welding speeds which are sometimes necessary to produce components at an economic rate. If weld quality can be guaranteed further economies arising from savings in labour costs, electrical power, and air usage are possible.

These considerations are of prime importance in multispot welding or transfer lines. The development of large fully integrated transfer lines over the last few years coupled with the general changes in economic considerations have led to the redesigning of welding equipment, particularly in the automotive industry. It is now considered that, in some instances, the use of large multispot machines is questionable from an economic

Dr Williams is Section Head — Welding at the Welsh Laboratory of the British Steel Corporation.

standpoint because such machines do not afford the required flexibility at medium and low production rates. In these circumstances it is essential for weld quality to be guaranteed, since any increase in the number of defective components could cancel out savings arising from the use of large units.

A further development which has to be considered is the increased use of robots for presenting the welding unit to the workpiece. To obtain maximum benefits from such systems it is again essential that weld quality is guaranteed. This necessitates the use of effective monitoring system or, ideally, a self-correcting system.

In recent years numerous techniques have been developed for monitoring weld quality during resistance spot welding. The various techniques so far proposed operate on the principle of recording one or more functions of the various parameters which are considered to have a direct relationship to weld quality. Parameters such as secondary voltage, secondary current, weld resistance, weld time, ultrasonic attenuation and pulse-echo recordings, the number of acoustic emissions, total weld expansion, initial expansion rate, and infrared emissions have all been used in various investigations. Many of these techniques have been developed into commercial monitors which have had a varied degree of success when used in production. It is now widely accepted that a particular monitoring system may be suitable only for certain specific applications and that much of the widespread scepticism on the part of production engineers (which still exists concerning monitors) stems from their use in areas which were clearly unsuitable for their application. The number of systems now available to the welding engineer are so numerous, and are of such a high degree of sophistication, that it was deemed necessary to develop an accepted standard practice for the assessment of any monitoring and feedback system. Such a procedure has been drawn up by the International Institute of Welding and is now used on an international basis.[1]

As a result of the increasing emphasis throughout the automobile and domestic appliance industries upon improved, consistent resistance spot weld quality, it is necessary to examine how the performance of such monitoring systems is influenced by steel quality. Perhaps, more importantly, such monitoring/ feedback systems would be most cost effective in welding coated steels, where electrode costs are relatively high and weld quality can either be inconsistent over a long production run or drop off rapidly at the end of the useful life of an electrode. The suitability of using the various systems developed has been investigated for different coating types and this Paper covers the investigation on one monitoring system based on voltage integration: the Digimetrics.

PRINCIPLES OF SYSTEM

The Digimetric monitor is a commercially available spot welding monitor and operates on the principle that the weld voltage when integrated over the weld period, i.e. the area under the voltage time curve, gives an indication of the heat input and thus the weld dimensions:

$$\text{Rate heat input, } \delta H = I^2 R\, \delta t = IV\, \delta t \quad [1]$$

where I = welding current
V = welding voltage
R = weld zone resistance
δt = weld time increment

$$\delta H \propto V \delta t$$

$$H_T \propto \int_0^T V \delta t \quad [2]$$

where H_T = total heat input and T is the weld time.

Ganowski[2] has utilised this principle in the development of an open-loop control system, known as the Normalpunkt NDT 500N. This control system operates on the same basic principles as the Digimetric NDT 500 in that pickup leads are attached to the welding electrodes and the weld voltage integrated with respect to weld time, this being displayed as a digital millivolt-second count. However, the Normalpunkt has the additional facility for automatic adjustment of the welding current for subsequent welds, if the millivolt-second count (Normalpunkt count) for a given weld falls outside preset upper and lower count limits. This is carried out via a phase shift heat control potentiometer in the Normalpunkt control circuitry, which is interfaced with the welding machine heat control potentiometer. Normalpunkt counts greater than the preset upper limit result in a stepwise reduction in welding current for the succeeding weld. Conversely, if a Normalpunkt count less than the preset lower count limit is obtained, the welding current is increased in a stepwise manner for the subsequent weld. The total number of increment steps and their magnitudes can be adjusted according to a predetermined programme.

Ganowski showed that, with this control system, it was possible to maintain weld quality within an acceptable range for a significantly greater number of welds than if no control system were used. Additionally, improvements in electrode life were reported on mild steel from 7000-10 000 to 50 000-70 000 welds under production conditions. With zinc-coated steel improvements in electrode life from 500-1000 to 7000-10 000 welds were also recorded.

In the present investigation the open-loop feedback control system was evaluated under conditions of electrode wear so that the reliability of the monitor in compensating for electrode tip deterioration could be assessed. Electrode tip deterioration is a major restraint upon fabricators when spot welding coated steels and, in general, increases in the welding current are required to compensate for the gradual loss of current density caused by electrode tip growth. The Normalpunkt monitor used was of the same basic construction as the Digimetric NDT 500 with the conventional digital display and upper and lower count limits. However, there were several additional features, including a phase shift heat control potentiometer and sixteen weld current levels by which the initial welding current could be selected. It was general practice to commence the electrode life tests on Step position 3 of the sixteen possible step positions. This meant that there were two lower welding current levels available in the event of 'splash' or hot welds, particularly in the early part of the life test. More importantly, thirteen higher welding current levels were available to allow for deterioration in weld quality during the latter part of the electrode life.

Setup procedures with the Normalpunkt consisted, firstly, of single weld trials at various welding current levels, current adjustment being made using the Normalpunkt phase shift heat control potentiometer. The voltage integral and weld size readings were recorded for welds ranging from 'stuck' to splash quality. In this way upper and lower voltage integral or count limits were determined for the subsequent electrode life tests. These were carried out in the conventional manner, except that the initial welding current was selected from various positions in the weldability lobe. It was considered that the conventional idea of using initial current settings chosen near the upper limit of the weldability lobe was not necessarily valid when using current stepper control techniques, since using an initially high current density could result in a more rapid deterioration of the electrode tip condition.

USE OF NORMALPUNKT NDT 500N TO INCREASE ELECTRODE LIFE

The relevant weldability lobe used for the various electrode wear tests on 0.9mm normal spangle Galvatite is shown in Fig.1. A weld time of 240msec was selected for the various electrode life tests and the welding current/weld size and Normalpunkt count/weld size graphs at this weld time are shown in Fig.2. The lower limit of the weldability lobe corresponded to a welding current of 9.8kA and a Normalpunkt count of 77mV-sec. Corresponding values for the upper limit of the range for satisfactory weld quality were 10.9kA welding current and 89mV-sec Normalpunkt count.

Variations in weld size and Normalpunkt count during a conventional electrode life test, i.e. no correction, 0.9mm normal spangle Galvatite are shown in Fig.3. This test was carried out using a welding current of 10.8kA, this being chosen in accordance with weldability lobe criteria at a predetermined weld time, Fig.1. The electrode life obtained in this instance was between 1000 and 1600 welds.

It can be concluded from this result that the Normalpunkt NDT 500N, when used in the noncontrol mode, gave a clear indication of the final loss in weld quality at 1300-1400 welds. Results obtained when the Normalpunkt NDT 500N was used in the open-loop control mode are shown in Fig.4. A starting welding current of 10.4kA was used in this test, this level of current corresponding approximately to the centre of the weldability lobe, Fig.1. The Normalpunkt limits were set at the same values as in the previous test, i.e. Normalpunkt counts of 77 and 89mV-sec were used to activate changes in welding current. It was observed that, during the first one hundred welds, the welding current was automatically reduced to approximately 9.2kA. This occurred because excessively 'hot' welds were produced during the period corresponding to the conditioning of the electrode. As a result the Normalpunkt count indicated was greater than the present upper limit of 89mV-sec and, consequently, the control system systematically reduced the welding current for the subsequent weld by one of the available sixteen steps built into the monitor circuitry. This process of current reduction continued until welds were being produced which gave Normalpunkt counts of between 77 and 89mV-sec. On those occasions when the control system reduced the current level to such a low value that welds of less than 77mV-sec Normalpunkt count were produced, the control system was observed to operate in the opposite manner thereby increasing the current level so as to produce welds of an acceptable Normalpunkt count. In this way the current step 'number' which was initially set at 'THREE', quickly fell to 'TWO', and then 'ONE' and stayed at this position for the first hundred welds. Such a reduction in current minimised brassing and hence the rate of electrode tip deterioration.

At the end of the initial electrode tip conditioning period it was observed that the welding current automatically increased to 10.0kA between 100 and 300 welds, this being necessary to maintain the integral count at above the preset minimum value of 77mV-sec, Fig.4. This process continued until, after 1600 welds, the welding current had increased to approximately 11.0kA. The process of gradual 'stepping-up' in welding current whenever a Normalpunkt count of less than 77mV-sec was obtained continued and at 2600 welds the weld size was larger than 4.0mm. From this point onwards deterioration of the electrode tips proceeded at a faster rate, and the 'stepping-up' of welding current occurred more frequently. In general, the extent of variation in weld size also increased. After 3000 welds the welding current had increased to 17.0kA corresponding to Step 12 on the monitor. By this time the electrode tips had grown to approximately 7.7mm as compared with a starting diameter of 4.8mm. During the next fifty welds the control system operated in an erratic manner since abnormally high Normalpunkt counts began to be observed. Consequently, in accordance with the preset, dialled-in programme, the welding current decreased until after 3100 welds; stuck welds were produced consistently. The final electrode life, based on measurements of weld size, was considered to be 3000, a value of 2-3 times that obtained under conventional static welding conditions for the configuration being assessed.

A further electrode life test was made using the same initial welding current, i.e. 10.4kA, but with a change in the value of the Normalpunkt count lower limit, this being increased from 77 to 83mV-sec. It was observed that, in the first test, only twelve of the sixteen available welding current steps were activated, thereby limiting the potential life of the electrodes. It was considered that further improvements could be possible by a greater utilisation of the available current steps. This could be achieved by setting the lower limit to a higher value. In addition, during the first test very large changes in welding current occurred over a relatively small number of welds, for example:

Weld no.	Current step, A
900	+ 1600
2000	+ 1500
2500	+ 1500
3000	- 2000

It was considered that with a large working Normalpunkt count range, the control system was not detecting and compensating for relatively large variations in weld quality at a sufficiently early stage. In considering the method of operation of the monitor with a wide working range, Fig.5a, the count and, indirectly, the weld size, can fall a considerable extent before going out of limits. The longer the delay before remedial action is taken the greater is the rate of fall R_1 in weld size. When the count does fall outside the lower limit a relatively large increase in weld current is required to compensate for this relatively rapid fall in weld size. By using a narrower working Normalpunkt count range, i.e. with a higher low limit, it was considered that, when the count fell outside limits, the rate of fall in weld size R_2 at that point, Fig.5b, would be correspondingly lower. In such an instance the adjustment in welding current would be more gradual, with lower overall current steps to compensate for the fall.

During the test, in which a narrower range between the upper and lower count was set, similar reductions in welding current occurred during the initial conditioning period of a few hundred welds, Fig.6. As the number of welds increased to 3000 the welding current was progressively 'stepped-up' in 100-200A increments. At this stage, weld quality was consistently good and the level of welding current was relatively low at 12.0kA. Between 3000 and 3400 welds the welding current increased rapidly in a series of steps totalling 1100A. Weld quality remained at a satisfactory level and the average weld size increased steadily throughout the test, in balance with an accompanying increase in electrode tip diameter. At 4800 welds the welding current had been increased to approximately 14.5kA, corresponding to Step 9, after which the weld size began to decrease rapidly from 7.0 to 5.0mm without count values of less than 83mV-sec being recorded. Consequently, the drop in weld quality was not sensed sufficiently early by the control system to allow a correction to be made to the welding current. The test was terminated at the stuck weld condition obtained at 5100 welds, only eleven of the sixteen current step positions having been utilised. An electrode life of 5000 welds was achieved during this test, a value four to five times greater than the corresponding value achieved using conventional static welding conditions.

DISCUSSION

Results indicated that an increase in electrode life can be obtained by using an open-loop control system, an increase of two or even three times the standard electrode life being observed for normal spangle Galvatite. These results were obtained by choosing a lower count limit, according to the minimum acceptable weld size criterion, here 3.8mm ≡ 77mV-sec based on single weld tests. However, the results indicate that further improvements are possible and that the real benefits of the control system can be derived only after a number of tests have been carried out to give a better choice of the preset limits. The performance of the open-loop control system depended principally on the choice of the lower integral limit. A factor which has to be considered is the choice of the initial welding current: for best results this should be midway along the current work range for the particular weldability lobe. This is contrary to normally accepted concepts used for welding with no feedback, where a welding current just below the upper limit of the weldability lobe is recommended. The choice of a lower Normalpunkt count limit which is only 5-6mV-sec below the upper integral limit is beneficial. The latter is important since, in a number of tests, the calibration of the monitor appeared to change with the condition of the welding electrodes. For example, electrode contamination or deformation can influence the value of the measured voltage between the electrodes as can the temperature of the electrodes.

Increases in the value of the measured voltage across the electrode tips for a given weld size are a function of the number of welds made. This is a characteristic feature of those open/closed-loop voltage-based control systems which use voltage connections located away from the electrode tips. The implication of this effect is that, in the Normalpunkt NDT 500N, although it will be able to effectively extend the working life of a pair of spot welding electrodes, it will not indicate the true end point of that extended life unless the low/high limit values are modified during the production run. In a production situation this could be facilitated by the use of the second available Normalpunkt count low limit.

It is considered that the Normalpunkt NDT 500N should be used in its open-loop control mode using the critically set lower limit for only a portion of the extended electrode life. For example, during the test shown in Fig.6 it would be desirable to use the initial lower limit up to about 4000 welds. At this point it would be advantageous to change to an increased lower limit value to compensate for the erroneous voltage component. The value of this second lower limit would be determined from previous tests using the same welding schedule/monitor limits. This is best carried out by obtaining the relationship between weld quality and Normalpunkt count for the stuck or minimum acceptable weld size to the splash weld condition using the worn electrodes. In this way it would be possible to define new high and low limits which will be different from the corresponding limits determined using new electrodes. It is the balance between the change in the high and low limit that has to be taken into account if optimum performance of the Normalpunkt is to be achieved. By using this higher low limit value it is likely that a full utilisation of the sixteen available current steps will be possible, thereby giving a further increase in electrode life.

In production, tests in which the Normalpunkt is used in the active control mode would have to be carried out to determine the point during the life of an electrode at which the correction system loses control, i.e. is unable to compensate sufficiently quickly for reductions in current density. A number of tests would be necessary to be able to take into account aspects such as differences in electrode or machine condition, variations in pressing or material quality, etc. Thus, meaningful limits need to be determined to cover operation on a long-term basis, i.e. shift to shift, because, once set up, the system would be left to its own devices to control weld quality without supervision.

In terms of application of the system a limiting factor is its cost, and it would not be feasible to install the control system on a large number of welding stations. However the principle is sound and can be applied in an alternative manner with very little increase in cost. The method of correction is basically a stepper system of increasing the welding current via a phase shift control. This can be carried out quite easily using a simple and cheap stepper control based on the number of welds made. Such systems are used extensively in the USA and Continental Europe but have found limited application in the UK. Investigations have shown that a x2 to x3 improvement is possible using automatic stepper control systems. Difficulties encountered with these systems are associated with the problem of determining the optimum programme for the stepper system. For example, incorrect setup can lead to a decrease rather than an increase in electrode life, Fig. 7. If current increases are made too early the splash condition may be reached too early in the life of an electrode. Conversely, if made too late an unacceptable weld size may be reached before correction can compensate. As soon as one of the current steps goes out of sequence, i.e. the automatic increase in current is not in phase with the actual or real changes in weld quality, all subsequent phase shift steps will be out of phase. Such systems change the welding current on a simple periodic basis, unlike the Normalpunkt which allows the welding current to be changed, if required, for every weld.

What is required is a careful precalibration of the precise sequence and timing of the current steps, i.e. the basic curve defining the changes in current density with the increasing number of welds made from a particular set of welding electrodes. The Normalpunkt can be used to determine the general trend for a particular welding station. The number of current steps and their magnitude would be determined, and a conventional stepper control can then be preprogrammed according to the calibration obtained. To date a limited amount of work has been carried out using this philosophy and the results obtained are encouraging.

Aluminium-coated steels pose a major problem in terms of electrode life, a typical electrode life for a particular welding station under severe welding conditions being in the region of 150 welds. Using the Normalpunkt it was possible to increase electrode life to approximately 700 welds, an increase of around x4.5. Results obtained, Fig. 7, using a stepper control setup in the conventional manner resulted in an electrode life of 400 welds, but preprogramming the stepper control based on the step up pattern obtained with the Normalpunkt, resulted in a further increase in electrode life to 600 welds.

Further work is necessary to evaluate the potential of this technique, particularly under production conditions. Advantage should also be taken of the characteristic of the Normalpunkt to step down during the first few hundred welds corresponding to the conditioning period of the electrode. During this period it is possible to move from the nonsplash to the splash condition after only a few welds, thereby causing considerable damage to the electrode. This situation is eliminated if the current is stepped down; consequently, any preprogramming of a stepper control should incorporate current reduction steps in the early stages.

In addition it is recognised that the monitoring of a single process parameter has limitations in controlling weld quality. For example, systems based on the integration of voltage across the electrodes are sensitive to shunt conditions, and the sensitivity therefore decreases at small interweld spacings. It is most probable that optimum results would best be obtained by a control system based on measuring two separate parameters. Such a technique has been recently developed by Eichhorn and Singh[3] who combined dynamic resistance and nugget expansion measurements and related them to weld quality. Similarly, combining voltage integration and nugget expansion could give excellent results in terms of increased electrode life. Voltage integration measurements could be used to control weld quality until the point is reached when instability sets in, after which any further control could be based on expansion. However the cost of such a system would have to be balanced against the financial return obtained in maintaining a high level of weld quality. Whichever technique is used, the best balance between satisfactory weld quality, optimum electrode life, and capital cost of equipment would appear to be obtained by using precalibrated stepper control techniques.

CONCLUSIONS

It may be concluded that marked increase in electrode life can be obtained by using techniques based on integration of the voltage between the spot welding electrodes to control weld quality. The commercial system based on this principle, i.e. Normalpunkt, can, in addition, be used to set up and preprogramme a simple stepper control system to achieve similar increases in electrode life.

ACKNOWLEDGEMENTS

The author wishes to thank Dr R. K. Snee for carrying out the experimental work reported.

This Paper is published by the kind permission of the Manager of Welsh Laboratory, British Steel Corporation.

REFERENCES

1 ANON, 'Recommended practice for the

evaluation of resistance spot welding monitor and correction systems'. Int'l Inst. of Welding Doc. III-457-72.

2 GANOWSKI, F. J. 'Technique for ensuring spot welds in a standard welding range'. Int'l Inst. of Welding Doc. III-505-74.

3 EICHHORN, F. and SINGH, S. 'Method of ensuring the maintenance of constant quality of spot welds'. Int'l Inst. of Welding Doc. III-564-77.

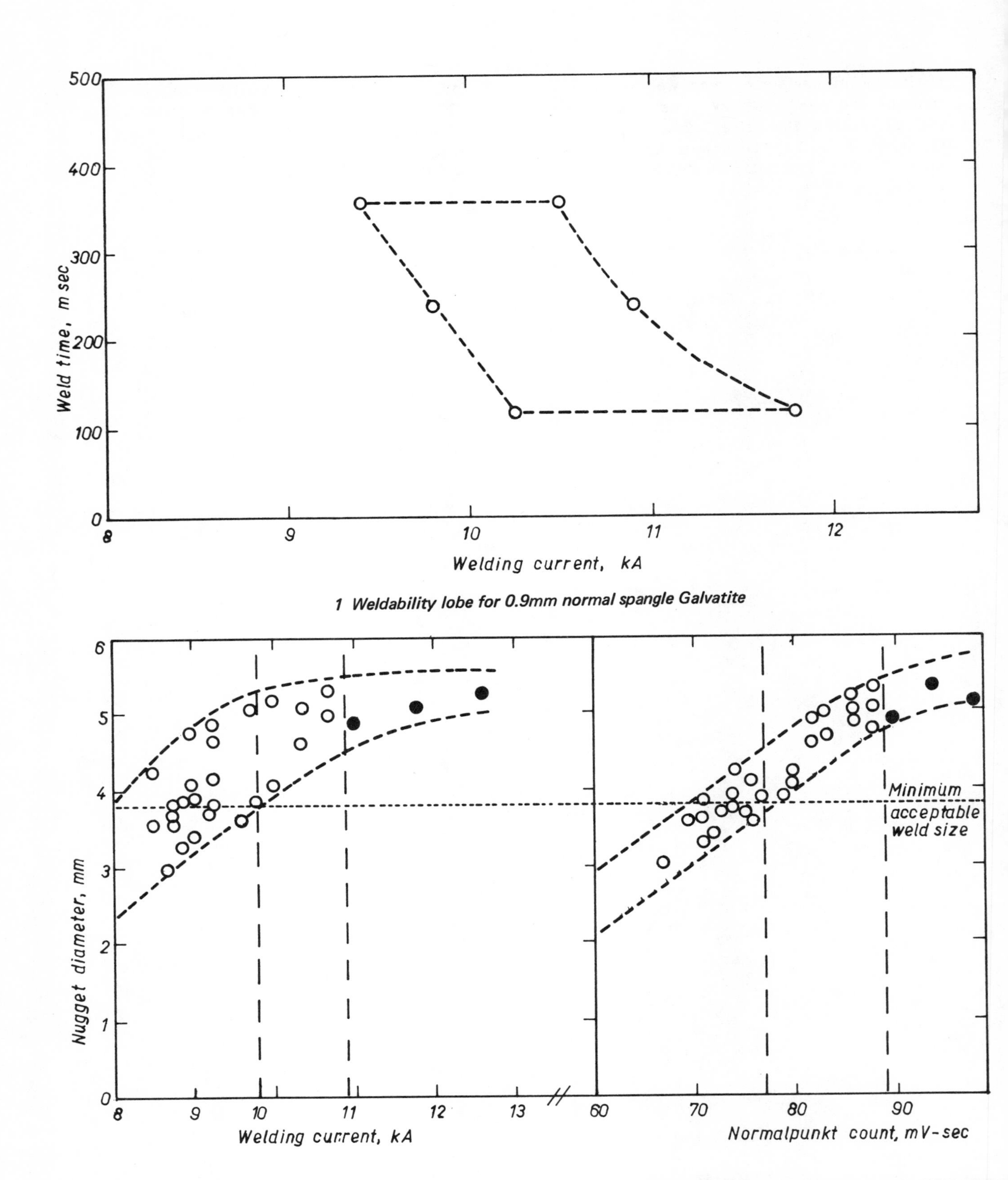

1 Weldability lobe for 0.9mm normal spangle Galvatite

2 Nugget diameter/welding current or Normalpunkt count for 0.9mm normal spangle Galvatite using Normalpunkt NDT 500N in nonactive control mode. ● splash welds

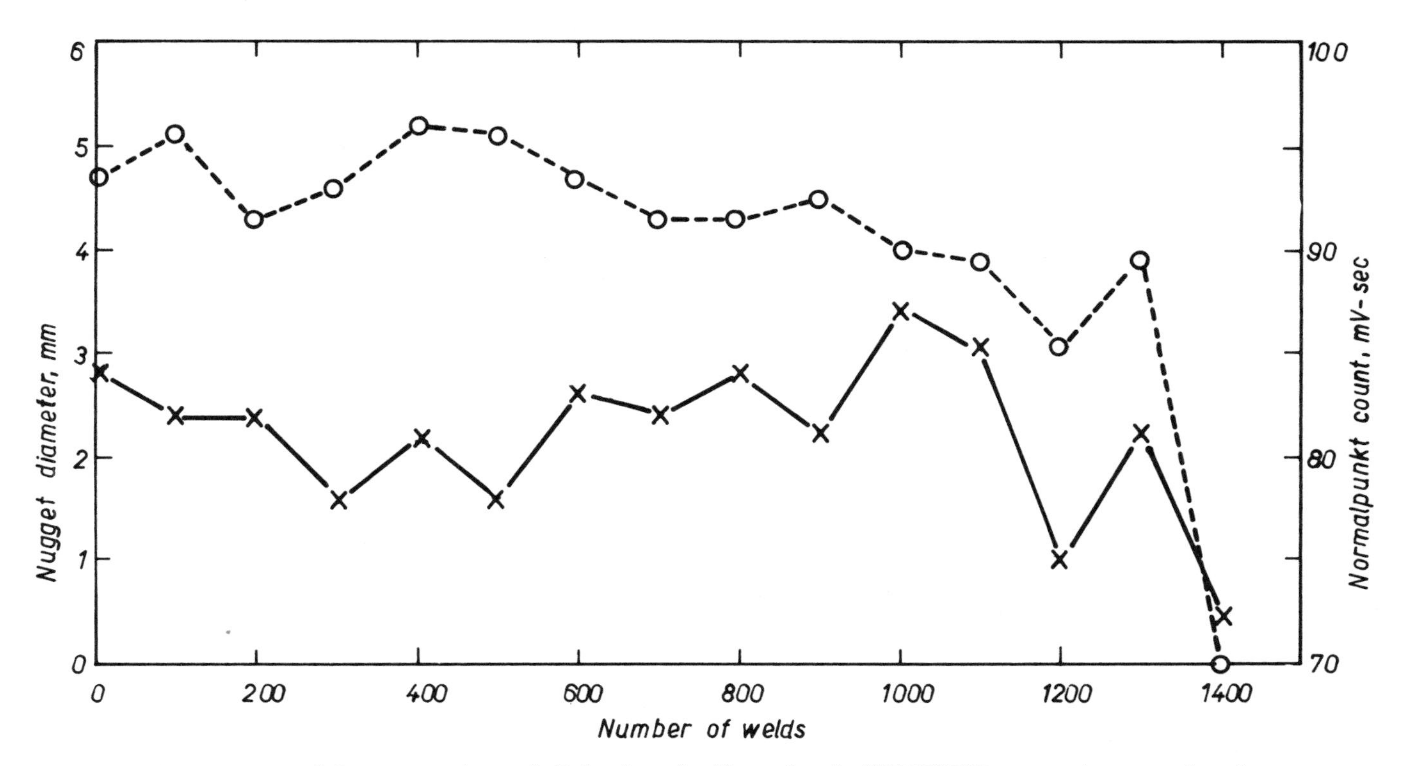

3 *Electrode wear test on 0.9mm normal spangle Galvatite using Normalpunkt NDT 500N in nonactive control mode. ---- nugget diameter; ——— Normalpunkt count*

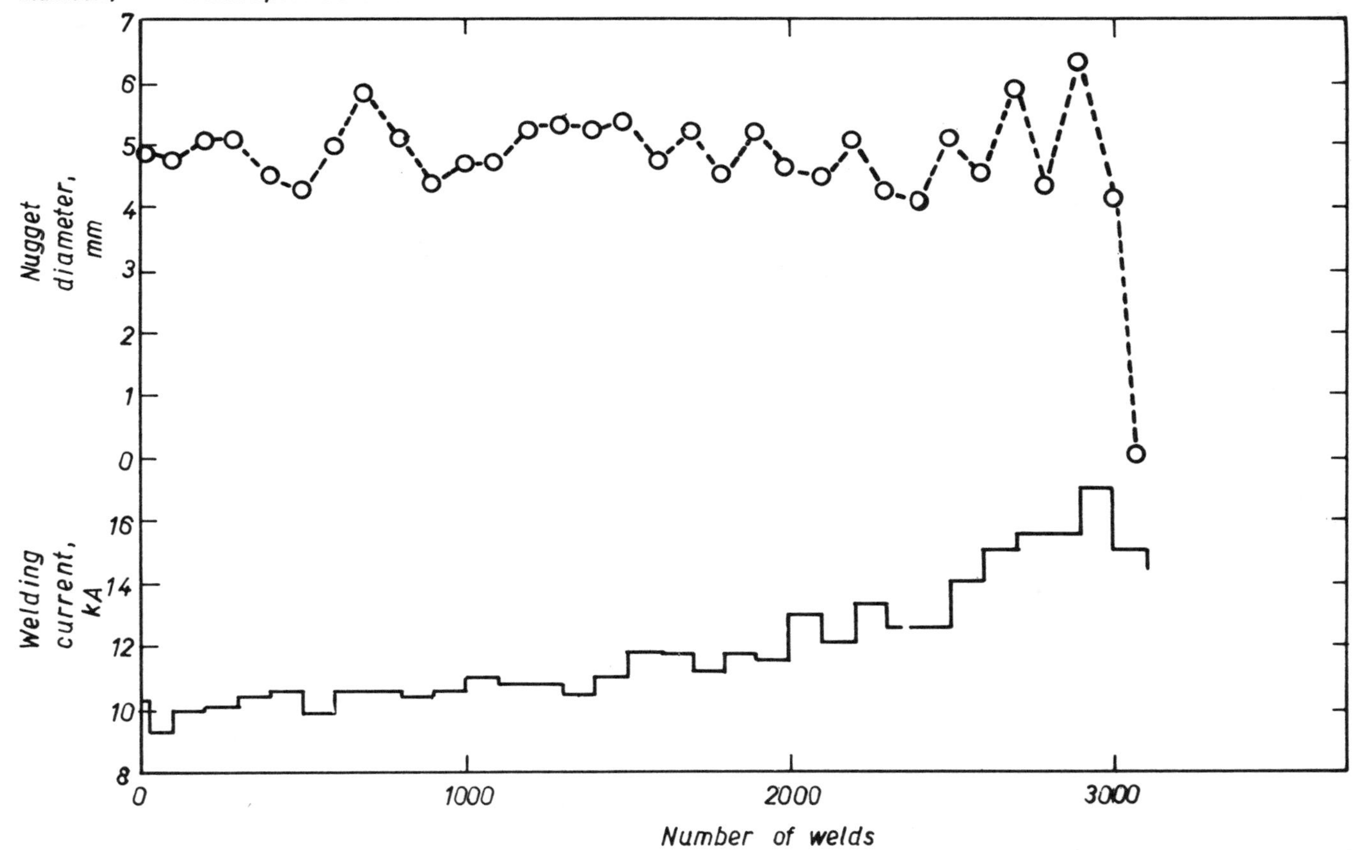

4 *Electrode wear test on 0.9mm normal spangle Galvatite using Normalpunkt NDT 500N in active open-loop control mode. Effective electrode life — 3000 welds; start-current setting — no.3 heat stage; high limit — 89mV-sec; low limit — 77mV-sec*

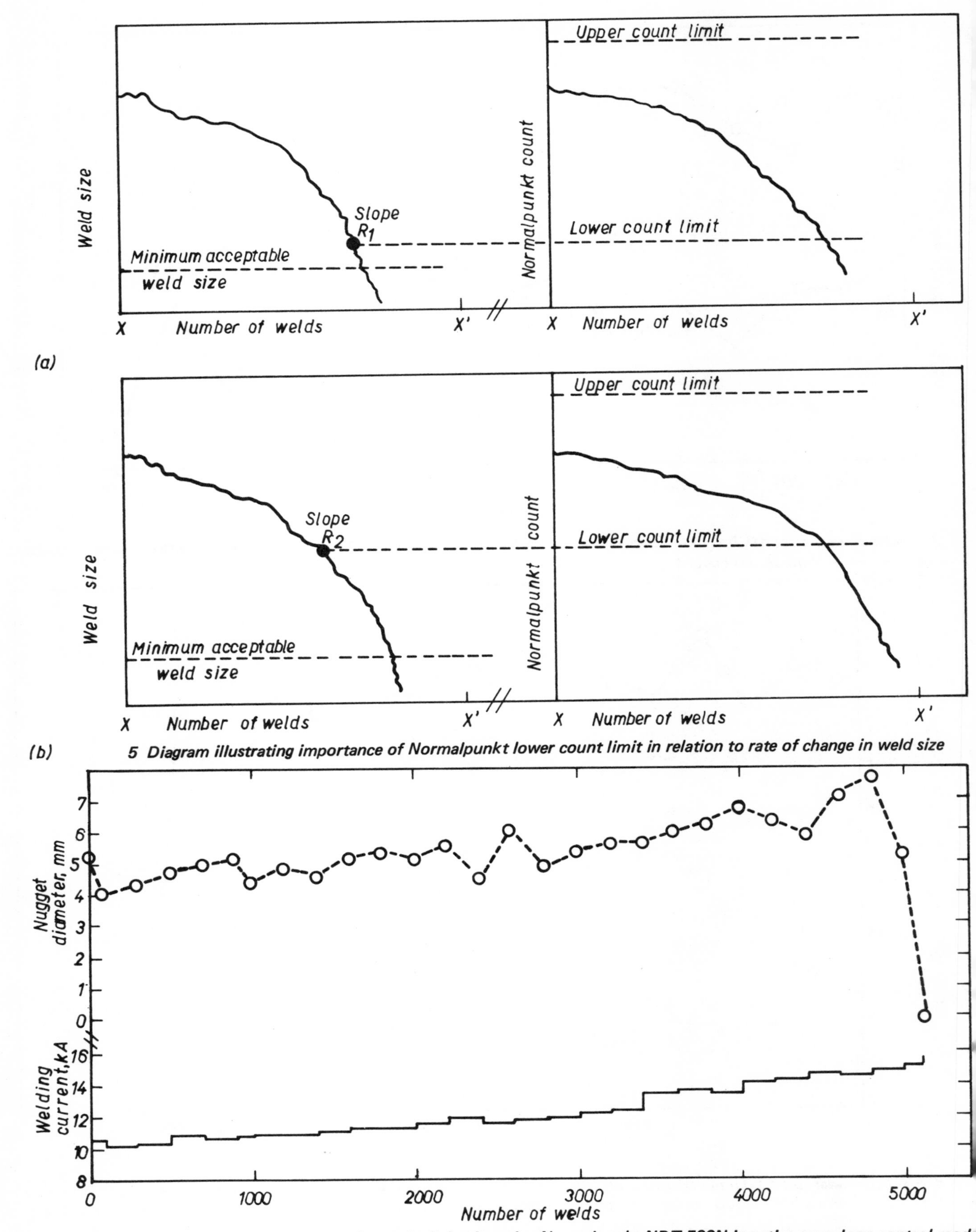

5 Diagram illustrating importance of Normalpunkt lower count limit in relation to rate of change in weld size

6 Electrode wear test on 0.9mm normal spangle Galvatite using Normalpunkt NDT 500N in active open-loop control mod Effective electrode life — 5000 welds; start-current setting — no.3 heat stage; high limit — 89mV-sec; low limit — 83mV-sec

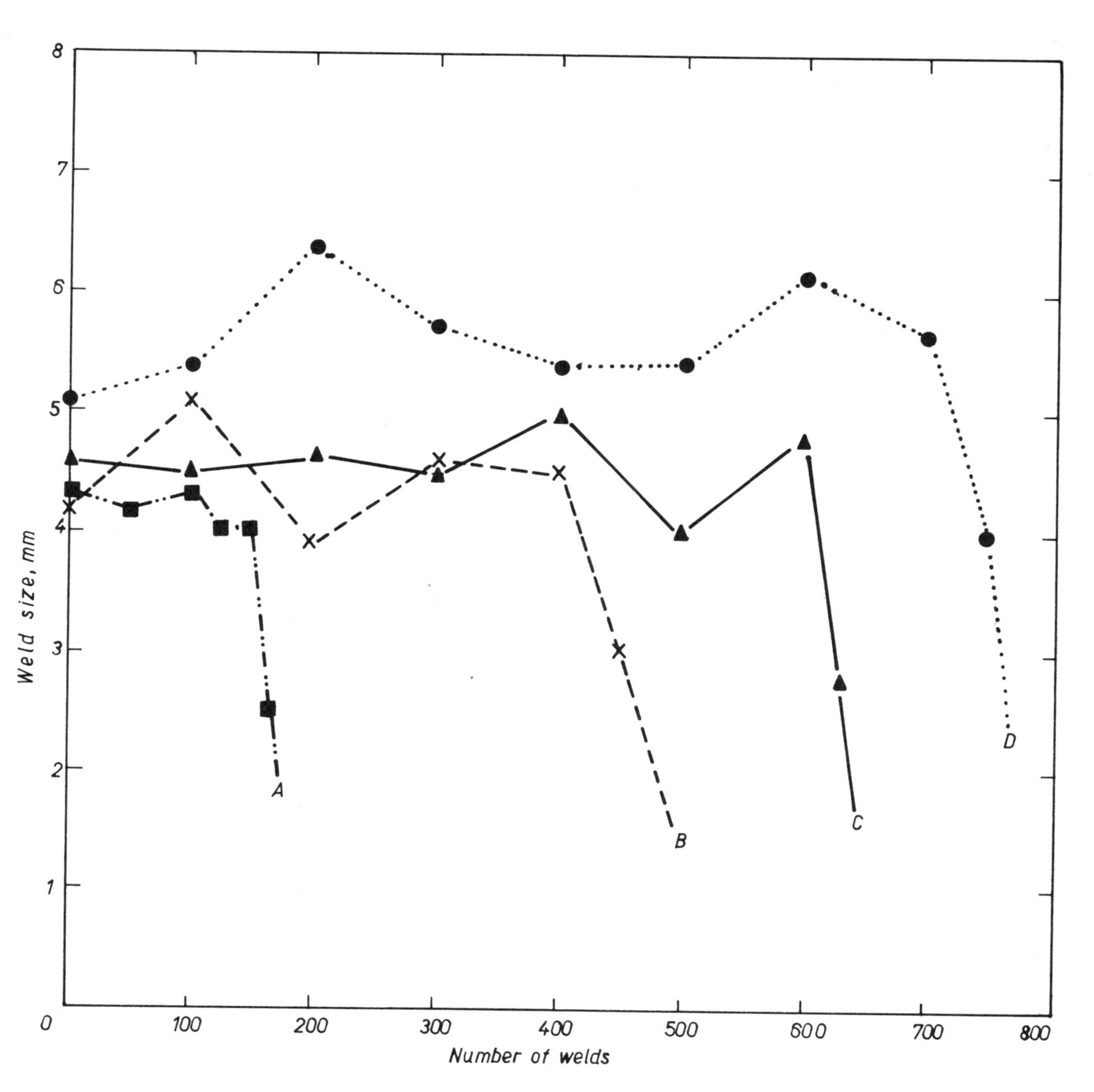

7 Influence of calibration on electrode life obtained with stepper controls — aluminium-coated steel. A — normal; B — stepper (uncalibrated); C — stepper (calibrated); D — Normalpunkt

Adaptive system of resistance welding control

M.Janota

INTRODUCTION

The selection of a variable which provides information on the controlled process appears to be the major problem in the development of a resistance welding control system. The system may be the more successful in practical application the closer the variable is related to the output of the process, i.e. a criterion of quality of the weld joint.

In resistance spot welding the most significant aspects (from the point of view of their results) are the phenomena connected with the weld nugget formation. A relatively complex information on the weld nugget growth can be obtained from the thermal expansion (TE) of the weld nugget with time. The heating and especially the melting of material in the weld joint manifest themselves in a force causing relative movement of the welding electrodes which can be measured.

This variable is connected mainly with the volume of molten metal and is thus an indication of the nugget growth.

The relationship of the TE to weld joint quality has been well known for many years.[1] Nevertheless, the known systems utilising this variable for monitoring or control have shown some shortcomings[2] on which some attempts have here been made to overcome them.

MEASURING THE THERMAL EXPANSION

Expansion characteristics

According to de Vos[3] the total TE comprises three components, for each of which a simple calculation can be set up:

1 Expansion coaxial with the electrode force, caused by linear TE of the welded material

$$\Delta S = \alpha \int_0^{S_1} T_1 d\, S \qquad [1]$$

where S is the change in thickness

α is the coefficient of linear expansion for the welded material

T_1 is temperature distribution in the sheets along the axis of the electrodes, Fig.1a.

2 The difference in thickness as a result of expansion perpendicular to the direction of the electrode force:

$$\Delta' S = d\, \bar{T}_2\, S_2\, (2 + \alpha T_2) \qquad [2]$$

where $\bar{T}_2 = \frac{1}{R_1} \int_0^{R_1} T_2\, d\, R$ is the mean temperature according to the radius R_1, Fig.1b.

3 Expansion as a result of increase in the volume of the molten nugget material:

$$\Delta'' = S_3 \delta / 100 \qquad [3]$$

where S_3 is molten nugget thickness

δ is percentage volume increase caused by the melted portion.

The proportional effect of the different components varies during the welding process. Components 1 and 2 above are dominant at the beginning and component 3 prevails towards the end of the process; it constitutes about 65% of the total TE.

Thermal expansion manifests itself by the relative electrode movement which again consists of two components:

1 The movement of the upper electrode relative to the pneumatic cylinder against the acting air pressure

2 The deformation of the arms of a welding machine owing to friction forces in the upper electrode head and pneumatic cylinder

Function of the system

The measurement of electrode movements, typically about 0.15-0.2mm in the time

Mr Janota is at The Welding Research Institute, Bratislava, Czechoslovakia.

interval 0.2sec, may be rather difficult when accompanied in operation by mechanical impacts and intense electromagnetic fields. Moreover, the deformations as a consequence of friction forces vary from one machine to another and may even vary from weld to weld. This causes great error when attempting to measure the absolute value of TE.

Therefore here it is preferred to control the process in terms of the characteristic features of the TE with time which do not invoke absolute values. As has been described previously,[4,5] two such features occur in the time course of TE which can be utilised for process control. The first is the maximum in the TE course which represents the optimum moment for the termination of the welding current. The second is the rate, or dTE/dt value, which is connected to the intensity of the heating process and thus the rms value of welding current.

Figure 2 shows the block diagram of the control system. The components which evaluate the TE course consist in principle of a differentiator and of detection and indicating circuits. The principle of the system function is illustrated in Fig.3. At the beginning of the process the rms value of the welding current rises with a certain preset slope. A signal from block 7, Fig.2, detects when the intensity of the current has reached the desired level, and the slope is then terminated. The process then continues, at approximately constant rms current value, until the TE reaches its maximum. Then the signal from block 8 switches off the current.

The parameters which remain to be chosen manually are then the diameter of the electrode, the welding force, and the process intensity, that is level 1, Fig.3.

EVALUATION METHOD FOR SYSTEM

The tests were performed in general accordance with the methods recommended in International Institute of Welding documents.[6] The material used was low carbon steel in thicknesses of 0.5, 1.0, 1.5, and 2.0mm. The results obtained with a conventional timer and with automatic control have been compared. For these tests two types of welding machine were chosen. The Sciaky E266H welding press with small throat depth to provide a very rigid machine, and spot welding unit BP 80.2 which possesses low rigidity and throat depth of 750mm. In the conventional mode of working both machines had synchronous timing and phase shift control.

During the tests the welding current, time, and electrode force were measured and recorded together with the TE course and control system response.

The quality of individual welds was evaluated by the peel test[6] and in certain aspects also by tensile shear testing. Every point in Figs 4, 5, 6, and 7 represents the mean value of five measurements.

Mains voltage fluctuation

The voltage fluctuations were simulated by means of transformer tap changes. The test results are shown in Fig.4a. By extrapolating the regression equations it can be estimated that a 25% decrease of the nugget diameter (from the nominal value of 5mm) has been caused not by a 15% decrease in voltage with conventional control but by a 35% decrease in the automatic control. The expulsions or splashed welds have arisen at 20% increase in voltage with conventional control compared with 45% increase under automatic control.

Fluctuations in electrode force

It can be seen from Fig.4b that over the whole range tested, i.e. 57 to + 86%, around the nominal value 70MPa, the resulting curve obtained with automatic control is very flat. Within the whole range a statistically significant decrease in the nugget diameter has not been found.

Influence of shunting

In this test the distance between two spot welds was changed and the diameter of the second weld nugget was measured, Fig.5. Both conventional and automatic control systems did not differ significantly when the distance between spots was larger than 3De; De being the diameter of the electrode. Below this value the nugget diameter as produced with conventional control decreases whereas in automatic control it increases. It seems that from this point of view the automatic control system is slightly overcompensated.

Distance from the sheet edge

The two systems, conventional and automatic control, did not show much difference with regard to this aspect, Fig.8. Nevertheless, one feature is the ability of the automatic control to stop the process immediately after the expulsion has occurred. This means at least a certain improvement in the operator's working conditions.

Compensation for different sheet thicknesses

The results are clear from Fig.6: in conventional

Table 1 Reproducibility of tensile shear strength tests, welding on machine E266H

	Automatic control	Conventional control	Dispersion ratio
Clean sheets	n = 9 $\bar{x}$ = 5956N v_x = 1.88%	n = 10 $\bar{x}$ = 6069N v_x = 0.99%	Significant
Rusty sheets	n = 9 $\bar{x}$ = 5580N v_x = 9.44%	n = 7 $\bar{x}$ = 6217N v_x = 0.79%	Significant
Difference in diameters	Nonsignificant	Significant	
Dispersion ratio	Significant	Nonsignificant	

n = number of specimens tested
$\bar{x}$ = mean value
v_x = variation coefficient

Table 2 Reproducibility of tensile shear strength tests for machine BP 80.2

	Automatic control	Conventional control	Dispersion ratio
Clean sheets	n = 10 $\bar{x}$ = 5850N v_x = 2.13%	n = 10 $\bar{x}$ = 5650N v_x = 1.86%	Nonsignificant
Rusty sheets	n = 5 $\bar{x}$ = 5634N v = 2.21%	n = 6 $\bar{x}$ = 5910N v = 5.51%	Significant
Difference in diameters	Significant	Nonsignificant	
Dispersion ratio	Nonsignificant	Significant	

n = number of specimens tested
$\bar{x}$ = mean value
v_x = variation coefficient

ontrol the preset parameters have not produced ny molten nugget for 2mm thickness, whereas he automatic control has been able to produce uite reasonable joints. A similar situation rises when the number of sheets to be welded as changed. This illustrates one of the reatest advantages of the new system, the o-called multiprocessing[7] ability, i.e. the ossibility of welding different combinations of sheet thickness and numbers without any changes in the setting of the welding parameters.

Reproducibility of tensile shear strength

The strength tests have been carried out on both the clean and slightly rusted sheet surfaces. Tables 1 and 2 show that in this respect the automatic control system gives slightly worse results on the clean surface, and, in the rigid

Table 3 Review of test results for machine E266H

Test no.	Feature tested	Characteristics of automatic control system
1	Voltage fluctuations	Better than conventional control
2	Electrode force fluctuations	-
3	Effect of shunting	-
4	Welding on sheet edge	Equal to conventional control
5	Different sheet thickness	Better than conventional control
6	Dispersion of tensile shear strength	Slightly worse than conventional control
7	Effect of rust	Worse than conventional control

Table 4 Review of test results for machine BP 80.2

Test no.	Feature tested	Characteristics of automatic control system
1	Voltage fluctuations	Better than conventional control
2	Electrode force fluctuations	Better than conventional control
3	Effect of shunting	Better than conventional control
4	Welding on sheet edge	Better than conventional control
5	Different sheet thickness	Better than conventional control
6	Dispersion of tensile shear strength	Equal to conventional control
	Effect of rust	Worse than conventional control
7	Bad fit-up	Better than conventional control

welding machine E266H, also the worst performance on rusty surfaces. This can be explained in terms of the control system functions. The increase in the welding current, i.e. slope up, at the beginning of the process (which has a more intensive effect in a rusty surface) can be terminated within one half-cycle (0.01sec) only, which apparently is too long a time interval.

Influence of fitup

This test is not included in the IIW document III-457-72,[6] however this feature is very important in practical application. The specimen used is shown in Fig.9.

In the first phase of the welding process the current is led through the shunt paths. Simultaneously the different rigidity of the work is simulated by the changing distance, S_s, in the test sample.

With the conventionally controlled process even the nominal value of the nugget diameter was not obtained, and below a value of S_s = 50mm the nugget diameter decreased further. The automatic control not only provided larger weld diameters but even certain overcompensation can be seen here again, since the diameter rises with decreasing S_s value, Fig.7.

CONCLUSIONS

The overall evaluation summarised in Tables 3 and 4 shows that the new control system gives better results in all the tests with the exception of welding rusted sheets. For welding near the sheet edge, the performance of the automatic control system is at least equal to that of conventional control.

Much better results were obtained in tests on the effect of voltage fluctuations, bad fitup, and especially for changes in sheet thickness and configuration.

On the other hand, it should be pointed out that a rigid spot welding unit with low friction forces is necessary for a proper function of the automatic control system. Application of the system to the portable gun type welding machines cannot apparently be successful.

REFERENCES

1 BALKOVEC, D.S. 'Control methods for spot welding'. Avtogenoe delo, (12), 1947, 9-14.

2 JOHNSON, K.I. and NEEDHAM, J.C. 'Quality control systems for resistance welding'. Paper presented at Conference 'Steel Sheet and Strip Welding', Kenilworth, 15-17 March 1972, 14-19.

3 VOS, E. de. 'Measurement of electrode movement as a means of spot weld quality control'. IIW Doc.III-283-66.

4 JANOTA, M. 'The relationship between thermal expansion and the growth of a resistance spot weld'. Paper presented at Conference 'Advances in Welding Processes', Harrogate, 7-9 May 1974, 21-6.

5 JANOTA, M. 'Control of the welding current and time on basis of the weld nugget expansion'. IIW Doc.III-521-75.

6 ANON. 'Recommended practice for the evaluation of resistance spot weld monitor for assessing resistance spot weld quality'. IIW Doc.III-457-72/IIIG-49-72.

7 SNEE, R.K. and WILLIAMS, N.T. 'A review of monitoring techniques for assessing resistance spot weld quality'. IIW Doc.IIIG-50-76.

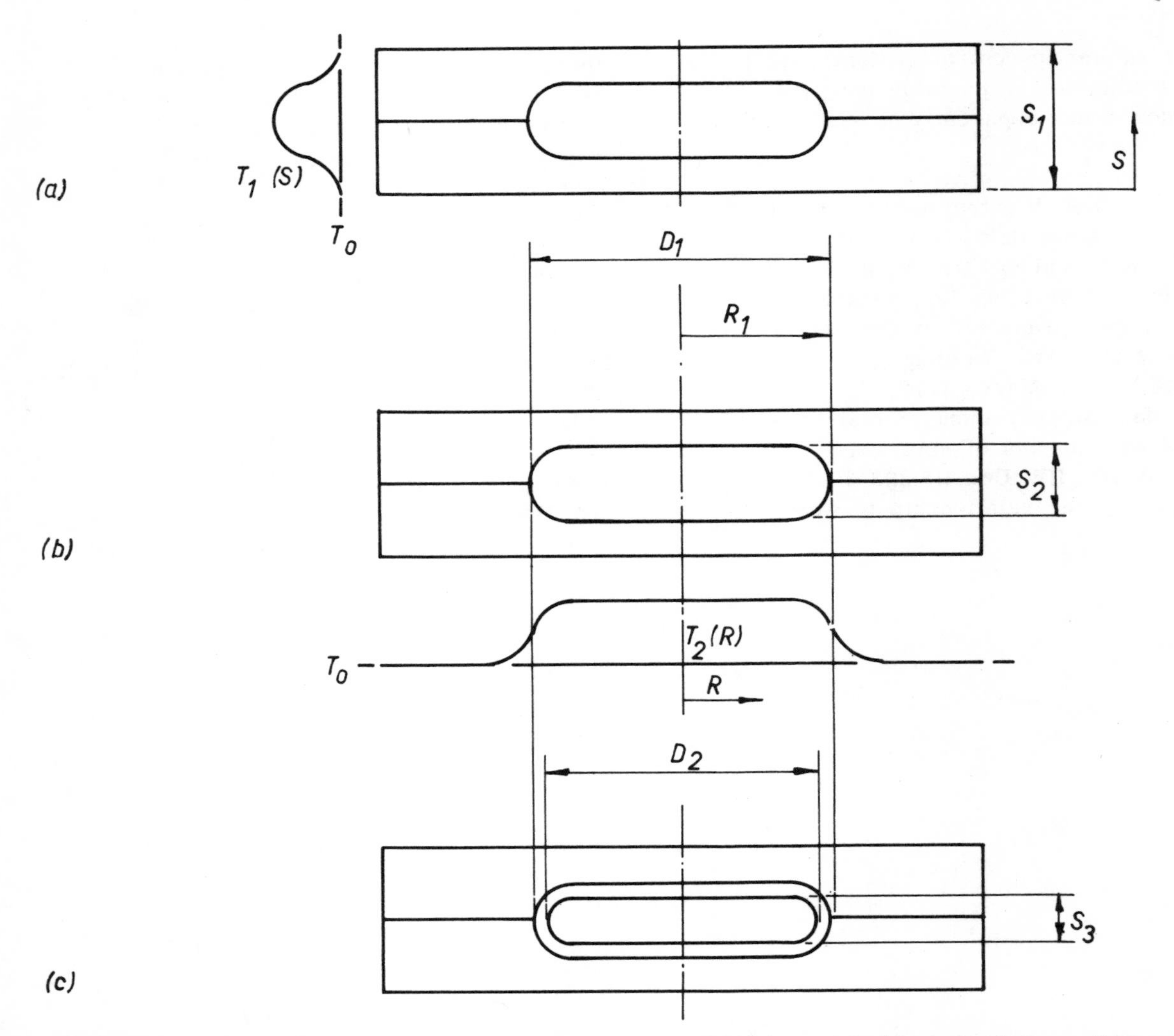

1 Schematic diagram with symbols to calculate components of TE: (a) coaxial with electrode force, (b) from expansion in perpendicular direction, and (c) resulting from material melting

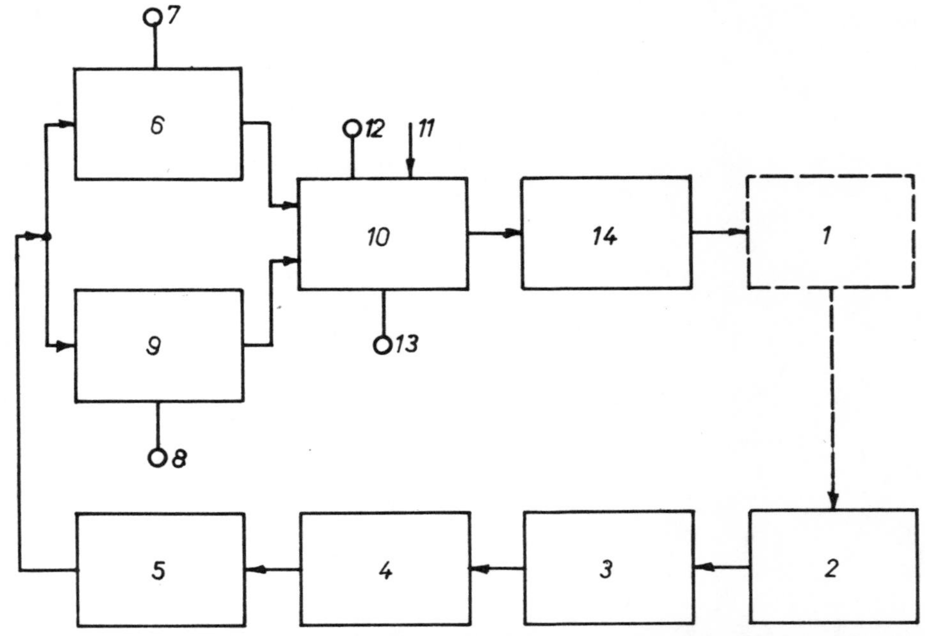

2 Block diagram of automatic control system. 1 — welding process; 2 — TE transducer; 3 — differentiator; 4 — filter; 5 — amplifier; 6 — current control circuit; 7 — potentiometer for setting slope termination; 8 — potentiometer for setting current termination (weld time); 9 — welding time control circuit; 10 — slope control; 11 — start condition; 12 — slope setting; 13 — initial current setting; 14 — thyristor contactor

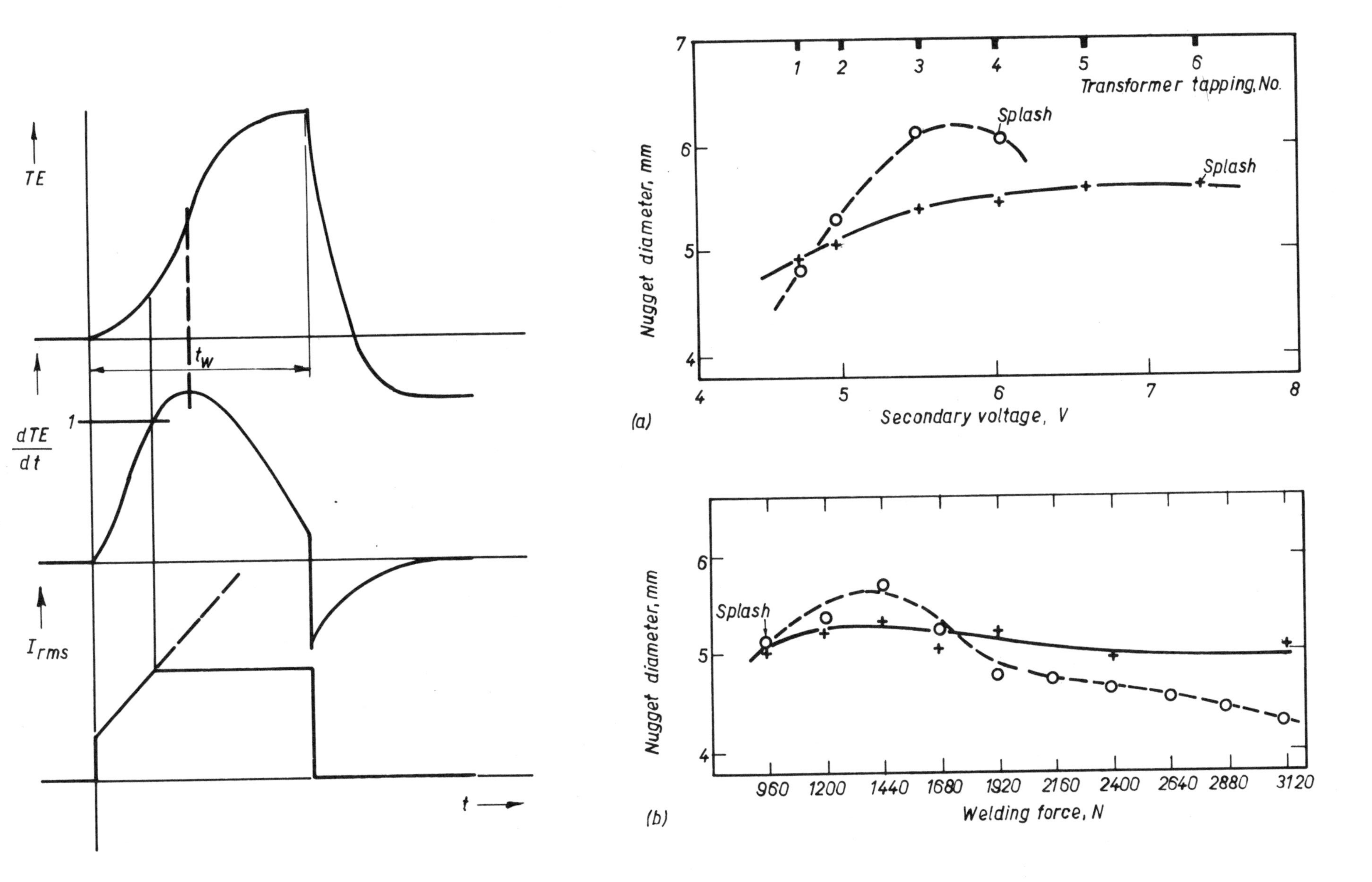

3 Function of control system. TE – course of TE; dTE/dt – course of first derivative of TE according to time; I_{rms} – rms current course; t_w – welding time; 1 – level for switch off of slope current

4 Influence of change on nugget diameter for machine E266H: (a) in transformer tapping, (b) in electrode force. ○ – conventional timer; + – automatic control

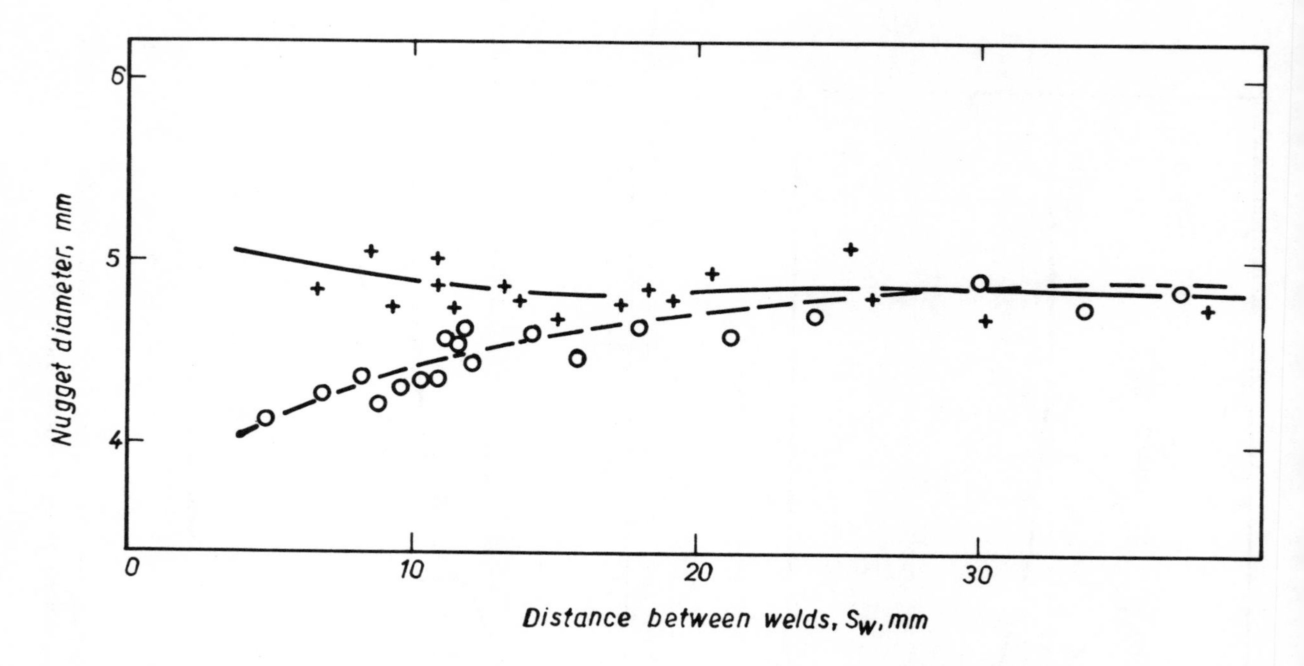

5 *Influence of shunting effect for machine BP 80.2. S_W — distance between spot welds; ○ — conventional timer; + — automatic control*

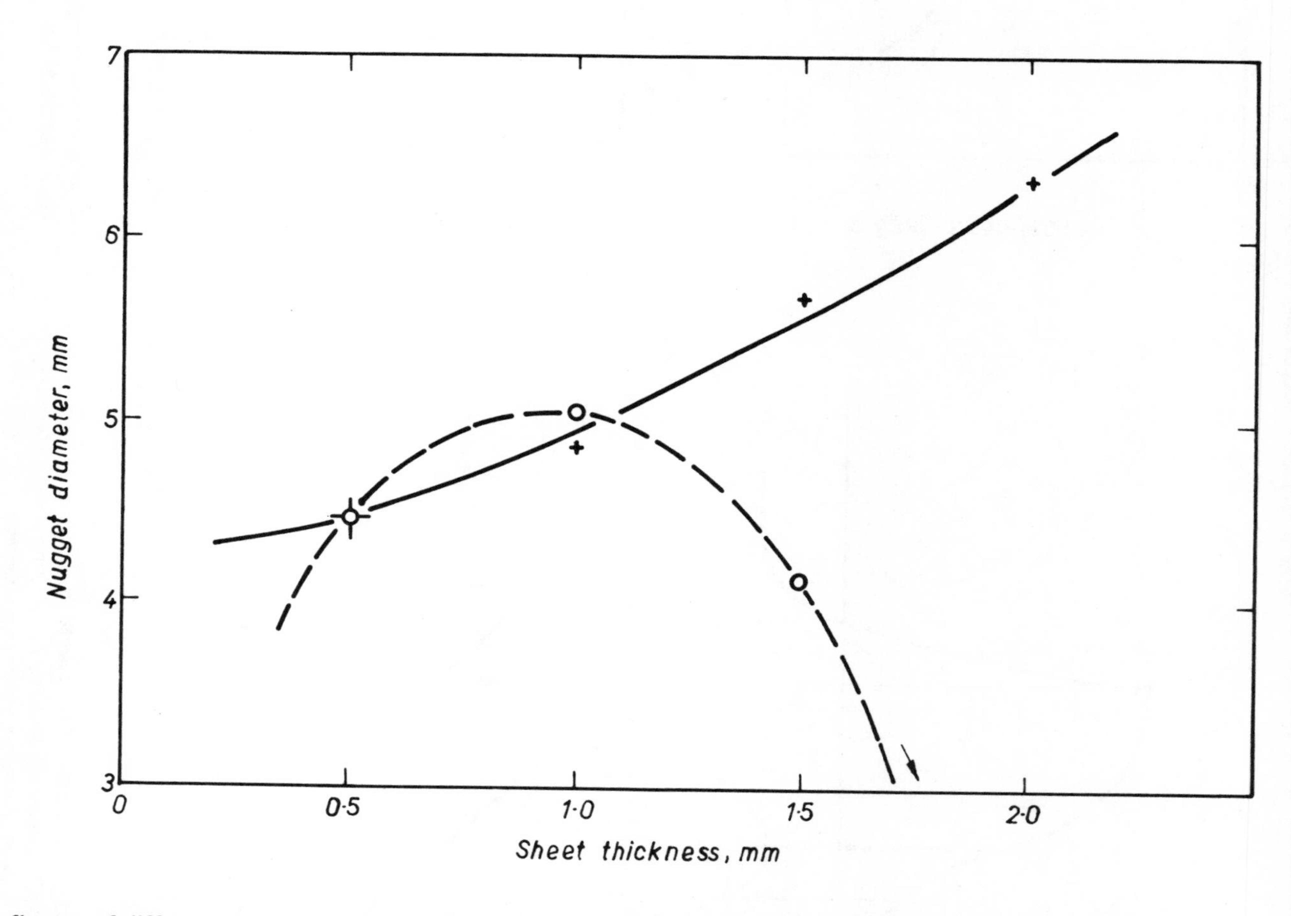

6 *Influence of different sheet thicknesses welded on machine E266H. ○ — conventional timer; + — automatic control*

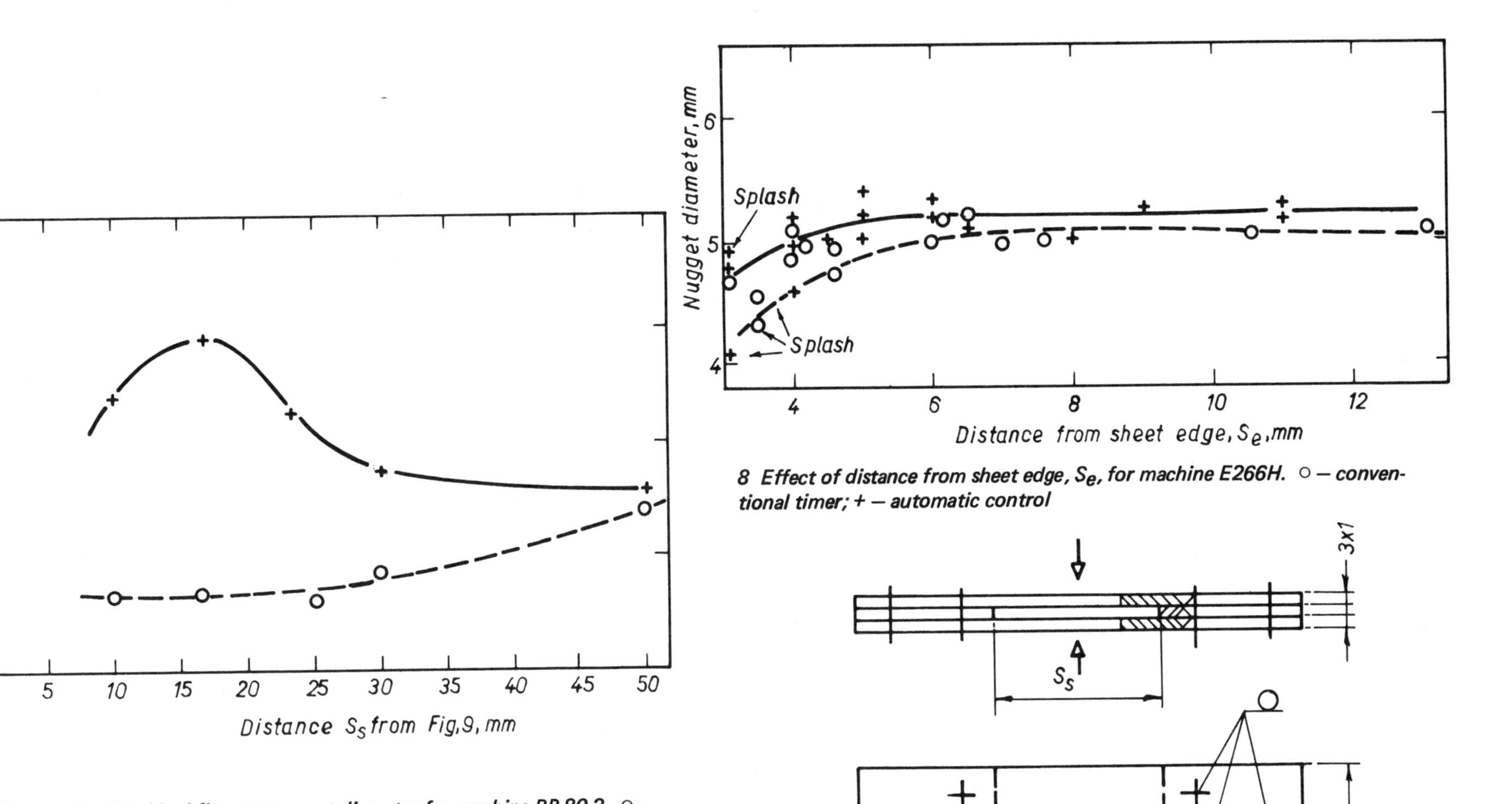

7 Effect of simulated bad fitup on nugget diameter for machine BP 80.2. ○ – conventional timer; + – automatic control

8 Effect of distance from sheet edge, S_e, for machine E266H. ○ – conventional timer; + – automatic control

9 Sample used for tests of bad fitup. Distance S_s is changed to 50, 30, 25, and 15mm

Welding with a magnetically moved arc (MBL welding): a new means of rationalisation

E.Schlebeck, Dr Ing

INTRODUCTION

Characteristically at present the technique for pipe welding in smaller sizes (up to approximately 150mm diameter) is that up to 90% of the circumferential seams are made by the manual process. Gas welding is clearly most widely used; it is, however, very time-consuming and many qualified welders are kept occupied. Also, a similar technological situation exists in the manufacture of round parts where, in addition to gas welding, mechanised fusion welding methods are used more and more.

At present the main fields of application of the process are in pipeline manufacture, the production of steam boilers, shipbuilding, the production of plastic processing machines, as well as the automobile industry and the component suppliers. Other enterprises such as the chemical industry, the production of agricultural machines, and ventilation and heating engineering are also included.

The technological development of the MBL welding technique is not yet finished. The possibilities for application to materials other than carbon steels, for larger and thicker weldments, as well as to parts with different material and geometric combinations, depend largely on whether, in the next few years, the energy input of the rotating arc can be controlled in a more specific manner and suitable specialised welding programmes and efficient quality-ensuring systems can be developed.

JTILISATION OF A KNOWN WORKING PRINCIPLE

n MBL-type welding a working principle is ısed which has been known in physics for more han 100 years, and from which two variants of he process have emerged:

a) MBL-P welding (welding with magnetically moved arc with pressure force: Schweissen mit magnetisch bewegtem Lichtbogen mit Presskraft). Figure 1a illustrates the principle of this variant where the arc is ignited between the two weldments and rotates between them. After reaching the necessary welding heat the joint is completed by means of forge pressure.

(b) MBL-H welding (welding with magnetically moved arc with auxiliary electrode: Schweissen mit magnetisch bewegtem Lichtbogen mit Hilfselektrode), as shown diagrammatically in Fig.1b. The arc in this variant is ignited between the weldment and the water-cooled auxiliary electrode and also rotates between them. The joint is effected without a pressure force. This variant is especially suitable for thin-walled sheet metal formed parts which are joined by different kinds of welding seam: flanged seam, edge joint weld, or corner seam

The Zentralinstitut für Schweisstechnik der DDR (ZIS Halle) has been engaged since 1969 with the theory and technological development and the application of MBL-type welding. Five years ago the first MBL welding machine was used in an industrial enterprise. At present altogether twenty-eight machines are being used in institutes and industrial enterprises in the German Democratic Republic and other parts of Europe.

TECHNOLOGICAL STATE OF DEVELOPMENT OF MBL-P WELDING

Figure 2 shows the types of joint for MBL pressure welding; instead of pipes hollow sections can also be employed. For the diameter range between 8 and 90mm and wall thicknesses between 1 and 5mm (up to 1000mm^2 welded cross-section) technologies and machines for industrial application are available, including pipe/flange joints up to 85mm nominal width.

More than fifty different weldments have been tested by ZIS for organisations in Germany and abroad.

Realistically, welding times lie between 1.2 and 12sec and fillers are not required. Moreover, for unalloyed steel additional shielding gas is not used. In comparison with the

Ir Schlebeck is at the Zentralinstitut für chweisstechnik der DDR, Halle (Saale), German emocratic Republic.

traditional pipe welding processes an increase of some 400 to 3000% in the working performance is obtained. The consumption of electrical energy in the MBL process amounts to only 50% in comparison with CO_2 welding, and to only 25% in comparison with flash butt welding. There are two approaches to welding by this process: on the one hand, with the so-called 'hard regime' high and constant welding currents are used with short times, or, alternatively, special welding programmes are employed. Thus the technique is adapted to meet the high and manifold quality requirements of the manufacturers of energy-generating equipment, steam boilers, and piping systems. MBL welded low and high pressure heat exchangers with operating pressures up to 25MN/m^2 and steam temperatures up to 230°C are operating in power stations in the German Democratic Republic.

The following examples, Fig.3, indicate that MBL pressure welding has already found many fields of application. Thus Fig.3a shows diagrammatically the carrying roller axle of conveyor equipment in lignite open-cast mining. This dynamically stressed joint between a thin-walled pipe and two round solid sections places high demands on the MBL welding technique. Fatigue strength tests on CO_2 and MBL welded parts have proved that the two welding processes are equally good for this application. Figure 3b shows an MBL welded CO_2 pressure cylinder for fire extinguishers for which the cycle time is 20sec. The test pressure amounts to 40MN/mm^2. Likewise Fig.3c shows, full size, an MBL welded pipe screw coupling for a hydraulic line with operating pressures up to 15MN/m^2. For about three years these mass-produced components have been manufactured on MBL equipment in shipyards and installation for the production of plastics processing machines and in the machine building industry. Finally, pipe/flange joints with the nominal widths 32 to 85mm are shown in Fig.3d. Such MBL welded pipe components are manufactured centrally in a pipeline factory of the Federal Republic of Germany and in a chemical combine of the German Democratic Republic. When fatigue strength tests on gas-welded and MBL welded pipe/flange joints were carried out the MBL welded seams resulted in higher strength values.

Some further applications which require high quality welds are illustrated in Fig.4. Thus Fig.4a shows a drive part for a motor-car consisting of a dissimilar material combination, i.e. carbon steel (0.45%C) and unalloyed structural steel. This component has been welded with a specific current programme. The latter technique is always used for MBL welding when materials or material combinations sensitive to hardening and cracking are to be joined.

In Fig.4b a part of the chain-type carrier of a beet harvesting machine is shown. The requirement here was to rationalise the production of the parts for the chain-type carrier from the previous two million to five million welds per year in a factory with constant welder personnel. Numerous welding processes were compared. In the end only CO_2 fusion welding and MBL pressure welding were considered satisfactory. However, for the CO_2 process thirty to forty automatic welding units would have been required. Therefore the decision was taken to use the MBL process. The whole annual production of five million welds will be carried out from 1978 onwards on three MBL automatic welding units. The direct weld time per catch amounts to 1.2sec.

Figure 4c shows some T butt joints of hollow sections, pipe/plate, and section/plate. The welding times amount here to between 0.5 and 1.2sec. The technological investigations are not yet finished, but the results obtained so far already permit the conclusion that such forms of joint can also be accomplished with the MBL welding technique. At the moment efforts are being directed to substituting both manual electrode welding and the CO_2 fusion welding processes, which were dominant until now, by the more productive MBL welding method.

MACHINES AND POWER SOURCES FOR MBL-P WELDING

At present there are two machine types in the German Democratic Republic which are used for quantity production. The first is the welding machine MBL-S 6.3, Fig.5a, which is designed on a modular assembly principle for the butt welding of closed hollow sections. Other component shapes can also be welded on it by changing the clamping jaw and magnet coil systems. The weldable cross-section for unalloyed steels lies between 120 and 1000mm^2, and for high alloyed steels up to 400mm^2.

The MBL welding equipment ZIS 786, Fig.5b, is provided especially for the use in prefabrication departments of workshops and construction sites. It exists in two variants: one for welding pipes and the other for welding pipe screw joints. The weldable cross-section amounts to 300mm^2 maximum.

The MBL process does not require special power sources, and the conventional electrical systems used for the fusion welding technique can be employed. Direct current welding current sources with a drooping volt-ampere characteristic can be used in principle. The no-load voltage must amount to at least 65V.

QUALITY OF THE MBL-P WELD JOINTS

For pipe/pipe and pipe/flange joints yield and ultimate strength of 100% of the standard values of the parent material of the pipe was verified in the tensile test. Up to the present no MBL welded seam has ruptured during burst tests. The uniformly formed upset metal has a height between 0.2 and 0.4 times the thickness of the pipe wall. The bend angle amounts, for flat samples (inside or outside in tension), to 100° at least. Comparison tests carried out with gas- and CO_2 welded pipes under dynamic loads have proved that the MBL welded joints are as good. In plate/pipe and/or plate/section joints a quality factor for the welded joint of 0.85 is reached.

TECHNOLOGICAL STATE OF DEVELOPMENT OF MBL-H WELDING

The MBL-H process is suitable to join flanged end seams, edge joint welds, and corner seams on components with closed seams, made of both unalloyed and high alloyed steels as well as oxygen-free copper, Fig.6. For weldment diameters up to 300mm production-suitable technologies have been developed in the ZIS Institute. Although this size is the upper limit for applications using the standard design of the MBL-H welding unit, Fig.7, it is not the upper limit of the MBL-H process. At present the maximum weldable thickness of sheet metal is 1.5mm.

The parts suitable for welding by the MBL-H process are to be found in large numbers in the sheet metal processing industry, consumer goods industry, machine building, the manufacture of agricultural machinery, and in the automobile industry. These parts are currently fabricated by means of gas, fusion, and resistance welding, but at relatively high expenditure because of the slow welding speeds.

Figure 8 shows an MBL-H welded fuel container. The welding time amounts to only 1.5sec. This example demonstrates that the process is not limited to welded sections which are rotation-symmetrical. These new developments are, however, not yet finished, but if production-suitable technologies are also successfully established for the automated manufacture of these complicated weld cross-sections, the production output in container manufacture can be increased by several thousand per cent.

SUMMARY

The MBL process with its two variants is a highly efficient means of fabrication in that it offers the possibility for major improvements in the welding of pipes, hollow sections, and round parts, which at present involve a very high expenditure of time and money.

The most important features of the new process are:

1. Extremely short welding times, and thus savings in manufacturing times of between 25 and 90%
2. Low energy consumption, and thus reduction in electrical demand by 50 to 75%
3. Simple preparation of the seam with no filler required
4. Optimum utilisation of the supporting cross-section
5. It can be mechanised and automated
6. Shorter training time because of the employment of semi-skilled welders

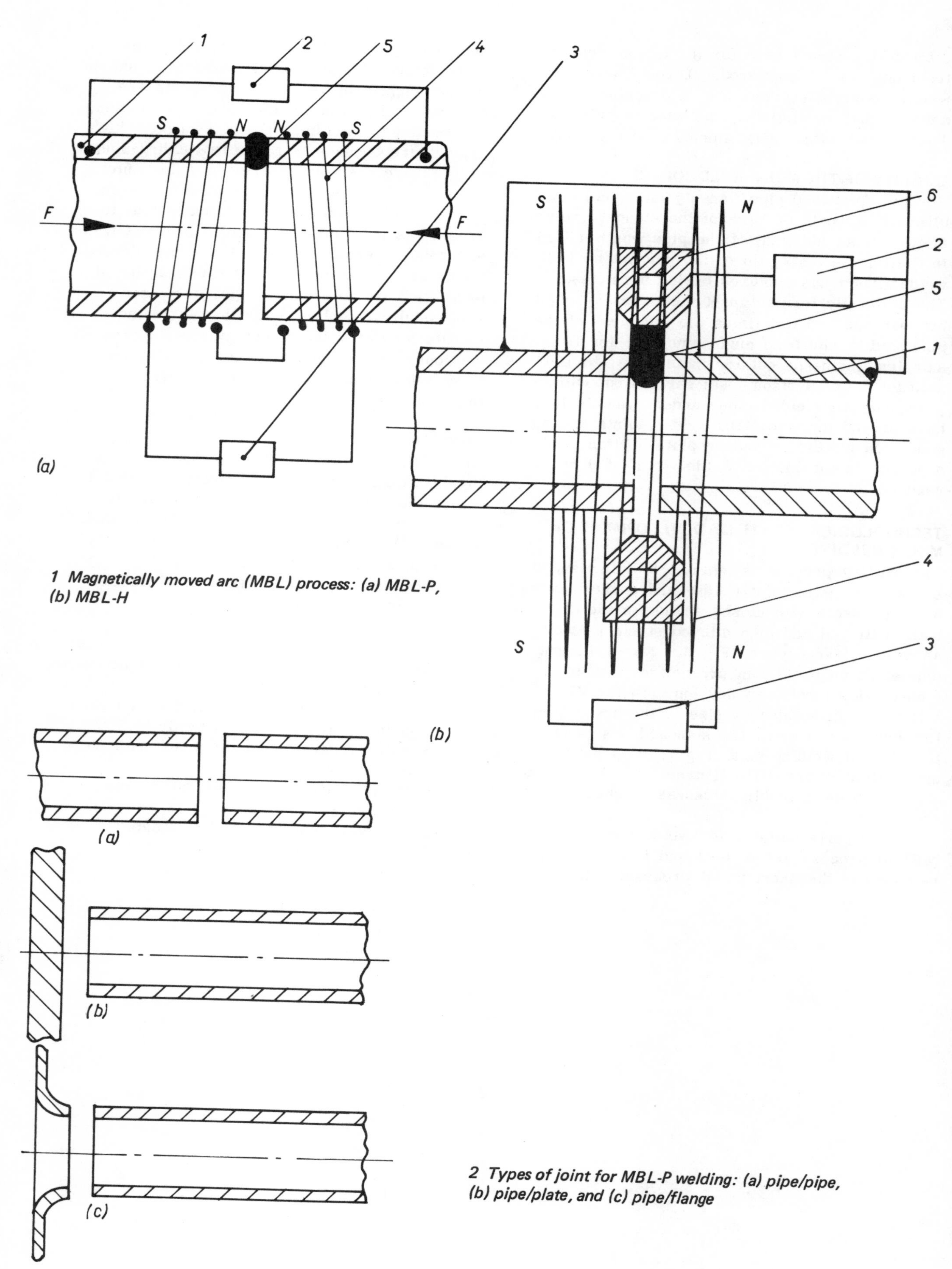

1 *Magnetically moved arc (MBL) process: (a) MBL-P, (b) MBL-H*

2 *Types of joint for MBL-P welding: (a) pipe/pipe, (b) pipe/plate, and (c) pipe/flange*

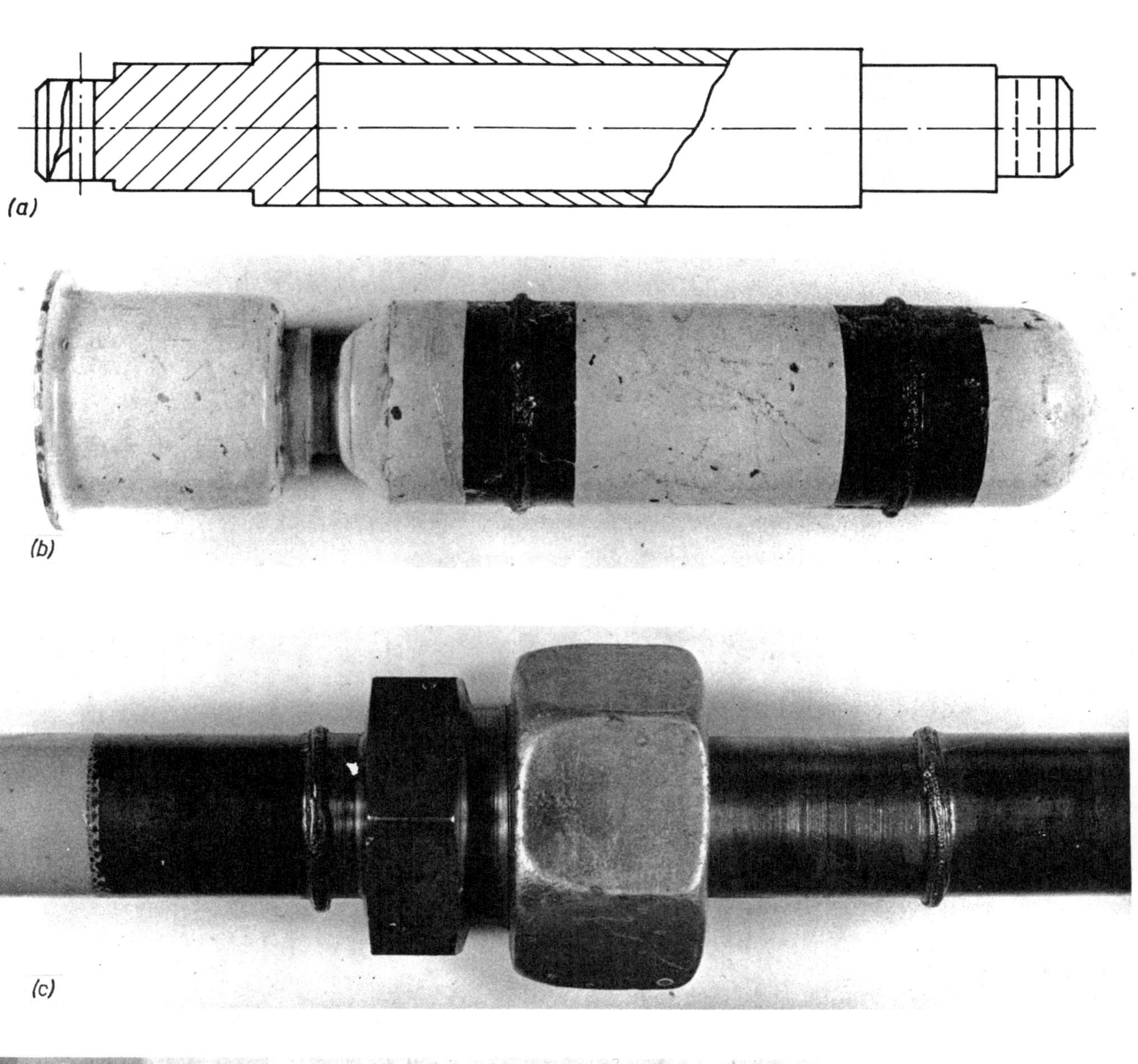

3 Applications of MBL-P process to engineering components: (a) carrying roller axle (schematic) for conveyor equipment, (b) CO_2 pressure cylinder for fire extinguishers, (c) pipe screw joint for hydraulic lines, and (d) pipe/flange joints of 32 to 85mm nominal width

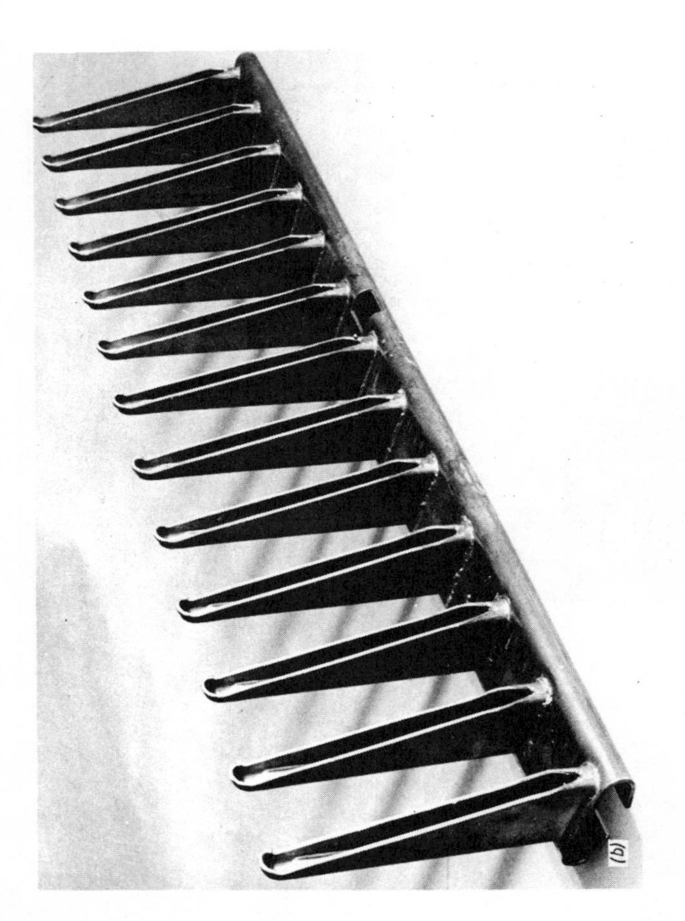

4 *Special applications: (a) drive part for motor-car, (b) part of chain-type carrier of beet harvesting machine, and (c) T butt joints of hollow sections, pipe/plate, and section/plate*

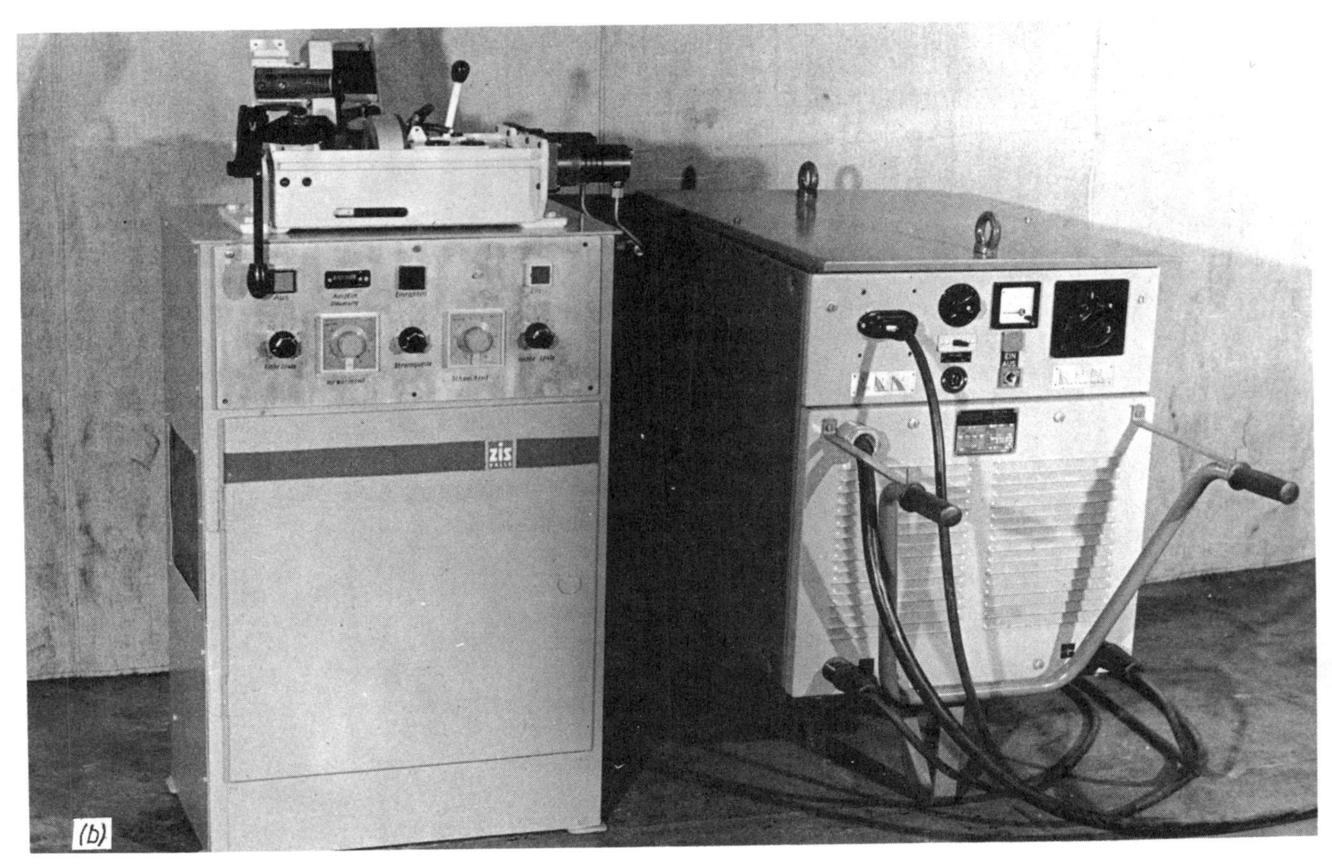

5 (a) MBL butt welding machine, Type MBL-S 6.3, (b) MBL workshop unit ZIS 786

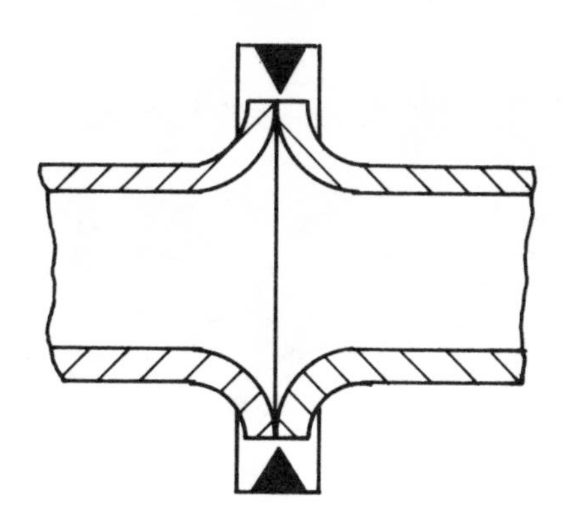

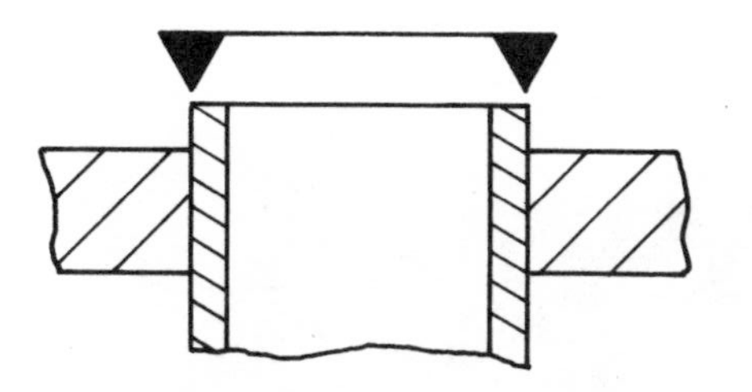

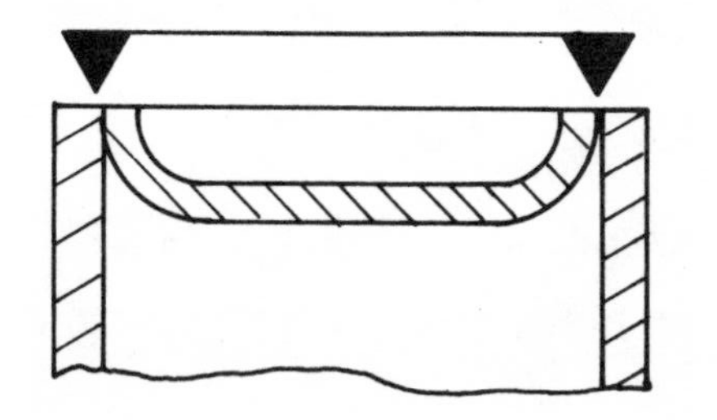

6 Types of joint for MBL-H welding

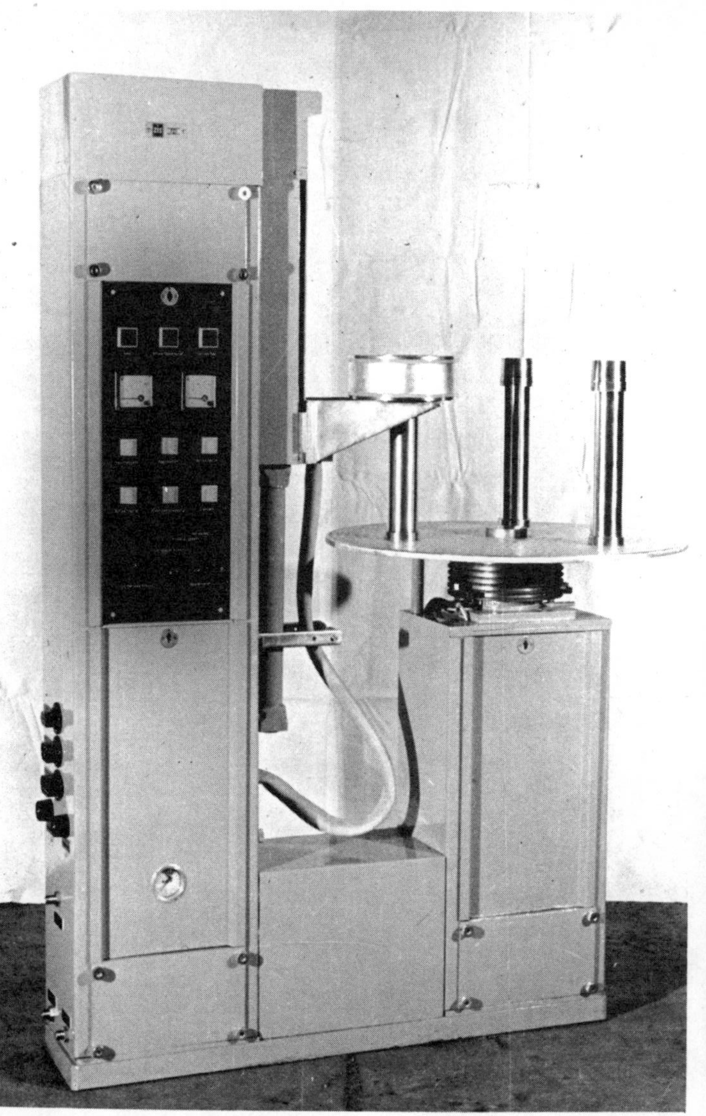

7 MBL-H welding unit ZIS 721

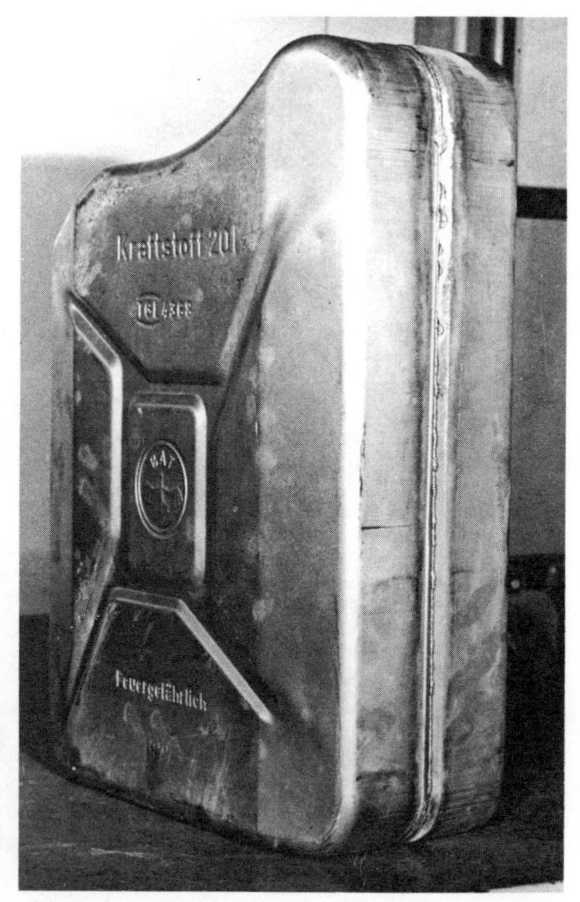

8 MBL-H welded fuel container, 5 litre capacity

Arc-augmented laser welding

M.Eboo, BSc (Eng), ARSM, W.M.Steen, MA, PhD, DIC, CEng, MIChemE, FIM, and J.Clarke, BSc (Eng), ARSM

It has been found that an arc can be successfully rooted to the hot spot created by a laser beam striking a metal surface. Currents of up to 130A have thus been rooted to the same small area which is struck by a 2kW laser beam. The result has been that bead-on-plate welds in tin plate have been made at four times the laser-alone speed.

The operating procedure for this laser-cum-arc-enhancement technique is described together with the resulting weld structures.

INTRODUCTION

Laser welding is an established process producing high quality welds.[1] The welds produced are comparable in quality with those from an electron beam with the possible advantages of being done without a vacuum causing a greater cooling rate and little degassing. The greater cooling rate allows less grain growth in sensitive materials such as titanium.[2] Both these processes weld by forming a 'keyhole'. This produces welds of high aspect ratio, e.g. greater than 1 : 1 depth : width with consequent low heat-affected zone (HAZ) and thermal distortion. However, as the welding speed is increased insufficient energy is deposited in a given area of surface to cause a keyhole to form and penetration is consequently lost. Figure 1 shows how Crafer[1] found penetration varying with speed. If more energy could be deposited in the interaction zone between the laser and substrate, penetration might be re-established and the welding speed consequently enhanced.

It was found that the fast moving laser beam lowered the anode work function of the hot spot it created to such an extent that an arc could be stably rooted there. This resulted, with sufficient arc current, in a fully penetrating weld at speeds when the laser alone would penetrate only a fraction of the thickness, and the arc alone would melt the substrate only in sporadic spots, Fig. 2.

Described here is a preliminary experimental survey of this combined interaction showing welding speeds increased by a factor of four.

Mr Eboo and Miss Clarke, Postgraduate Students, and Dr Steen, Lecturer in Applied Metallurgy, are all at the Department of Metallurgy and Materials Science, Imperial College of Science and Technology, London.

EXPERIMENTAL

A BOC Model 901 2kW CW CO_2 laser was used together with a BOC Model THF/TCM 450 water-cooled 250A TIG-welding set. The experimental arrangement is shown in Fig. 3.

The laser produced a near-parallel beam of 10.6μm radiation 10mm in diameter. The power distribution within the beam approximated the fundamental Gaussian spread known as a Tem_{oo} beam. The beam was focused through a 3mm diameter copper orifice using a polycrystalline KCl lens of 75mm focal length. The resulting spot size, measured by acrylic prints, was around 250μm diameter. Helium shielding gas was passed through the copper orifice coaxially with the laser beam. Because of the high speed of the welding process extensive trailing shielding was also required. The nozzle/substrate distance was maintained approximately constant at 3mm.

The arc was mounted either above or below the substrate. The tungsten electrode was always kept negative while the workpiece was positive. Arc initiation was triggered by a high frequency signal of 450kHz. The undersurface of the plate being welded was shielded with extensive shielding plates of argon.

The specimens were 300 x 300mm square

sheets mounted horizontally which were rotated and moved linearly beneath the arc/laser system. This movement produced spiral bead-on-plate melt runs. The point at which penetration ceased could be easily located and a simple calculation revealed the welding speed at that point. The tin plate was supported by thin cross-struts of metal mounted above and below. This gave the added advantage of producing a spiral weld (when the arc was below and the laser above) with alternate zones of laser alone, arc and laser, and arc alone, with varying velocities across the field of the plate. The speed range across a single sheet was 1 to 1.7.

Each specimen was sectioned at various points around the spiral. The cross-section was then polished and etched in 2% nital to reveal the fusion and HAZs.

EXPERIMENTAL RESULTS AND DISCUSSION

Figure 4 shows how the fully penetrating bead-on-plate welding velocity is increased by increasing the arc augmentation current for two materials, titanium and tin plate. For the former the arc is on the same side of the substrate as the laser. For the 2mm thick material increasing the current gives a small increase in speed; however, a similar power increase in the laser radiation gives a greater rise in velocity. On the other hand, for 0.2mm thick tin plate, but with the arc beneath the substrate and the laser incident above, quite a different event is shown, i.e. a fourfold speed increase for the addition of only a 25A arc (equivalent to $\sim$ 200W).

The penetration of the laser radiation has been found to be partially blocked by the arc at low arc currents (<30A) when both are acting on the same side of the material, Fig. 5. Such absorption or reflection of the laser radiation by the arc plasma might be expected. In addition, the anode spot size is much larger than the focused laser beam, thus tending to create a much wider fusion zone at the top. This may account for the different success between the two examples in Fig. 4. However, Fig. 6 shows runs made on 2mm titanium strip at 200mm/sec and with an arc current of 50A. The much higher arc current causes a substantial increase in energy input so that in this instance penetration is achieved. The three runs are of laser alone, laser with arc augmentation, and arc alone respectively. The central run shows clearly how arc augmentation to the laser increases the energy input to the plate and actually causes penetration; see Fig. 6a and b for the upper and undersurfaces, respectively, of the metal.

The macrostructure of a tin plate weld made with the arc and laser on opposite sides is shown in Fig. 1. The effect of the laser alone is shown in Fig. 1a with only partial penetration, and Fig. 1b is the result of the arc alone which shows no perceptible penetration but only a HAZ. The combined event, however, Fig. 1c, exhibits a relatively large fusion zone with the same laser power and welding speed as for the run in Fig. 1a. It can thus be seen that the combined event is greater than either separately or even that of the two separate events added together.

DISCUSSION

Arc rooting

It is surmised that the arc and laser lock together in a useful way. To establish if such a locking mechanism exists, experiments were carried out on 2mm steel plate 300mm square. By carefully measuring the arc column length, arc current, and arc voltage on a rotating plate the following two factors were established:

(a) the effect of arc length on column resistance, by the production of a circular trace on a tilted plate at constant velocity, and

(b) the effect of velocity on column resistance by the production of a helical weld on a flat plate

The column resistance was thus found as a function of arc length and velocity, Fig. 7. It is possible to combine these two sets of lines by calculating the angle of the arc column to the normal of the plate, shown as the 'drag' angle' in Fig. 8, to produce a universal curve (for all currents) of drag angle as a function of velocity. This curve, however, has a series of arc current stability ranges and is valid for only an initial arc length of 2mm. Different initial arc lengths would result in a displacement of the curve.

Figure 8, however, emphasises that the arc column extends itself with increasing velocity because it has a favoured root at the arc-generated hot spot, a point observed by Reeves-Saunders.[3] However, there is a velocity limit associated with a particular current beyond which the arc will become unstable and can no longer hold a stable position. This limit is reached when the resistance of the arc column to the hot spot exceeds that required to initiate a new hot spot directly below the electrode. At low currents, this limit is reached more quickly.

At the hot spot the anode work function is diminished making the anode potential drop lower here than elsewhere and hence the path of least resistance. This conforms with the Steenbeck minimum principle which states that an electric arc will operate at the lowest potential possible under the conditions of its maintenance.[4,5] This feature of rooting to a hot spot was confirmed by making a circular arc run on 2mm mild steel plate. A lower arc resistance was observed when the arc was struck on to a preheated substrate, under exactly the same condition as had previously been used with a cold substrate; see data Fig. 7a at 50A. Consequently, if a localised hot spot can be created within the maximum arc length away from the electrode which provides a path of lower resistance for the arc than that to the cold anode plate directly, the arc will favour that path. More significantly, experiments with the laser and arc together on tin plate showed stabilisation of the arc on the laser-created hot spot. The intermittent rooting of the arc ceased in the presence of the laser hot spot. The calculated temperature difference for this was only 300degC.

It is postulated that the hot spot generated by the laser acts as an effective root for the arc and partly explains their mutual action in welding. Figure 9 with both the laser and the arc above the plate shows the laser-generated plasma acting as a favourable rooting zone for the arc.

At high arc currents, however, the arc appears to be self-stabilising and this mutual locking of arc and laser might not be so pronounced. Figure 7a shows that at high currents — 130A — there is little change in column resistance with welding velocity.

At this preliminary stage of the work it is reasonable to consider theoretically whether there is a maximum welding speed for laser/arc welding. If there were it would be expected to depend upon adequate stable arc rooting and the insertion of sufficient power into the substrate.

From the previously mentioned experiments, stable arc rooting acting with the laser would require the temperature of a laser-generated hot spot on the undersurface of the substrate to be greater than 300°C. A mathematical model[6] of a moving, surface disc, heat source having a Gaussian power distribution was used to calculate the undersurface temperature. This showed that there would be a speed for a given laser power or substrate thickness at which the undersurface temperature would be insufficient for the arc to root. This would be expected to be the upper speed limit for this process. For a laser power of 1320W as used in these preliminary experiments this limit is calculated to be around 1m/sec.

To extend this model to simulate the arc-augmented example, to calculate what current would be required for penetration welding, requires:

(a) an input of the area over which the arc falls
(b) the area heated by the arc radiation
(c) the internal flow of current within the substrate, together with any convection terms within the molten pool owing to the magnetic and electric forces, and
(d) convection of heat from the hot arc gases, including latent heat from the arc plasma

This work is currently in hand.

CONCLUSIONS

What is described here is a new method of using a laser and an arc to produce a welding process of far greater joining rate than any fusion process.

So far, welding rates in 0.2mm tin plate have exceeded 0.9m/sec for bead-on-plate welds. The maximum rate of welding is probably dependent upon the laser power and substrate thickness, and the arc current decides whether a fully penetrating weld would occur.

REFERENCES

1 CRAFER, R. 'Welding with the 2kW CO_2 laser'. Welding Inst. Research Bull., 17 (2), 1976, 29-33.

2 MAZUMDER, J. and STEEN, W.M. 'Laser welding of titanium'. Munich, 1977.

3 REEVES-SAUNDERS, R. 'Observation of a new transition in low-current DC arc properties'. J. Phys. D: App. Phys., 6 (2), January 1973, 212-23.

4 LUDWIG, H.C. 'Current density and anode spot size in the gas tungsten arc'. Int'l Conf. 'Gas Discharges', London, 15-18 September 1970. Inst. Electrical Engineers, 498-502.

5 STEENBECK, M. Physik Z., 33, 1932, 809.

6 MAZUMDER, J. PhD Thesis, London, 1977.

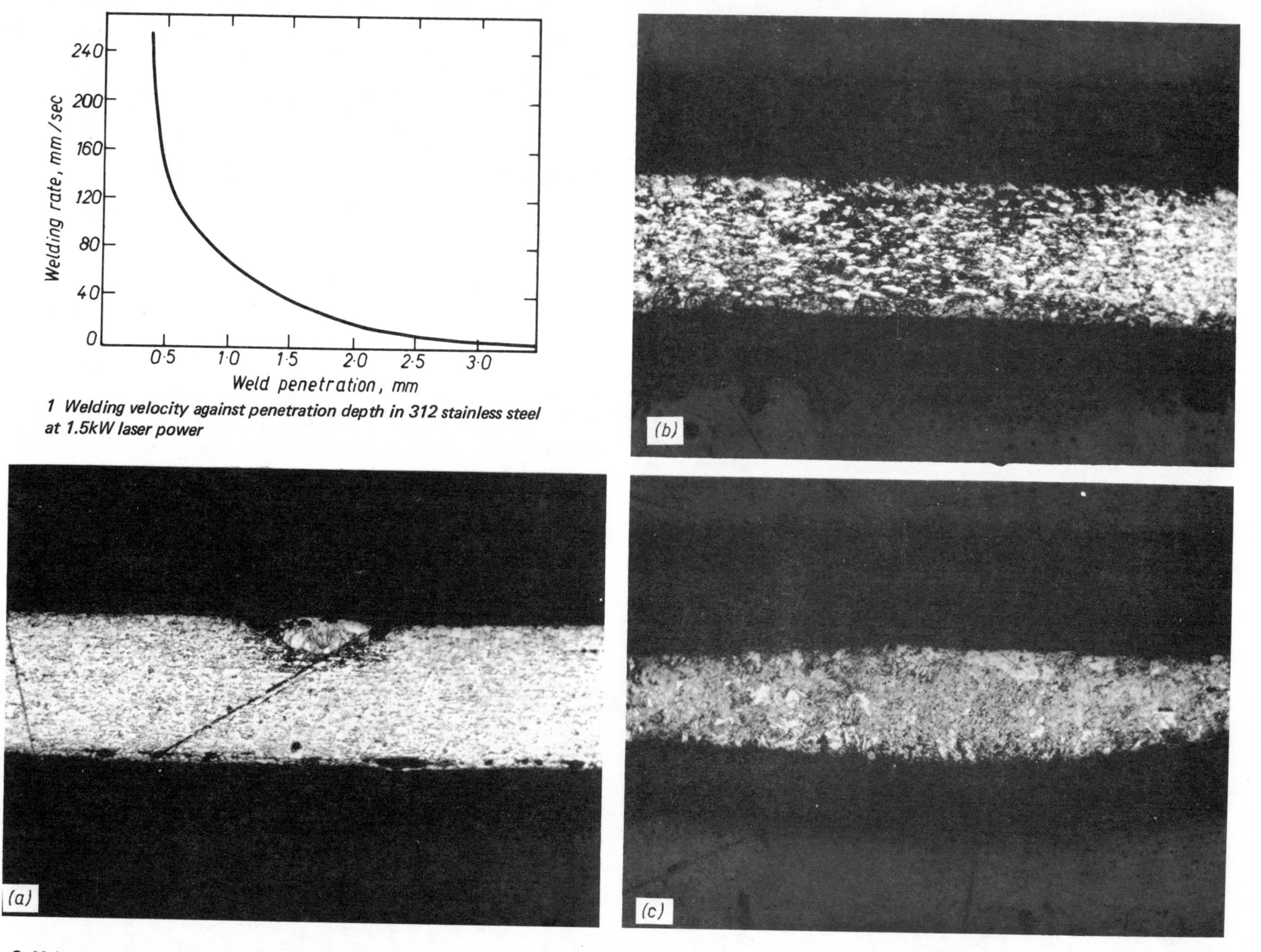

1 Welding velocity against penetration depth in 312 stainless steel at 1.5kW laser power

2 Melt runs in 0.2mm tin plate (etchant 2% Nital): (a) laser alone at 800mm/sec, (b) arc alone at 650mm/sec; note: only HAZ observed with no melting, and (c) laser/arc combination at 800mm/sec with laser incident on top surface x 115

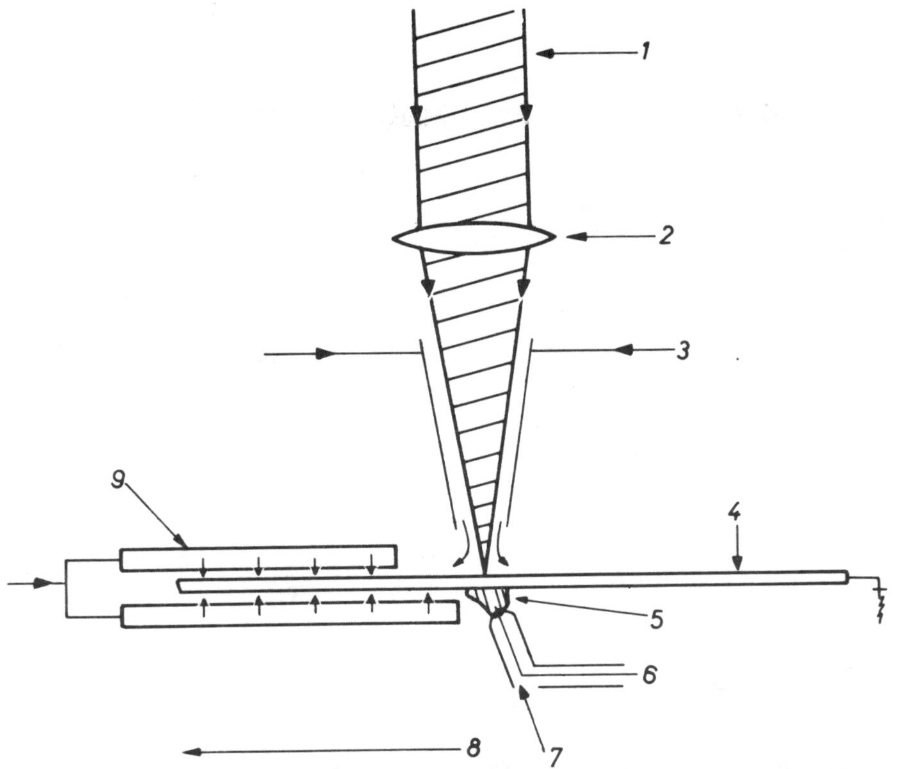

3 Schematic diagram of laser/arc welding system. 1 — laser beam; 2 — 75mm KCl lens; 3 — He shield; 4 — +ve workpiece; 5 — arc plasma; 6 — tungsten cathode; 7 — argon; 8 — welding direction; 9 — argon shielding jets

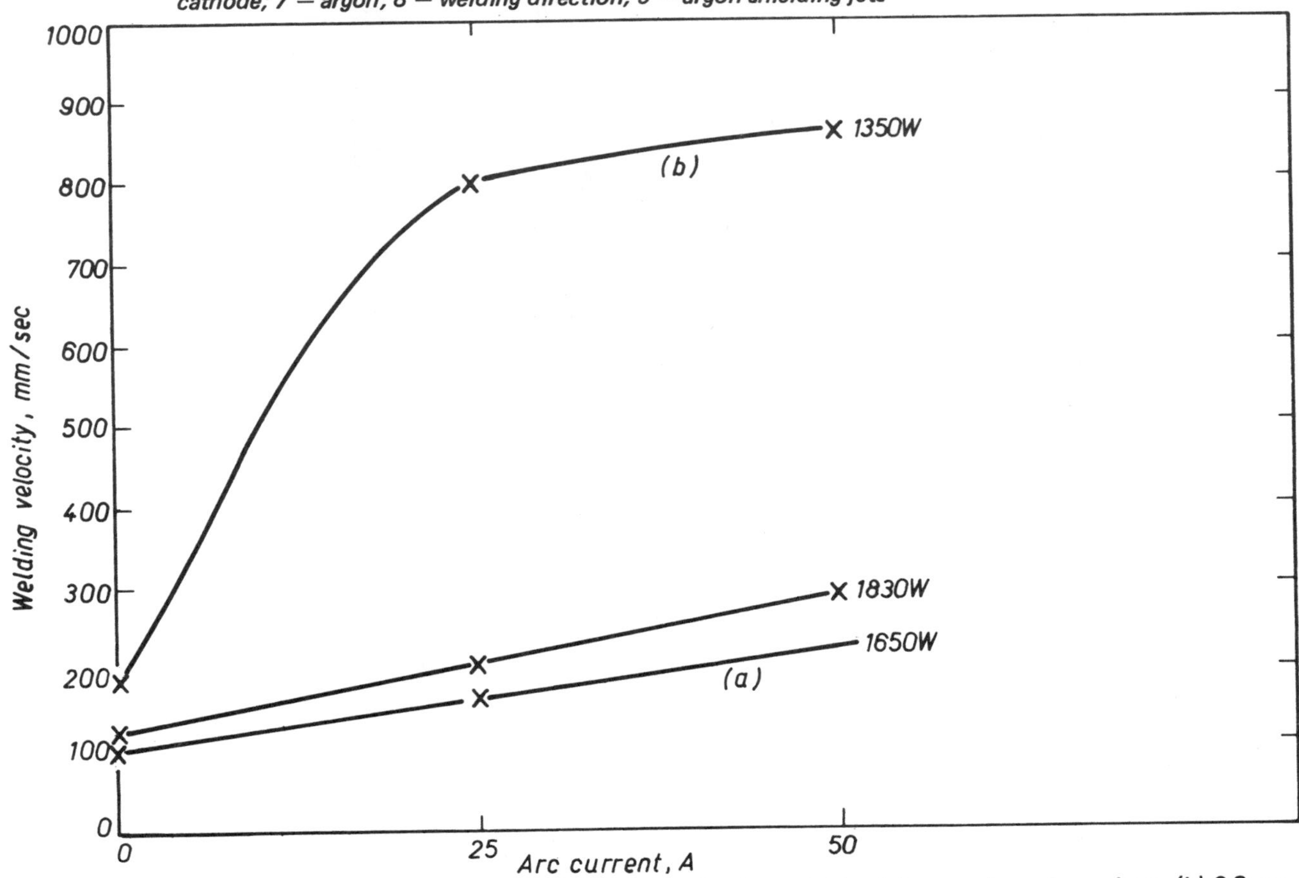

4 Graph of welding velocity against arc current: (a) 2mm cp titanium with both laser and arc above sheet, (b) 0.2mm tin plate with laser above and arc below

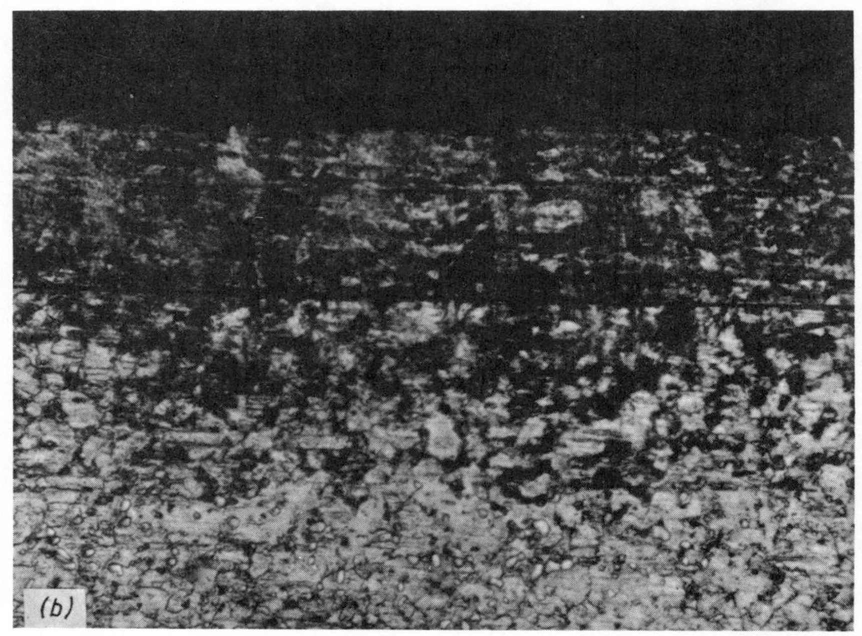

5 Fusion zones in 2mm mild steel: (a) laser alone, (b) laser/arc at 20A; note: fusion zone is spread out and penetration appears to be diminished x 130

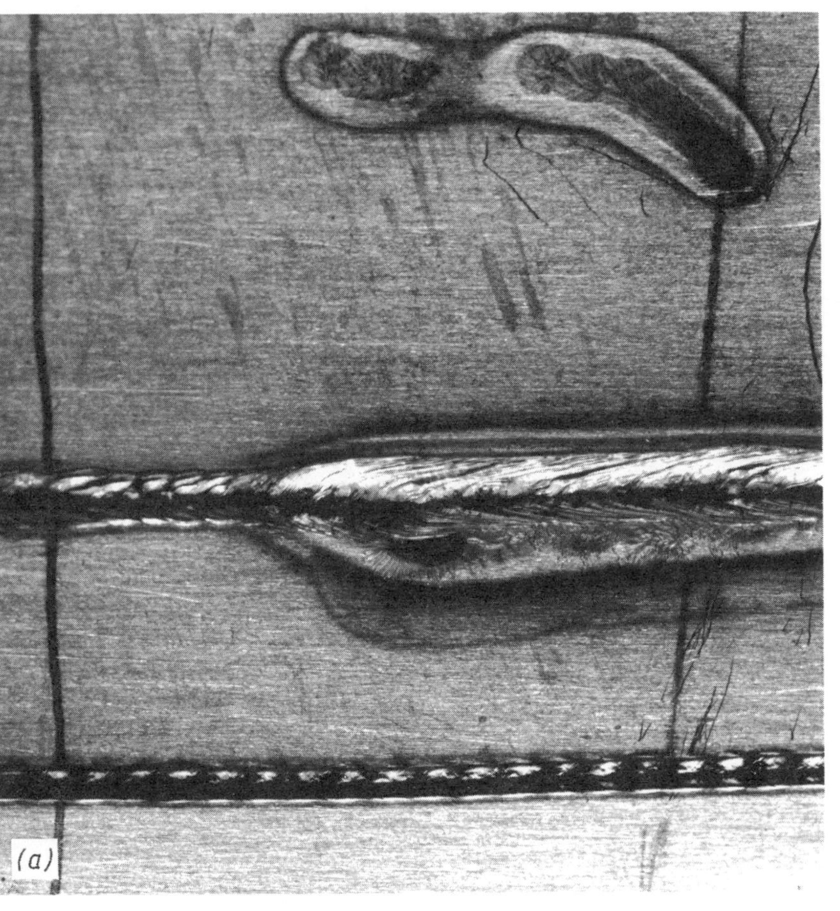

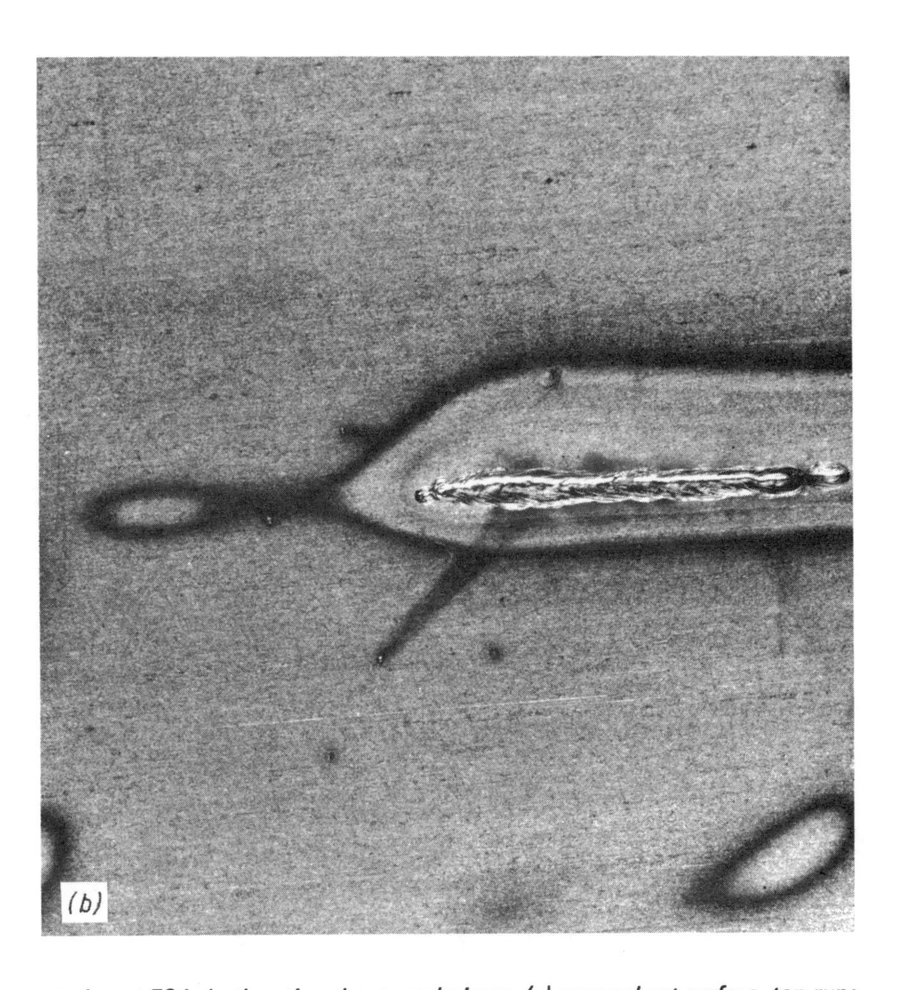

6 Melt runs in 2mm titanium sheet at 250mm/sec using laser at 1830W with arc augmentation at 50A, both acting above workpiece: (a) upper sheet surface, top run: laser alone; centre run: laser plus subsequent arc augmentation; bottom run: arc alone, (b) undersurface of sheet; note: penetration caused by arc augmentation x 10

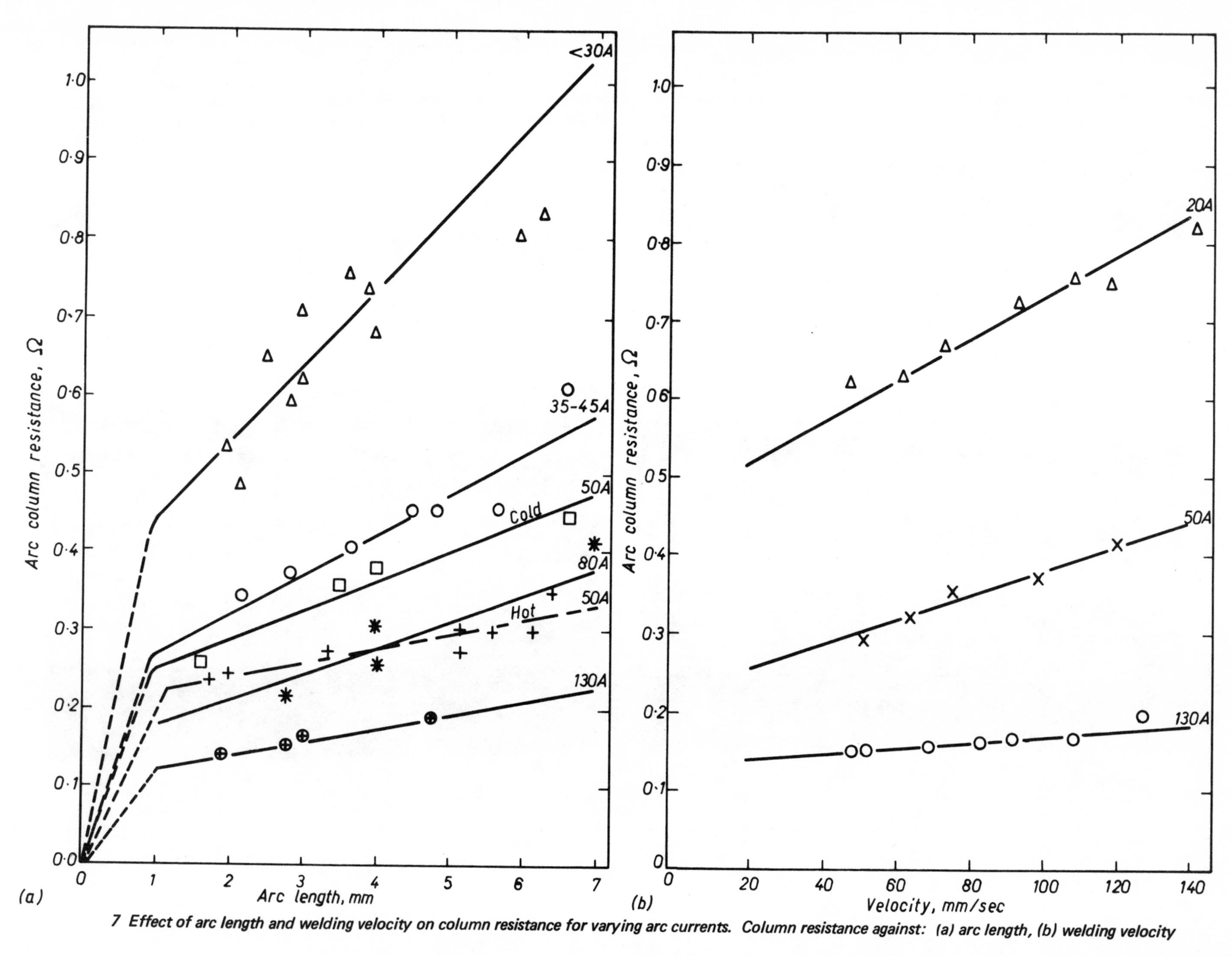

7 Effect of arc length and welding velocity on column resistance for varying arc currents. Column resistance against: (a) arc length, (b) welding velocity

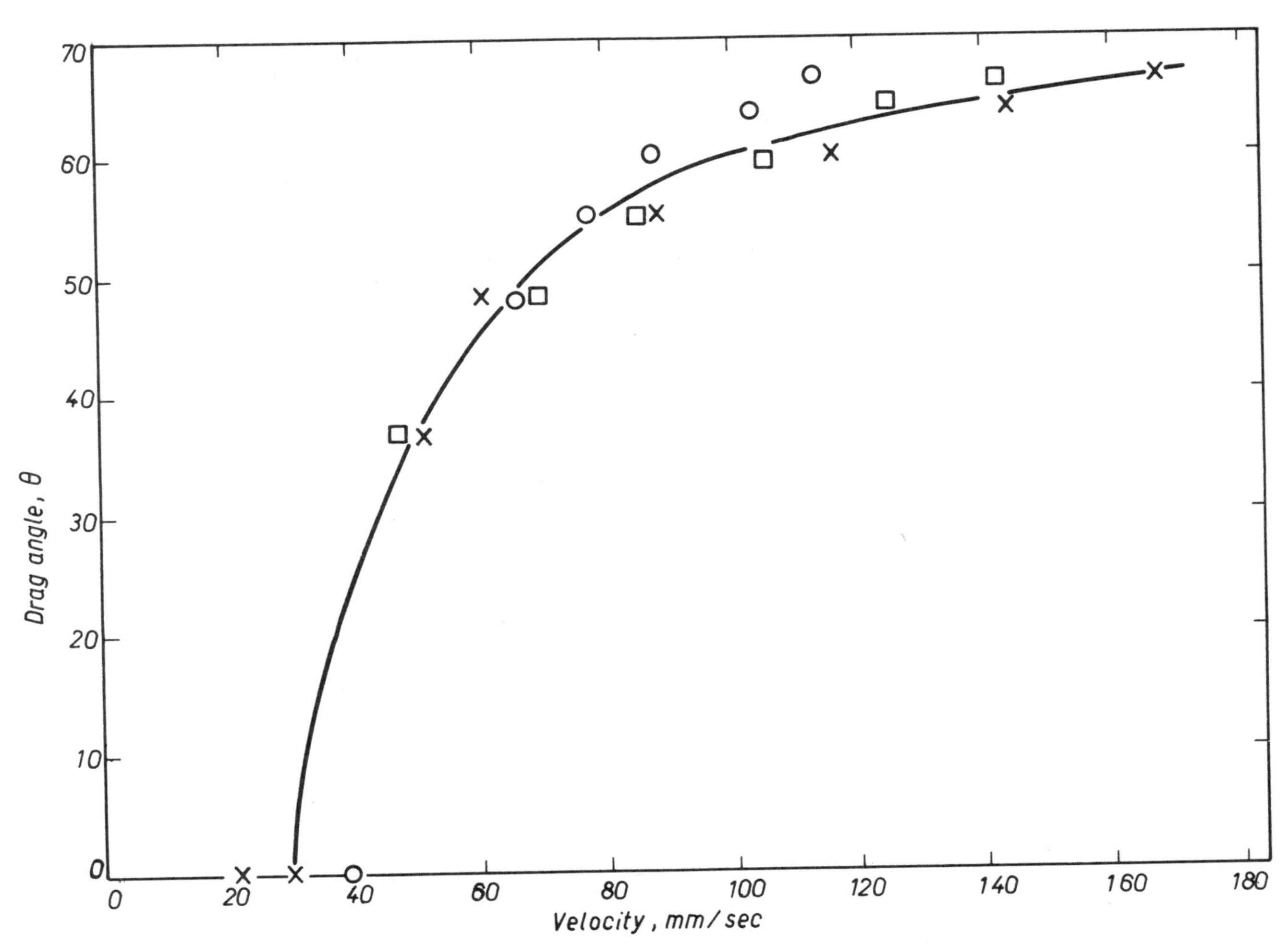

8 Graph of drag angle, θ, against welding velocity. X – 130A; ○ – 50A; □ – 20A

9 Arc rooting into laser plasma x 2.6

Improved welding performance from a 2kW axial flow CO_2 laser welding machine

R.C.Crafer, MA, PhD, MInstP

INTRODUCTION

Recent developments in the generation and handling of laser light have led to dramatic improvements in welding performance. For example, the welding of this gauge engineering materials (<3mm thick) is now comparable in performance to electron-beam (EB) welding and, power for power, thicker materials (up to 6.5mm) are welded faster and deeper than before. High conductivity materials such as aluminium alloys can also be welded at relatively low powers, such as 2kW, whereas previously powers of around 5kW were required.

This Paper describes the laser modifications involved together with details of the improved performance, giving comparisons where possible with other published data. These comparisons show that the axial flow 2kW welding machine[1] now out-performs lasers of more than double this power, particularly for materials up to 4mm thick, and further that it remains competitive in certain materials for thicknesses up to 6mm.

EQUIPMENT

Power supply

The original 30kV power supply has been superseded by a modified EB supply capable of delivering 40kV at 1A. Unlike the original supply, which was operated manually, the new unit incorporates a motor-driven autotransformer to give remote control of power supply voltage.

Control unit

All the laser controls and operating logic have been centralised in one semi-automatic control console, Fig. 1. The system has four states of operation: OFF, VACUUM ON (system evacuated, leak test completed), BLOWER ON (correct gas mixture in system, recirculation operational), and LASER ON.

Logical chain operation is employed. To select a particular state the appropriate button on the console is simply depressed. If the previous state is not already operational it will be brought into operation automatically. Any state can be cancelled by selecting a previous state, e.g. the laser is switched off by selecting OFF, VACUUM, or BLOWER. (The only exception is LASER ON, which is decoupled for safety reasons.) Alongside the state selectors are facilities energising the primary of the autotransformer and for raising and lowering the supply voltage. In addition to these controls there are override facilities to give complete flexibility of operation when required.

Information on laser operating parameters is also available at the console:

1 Discharge gas pressure, Torr
2 Injection gas pressure, Torr
3 Pressure ratio computed automatically and continuously from 1 and 2
4 Traverse velocity, mm/sec
5 Laser output power, kW
6 Transformer primary voltage (arbitrary scale)
7 CO_2 partial pressure, psi } in gas mixing manifold
8 N_2 partial pressure, psi } in gas mixing manifold
9 He partial pressure, psi } in gas mixing manifold

In addition, several indicators yield information on gas pressures, flow rates, valve openings, and other conditions of interest. Safety interlocks are used throughout the logic system to prevent the occurrence of hazardous situations. For example, the failure of any of the services (gases, air, coolant, etc.) will cause the laser to revert to a nonoperational state, e.g. VACUUM ON. As an additional safety feature, full operation can be resumed only on manual command.

Remote control facility

For ease of operation duplicate work handling controls, shielding gas switches, state selectors, and power supply voltage adjustments are mounted on the portable hand held unit shown on the right-hand side of Fig. 1. This allows normal laser use within recourse to the main console except for initial setting up.

Dr Crafer is a Senior Research Physicist in the Process Operation and Control Department of The Welding Institute.

Excitation

Two modifications have been made to the excitation system. One of these involves a change in the method of mounting the isenthalpic expansion nozzles. Previously these had been bonded to their respective discharge tubes with the result that initial distortion caused misalignment of the annular nozzle apertures. The new system allows the nozzles to be aligned independently of the discharge tubes, permitting a more efficient gas flow through the more symmetrical aperture.

The second modification involves reshaping the aluminium alloy cathodes to reduce downtime caused by tube replacement. Since fitting the new cathodes, the laser has operated for more than two years with tube replacement required only during routine maintenance.

Resonator

The laser's internal optical system, or resonator, has been modified in two respects. The major modification has been a complete redesign of the rooftop fold assembly to incorporate mirrors of a larger diameter and thickness to eliminate aperturing within the mode volume and to minimise mirror distortion, both mechanically and thermally.

The other modification has allowed larger diameter output windows to be used. The advantages are twofold:

1 Elimination of beam aperturing at the output window mount thereby improving efficiency and output power
2 Reduction of back reflection from the window mount into the laser resonator causing temperature changes in critical components

IMPROVED PERFORMANCE

Beam characteristics

The outcome of these modifications has been a greatly improved laser performance. The laser power has been upgraded to a routine 2.5kW (3.0kW maximum) with a resonator efficiency approaching 25%. (The efficiency is calculated from the measured tube currents, total supply voltage, and the known value of ballast resistance.) This high value has been maintained in subsequent operation.

Figure 2a shows a near field acrylic print of the unfocused laser beam. The beam appears to be axisymmetric and predominantly Gaussian, although the beam diameter of approximately 20mm is somewhat larger than expected for the dimensions of this resonator. Further work is currently in progress to elucidate this point.

The focal profile at the workpiece position is shown in Fig. 2: Fig. 2a shows a tracing of the original photon drag oscillogram, and Fig. 2b shows the intensity profile deconvoluted by a method of Sanderson.[2] The measured focal spot diameter at the 1/e intensity points is 0.14mm. This is consistent with diffraction-limited focusing of a single mode of the measured (not theoretical) beam size when lens aberrations have been taken into account.

Stainless steel welding

Although stainless steels do not constitute a large percentage of welded materials they do possess two attributes of special interest with respect to laser welding:

(a) they provide a reproducible laser welding response
(b) they are relatively easy to weld, even with indifferent lasers and primitive gas-shielding conditions

Hopefully as a result of the first attribute, but probably because of the second, welding performance with stainless steel has become widely accepted as a yardstick of industrial welding laser performance. Although the author does not necessarily approve of the use of this yardstick, it is the currently accepted method of comparison and for this reason the following stainless steel data are presented.

Maximum welding speed

The maximum welding speed for complete penetration in 18/8 type stainless steel for the modified laser is shown plotted against material thickness in Fig. 3, curve A. The net laser power incident at the workpiece is 2.2kW, corresponding to a gross output power of 2.5kW* (unless stated otherwise, the term 'laser power' should be interpreted as net power measured at the workpiece). Curves B, C, D, and E are performance curves for other lasers at as similar a power level as possible, plotted from data available in the open literature.[3-6] Curve C (UKAEA, Culham) is the most competitive at an identical power level to The Welding Institute data, but is a factor of 1.7 slower in speed for thicknesses between 1.6 and 4.6mm. At constant speed, i.e. 50mm/sec, the depth of penetration is nearly 1mm less. Results from United Aircraft at 10% less power than for

* The difference arises from reflection and transmission losses in the beam-handling optics.

the modified 2kW laser are a factor between 2 and 3 slower in welding speed, or 1.5mm less in penetration, between 2.6 and 3.8mm thickness.

The remaining curves are at more scattered power levels and are included for completeness rather than direct comparison. Curve B (AVCO) corresponds to 3kW but still appears less competitive below 4mm thickness. Curve E (Sylvania) is from their model 971 laser, the welding performance at 1.35kW laser power for very thin material appears promising but the performance at greater thicknesses is uncertain.

Curve F is an EB result plotted from data using a standard commercial 6kW machine. Below 3mm thickness the EB results are identical with the laser. Above this thickness the electron beam can achieve penetration at significantly higher traverse speeds.

Weld profile

To produce welds acceptable in an engineering context, the welding speed is generally reduced from the values given in Fig.3 to ensure excess penetration and improve weld profile. Figure 4a-e shows a series of welds in 18/8 stainless steel for thicknesses ranging from 1.6 to 5.2mm using conventional shielding techniques.

Satisfactory profiles were obtained in thicknesses up to 3.4mm simply by reducing welding speeds by about 20% from those required at the limiting penetration. Greater thicknesses suffered a degradation in profile even with speed reductions of up to 60%. An extreme example of this is the 5.2mm weld shown in Fig.4d with incipient weld metal dropout.

To improve the profile and increase the penetration, a plasma-control attachment[7] using helium gas has been developed experimentally in which a narrow high velocity gas jet interacts with the metal vapour plasma. Figure 4e shows the improved performance brought about by this attachment in 6mm material. The improvement in surface profile over that in Fig.4d is obvious and also significant in that a greater penetration was achieved at a higher welding speed. Work is in hand to modify the profile at this thickness and already further improvement has been achieved. Figure 5 shows a complete welding head incorporating the plasma-control attachment.

At the lower end of the thickness range the plasma-control attachment is not necessary as the weld profile can be adjusted by means of welding speed and focal position, see for example the series of welds in 1.6mm thick 18/8 stainless steel, Fig.6, for welding speeds between 30 and 100mm/sec. In all of these the focal position was optimised to obtain the most parallel-sided weld possible.

For moderate thicknesses, say from 2 to 4mm, the plasma-control attachment plays a special role. Although not necessary to achieve penetration it can have a considerable effect on weld profile. An example is shown in Fig.7 in which the laser power and welding speed have been kept constant. The first section, Fig.7a, shows a conventional weld in 3.7mm thick martensitic stainless steel without the benefit of plasma control. The reader's attention is drawn to the distortion-inducing weld profile asymmetry and the presence of root porosity commonly associated with nonpenetrating or marginally penetrating welds. With partial plasma control, Fig.7b, the penetration finger has become wider at the expense of the nail head. With full control, Fig.7c, the nail head has completely disappeared resulting in a much more parallel-sided weld, with a complete absence of root porosity.

APPLICATION TO ENGINEERING MATERIALS

Mild steel welding

Mild steels are now weldable with acceptable profiles in thicknesses up to 4.5mm. Figure 8a shows the variation of maximum welding speed with material thickness for a low carbon mild steel, in comparison with the stainless steel data of Fig.3, together with the full penetration profiles for mild steel, Fig.8b. The effects of simple shielding, i.e. no plasma-control attachment, are shown and up to 3mm the welds are satisfactory. At 4mm, however, the weld without plasma control reverts to a classical 'nail' shape with a considerable amount of porosity both in the root and elsewhere. The use of a plasma-control attachment on a 4.5mm specimen is shown in Fig.8c. The profile has become more parallel-sided and the porosity has almost entirely been eliminated, though at the expense of weld speed which has dropped from 17mm/sec at 4mm thickness, Fig.8b, to 8mm/sec at 4.5mm thickness, Fig.8c. In terms of weld profile and its implications for distortion, the latter is by far the better weld even at the lower speed.

To demonstrate the effect of weld speed upon profile Fig.9 shows a series of welds in 3mm thick mild steel, this time without plasma control. All the profiles are very similar over the speed range 17-38mm/sec with the exception of a small decrease in root width as speed increases. The only feature to change considerably with speed is the width of the parallel-sided heat-affected zone which varies from being equal to the material thickness at 17mm/sec, to a factor of two smaller at 38mm/sec.

This tolerance to weld conditions augurs well for the future of laser welding machines in light engineering production processes.

WELDING ALUMINIUM ALLOYS

Until very recently aluminium alloys could not be welded at laser powers less than about 5kW. The reason for this restriction lies in the nature of the welding process itself. Laser welding proceeds via the intermediary of a keyhole — a vapour-filled cavity surrounded by a wall of molten metal — extending the full depth of the weld. In butt welding applications the keyhole usually extends the full depth of the material, resulting in a guaranteed complete penetration weld. In these respects the laser process bears similarities to EB and plasma welding used in a keyhole mode.

The first stage in the formation of a laser weld is absorption of laser light at the surface of the workpiece. At the CO_2 laser wavelength of 10.6μm (in the infrared region of the spectrum) most cold metals are excellent reflectors, reflecting between 90 and 99% of the incident light, the exact value depending on composition and cleanliness. For example, most common steels, in clean condition, reflect about 92% of the incident light and the initial heating rate is therefore low. As the temperature rises the reflectivity gradually falls so that the heating rate tends to accelerate. By the time the metal at focus has reached its boiling point, which may take only a small fraction of a millisecond, the reflectivity will probably have fallen to around 70 or 80%. Metals with higher reflectivities will approach boiling point at progressively slower rates until, for a given laser beam, a situation will be reached where it cannot be achieved. For this reason highly reflecting metals such as gold, silver, and copper cannot at present be welded at the 2kW power level, although welding has been reported at higher powers.

When the metal boils the emitted vapour becomes highly absorbent to the laser beam, and the greatly increased boiling rate causes local surface depression and the formation of a keyhole in the metal surface. If the laser power is sufficient the keyhole will penetrate completely through the metal.

Recently, however, certain modifications to the laser resonator have enabled a beam of such high intensity to be generated ($>50kW/mm^2$) that aluminium alloys can be welded at powers of only 2kW. Preliminary studies on H15 alloy show weld speeds for limiting penetration comparable to those for 18/8 stainless steel for thicknesses up to 3mm. As an example of weld quality obtainable, Fig.10, shows a section of a laser weld in 1.6mm thick H15 alloy at a power of 2.2kW and a speed of 46mm/sec. Features to note are the smooth top and bottom beads and complete lack of porosity: qualities traditionally difficult to achieve in aluminium alloys, and particularly in H15, which is not recommended for welding by normal processes.

Joining rate

As an index of overall performance Fig.11 shows process joining rate in mm^2/kJ plotted against material thickness for stainless steel, mild steel, and aluminium alloy H15. For all these materials the improved 2kW laser clearly exhibits an optimum thickness range of between 1 and 3mm. The broken lines show the range of extended performances consequent on plasma control; in stainless steel the improvement is mainly in speed and penetration, but in mild steel it is directed more towards weld profile and lack of porosity at the greater thicknesses. Representative 2-3kW data on stainless steel from other laboratories are also included for comparison.

CONCLUSIONS

The improved performance described above demonstrates some of the possibilities of the upgraded 2kW laser system. Mild and stainless steels have been welded faster and deeper than previously was thought possible, and high conductivity materials such as aluminium alloys have been welded at lower power levels than hitherto considered feasible.

All three materials exhibit an optimum joining rate of $\sim 60\text{-}70mm^2/kJ$ in the thickness range 1-3mm, with extended thick section performance in the steels as a result of efficient control of the metal vapour plasma by means of the plasma jet attachment.

ACKNOWLEDGEMENTS

The author is indebted to Mr J.D.Russell for advice and to Mr D.G.Staines for technical assistance during the course of the research described in this Paper.

REFERENCES

1 CRAFER, R.C. 'A 2kW CO_2 laser system for welding sheet material'. Welding Inst. Conference 'Advances in Welding Processes', Harrogate, 7-9 May 1974, Paper 26, 178-84.

2 SANDERSON, A. 'Electron-beam monitoring techniques and probe trace analysis'. Welding Research Int'l, 7 (2), 1977, 144-74.

3 AVCO HPL10 Metalwork Laser. Commercial Information Sheet, AVCO Everett Corporation, Revere Beach Parkway, Everett, Mass 02149, USA.

4 ALLEN, T.K. et al. 'The current status of lasers as industrial tools'. Royal Society Conference 'Lasers in Industry', 1975.

5 BANAS, C.M. 'Laser welding developments'. CEGB Conference 'Welding Research Related to Power Plant', 17-21 September 1972, Paper 41, 565-73.

6 ENGEL, S.L. 'Kilowatt welding with a laser'. Laser Focus, February 1976, 44-53.

7 Patent applied for.

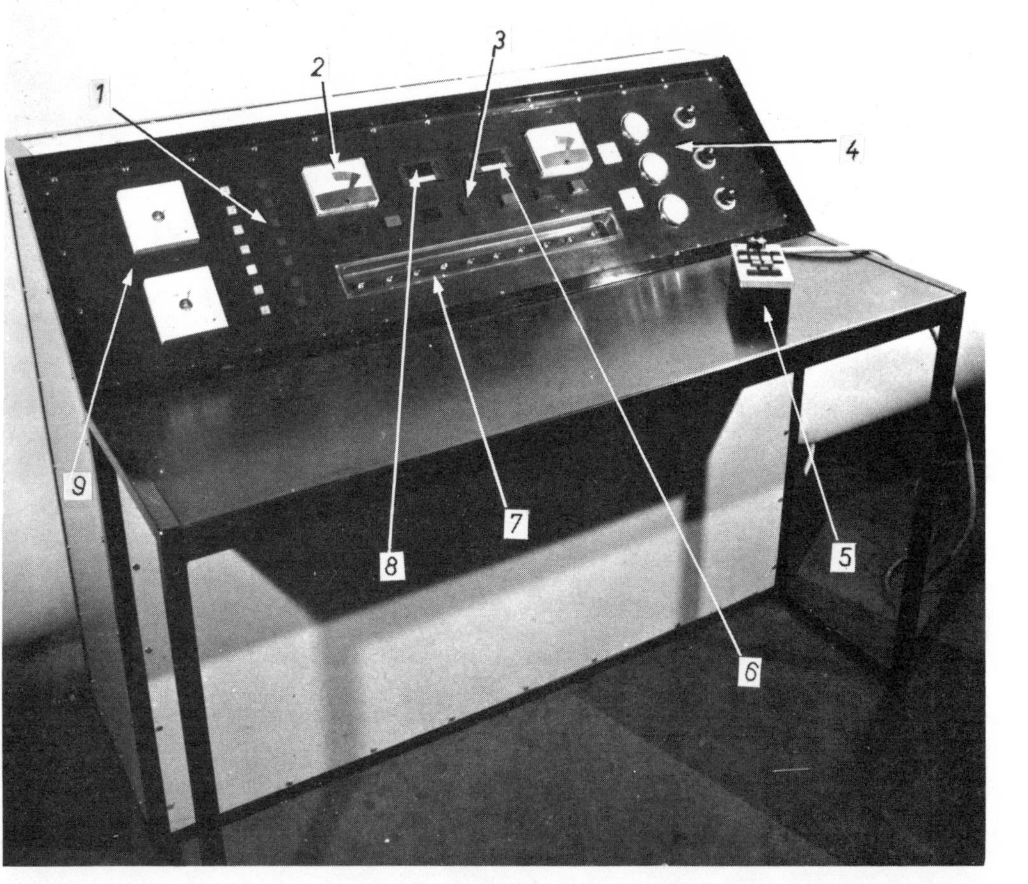

1 Centralised laser control console showing remote hand-held unit on right-hand side of work surface. 1 – indicators; 2 – pressure ratio; 3 – state selectors; 4 – gas mixing supply; 5 – remote control; 6 – laser power; 7 – override facility; 8 – welding speed; 9 – pressure gauges

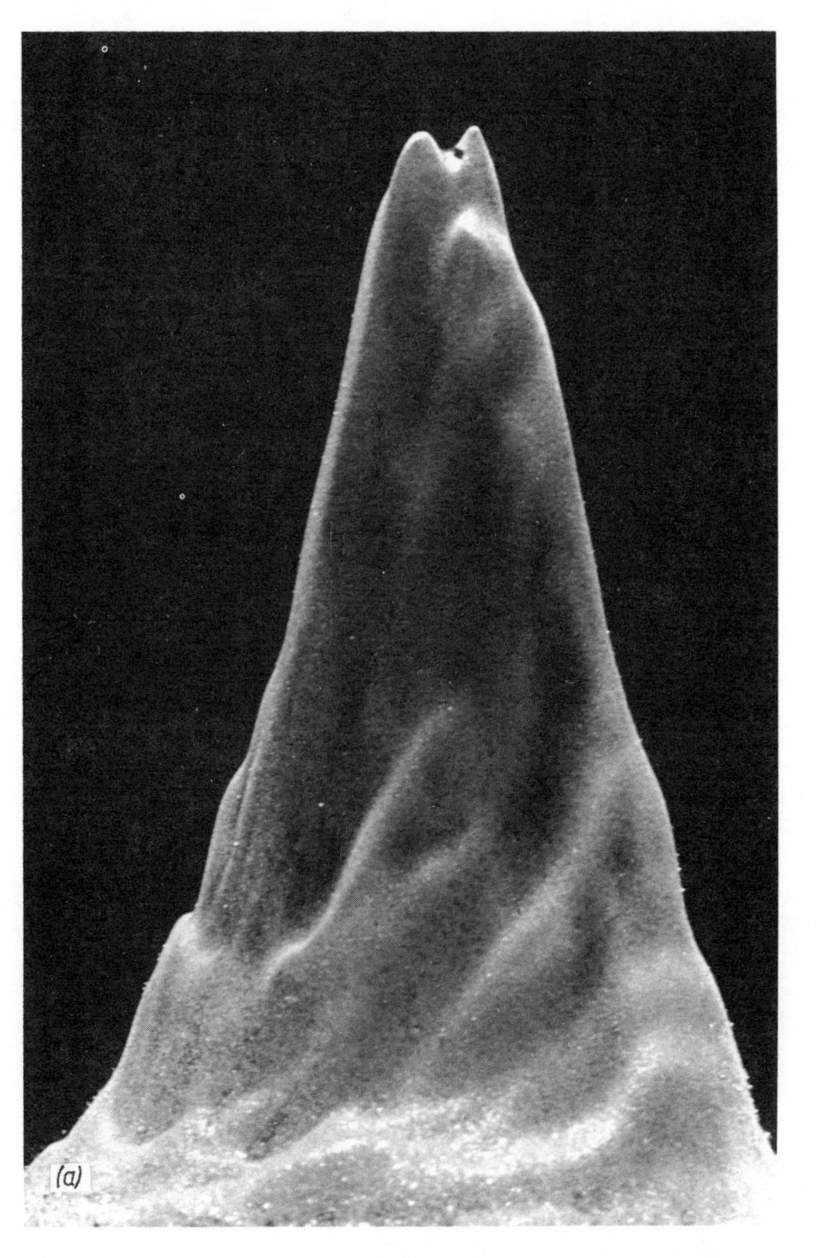

2 Laser beam analysis: (a) near field acrylic print of unfocused beam,

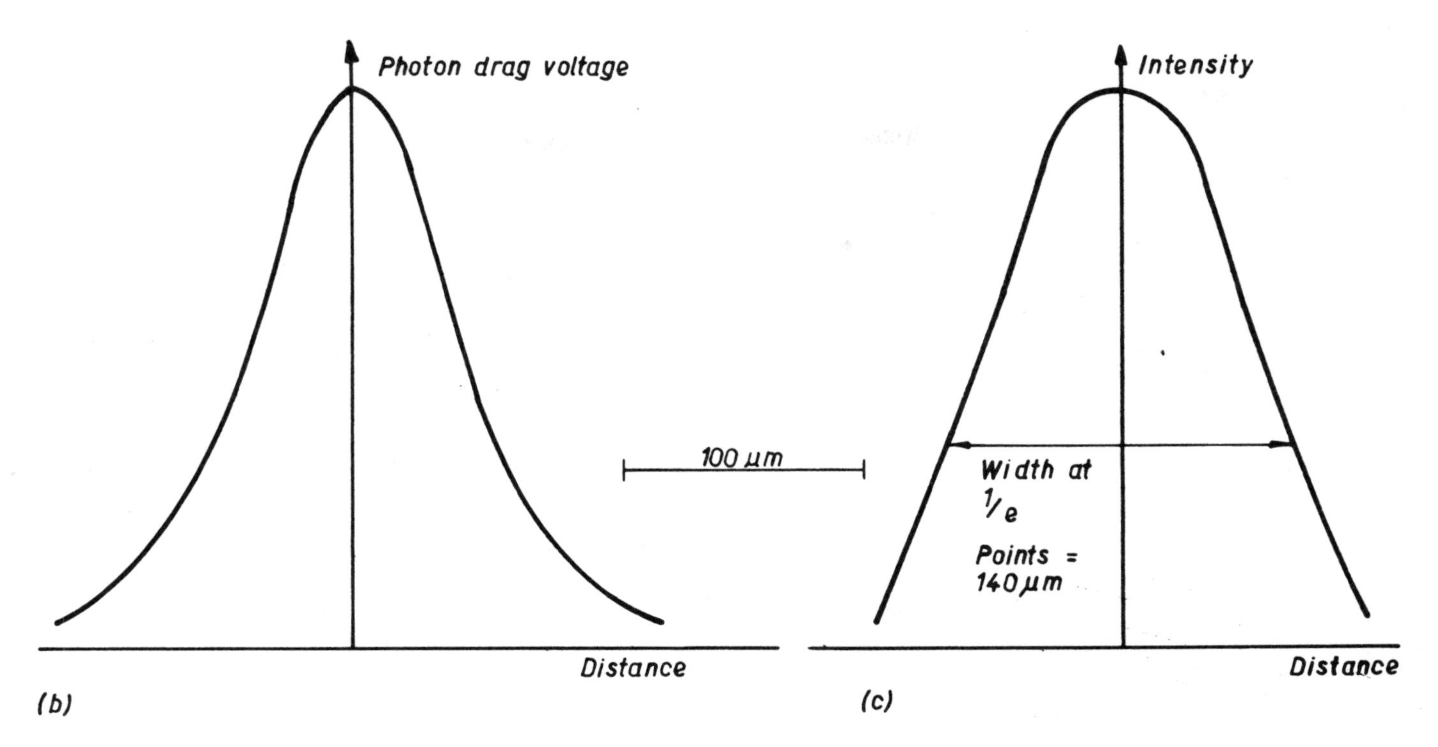

2 cont. (b) tracing of focused beam photon drag oscillogram, and (c) deconvolution of (b) according to Ref.2

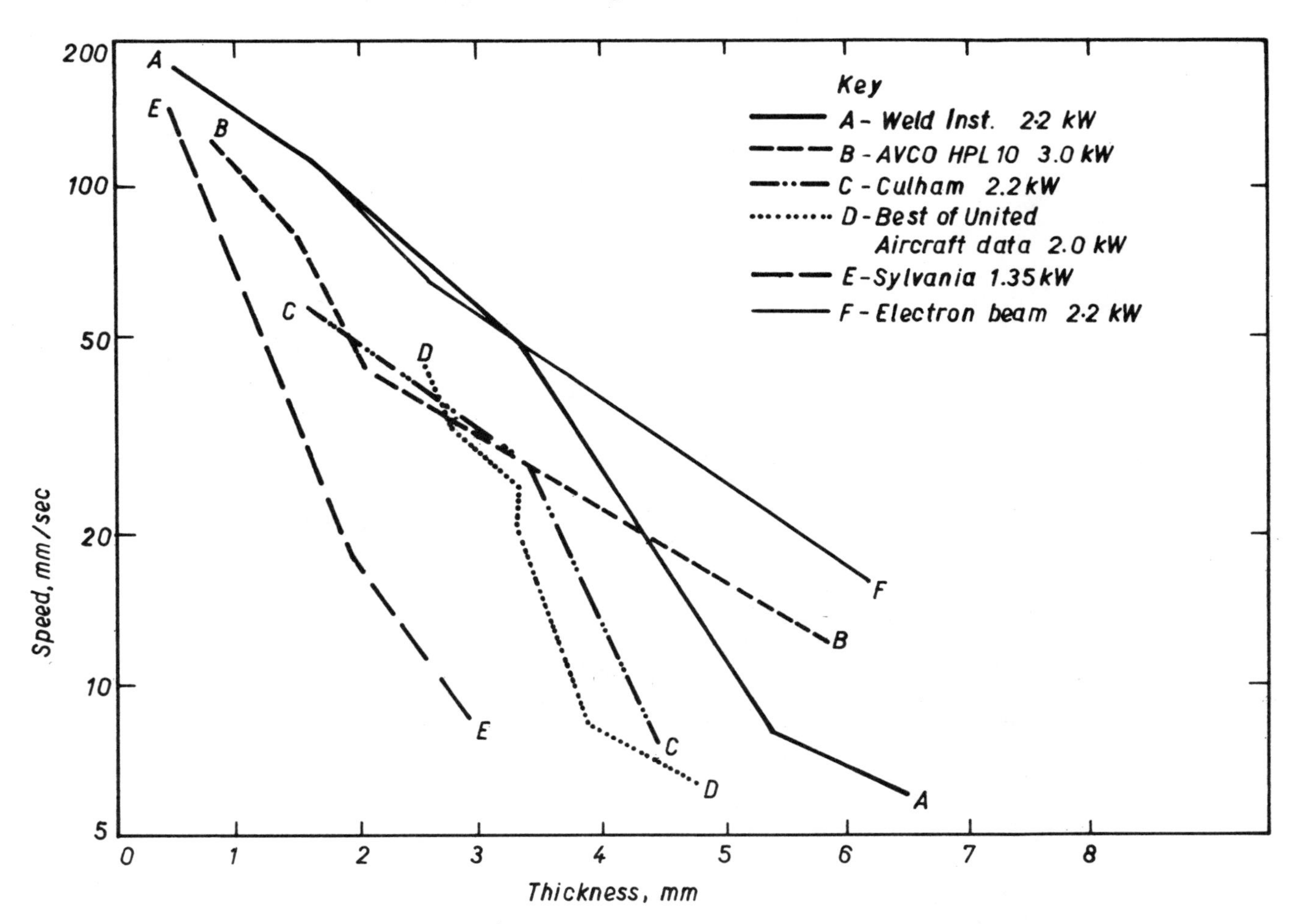

3 Laser welding data for 18/8 stainless steel. Weld depth plotted against welding speed

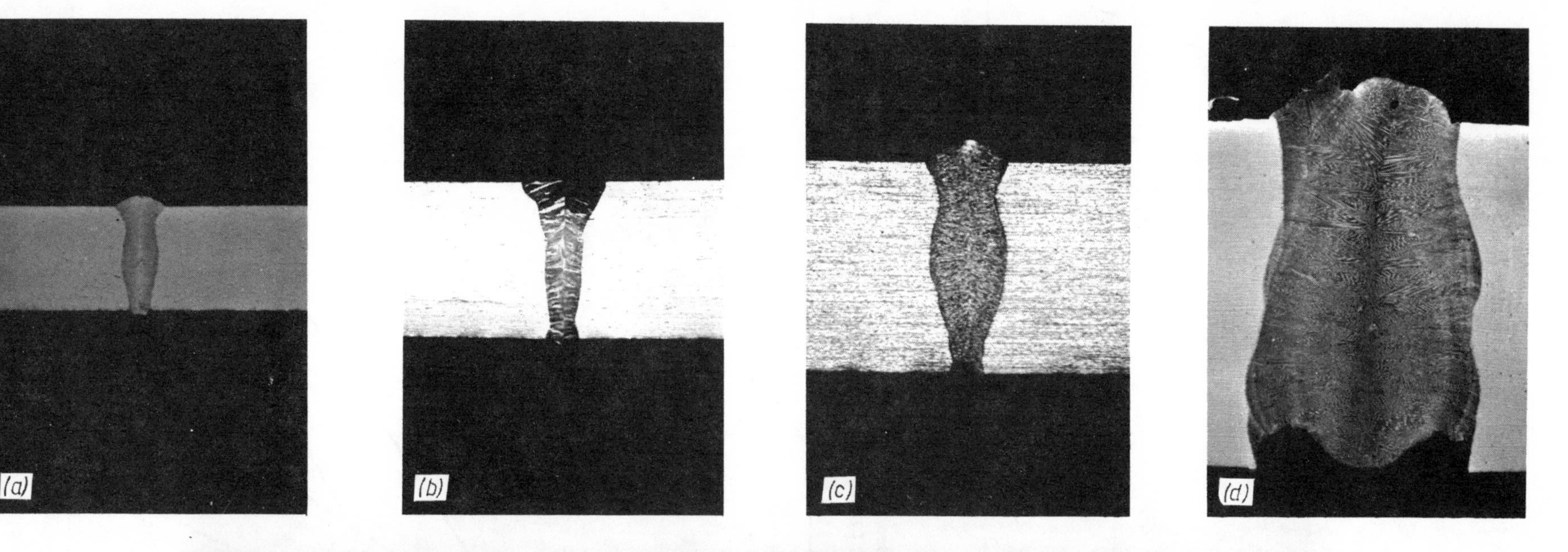

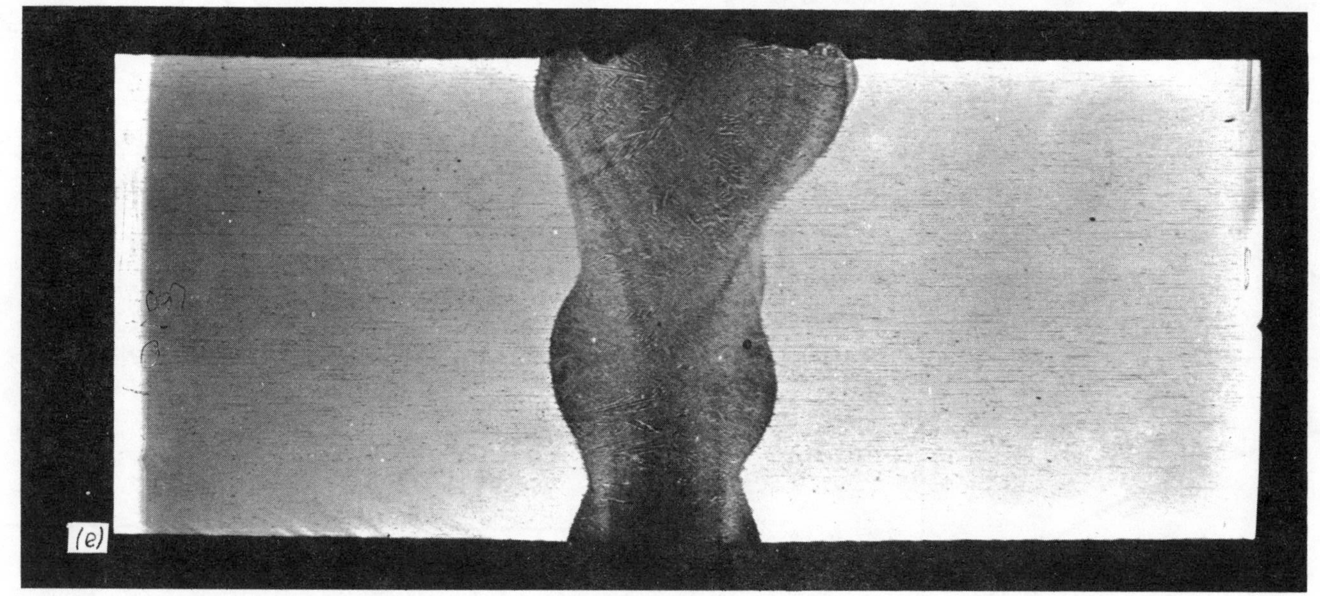

4 Effect of material thickness on weld profile in 18/8 stainless steel for marginal overpenetration. Power 2.2kW, simple shielding: (a) 1.6mm thick, 100mm/sec, (b) 2.7mm thick, 45mm/sec, (c) 3.4mm thick, 38mm/sec, (d) 5.2mm thick, 5mm/sec, and (e) power 2.0kW, plasma control, 6.5mm thick, 6mm/sec x 10

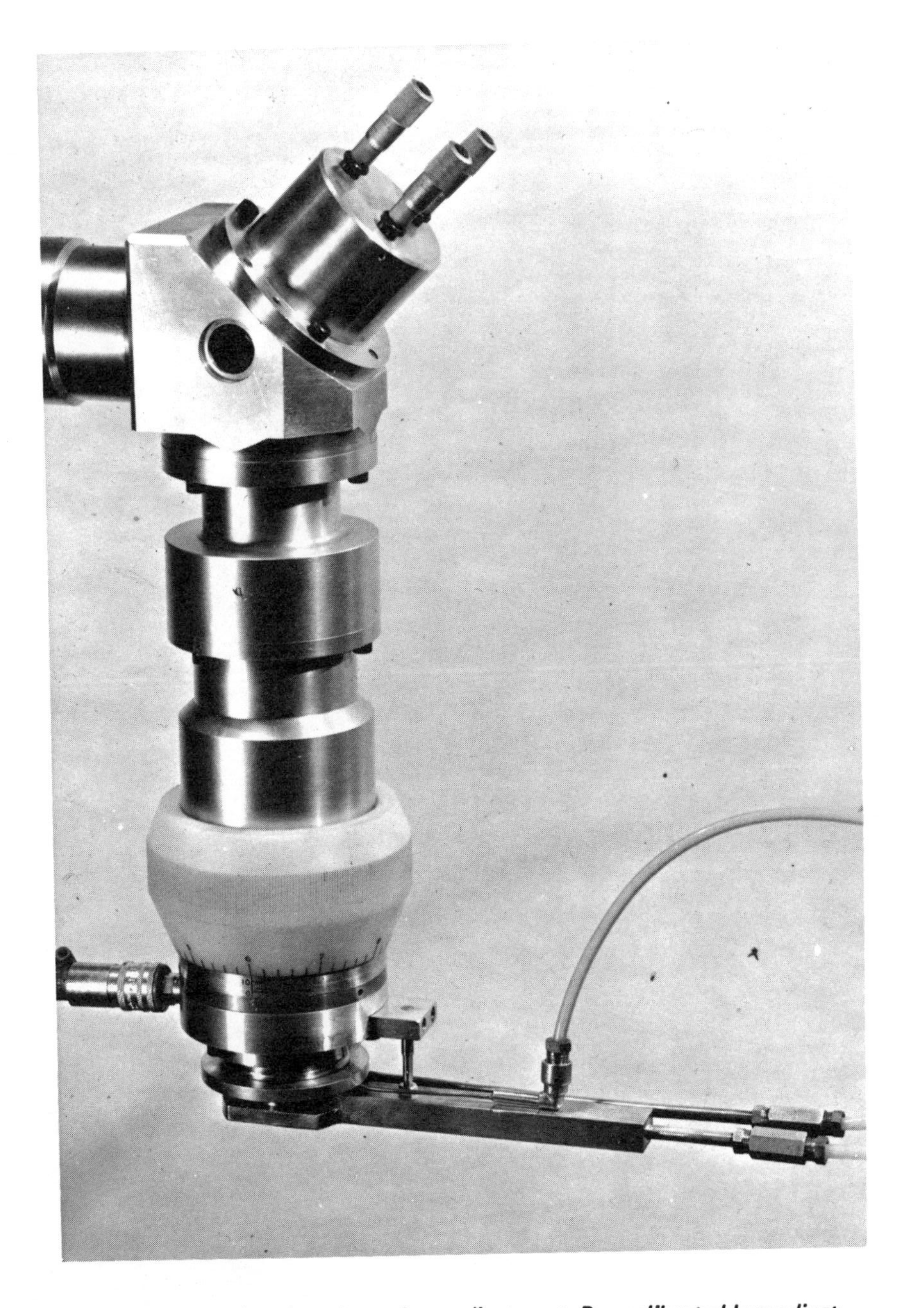

5 Laser gun showing: A — mirror adjustment; B — calibrated lens adjustment; C — gas shield and plasma-control assembly

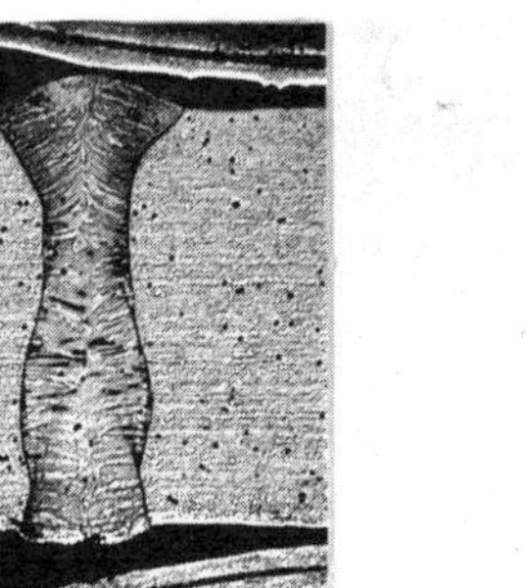

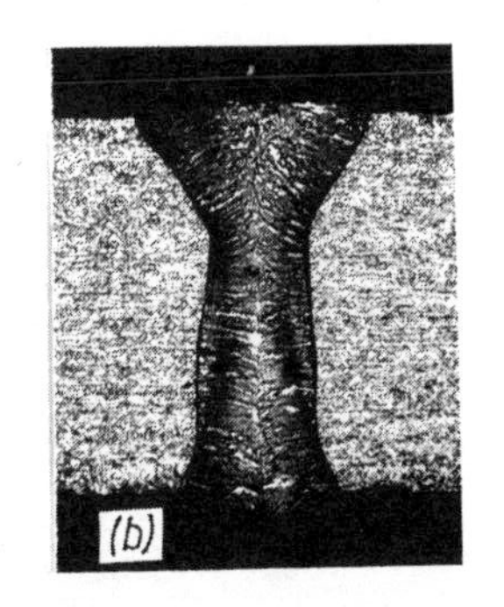

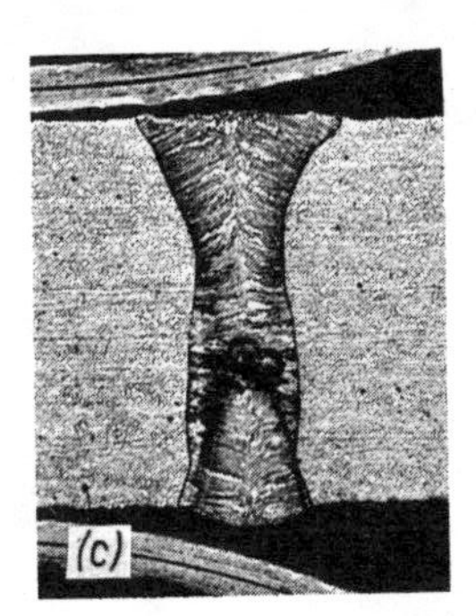

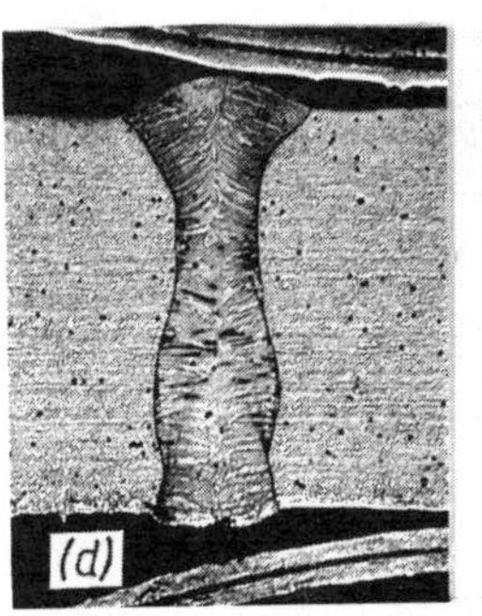

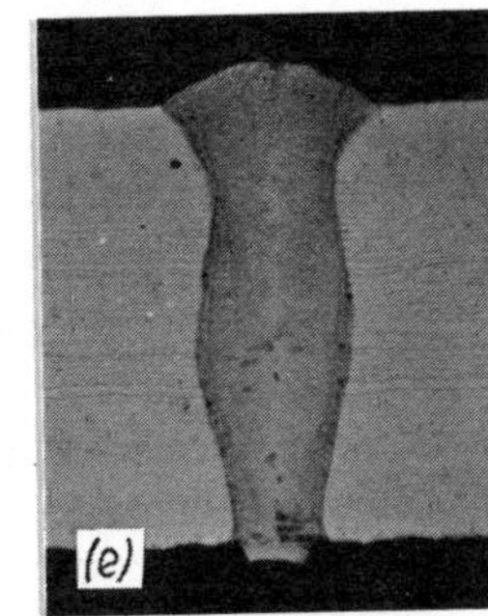

6 Effect of weld speed on weld shape for Type 304 18/8 stainless steel 1.6mm thick, 2.2kW at workpiece: (a) 30mm/sec, (b) 45mm/sec, (c) 70mm/sec, (d) 85mm/sec, and (e) 100mm/sec x 20

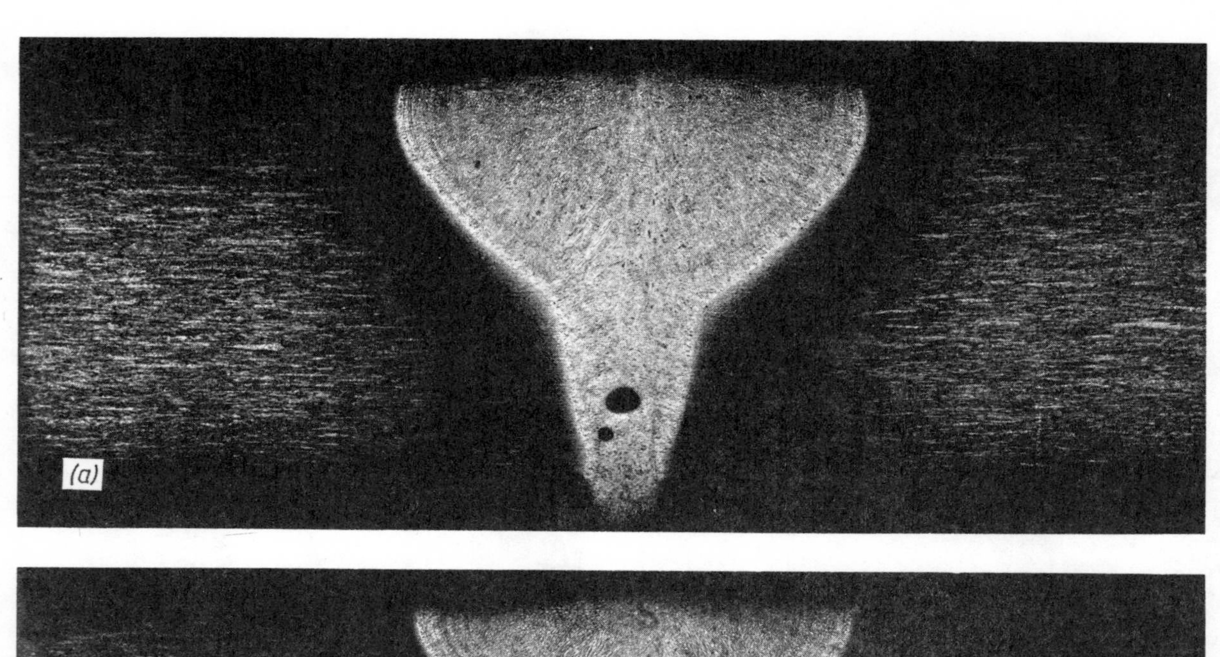

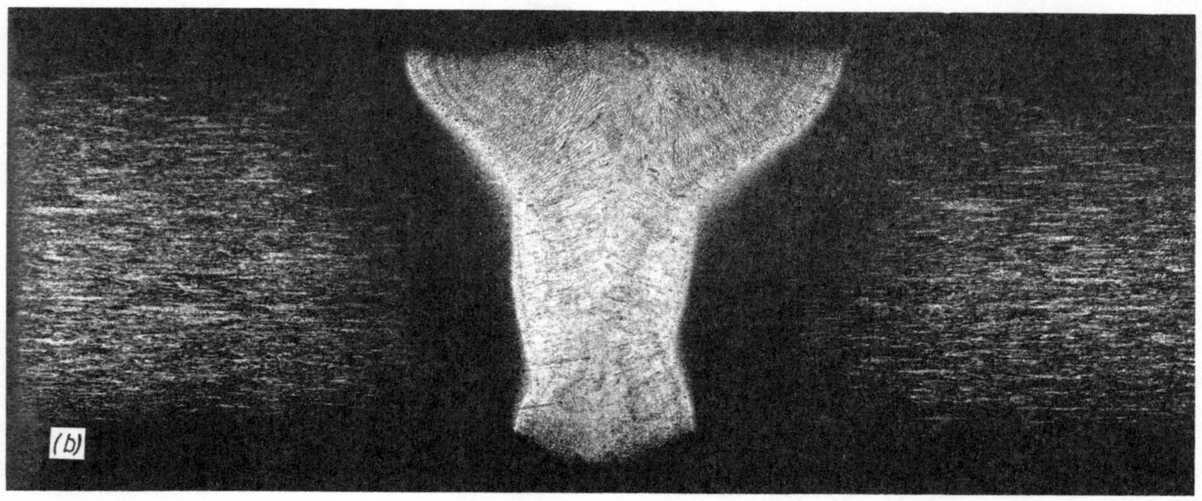

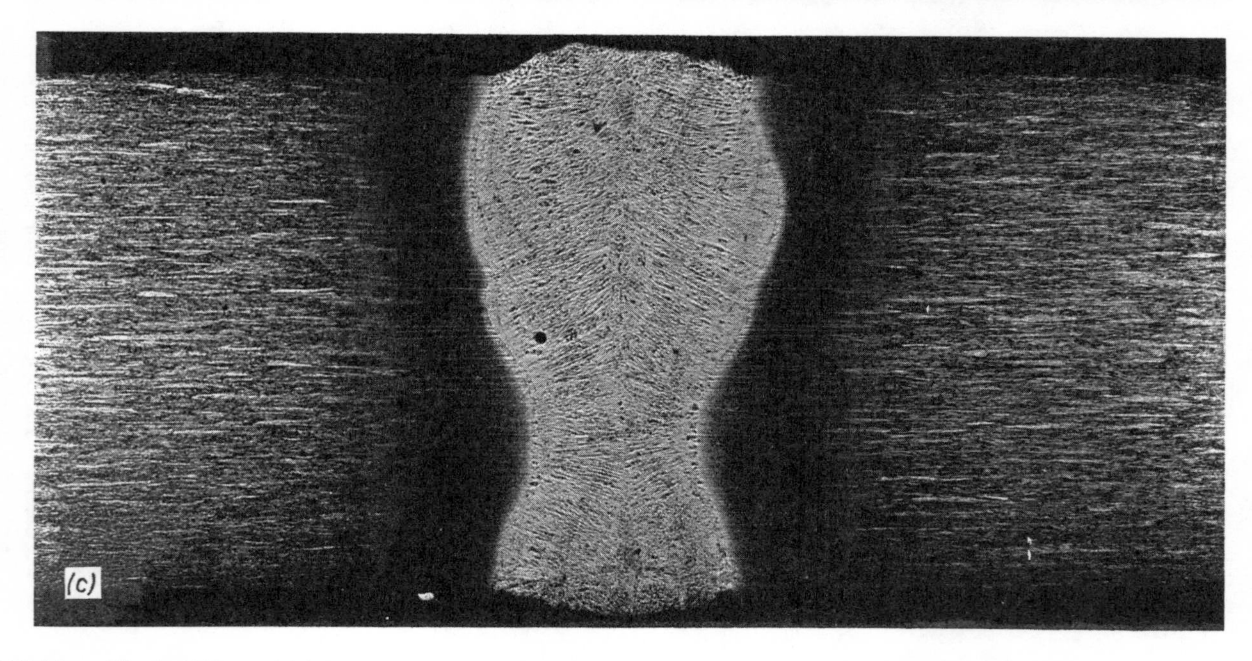

7 Weld profiles in 3.7mm thick high alloy steel showing effect of plasma control. Laser power — 2.5kW, welding speed — 8mm/sec, focus at surface: (a) no plasma control, (b) partial control, and (c) full control x 15

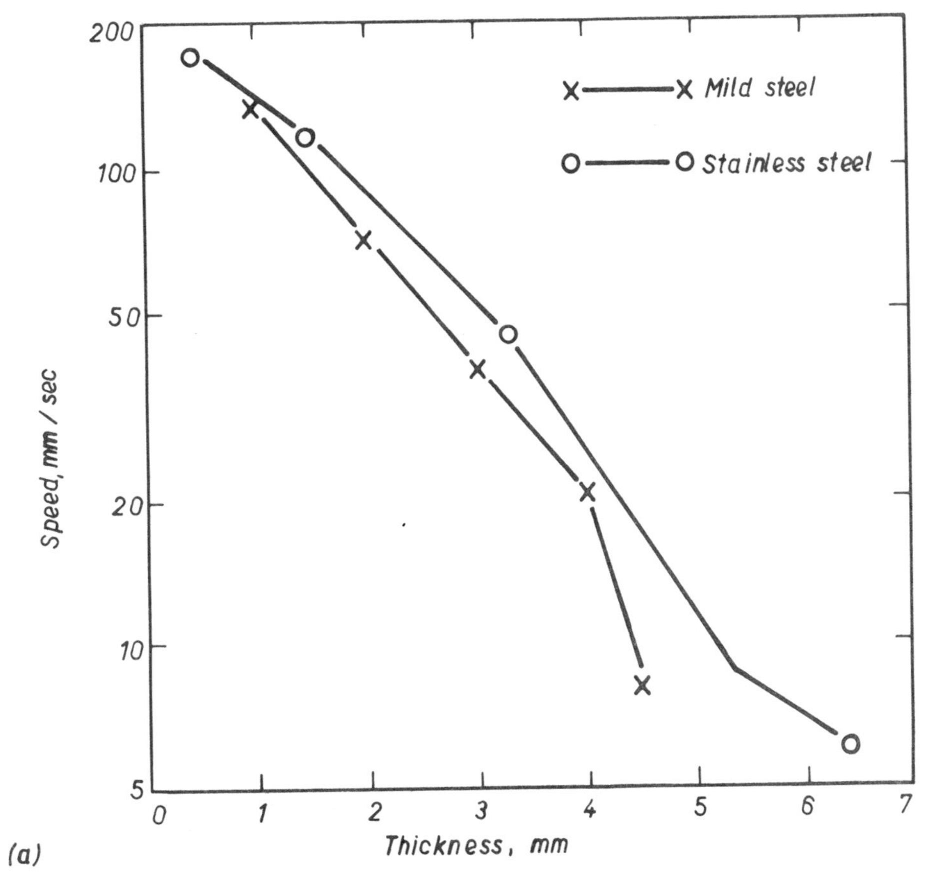

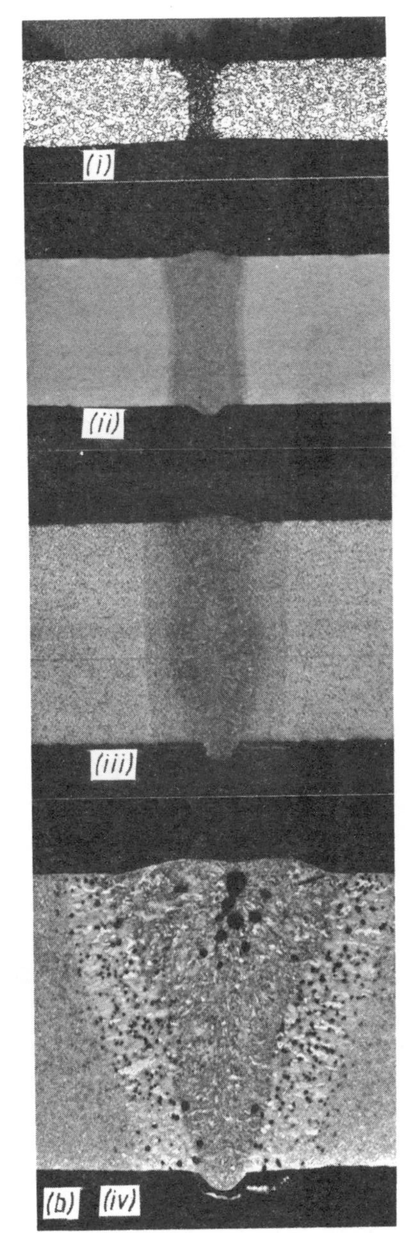

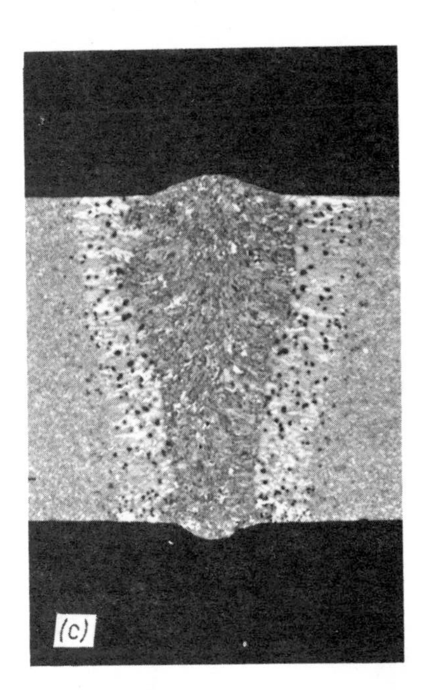

8 Laser welding of mild steel: (a) comparison of welding data for mild steel and stainless steel; laser power – 2.2kW, (b) effect of material thickness on weld profile in mild steel for marginal overpenetration; power – 2.2kW. Simple shielding: (i) 1.0mm thick, 135mm/sec; (ii) 2.0mm thick, 70mm/sec, (iii) 3.0mm thick, 31mm/sec, (iv) 4.0mm thick, 17mm/sec. Plasma control: (c) 4.5mm thick, 8mm/sec x 10

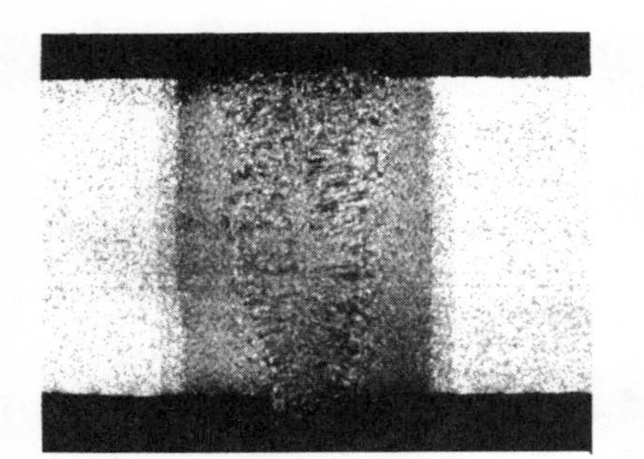

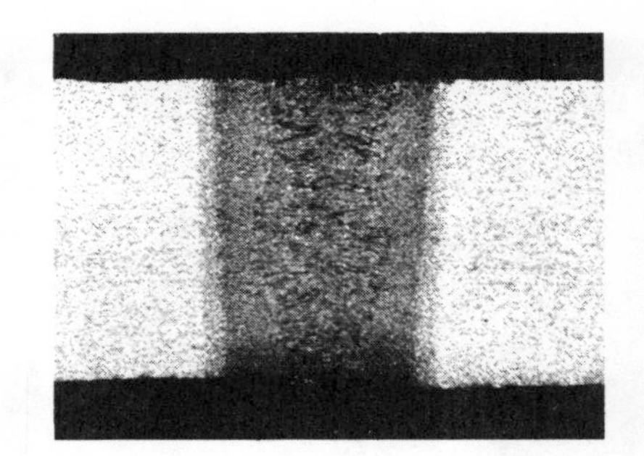

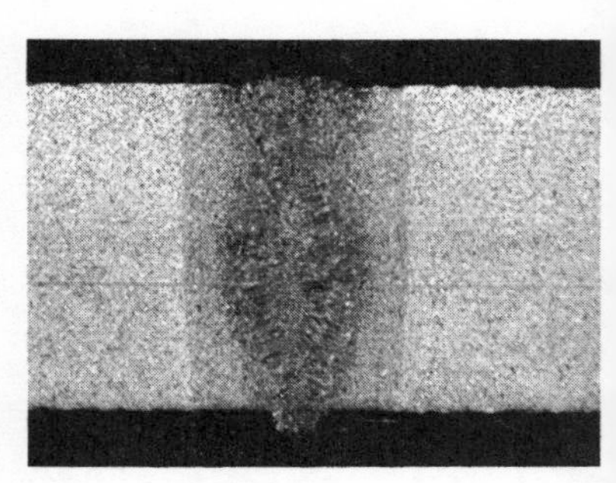

9 Effect of weld speed on weld shape for mild steel 3.0mm thick; laser power — 2.2kW: (a) 17mm/sec, (b) 24mm/sec, and (c) 38mm/sec   x 10

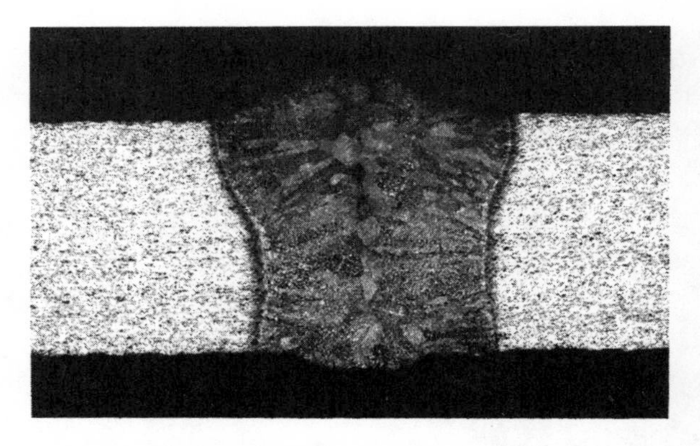

10 Laser weld profile in 1.6mm thick H15 aluminium alloy; laser power — 2.2kW, speed — 46mm/sec

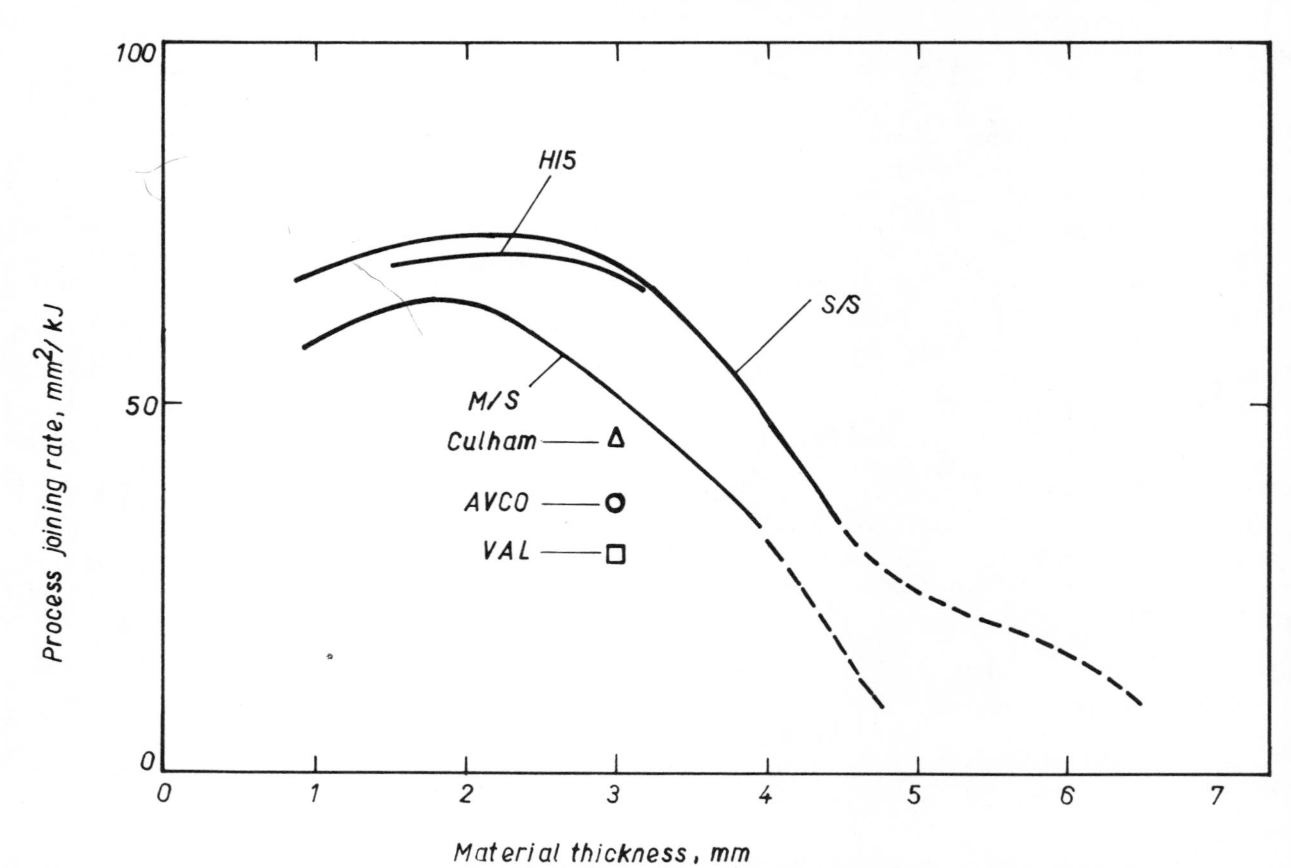

11 Process joining rate v. thickness for improved 2kW laser, S/S — 18/8 stainless steel; H15 — H15 aluminium alloy; MS — mild steel; ----- extended performance consequent on plasma control: △, ○, □ are typical data from UKAEA, Culham, at 2.2kW, AVCO at 3kW, and UAL at 2kW

A precision electron-beam welding machine for volume production

B.L.Miles, BSc

INTRODUCTION

The exploitation of electron-beam (EB) welding for the fabrication of volume production assemblies depends upon the availability of high precision, reproducible, and reliable welding equipment at a reasonable price.

Experience within Joseph Lucas Limited has indicated that the high power and flexibility usually associated with expensive high voltage equipment was not necessary for the majority of volume production applications. A machine with a restricted working distance and limited power level would be acceptable, provided that the beam positional and power stability were good and that high machine utilisation could be achieved during shopfloor operation.

This Paper describes the design, manufacture, and testing of a prototype EB welding machine which has been built to meet these requirements. It is concluded that, providing sensible limits are accepted for maximum penetration and working distance, it is possible to build a low voltage EB welding unit with acceptable penetration characteristics and good beam stability.

MACHINE SPECIFICATION

The heart of any EB welding unit is the beam generator which consists of the EB gun and its associated power supplies. The generator specification was compiled by considering the:

1 Requirements of the various applications that had been identified
2 Need to eliminate dependence upon the skill of the operator, and
3 Importance of maintaining a high machine utilisation and low capital cost

Power level and working distance

High power levels and long working distances were not considered to be necessary for the majority of volume production applications envisaged. Most assemblies were manufactured from steel with a maximum joint depth of 5mm. Experience with high voltage equipment had indicated that to obtain 10mm/sec welding speed a power level of about 1kW would be required. The maximum working distance could also be restricted. The majority of assemblies could be welded with the joint line positioned within 150mm of the focusing coil, and a working distance of 100mm would usually be acceptable. It was also considered unnecessary to make allowances for an optical viewing system as this facility would not be required on a machine with good beam positional stability.

Power density and accelerating voltage

Electron-beam welding is usually selected because of its ability to produce a weld geometry with a high depth to width ratio. This ensures that the overall heat input is low, thus reducing distortion and increasing the possibility of producing acceptable welds in alloys normally considered unweldable by other processes. It is therefore important to achieve a high power density, but at minimum cost. Bearing in mind the limited working distance required, it was decided that an accelerating voltage of 30kV should be evaluated. This would limit the level of electrical insulation and X-ray shielding required and help to reduce overall manufacturing costs.

Beam stability

It was envisaged that the machine should be capable of operation on the shopfloor for long periods, controlled by unskilled or semi-skilled operators. Once welding parameters had been established for a particular assembly it would be essential that the machine could be set and production continue without the need for frequent adjustments to weld settings. For this to be achieved, the following characteristics would be required:

1 Good long-term stability of beam power density to ensure reproducible weld penetration and geometry
2 Excellent beam positional stability
3 A simple procedure for filament changing, without the need for beam re-alignment

---

Mr Miles is a Senior Research Officer at the Lucas Group Research Centre.

These features would depend upon careful attention to gun and column design and the specification of stable electrical supplies incorporating closed loop to ensure precise control of electrical parameters.

DESCRIPTION OF THE PROTOTYPE MACHINE

Gun and column assembly

A diagram of the electron gun and column is shown in Fig.1.

The most significant feature of the overall design is the interchangeable gun, enabling a used gun to be replaced by a serviced assembly in a short period of time. Filament changes and electrode cleaning and alignment are therefore carried out away from the machine, limiting equipment downtime to a few minutes.

Each gun is a complete anode, cathode, and bias cup assembly. During servicing the electrodes are aligned using a special-purpose tool. This maintains diameters D3, D4, and D5 concentric to $\pm$0.01mm and spacing Z2 to $\pm$0.025mm. The cathode consists of a directly heated tungsten ribbon filament which is shaped to produce an emitting area of 1.5 x 1.5mm. This is again aligned using special-purpose equipment to ensure that its centre is concentric with D4 and that the filament set back Z1 is held to $\pm$0.025mm.

During operation the bias and cathode support assembly rise in temperature as a result of thermal emission from the cathode. Equalisation temperature is 300°C and it is reached after 2$\frac{1}{2}$hr operation. Throughout this period it is important to maintain both the concentricity and spacing of the electrode system. Failure to do this could result in movement of the spot position and change in beam power density. To minimise the effects of thermal expansion the gun assembly is manufactured from materials with matched expansion characteristics. It is also symmetrical in design to ensure that the electrode surfaces remain concentric during warm up.

A most difficult area of design is that relating to the cathode support assembly. The cathode alignment is accurately maintained by means of a ceramic disc which is positioned around the filament. This close control of the cathode/bias relationship ensures minimum movement during gun warm up.

Each gun is accurately located within the column by diameter D1 with a fit of K6/f8 (BS 308 : 1972 : Pt 3). The electromagnetic lens is installed with a concentricity of pole piece diameter D2 to gun location diameter D1 of $\pm$0.1mm. The column dimensions are generous and wherever possible all internal surfaces have a smooth profile to reduce the likelihood of flashover. Other features include, a side entry high voltage leadthrough to facilitate gun interchange and a venturi and column valve to permit the use of partial vacuum for welding.

Electrical supplies

The high voltage supplies are generated in an oil-filled tank and closed loop control is used throughout. The series regulator maintains the high voltage within close limits, and provides a fast response to fluctuations in beam current. Saturable reactors are used to control both filament and bias supplies. The focus current is provided by a highly stable constant current DC source.

Complete machine

A photograph of the prototype machine is shown in Fig.2. The column assembly is mounted on a 460 x 300 x 300mm work chamber and is pumped by a 100mm diffusion pump. The front loading chamber is fitted with a full view leaded glass door and is pumped by a rotary and booster pump in series. The chamber houses X-Y and rotary motions. These are powdered by an externally mounted constant speed motor with variable speed gearbox.

MACHINE PERFORMANCE

Penetration characteristics

Configuration that the beam displayed acceptable penetration characteristics was obtained by assessing the effect of variations in parameters on the penetration of melt runs in bright drawn mild steel. All runs were produced with the beam at optimum focus and a chamber pressure of 3 x 10^{-2}mbar. Figure 3 shows the effect of variations in beam current, working distance, and welding speed on penetration at otherwise constant conditions. Figure 3a and b indicates that the required penetration of 5mm was achieved at a power level of 900W and a working distance of 90mm.

Figure 4 compares the weld sections produced by the prototype machine and a commercially available high voltage machine. Both were produced at the same conditions with the focus current optimised to give maximum penetration. The weld geometry produced by the prototype machine compares favourably with that of the high voltage machine.

Effect of variations in electrode geometry

Small variations in electrode spacing and concentricity can occur because of machining tolerances during gun manufacture, incorrect use of the gun alignment tool, and thermal expansion during gun warm up. For a machine to be suitable for operation in a volume production environment it is important that these variations have minimum effect on weld penetration and position. Closed loop control can be used to counteract changes in beam current but it is difficult to vary the focus current automatically to combat changes in beam power density, and often impracticable to adjust the joint position following movement of the spot.

Variations in electrode spacing

The effect of variations in electrode spacing was determined by measuring the penetration produced by three gun assemblies over a range of focusing currents. Table 1 lists the variations in spacing evaluated and Fig. 5 shows the corresponding penetration produced. The melt runs were again produced in bright mild steel and penetration was assessed by examining longitudinal rather than transverse sections. This helps to reduce inaccuracies caused by the irregular penetration (spiking) of non-penetrating welds in mild steel. Throughout the exercise beam parameters were maintained at 30kV and 20mA, 10mm/sec welding speed, and chamber vacuum at 3 x 10^{-2}mbar. Figure 5 shows that, in each one, maximum penetration occurs at a focus current of 436 or 437mA, and that penetration lies within a ±5% tolerance band throughout the range evaluated. No correlation between specific variations in electrode spacing and penetration can be observed. This exercise demonstrates that, as long as the electrode spacing remains within the range evaluated and the optimum focusing current is sensibly constant, the penetration is within ±5% for mild steel.

Variations in filament concentricity

Variations in beam position were measured using the apparatus shown in Fig. 6. A copper disc with a small central hole was positioned

Table 1 Variations in electrode spacing evaluated

Test series	Gun no.	Variation on Z1, mm	Variation on 32, mm
R	2	+0.1	Zero
S	1	-0.025	-0.125
T	1	-0.025	Zero
U	3	Zero	+0.1
V	2	-0.075	+0.025

at a working distance of 100mm with a Faraday cup below. The beam was aimed through the hole, collected by the Faraday cup, and its current measured. Small movements of the beam position caused it to be partially intercepted by the disc and resulted in significant variations in the current measured. The deflection coils were calibrated so that the relationship between coil current and deflection of the spot at the working distance was known. Movement of the spot could therefore be determined by noting the deflection coil current required to restore the spot to its original position, as indicated by the current collected by the Faraday cup. The technique was sufficiently sensitive to measure beam movements down to 0.01mm.

The procedure was used to determine the effect of variations in filament position on spot position. The gun was modified to enable the filament to be adjusted in both X and Y directions. The actual filament position with respect to D4, Fig. 1, was measured using a travelling microscope. It was established that the ratio between beam movement and filament movement was 1:4. It was also shown that, when the special-purpose loading tools were used for filament changes, the movement in spot position was less than 0.01mm.

Stability during use

The previous exercises indicated that, provided electrode spacing and filament concentricity were maintained within the prescribed tolerances, mean penetration and position could be held ±5% and ±0.01mm respectively. The ability to maintain stable penetration and beam position during gun use with its associated thermal expansion and filament degradation were then examined using a variety of procedures, of which two are briefly described below.

Confirmation that control of penetration could be maintained was obtained by producing melt runs in bright drawn mild steel after the gun had been in use for many hours, and comparing the results with those obtained with the gun cold. Figure 7 shows the penetration produced at four focus settings. The ±5% tolerance band from Fig. 5 is superimposed and it can be seen that penetration with the 'hot' gun falls within this band. A thorough assessment of penetration reproducibility during prolonged use is planned.

Stability of spot position during use was assessed using the equipment previously described, Fig. 6. The beam was deflected away from the copper disc on to a water-cooled block and cycled on and off for a number of hours. It was then returned to the undeflected position where no change in position, i.e. Faraday cup current, could be detected. A similar procedure was also used to confirm that negligible movement occurs during a filament and gun change.

An important aspect of gun performance not already mentioned is the need for freedom from 'flashover' where the breakdown of bias voltage results in the flow of a high uncontrolled beam current. This was evaluated by cycling the beam 'on' and 'off' on a water-cooled block over a period of five days. The filament was changed at the begining of each day and 31hr of operation were accumulated. During this period only one flashover was recorded.

Overall stability of the EB generator

It has been shown that the electron gun achieves reproducible penetration to within ±5% providing the accelerating voltage, beam current, and focus current remain constant. In practice, variations in these parameters do occur and the tolerance on penetration is widened accordingly.

For example, referring back to Fig.5, a change of focus current by plus or minus 1% of the optimum setting increases the variation in penetration to around ±15%. Clearly, if close control of penetration is to be obtained during extended use, highly stable electrical supplies are required. By careful design of electrical circuitry, incorporating closed loop control and temperature compensation, the following limits for long-term stability have been achieved:

Beam current:	±0.5%
Accelerating voltage:	±0.5%
Focus current :	±0.2%

The effect on penetration of small variations in power (beam current x accelerating voltage) is insignificant when compared with small changes in focus and can therefore be ignored. The effect of accelerating voltage on beam focus can be considered by examining its relationship with focus current. Figure 8 shows the focus current required to keep the beam in focus at a constant working distance over a range of accelerating voltages. It can be seen that a ±0.5% change in voltage (±150V) produces the same effect as a ±1mA change in focus current. The combined effect of the variations in accelerating voltage and focus current (±0.2% at 436mA $\simeq$ ±1mA) would therefore be equal to a total change in focus current of ±2mA. Referring again to Fig.5 it can be seen that this would result in a variation in penetration of up to ±9%.

The prototype machine has been used for the development of a number of production applications involving the welding of large batches of components by unskilled operators. Equipment performance has been good with acceptable penetration reproducibility, a neglible flashover rate, and no problems with beam positioning.

CONCLUSIONS

This work has shown that, providing sensible limits are accepted for working distance and maximum penetration, it is possible to design and manufacture a precision 30kV EB welding machine with characteristics that are needed for volume production. The following attributes are particularly significant.

1 Penetration characteristics that are comparable with commercially available high voltage equipment at the power level examined

2 Potential for high machine utilisation, achieved by the use of interchangeable guns for filament changes and gun cleaning, and the absence of beam alignment procedures

3 Penetration performance that is insignificantly affected by small variations in electrode spacing

4 Penetration reproducible to ±9% and beam position stable to ±0.01mm, both during use and after gun changes, without adjustment of machine settings

5 Low flashover rate resulting in neglible component scrap

ACKNOWLEDGEMENTS

The author wishes to acknowledge the help and encouragement given by Mr L.N.Sayer who was responsible for initiating this project, and the contribution made by colleagues at the Lucas Group Research Centre. Finally, thanks are due to the Directors of Joseph Lucas Limited for permission to present this Paper.

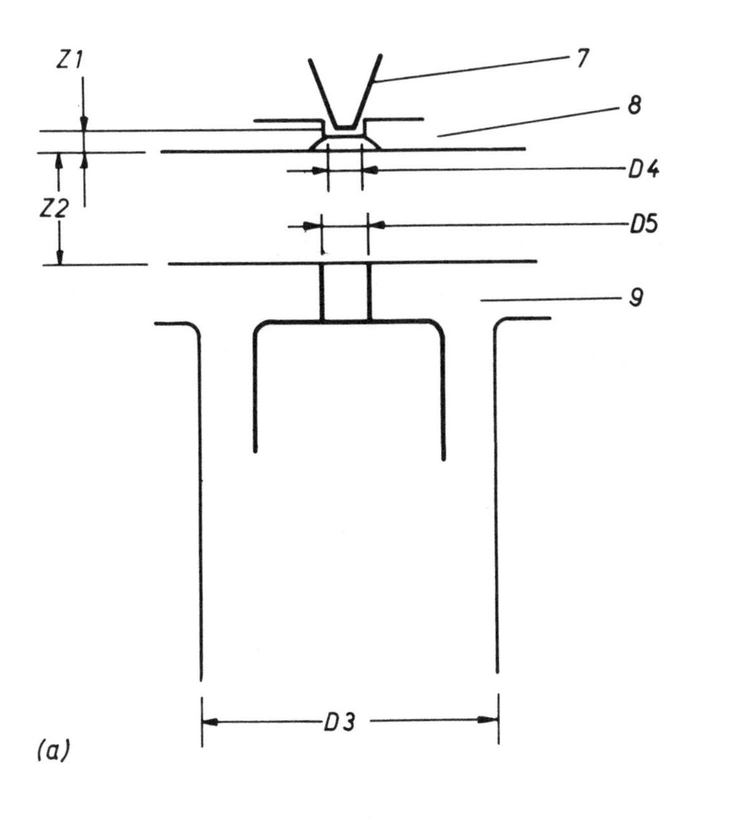

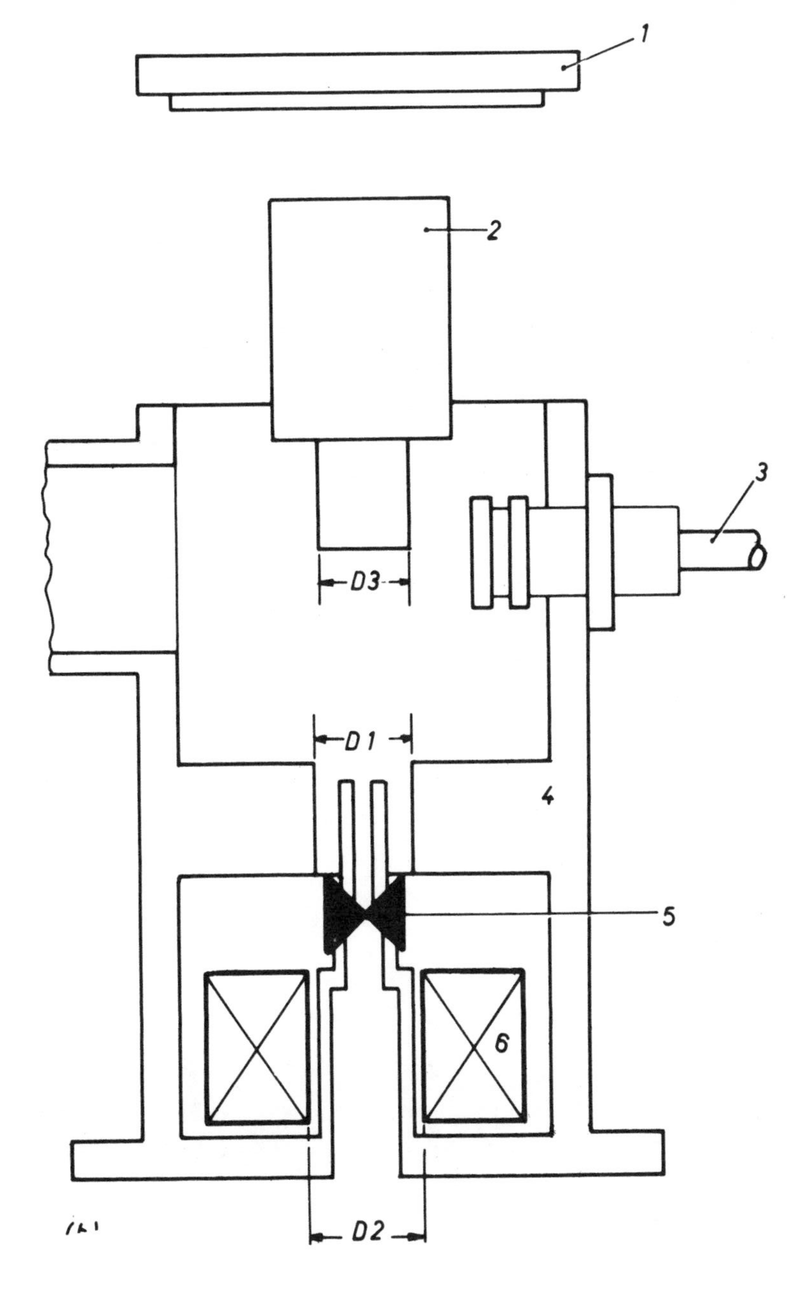

1 *Diagram of electron gun (a) and column (b). 1 — column top; 2 — interchangeable gun; 3 — high voltage cable; 4 — column; 5 — valve; 6 — lens; 7 — cathode; 8 — bias; 9 — anode*

2 Prototype EB machine

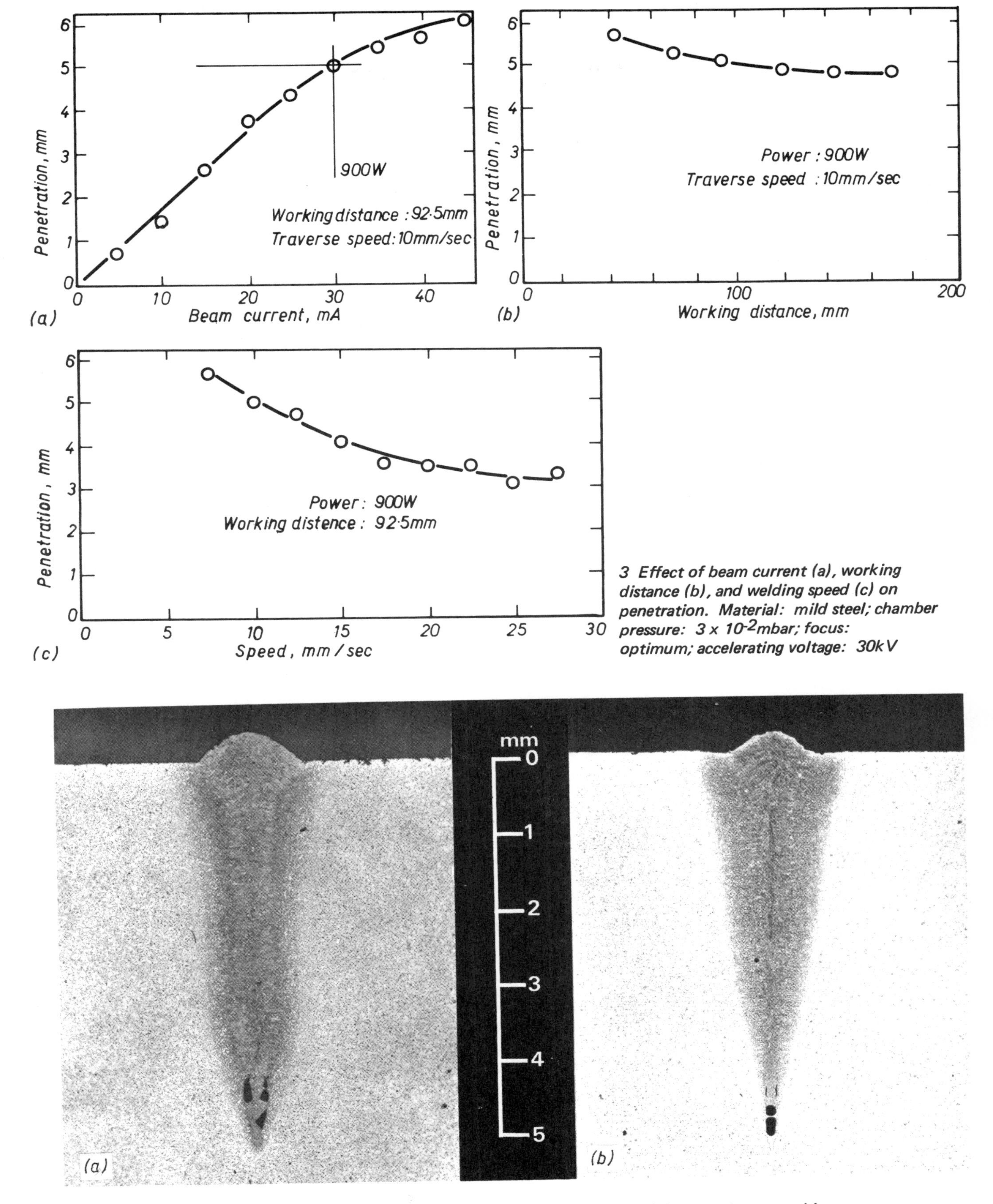

3 Effect of beam current (a), working distance (b), and welding speed (c) on penetration. Material: mild steel; chamber pressure: 3×10^{-2}mbar; focus: optimum; accelerating voltage: 30kV

4 Comparison of weld sections in mild steel: (a) prototype, (b) high voltage, machines

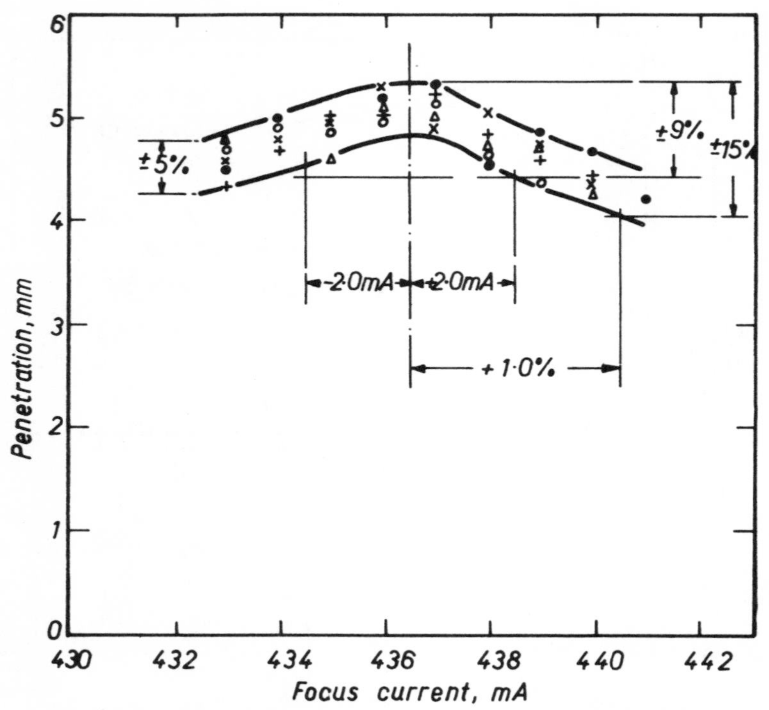

5 *Penetration v. focus current at various electrode spacings. Series: + — R; ○ — S; Δ — T; χ — U; ● — V*

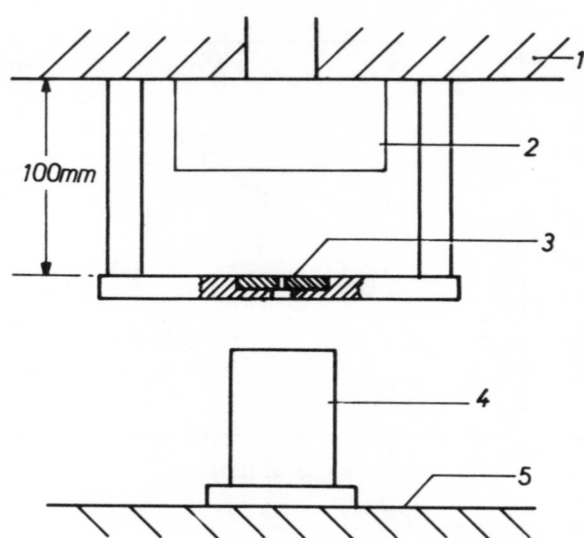

6 *Schematic design of apparatus for measuring beam movement. 1 — column base; 2 — deflection coil; 3 — copper disc; 4 — Faraday cup; 5 — chamber base*

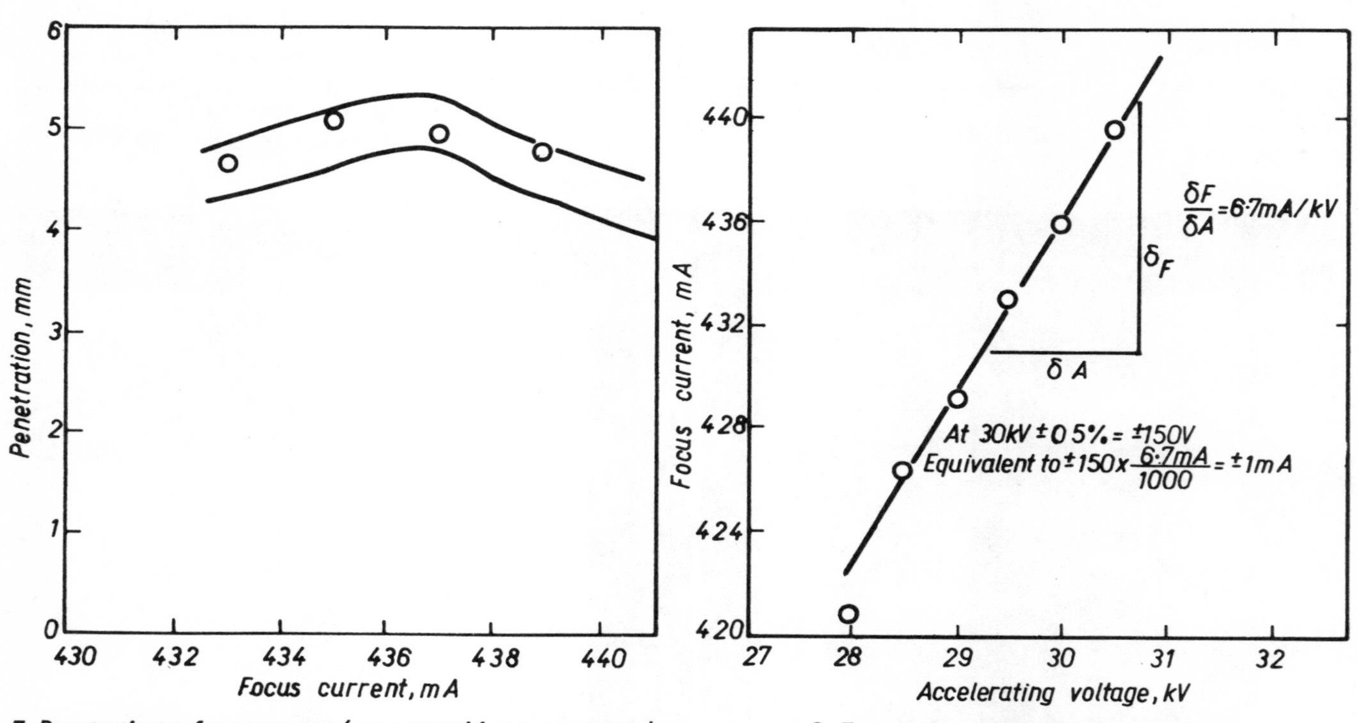

7 *Penetration v. focus current (gun at working temperature)*

8 *Focus current v. kV for focused beam*

A novel method of continuously and rapidly transferring components through an electron-beam welding machine

R.E.Trillwood

INTRODUCTION

One of the so-called 'limitations' of electron-beam (EB) welding equipment is the need to perform the welding in a vacuum (to prevent dispersion of the electrons). Only the larger, complex, and expensive high voltage (150kV) machines are capable of firing the beam through a pumped orifice and performing welding 'in air'. Even then, the performance of the equipment is degraded by beam collision with the air (or more frequently the helium-argon shielding gas). Beam throw is limited to around 5 or 10mm, thus losing one of the useful features of being able to project the beam into inaccessible welding areas. Special precautions to protect the operator from X-ray radiation have to be incorporated, thus adding to the work-handling complexity.

The limitation of the vacuum results from having to place the component (or many components) in a vacuum chamber, evacuate, weld, and then release the vacuum. The use of a vacuum reservoir is not a practical proposition since most of the evacuation time is spent in reducing the pressure in the high vacuum region (10 to 10^{-2} torr).

For small parts, close-fitting chambers have been produced in an effort to minimise the evacuation time. There are, however, limits to this minimum evacuation time determined by the:

(a) volume of the pumps (especially the Roots type)
(b) outgassing rate of the small chamber, and
(c) operating time and location of the vacuum valve

Moreover, each component requires the machine to go through an evacuate/weld/vent cycle of finite duration placing repeated stress on the valving and switching circuits.

To take advantage of the high welding speeds, typically 750-3000mm/min, it is essential that evacuation and indexing times are minimised. For example a 25mm diameter component welded at 1500mm/min has a welding time, with overlap, of only about four seconds.

Loading a number of parts into the chamber, for example on a multispindle fixture (carouselle), Fig.1, and indexing each part under the welding beam is a proven and satisfactory production method. However, the fixture must locate components with a high degree of precision and there is a need for a duplicate fixture if downtime is to be minimised during reloading.

The idea of creating a vacuum around the component before moving to the welding zone is not new and has been pioneered in the USA for the automotive industry. However, the usual method involves the use of face-sliding seals, which are prone to leaks and require frequent maintenance. Such systems have usually involved high development costs.

PRINCIPLE OF THE RAPID TRANSFER SYSTEM

This system* uses the well-tried and proven sealing methods for pistons into cylinders: 'O' or rectangular section neoprene sealing rings are placed at either end of a component carrier which is shaped like a cotton reel. The component to be welded is mounted between the seals.

A series of carriers is placed head to tail and pushed down a tube, the ends of which are at atmospheric pressure, and the centre at a welding vacuum of around 10^{-2} torr. As the carriers progress they act as their own valves passing through a series of pumping stages before reaching the machine vacuum chamber which is conventionally pumped. The component is rotated or moved linearly, or the beam is swept over the joint to perform the weld. On completion of one weld a push rod moves the next carrier under the beam and the next weld is performed. As the carriers move further along the tube they uncover a venting port and are returned to atmospheric pressure.

The advantages of the rapid transfer system can be summarised thus:

Mr Trillwood is Managing Director of Wentgate Engineers (1976) Limited.

*UK Patent Application no. 49162/75

1. For small parts several component carriers can be pre-evacuated before being moved sequentially into the welding chamber. This overcomes the limitation of the 'minimum outgassing time'

2. Pre-evacuation is being performed while welding. This is not possible with conventional machines

3. The mechanical engineering is simple and the rapid transfer system can be added to existing EB welding machines with vacuum chambers. This gives the ability to change the machine application with minimum investment unlike the dedicated, purpose-built welding machine

4. Tooling costs can be comparable or lower than carouselle jigs

5. The vacuum seals, being of a standard design and being part of the removable carrier, can be quickly replaced in the event of wear or damage

6. The system is a continuous flow line process, unlike the carouselle workholder, and lends itself to automatic loading and unloading.

DESCRIPTION OF OPERATION

The equipment is shown in detail in Fig.2. The components to be welded (1) are loaded into a series of cylindrical carriers (A-M). These are pushed through the EB welding chamber by an actuator (2) so that they index one after the other under the electron beam (3) by passing down the open-ended tube (4). This tube is sealed around its circumference at the point of entry and exit of the vacuum chamber (5) by an 'O' ring at point 6.

Initially, the complete system is at atmospheric pressure. To start the process the vacuum chamber is evacuated via its own port (7) and the rapid transfer system is evacuated via port 8. This creates a vacuum in carriers E, F, G, and H. A rotary drive shaft (9) is raised to engage the bevel gears (10). These rotate the component to be welded (1). On completion of the weld, shaft 9 is lowered, actuator 2 pushes the series of components to the right, and component in carrier G takes up the same position under the electron beam. During this sequence of events carrier F is sealed and isolated from port 8 before carrier D (which is at atmospheric pressure) is opened to the pumping port 8. This obviates pressure rise in carrier F. At the same time carrier K, which will carry a completed component, becomes opened to port 11. This tends to create a partial reduction in pressure which is a useful pre-evacuation in the carrier D via the connecting tube 11. On the next operation carrier K will move to the position occupied by carrier M which will vent the carrier to atmospheric pressure via the hole (12).

With this sequence, it is, therefore, possible to move components in these cylindrical carriers continuously through a standard vacuum chamber.

In the event that the components require a longer evacuation time than is occupied by one indexing position, port 8 can be extended to cover a number of carriers.

CONCLUSIONS

The system described has many theoretical advantages over present methods of high production EB welding machines, but has yet to be fully tried and tested.

Undoubtedly, there will be other non-welding areas where it can be used to good effect for the transfer of parts through vacuum chambers or other enclosures where gas pressure differentials are present.

1 Typical multispindle carouselle used in EB welding machines

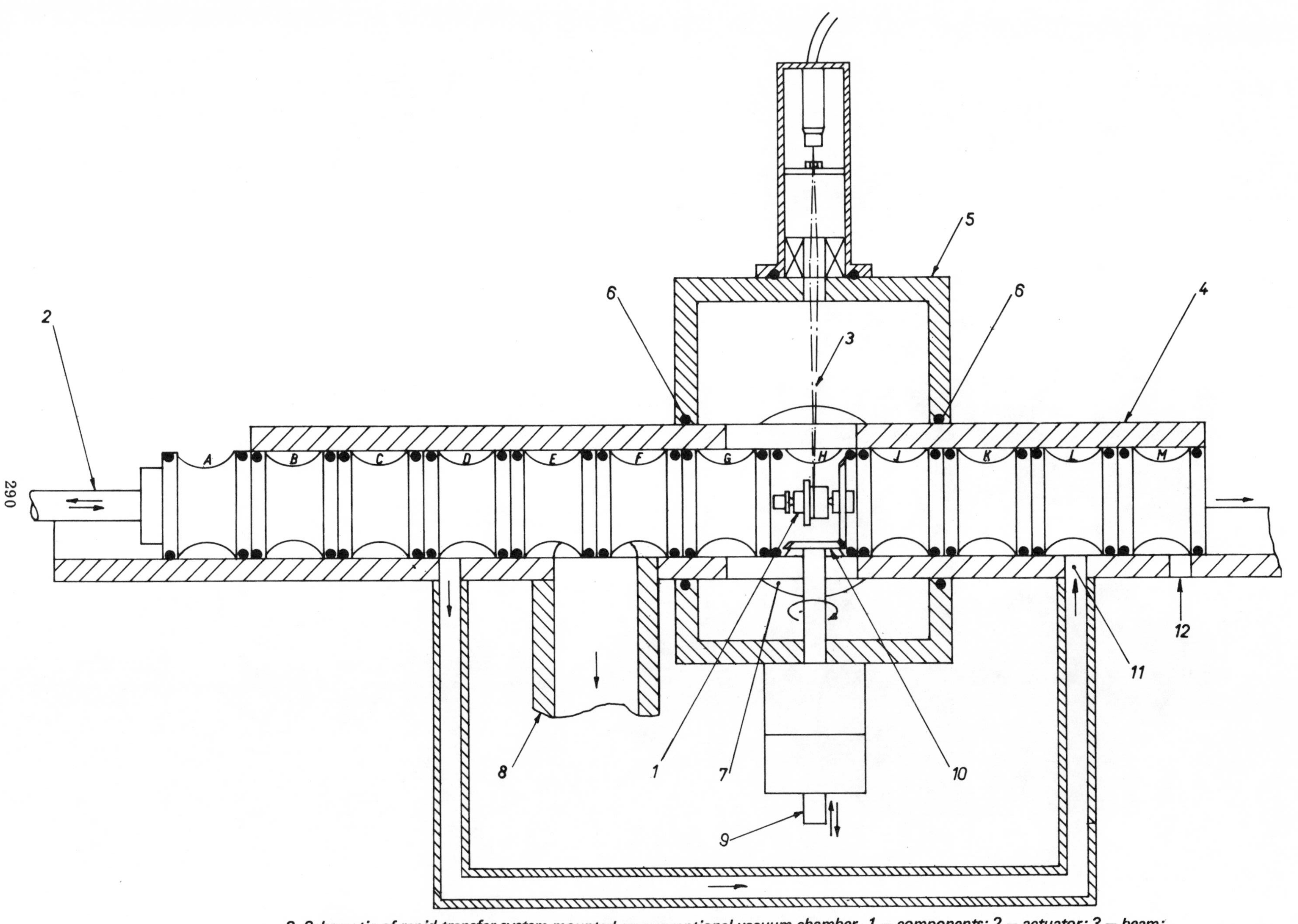

2 Schematic of rapid transfer system mounted on conventional vacuum chamber. 1 – components; 2 – actuator; 3 – beam; 4 – tube; 5 – vacuum; 6 – 'O' ring; 7 and 8 – ports; 9 and 10 – gears; 11 – port; 12 – hole

New realisations of electron-beam welding equipment to join tubes to tubesheet

Ph Dard, DPE, R.Roudier, and G.Sayegh, DSc, DEng

During the past few years a major effort has been undertaken by welding equipment manufacturers to introduce electron-beam (EB) welding into heavy industrial production. This Paper presents some recently developed equipment for welding tubes to tubesheet where a local vacuum chamber is used to create a primary vacuum around the joint to be welded.

The first equipment comprises a portable EB gun for welding tubes to tubeplate in heat exchangers in which a fixed gun is centred on the tube to be welded. The circular weld is realised by a double deflection of the electron beam. A reflection system detects the joint line and adjusts the deflection of the beam to the desired diameter. The welding cycle, which is automatic, can ensure a production of 120 to 150 tubes/hr. Tubes welded by EB can easily be inspected by ultrasonic techniques, which can be adapted to the welding head.

The second equipment, a very small electron gun, is capable of butt welding tubes to plate. This miniaturised gun (total length: 70mm, OD: 40mm) can be introduced into tubes of 100mm ID to butt weld them to the plate. Beam power as high as 5kW can be produced by the gun, which is capable of penetrating more than 20mm of steel.

The third equipment can realise circular axial welds for tubes of more than 70mm diameter in tubeplates by mechanical displacement of the gun over the joint line. When equipped with a 30kW EB gun penetration as deep as 70mm can be obtained, which could be very interesting to the welding of nozzles.

Some experimental results obtained by this equipment, together with the prospects for industrial use, are discussed.

1 INTRODUCTION

The weld quality of joints in tubes to tubesheets for power stations has always been of prime importance: any leakage through a weld causes considerable downtime of the boiler plant. When the power station is nuclear such leakage becomes catastrophic: vide the recent incidents at some nuclear power stations and the consequences.

Electron-beam welding can bring some very interesting solutions to getting high integrity weld quality, but vacuum, needed in EBW, could appear to be an obstacle which is difficult to overcome when very cumbersome components are considered, such as those met in the power station industry. This Paper describes some EBW equipment for welding tubes to tubesheet which employs the principle of 'local vacuum chamber EBW'. This means that vacuum is obtained only around the joint to be welded, without putting the whole component in a vacuum chamber. Two types of equipment are considered:

1. Welding tube to sheet for heat exchangers where the diameter of the tubes does not exceed 40mm
2. Welding tubes of more than 100mm diameter to tubesheet or to a shell with either internal welding of tubes to tubeplates (radial welding) or axial welding of tubes to tubeplates

The last two systems were developed to meet practical requirements in the assembly of

Mr Dard, Electron-beam Laboratory, Dr Roudier, Application Engineer, and Dr Sayegh, Manager — Scientific Department, are all with Sciaky SA, Vitry-sur-Seine, France.

cooling tubes in certain types of nuclear reactor.

2 WELDING TUBES TO TUBESHEET IN HEAT EXCHANGERS

The tubes, which can amount to several thousand on a sheet, are characterised by having a diameter smaller than 40mm. Currently, the tubes are welded by TIG and the penetration obtained is thus very limited.

The use of EBW can bring the following advantages:

(a) deeper welds
(b) automatic and reproducible welding cycle
(c) facility for inspection after welding or repair

The first EBW unit[1] for use in this application was constructed in 1965. It consisted of a machine with a portable welding head covering about 300 tubes at a time and forming (with the tubes to be welded) a local vacuum chamber which was pumped to the desired vacuum, Fig. 1. The EB gun was made to slide on the table comprising the welding head, and a reflectron was used to detect the joint to be welded and to align, with high precision (±0.01mm), the gun axis with the tube axis. Welding of the tube was then carried out by circular deflection of the beam, after adjustment of the welding diameter according to an automatic control. This highly sophisticated machine with its high production (up to 3000 tubes/hr) capability is quite expensive.

Some years ago other equipment manufacturers[2,3] developed simpler equipment to weld one tube at a time. The principle used was to move the gun mechanically to achieve the circular weld. The mechanical displacement of the gun, however, introduced technological difficulties with the high voltage feed and vacuum sealing. Therefore a simpler machine has been developed by Sciaky SA where no mechanical movement of the gun is involved since all displacements are ensured via electromagnetic deflection of the beam.

2.1 Principle of the beam-deflection equipment

The machine is composed of a centering mandrel with appropriate sealing which permits a soft vacuum to be established only around the joint to be welded. The electron gun is fixed and the weld made, as previously, by a double-deflection system operating with a reflectron which detects the joint and adjusts the deflection of the beam to the desired diameter.

Figure 2 illustrates the equipment diagrammatically:

(a) a special centering mandrel which ensures vacuum sealing in the tube
(b) a fixed electron gun (2)
(c) a local vacuum chamber (3) providing primary vacuum around the joint to be welded but which is separated from the gun (which is at secondary vacuum) by a 'valve' (4)
(d) a deflection system comprising two electromagnetic deflecting coils (5), supplied by equal and opposite currents, in such a way that, after the double deflection of the beam, its axis is parallel to its original position. Each deflecting coil has two pairs of poles which, when supplied by $i_1 = I_o \cos \omega t$ and $i_2 = I_o \cos \left(\omega t = \frac{\pi}{2}\right)$, produce a circle. The diameter of the circle is proportional to the deflecting angle, α, and the distance, d. Thus the control of I_o will control the diameter of the beam

2.2 Equipment performance

Based on this principle, equipment has been developed, Fig. 3, with the following performance:

1 Welding diameter range: 10 to 40mm
2 Reduced dimensions to permit welding of the peripheral tubes located not less than 60mm from the shell extending beyond the outside face of the tubesheet
3 Simple mechanical equipment for high reliability:

   (a) stationary gun with no rotating high velocity feedthrough[2]
   (b) no water cooling because of the optimisation of the gun components
   (c) all electrical connections are placed in atmosphere to avoid excessive heating

4 Flexibility of the tube centering through the use of the reflectron for seam tracking
5 Penetration of the beam (depending on the thickness of the tube) from 2 to 10mm (beam power: 4kW)
6 Production rate of 100-120 tubes/hr, which can be increased by using several welding heads with a centralised computer control or CNC equipment
7 Automatic welding cycle used to obtain the vacuum, perform the weld with preset parameters, slope up and down the power in the overlapping zone, and finally make a cosmetic pass

Figure 4b shows a macrograph of a weld obtained with the equipment in 1.2mm thick stainless steel tube. In addition to the deep penetration obtained, there is also the possibility of repairing a defect by a second pass of the electron beam. Furthermore, the NDT inspection of the weld could be adapted very simply to the equipment.[4] As a matter of fact, the weld is very easily inspected by replacing the electron gun by an ultrasonic emitter.

The different possibilities of such equipment (in addition to the advantages brought by EBW) are decisive factors in the reduction and elimination of leaks in heat exchangers. This is most important since such leaks are potentially disastrous and extremely costly when the exchangers are in service.

3 WELDING LARGE DIAMETER (OVER 100mm) TUBES TO TUBEPLATE

When the diameter of the tube is greater than 100mm the deflection system described in Section 2 becomes less practical because of the beam aberrations created by large deflections. Furthermore, the penetration needed in such applications necessitates the use of higher power beams, hence a bigger electron gun.

In other types of joint circular radial welds are needed which can be obtained only by welding from inside the tube. Moreover, in some nuclear components hundreds of tubes are welded to two thick plates, with one circular axial weld (A) and the other weld (B) circular radial, Fig. 5. Welding these tubes by conventional methods causes serious deformation of the bore and the plate, requiring considerable machining. This has heavy consequences from both the economic and execution time standpoints. When used for this application EBW substantially reduces the deformation of the plate, particularly angular deformation, thus avoiding remachining of the piece.

Two separate machines were developed to accomplish these two welds:

1 circular radial welding of tubes to tubeplate
2 circular axial welding of tubes to tubeplate

3.1 Circular radial welding of tubes to tubeplate

Attempts to use an electron gun with an axis aligned with the tube axis, and deflecting the beam 90° to reach the weld, proved to be unsatisfactory for industrial production. Effort was therefore oriented towards the design of a very small EB gun, Fig. 6, with an axis perpendicular to that of the tube and capable of accomplishing the weld directly without any deflection of the beam. The equipment developed for this application is a mobile unit of small dimension, easily transportable, and comprises:

(a) a small rotating EB gun which is inserted entirely inside the tube to be welded[1]
(b) mechanical elements to rotate the gun
(c) a device for centering the machine mechanically on the bore and obtaining vacuum sealing
(d) a welding area sealed (on the plate end) by the electron gun housing and a temporary seal inside the tube. This area is separated from the gun, which is under hard vacuum, by a valve
(e) the equipment can be adapted to various diameters and geometries

The different components of the electron gun were optimised by using a computer program. Experimental results were in good concordance with the theory. A 5kW beam (20kV, 250mA) was obtained from a very limited volume. The total length of the gun (70mm) includes the cathode assembly, the electrostatic part of the gun, focusing and deflection coils, as well as the reflectron system, Fig. 7.

Regulation of the beam current is obtained by temperature control of the filament, independent of the high voltage control, which can be adjusted to ±1%.

The welding performances of the gun are:

(a) range of operating distance: 10 to 35mm
(b) welding speed: 100mm/min to 3m/min
(c) independent electromagnetic focusing
(d) deflection coils in the X and Y axes
(e) beam location by reflectron system
(f) minimum diameter of tube bore to be welded: 100mm
(g) maximum penetration in stainless steel: 20mm

Sample welds, Fig. 8, have been produced in the laboratory and are currently under evaluation. The next step will be the realisation of a smaller scale model for practical test.

One of the most important advantages of this technique, especially when large numbers of tubes are installed on a large tubeplate, is the very slight modification of the pitch between tubes which can occur after welding. Thus there is no need to remachine the plate. It should be noted that, in conventional welding, remachining the plate and bore is a necessity and that the operation is long and costly.

3.2 Circular axial welding of tubes to tubeplate

The difference between this application and that

concerning welding tubes in heat exchangers lies in the fact that the tube diameter is relatively too large to permit welding by beam deflection. Indeed, when the deflection angle is large, aberration of the beam could be so high as to produce an unacceptable spot diameter.

The principal difference with this equipment is the provision of a local vacuum chamber around the joint to be welded. Thus a mobile chamber is clamped on to the workpiece to enclose the joint, forming the welding zone. In this way only a small volume vacuum is required surrounding the joint to be welded, Fig. 9.

The displacement of the beam along the joint is obtained by mechanical rotation of the gun around the tube axis. Furthermore, the gun can be tilted to obtain conical welds which could be useful to certain weld configurations. A mandrel ensures centering of the equipment on the tube. The use of positioners especially adapted to the application can ensure quick handling and hence make the machine suited to mass production operations.

The characteristics of the equipment are:

(a) range of diameters to be welded: 70-270mm
(b) tilting of the gun: from 0° to 20°
(c) welding speed: 100mm/min to 3m/min
(d) maximum penetration with a 30kW gun: 70mm
(e) vacuum in the welding chamber: $\sim 2.10^{-3}$ torr
(f) production rate: 4-10 welds/hr according to the nature and geometry of the workpiece
(g) dimensions of the welding chamber: diameter = 900mm; height = 1900mm
(h) weight of the welding head: 1500kg

In association with the circular radial welding equipment these two units can be used to assemble many configurations of tubes to tubeplates, not only in nuclear plant but also in conventional heavy industries such as petrochemical and power plant.

CONCLUSIONS

Electron-beam equipment especially adapted to weld tubes of various diameters to plates has been developed which does not need the whole component to be introduced into a vacuum chamber. These equipments use a local chamber where the vacuum is created in a limited volume around the joint to be welded.

Welding tubes to tubesheets in heat exchangers can be accomplished by a stationary EB gun integrated in relatively simple equipment. In addition to the specific advantages of EB in this application, there is also the possibility of easy control by NDT of the deep penetrant welds.

Two machines are utilised to weld tubes of larger diameter:

1 One for circular radial welds with an internal gun characterised by its miniaturisation. The gun with its focusing and deflection coils as well as its reflectron system is housed in a tube less than 75mm diameter

2 One for circular axial welds with an external EB gun which is displaced mechanically along the joint to be welded

Potential applications of this equipment can be foreseen not only in nuclear plant but also in conventional industries such as petrochemicals where high integrity welds are required.

REFERENCES

1 JAMES, H.A. 'EBW equipment: process parameters, limitation, and controls'. EB Met. Procs Seminar, Oakland California, February 1971.

2 PEYROT, J.P. 'Appareillage portable pour soudage par bombardement électronique'. Coll. 1st Int. Soudage et Fusion par Faisceaux d'électrons, Paris, 1971, 717-31.

3 GOUSSAIN, J.C. and PENVEN, Y.le. 'Etude de soudage par faisceau d'électrons avec pistolet portable de tubes sur plaques d'échangeurs'. Soudage et Tech. Connexes, 30 (7/8), 1976, 281-93.

4 PENVEN, Y.le and DUBRESSON, J. 'Contrôle par ultrason de la profondeur de pénétration des soudures executées par faisceau d'électrons dans l'assemblage de tubes sur plaque d'échangeurs'. Soudage et Tech. Connexes, 31 (9/10), 1977, 371-6.

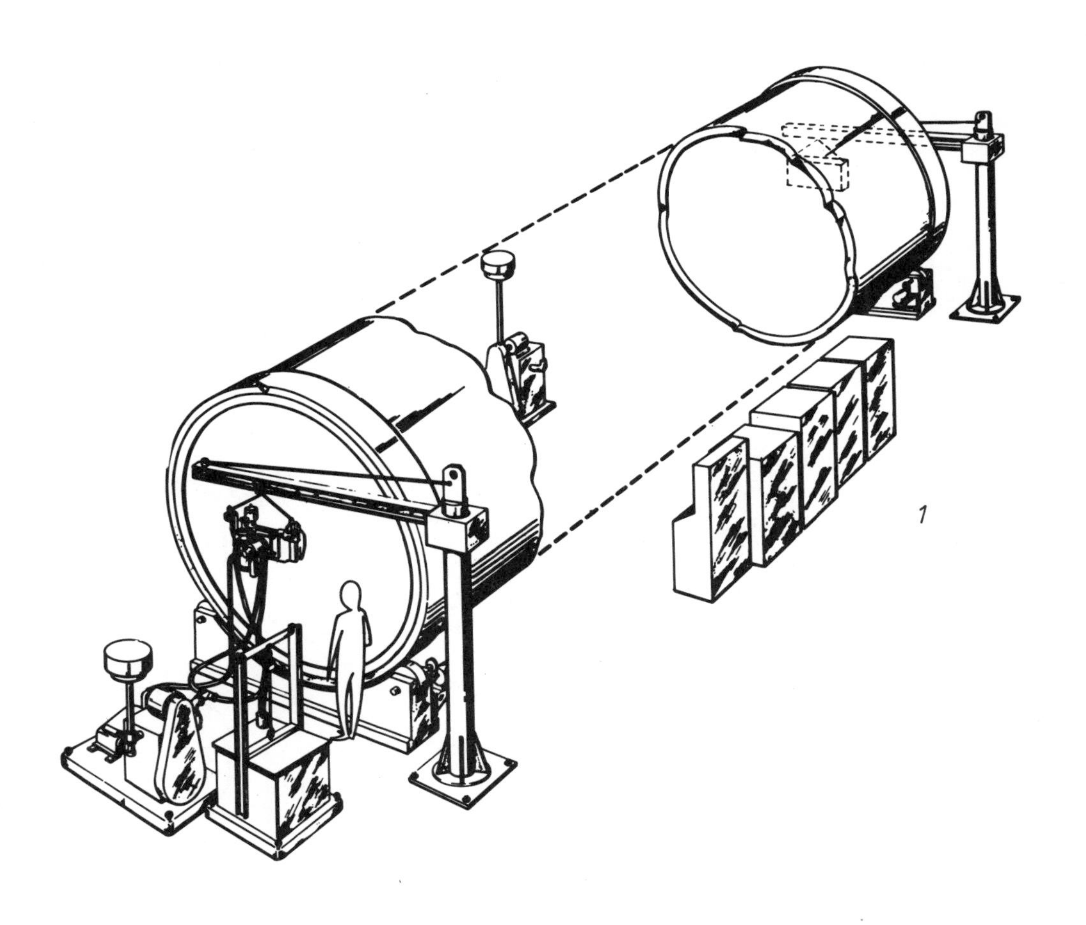

1 Schematic of first EBW equipment to weld tubes to tubeplate, circa 1965. 1 – sequence control Sciakydyne beam scanner

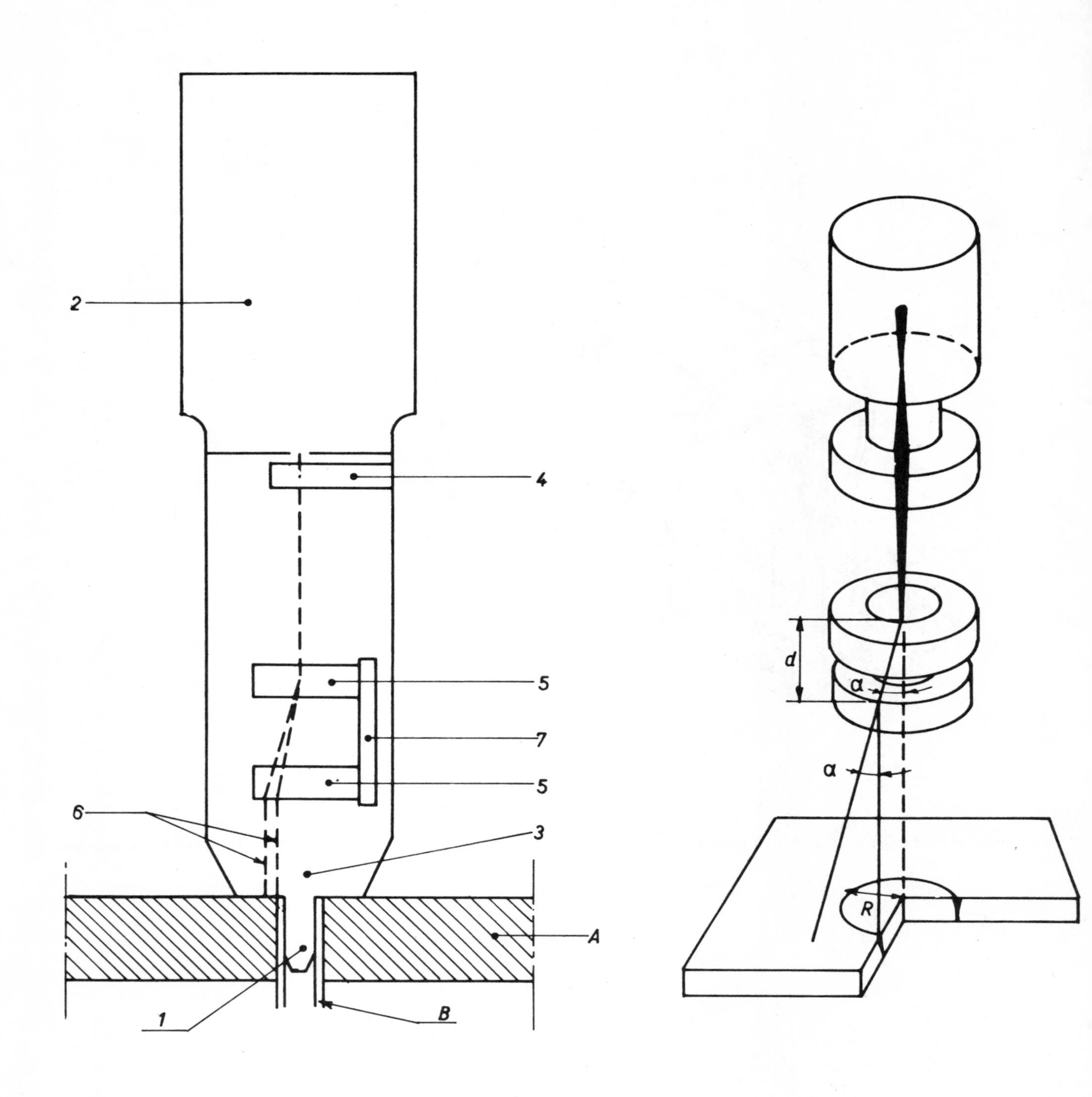

2 Principle of new system for welding tubes to tubeplate (annotations in text)

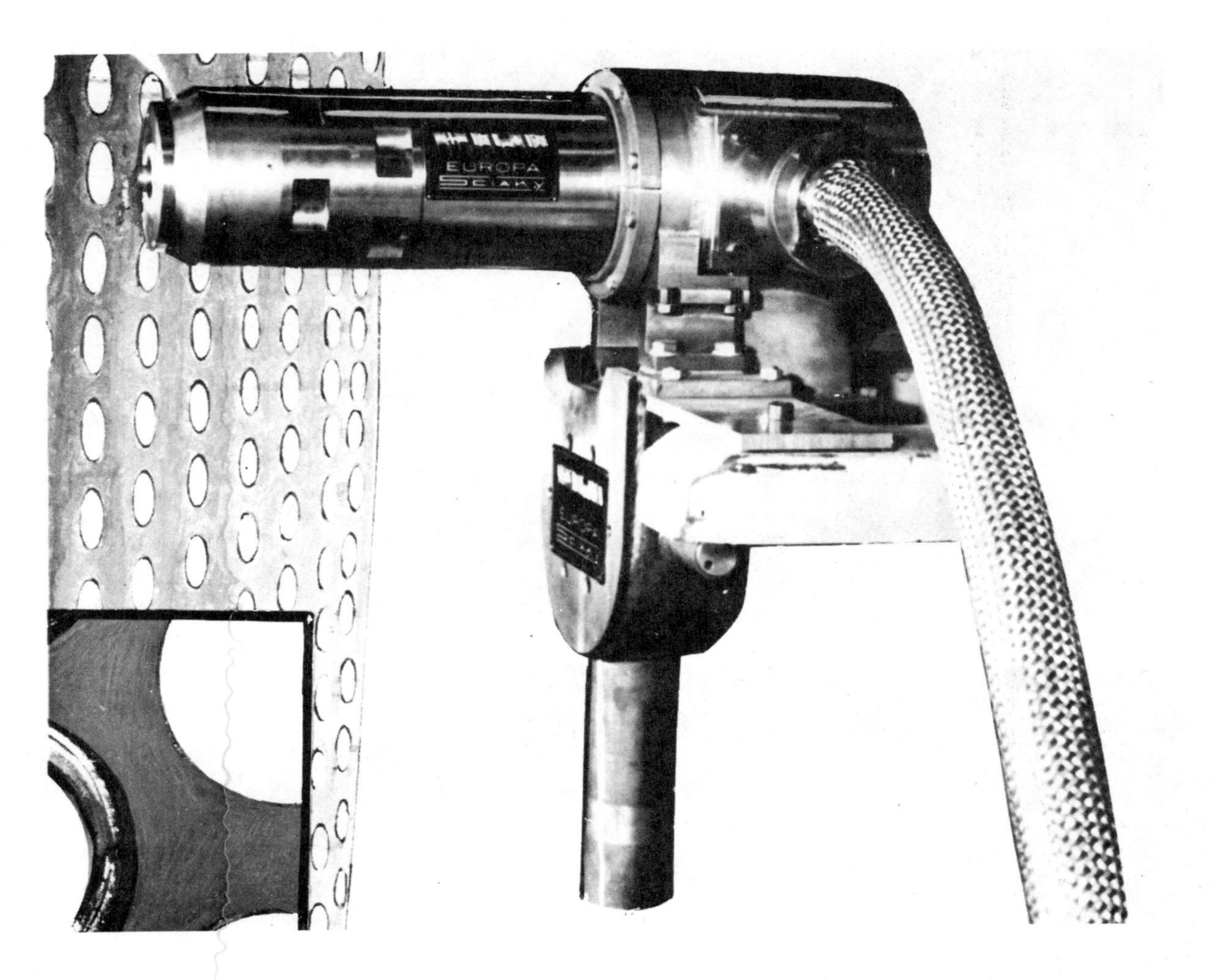

3 Complete equipment for use in welding tubes to tubesheet in heat exchanger

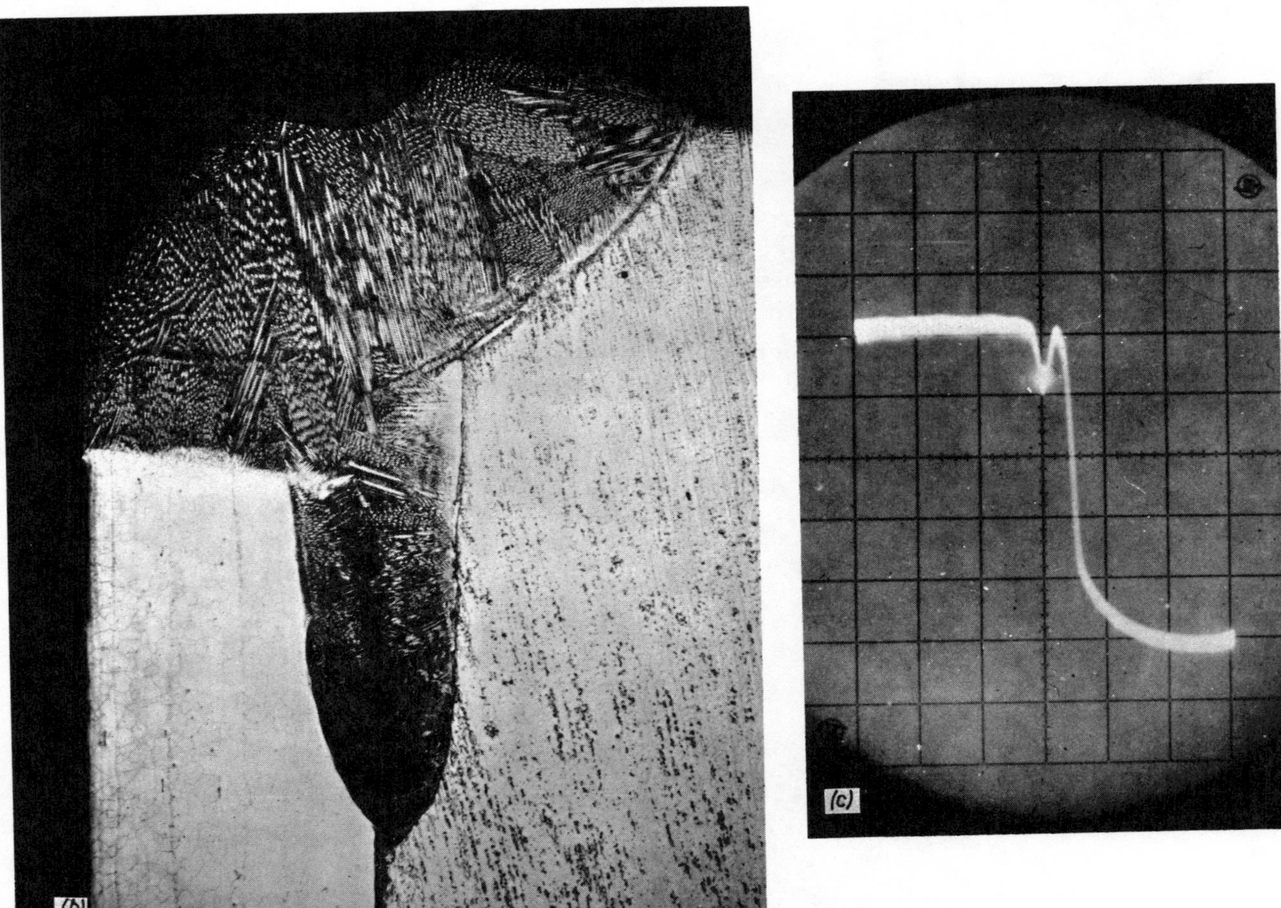

4 Tube to tubeplate welding: (a) examples of welded tubes by equipment, (b) macrosection of weld, and (c) detection of joi line by reflectron system

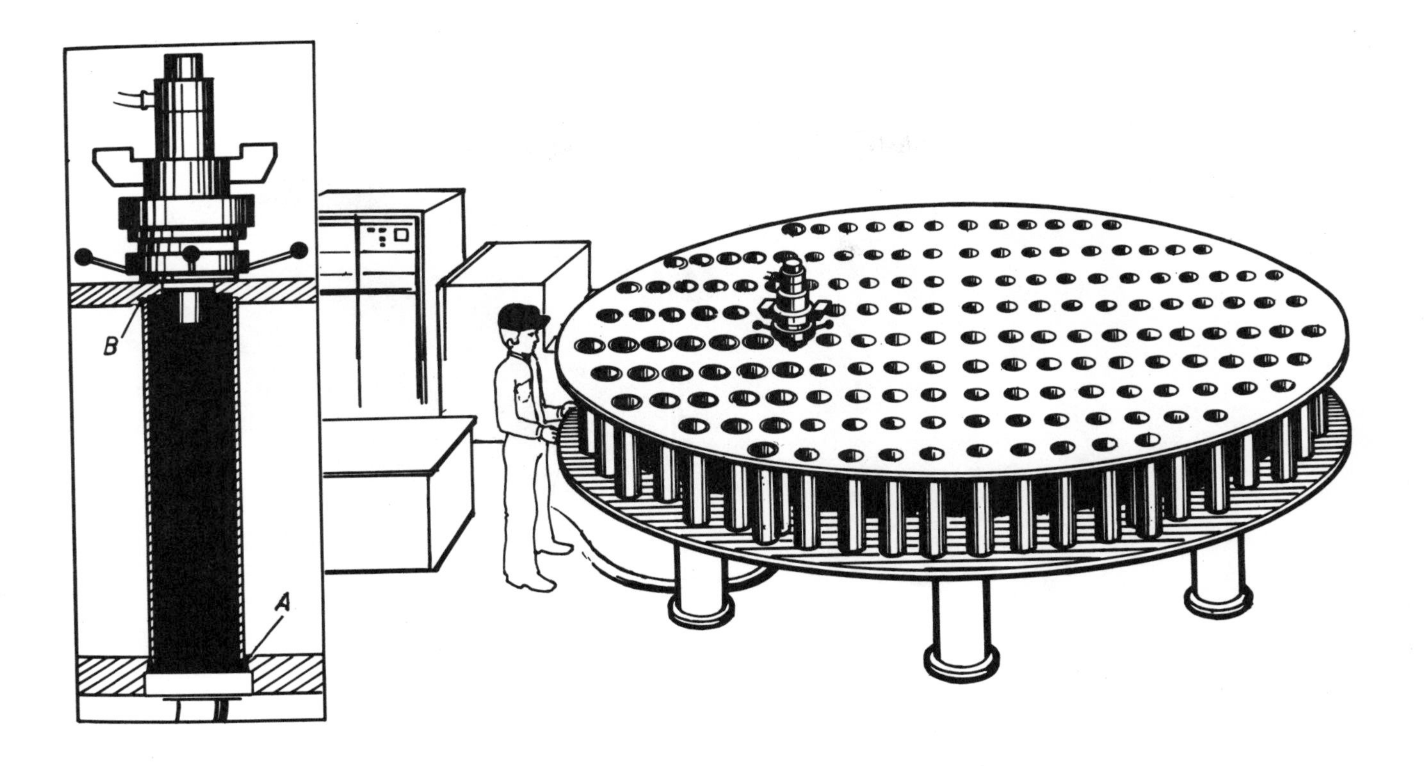

5 Circular axial welds (A) and circular radial welds (B) of large tubes (>100mm diameter) on tube sheet

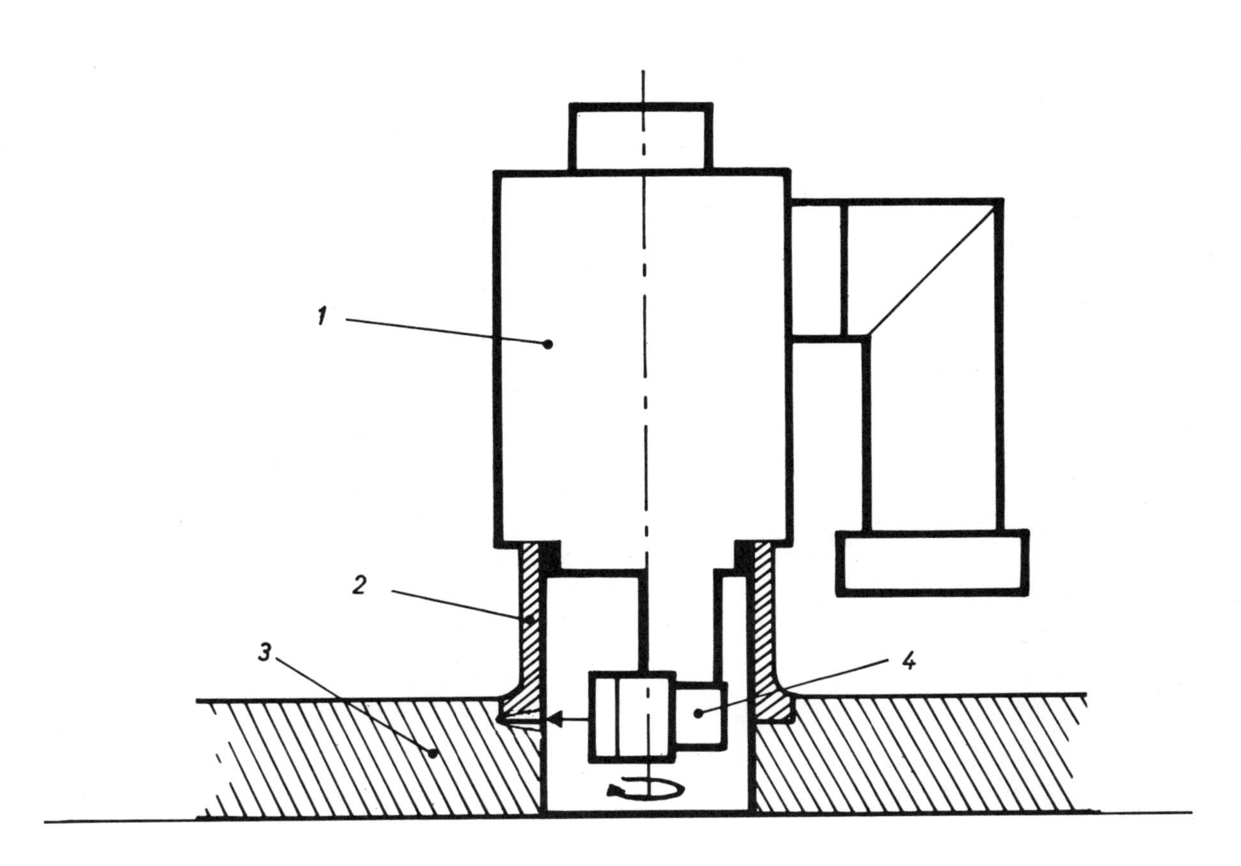

6 Principles of circular radial welding of tubes to sheet. 1 — local vacuum machine; 2 — electron gun

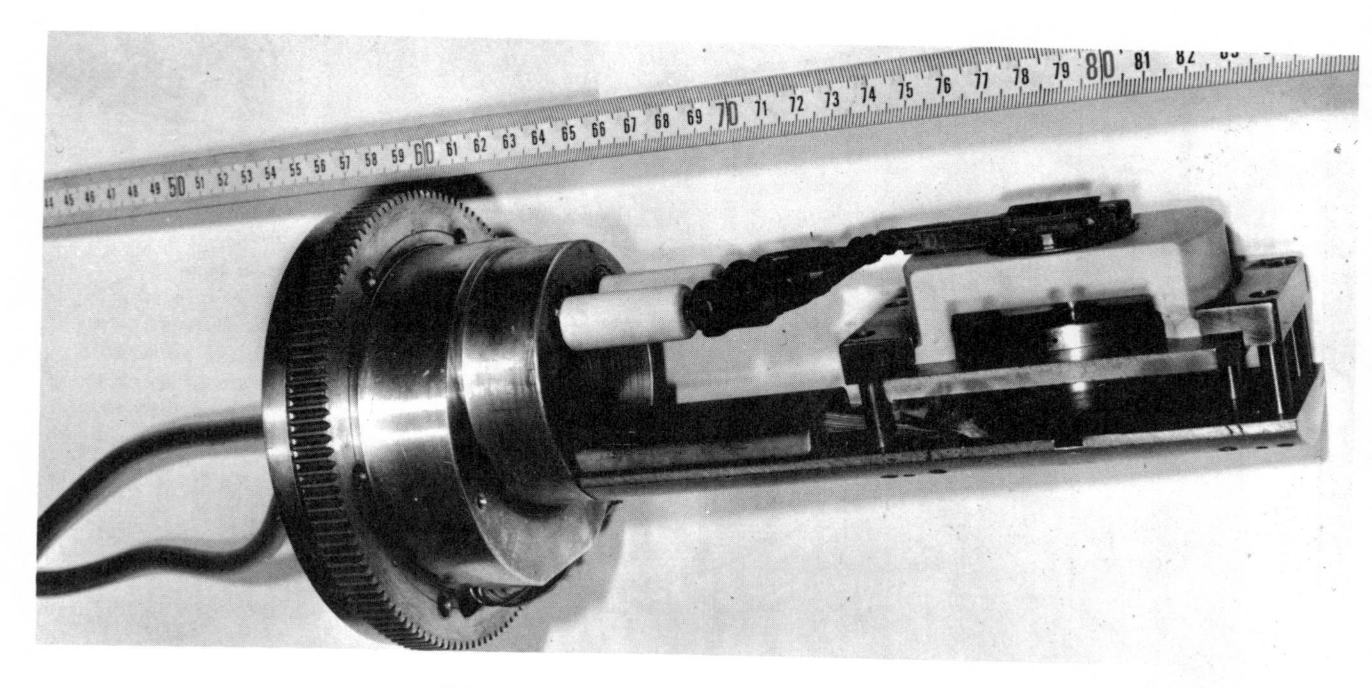

7 Small rotating gun got circular radial welding

8 Arrangement for radial tubesheet welding. Note: circular radial weld of tube to plate for which gun is located inside tube and electron beam comes through hole

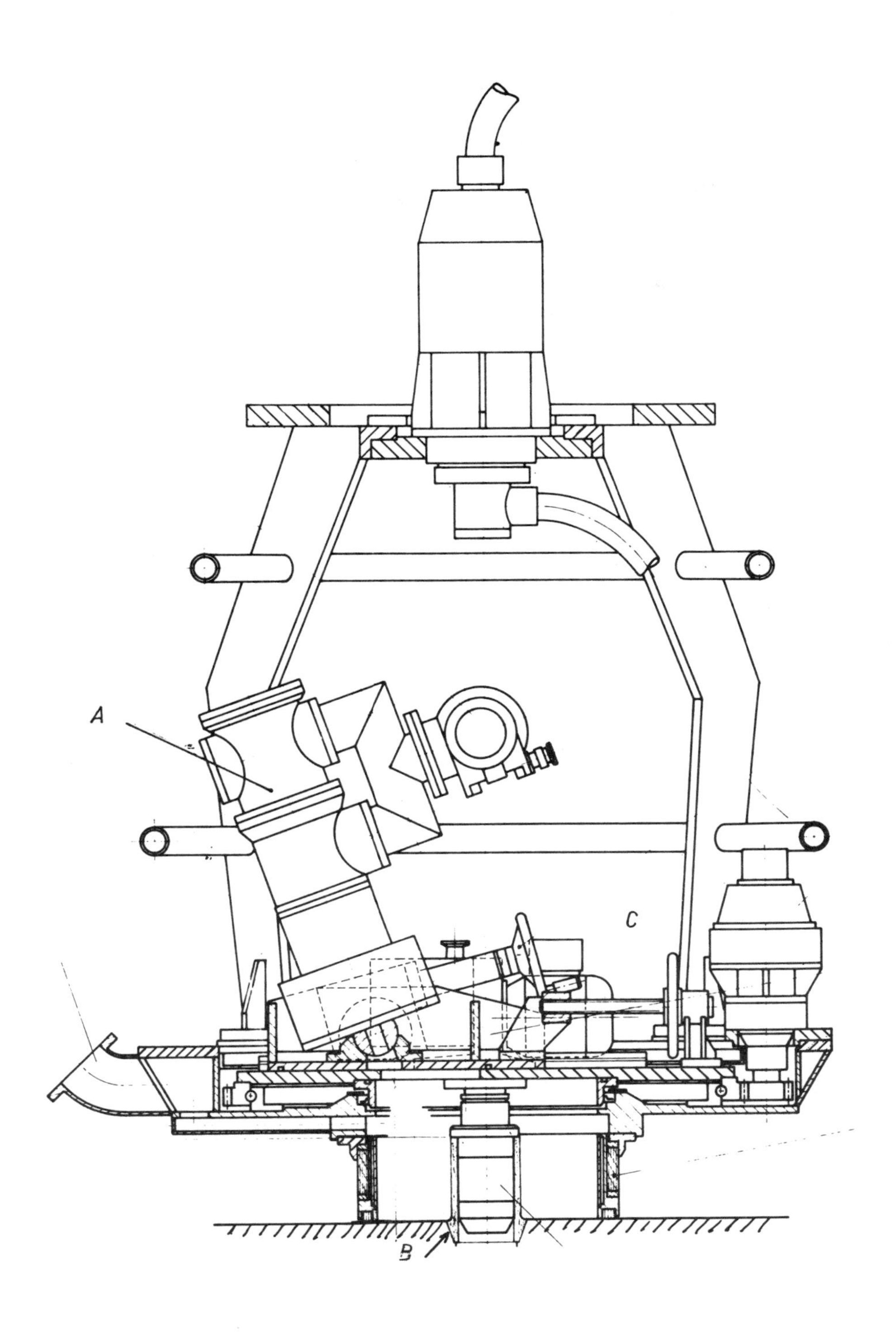

9 Electron beam equipment for circular axial welds: A – electron gun; B – EB weld; C – primary vacuum chamber

Magnetic trap devices for the elimination of high voltage breakdown in electron-beam welding equipment

A.Sanderson, MIM, AMCST, MWeldI

INTRODUCTION

One problem which has plagued the electron-beam welding process since its inception is flashover in the gun region and associated high voltage equipment. Flashover in its mildest form appears as a microdischarge which is of insufficient magnitude to trip the power supply overload circuitry but which may disturb the weld capillary, causing weld bead irregularities and trapped porosity. In severe instances, where microdischarges lead to a high current arc and protective shutdown of the power supply, the complete interruption of the welding process invariably creates huge weld cavities and crater cracks. Usually the latter are not easily repaired and in the extreme may result in a scrapped component.

High voltage breakdown of the main EHT standoff insulator can be avoided by careful control of the internal and surface voltage stresses. Similarly, discharge between gun electrodes can be controlled to a large extent by correct shaping and choice of materials. However, even the best electrode designs which may be discharge-free in the absence of an electron beam are often prone to breakdown during welding. It has been suspected for a long time that this form of discharge results from vapour or particles generated in the weld capillary entering the gun region. The exact nature of the offending particles was not clear, although gun discharge was known to be aggravated by inherently gassy or volatile workpiece materials. Discharges also occur when using large amplitude beam oscillation which possibly results from increased vapour generation.

To reduce if not eliminate the occurrence of major and microdischarges arising from weld pool particles, the possibility of introducing a suitable trap along the beam path has been investigated. Various techniques have been recommended in the past to prevent charged particles and metal vapour, originating from the workpiece, from entering the gun region. In one design the gun grid cup and anode are tilted relatively to each other and to the work, Fig.1a. Although this approach is suitable for reducing ion damage to the cathode in heating and melting devices, the asymmetry of the electrode structure would lead to several forms of aberration: electron emission would be asymmetrical across the cathode surface and the acceleration field region would cause non-circular beams to be generated.

In an alternative arrangement,[2] Fig.1b, a single bend of approximately 25° was introduced between the anode and lens. Such an approach, although probably fairly free from beam aberration, would be difficult to set up and adjust correctly. This may account for the lack of its widespread use. Moreover, any variation in accelerating voltage or deflection field strength would cause large changes in beam tilt angle, moving the beam off-axis in the lens region, with a consequent increase in beam astigmatism. Even more important is the displacement of the beam on the workpiece which would be caused. Some of these problems could be reduced by linking the magnetic field to the accelerating voltage electronically, but this would be at the expense of simplicity and flexibility.

This Paper outlines the development of more advanced trap devices which avoid the difficulties briefly indicated above.

NEW MAGNETIC TRAP DEVELOPMENTS

Double deflection system

The new concept[3] is based on a multiple deflection coil system in which the beam is bent off-axis by a first magnetic field and subsequently returned parallel to the original axis by a second field of equal strength and opposite sense. The essential features of this approach are shown schematically in Fig.2a. In a further development of the concept as shown in Fig.2b the beam is returned to the original axis. This latter arrangement also allows the beam to be used in the undeflected

---

Mr Sanderson is a Principal Research Physicist in the Process Operation and Control Department at The Welding Institute.

mode through a straight channel when conditions do not warrant the use of a trap.

Several prototype polepiece designs were built and tested in a 150kV, 6kW, Hawker Siddeley Dynamics machine to assess the deflection characteristics and beam aberration. A compact coil arrangement, Fig. 3a, was mounted between the focusing lens and the work. Beam aberration was studied at low beam currents by observing the spot focused on a smooth target (positioned 250mm from the lower coil) through the machine telescope.

For one set of polepieces (single deflection) it was found that beam aberration became excessive at beam angles in excess of 20°, whereas at angles below 15° sharp focusing of the beam was possible and melt runs made with deflected and undeflected beams were very similar in both depth and width.

Comparison experiments were carried out using the double coil arrangement, Fig. 3b. The curves in Fig. 3 illustrate the effect of coil current on overall beam deflection angle for a single, Θ, and for double coil deflection systems, $\emptyset$. It can be seen that a change in coil current from 100 to 150mA causes an increase in beam tilt angle of 6.5^o for a single coil, whereas the same change in coil current causes a deviation of less than 0.05^o for the double coil system. This corresponds to a maximum shift in beam position, at a working distance of 250mm, of 30mm for a single coil compared with less than 0.2mm for the double coil system, i.e. an improvement of 150:1. Conversely, for constant coil deflection currents the effect of change in accelerating voltage on beam deflection angle is shown in Fig. 4. For a single coil, varying the accelerating voltage from 140 to 120kV causes a change in deflection angle of 2.2^o compared with about 0.1^o for a double coil system. Based on the above data, a double deflection trap was designed.

Figure 5a shows the polepiece design with one deflector coil element in place, and Fig. 5b indicates the general arrangement of the prototype double deflection coil system for use with a standard electron gun. The angles of deflection are nominally 13° and the displacement of the entrant beam off the lens axis is 8mm. The set of four coils is connected in series, and with a working coil current of 94mA (for a beam accelerating voltage of 150kV) a uniform field is created across the polepieces. The total power dissipation of the unit under these conditions is only 0.62W. To restrict the flow of ions and other particles from the workpiece the drift tube comprises a series of copper blocks with decreasing boreholes. A general view of the prototype trap and its assembly ready for fitting to an electron gun column is presented in Fig. 6.

Performance

Extensive welding trials have been carried out. In general the double coil trap has been found to reduce gun discharges significantly, particularly for steel and copper welding at relatively high power levels. However for very gassy thick section materials, where spin amplitudes of above 3.5mm are necessary, some gun instabilities have occasionally been observed. Beam aberration on the other hand has been found to be minimal. This is confirmed by Fig. 7 which shows typical sections of melt runs made in 9.5 and 130mm thick steels at 4.5 and 37.3kW respectively. The melt runs were made with the travel direction in the same plane as the ion trap deflection plane and in the horizontal-vertical mode.

It can be seen that there is no significant aberration at least for this welding direction, and both the high speed low current melt run, Fig. 7a, and the high current low speed melt run, Fig. 7b, are relatively narrow, i.e. 0.5 and 3 mm respectively. In addition the total beam current intercepted by the copper drift tube, using a 4mm diameter anode aperture, was found to be less than 0.3mA over a beam current range 0-500mA at 150kV. The maximum pickup occurred in the range 50 to 100mA which coincides with the largest beam diameter and divergence for the gun used. Furthermore, the beam position is only slightly affected by relatively large changes in accelerating voltage and coil current. For example, in one application it was advantageous to fade out the beam power by decaying the accelerating voltage over the range from 150 to 100kV. This was satisfactorily carried out without any excessive deviation from the weld joint, even with a constant coil current, i.e. with no correction applied to the trap system.

Triple coil system

For power levels above 30kW, and in particular for welding very thick section aluminium alloy, it was thought further improvement in trap performance could be obtained by increasing the local beam deflection angle to approximately 45° with a maximum displacement from the nominal axis of about 25mm. Moreover an additional bend was incorporated to return the beam to its original axis, Fig. 8a. A 'straight-through' beam channel was also incorporated to facilitate initial column alignment and to allow the device to be switched off when not required.

Furthermore, ferrite coil cores were employed in an attempt to reduce the residual magnetic fields (which had proved to be a problem with the soft iron used in the double bend unit). In this situation the copper drift tube was made in two halves, Fig. 8a, which were insulated from the chamber walls and connected via a 390kΩ resistor to earth. This allowed the pickup current to be monitored and the system shut down if the latter became excessive. Also, neon bulbs were connected in parallel with the resistors to protect the drift tube insulation and the current pickup monitor amplifier by limiting the maximum voltage rise. A general view of the prototype equipment is shown in Fig. 8b.

Beam trajectories

Approximate theoretical beam trajectories were determined using a two-dimensional magnetic field program which was run on an Olivetti P652 microcomputer and companion P600 X-Y plotter. Magnetic flux levels were measured at the intersections of a mesh consisting of concentric circles and radial lines, Fig. 9, using a Hall effect transverse probe and fluxmeter. This was calibrated against a 0.1 Tesla standard magnet, and zeroed in a mumetal pot. The angular and radial increments were 15° and 10mm respectively.

Figure 9a shows electron plots for both central and outer trajectories of a 3mm diameter parallel beam. In this example the coils were connected in series and a current of 228mA (measured on a calibrated DVM) was applied. For an accelerating voltage of 150kV it will be seen that the maximum beam displacement is 25.5mm and the trajectories, at least for low currents, cross over in the region of maximum displacement of the beam. The effect on the beam of varying the accelerating voltage (for a constant common coil current of 228mA) is shown in Fig. 9b where the plots indicate that the system is quite tolerant to relatively large changes in voltage. The exit angle of the beam was maintained within ±0.05° for an accelerating voltage change of 120 to 150kV, although the beam was displaced laterally ±0.75mm. However, over a voltage range 130 to 140kV the displacement was only ±0.02mm. For higher accelerating voltage (170kV) the beam exit angle stayed within 0.4° of axis but the lateral displacement of 2.5mm with respect to the 150kV trajectory occurred. Trajectories for accelerating voltages of less than 120kV were similarly displaced, but the angle of exit caused the beam to intersect the drift tube at its outer wall. The effect of varying coil current for constant accelerating voltage was also studied; the system was found to be tolerant to approximately half the percentage change compared with accelerating voltage change, as would be expected from the approximate square root relations between these parameters and deflection amplitude.

Performance

Using the triple bend magnetic trap, current pickup on the copper drift tube for a 150kV beam was found to be less than 0.03mA for the current range 0 to 500mA for an optimised coil current (with all coils connected in series) of 204mA. Fully penetrating welds were made in a range of materials including 150mm thick mild steel, 125mm thick copper alloy, 50mm thick aluminium alloy with a 0.25mm wide magnesium shim placed at the joint interface, and 470mm thick aluminium magnesium alloy. It was found that no gun discharging or grid voltage breakdown occurred when the trap stop was closed, Fig. 8a, even when large amplitude beam deflections were employed. However, when the trap stop was opened and the beam still deflected, discharge occurred particularly in 470mm aluminium alloy welding. The triple magnetic bend caused some distortion of the focused spot as observed by visual optics but no adverse effects were noted in welding performance. Figure 10a and b shows a comparison of fully penetrating melt runs in 470mm aluminium alloy and corresponding UV beam current records for 'before' and 'after' trap installation. Without the trap periodic gun discharge and occasional power supply shutdown led to gross cavitation defects, Fig. 10a. With the trap in operation no discharges were observed on the beam current record and the fusion zone profile was free from cavities, Fig. 10b. After de-energising the deflector coils, a residual field of approximately 0.2mT was measured which gave rise to some distortion of the focused spot: however, subsequent bench experiments have shown this can be reduced to less than 0.05mT by purging with a collapsing 50Hz supply.

DISCUSSION

Significant improvement in breakdown performance has been achieved by the use of a small angle magnetic double bend trap device. The device illustrated in Fig. 6, in which the line of sight from the workpiece to the gun is blocked, has been found to be quite successful although occasional gun discharges can be induced at high power levels by very gassy materials or the use of excessive beam spin amplitudes.

The use of larger deflection angles and displacements in the triple bend device has been found to eliminate all discharge stemming from weld pool particles. Welding 470mm thick aluminium alloy, Fig.10, with 'beam-on' times in excess of 10min with no detectable discharge more than justifies this statement.

At the same time the design used eliminates the shortcomings of other systems in that beam movement as a result of small variations in accelerating voltage or magnetic field is minimised. Furthermore, beam aberration is minimal.

It is interesting that gun discharge still occurs with this device when the line of sight to the cathode is not blocked. The latter suggests that the particles which then cause arcing are uncharged clumps, atoms, or fast positive ions generated at the workpiece, since slow positive or negative charged particles would be separated out by the strong magnetic fields.

The double bend system requires the electron gun to be offset with respect to the column axis and therefore once installed it is not possible to align the gun electrode independently of the deflection system or revert quickly to an undeflected beam. On the other hand, the triple bend system allows gun electrode alignment without magnetic deflection by means of the undeflected beam channel, Fig. 8a. The latter is also useful since it permits the gun column to be operated conventionally when conditions do not warrant trap operation and where beam aberration needs to be minimised to achieve the smallest possible focused spot diameter.

It has been shown that, by virtue of the self-compensating characteristics of multiple deflection coil arrangement, beam positional stability is inherently far superior to single deflection systems to the extent that all reasonable fluctuations in accelerating voltage can be accommodated. Moreover large changes in accelerating voltage can be accommodated merely by reducing deflection coil current, providing the beam diameter and divergence is suitable for transmission through the drift tube assembly.

Minimum beam aberration would be expected for a parallel entrant beam, although a small angle of divergence or convergence is probably permissible, e.g. semi-angles up to approximately 2°. In addition it is expected that the same basic coil and drift tube components for the triple bend would be suitable for installation in a wide range of both low and high voltage machines, although the configuration of the supporting structure would need to be tailored for each machine type. However, to minimise the overall size of the device it is essential to site the coils as close to the gun anode as is feasible since the beam diameter is generally a minimum in this region in most equipments.

The results to date have indicated that the triple bend system eliminates all discharge caused by particles generated in the weld capillary even for ultrathick aluminium alloy welding, yet beam aberration can be held to levels which would be acceptable for thin section components. It is also expected that cathode lives will be significantly increased by elimination of ion and gun discharge damage and it is anticipated that this device will be suitable for a very wide range of electron-beam welding equipment eliminating one of the shortcomings of the process which has plagued its development since inception.

CONCLUSIONS

1 Gun discharge in the presence of a powerful welding electron beam which appears to be caused by neutral or fast positive particles originating from the weld pool can be eliminated by the use of magnetic trap devices

2 Two novel magnetic trap systems with counteracting magnetic beam bending fields have been devised which prevent gun breakdown without introducing excessive beam aberration. These devices are simple to set up and are extremely tolerant to accelerating voltage and coil current changes

3 Using the triple bend trap, gun discharge can be eliminated even for welding of highly volatile materials such as magnesium, for thick section aluminium, copper, and steel welding and for welding with large amplitude beam spinning

REFERENCES

1 BAS, E.B. et al. 'A 30kV electron-beam welding machine with indirectly heated tungsten cathode'. Procs 1st Conference 'Electron and Ion Beam Science and Technology', 3-7 May 1964, Toronto, Ed. R. Bakish. New York, John Wiley, 1965.

2 MEIER, J.W. 'New development in electron-beam technology'. Welding J., 43 (11), 1964, 925-31.

3 SANDERSON, A. British Patent 1485367.

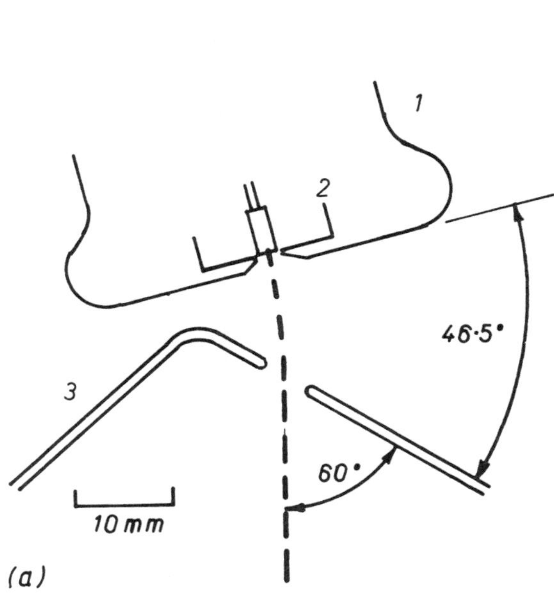

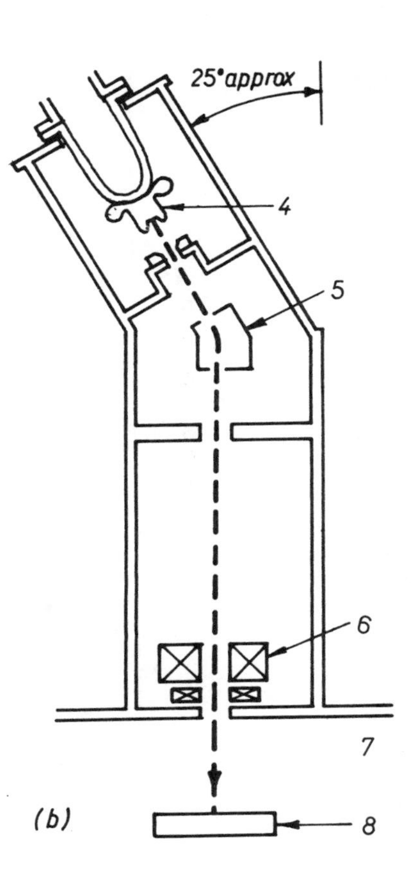

1 Schematic arrangement of two conventional techniques for ion separation: (a) tilted grid cup and anode (Bas et al[1]), (b) bent column (Meier[2] Hamilton Standard). 1 – grid cup; 2 – bolt cathode; 3 – anode; 4 – electron gun; 5 – beam deflection coil; 6 – magnetic lens; 7 – vacuum chamber; 8 – workpiece

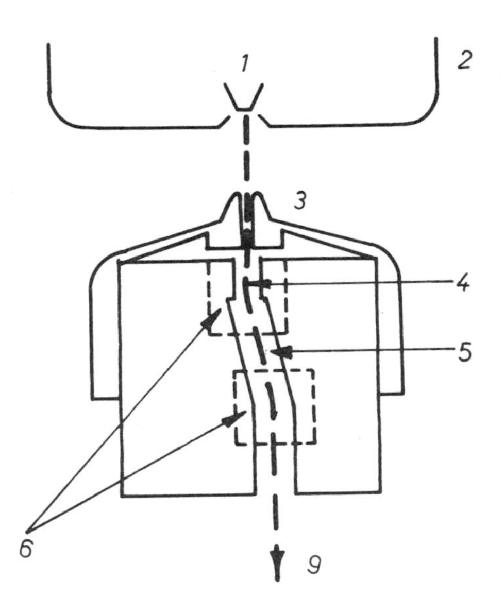

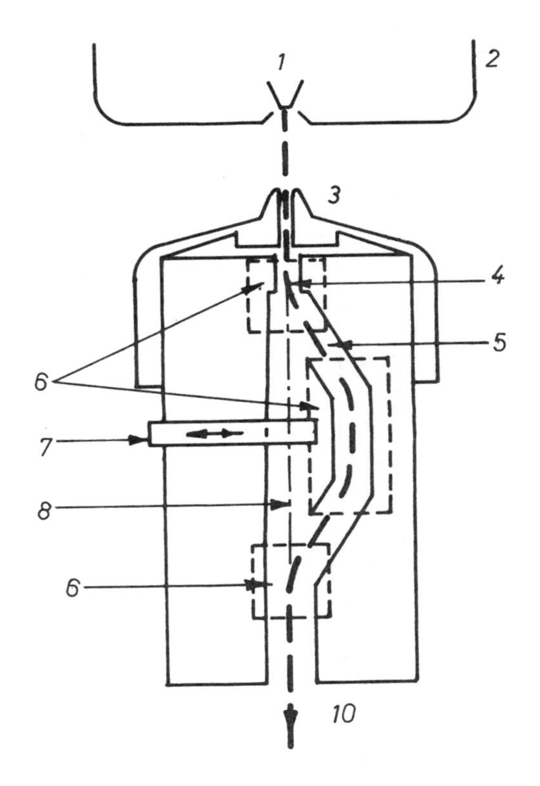

2 New magnetic trap deflection coil devices: (a) double bend, (b) triple bend. 1 – filament; 2 – grid cup; 3 – anode; 4 – deflected beam path; 5 – drift channel; 6 – deflection coils; 7 – sliding particle stop; 8 – undeflected beam drift channel; 9 – electron beam; 10 – axial electron beam

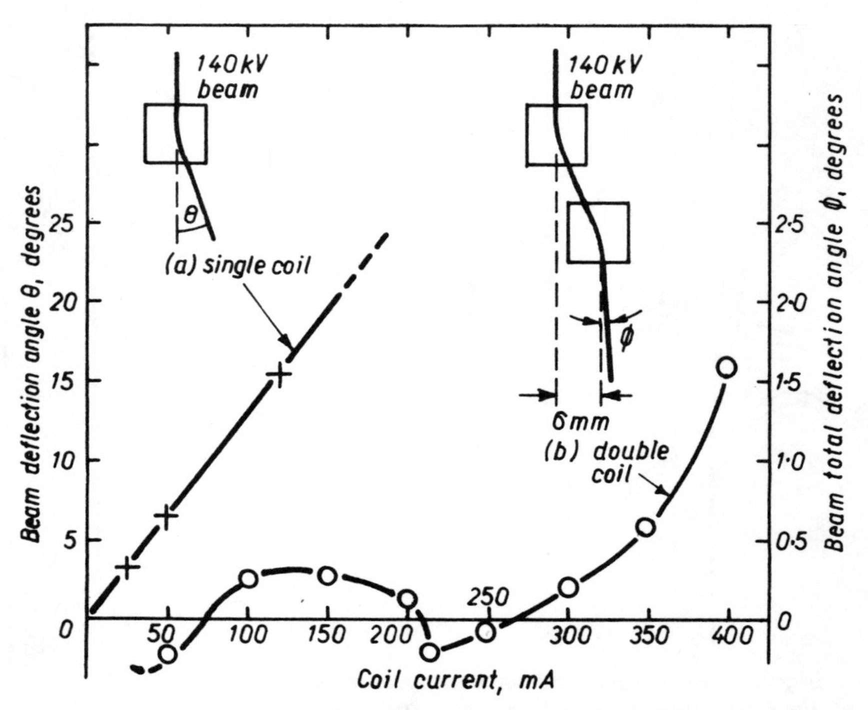

3 Effect of coil on deflection angle for: (a) single coil (LHS scale), (b) double coil (RHS scale)

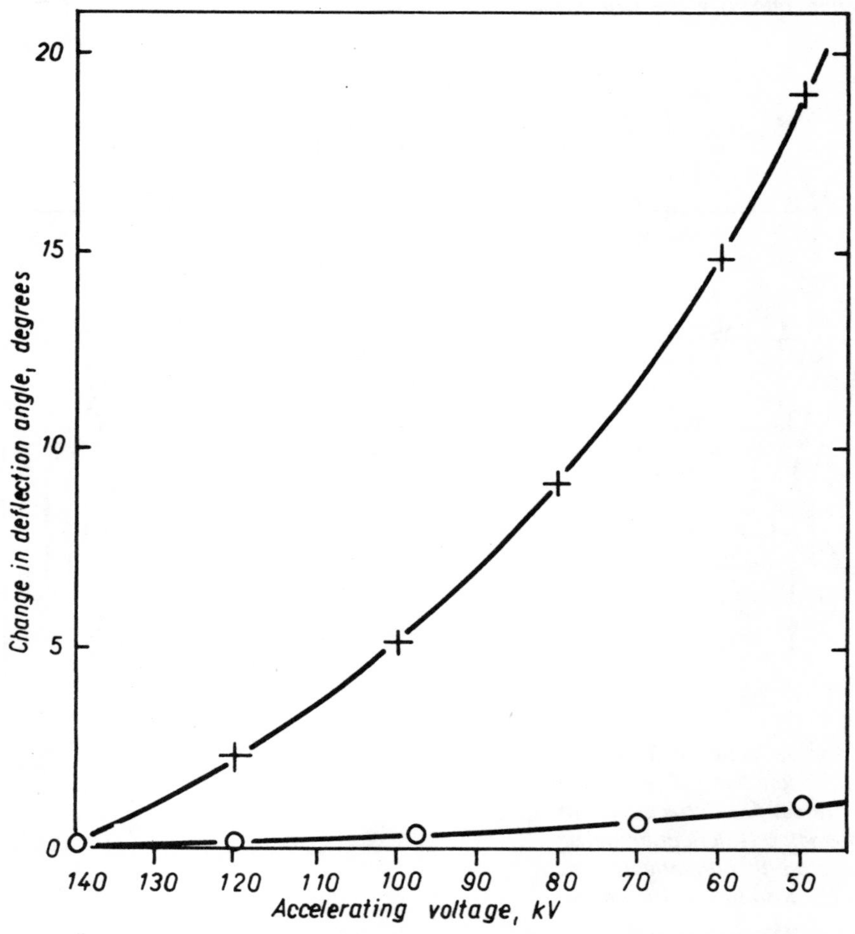

4 Effect of changes in accelerating voltage on beam deflection angle (coil current 215mA). + — single coil; O — double coil

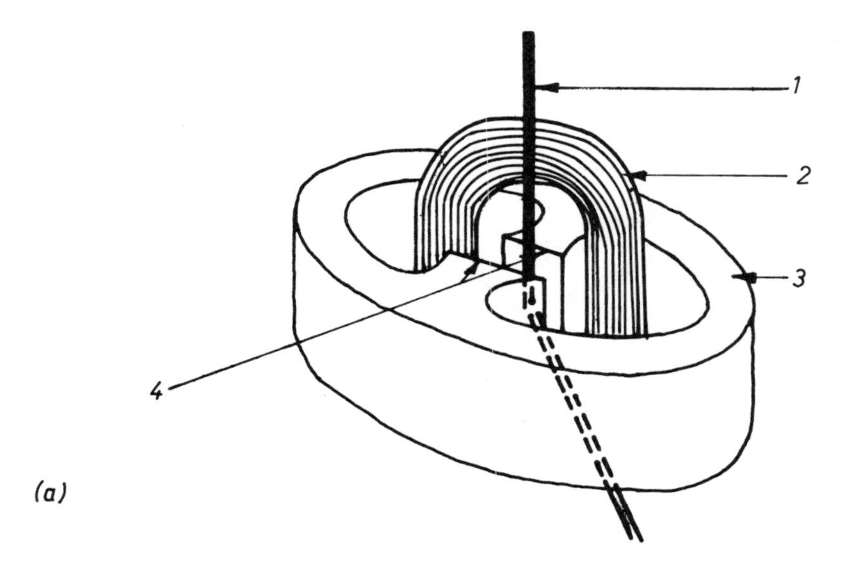

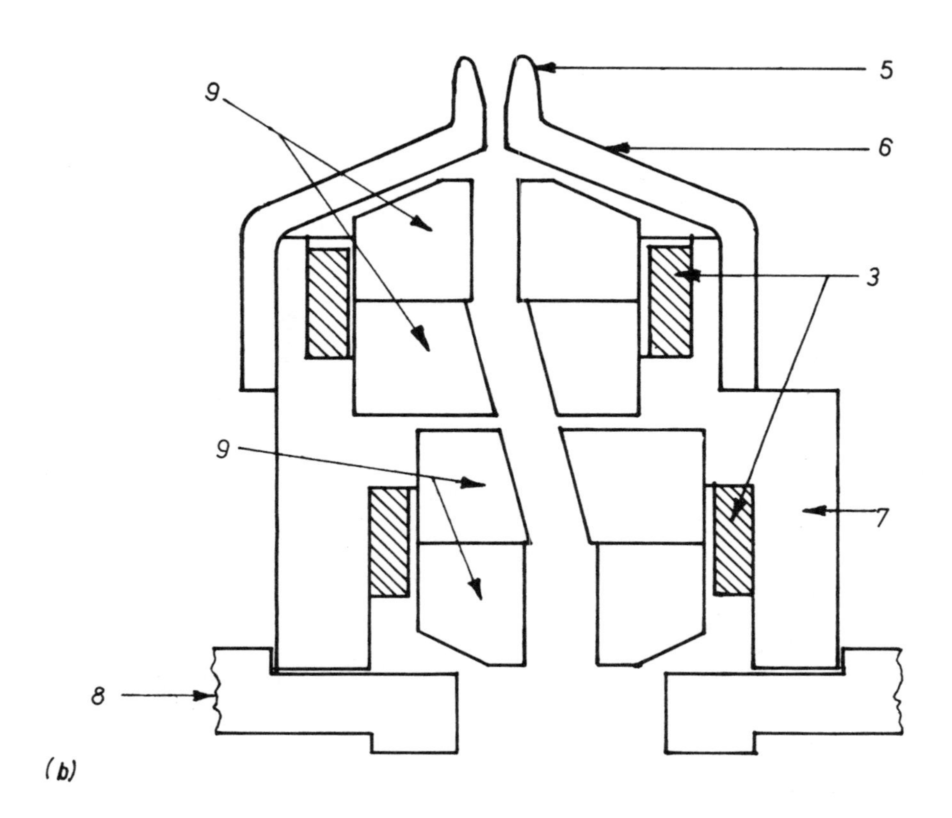

5 Double bend magnetic trap: (a) deflection coil element (one coil removed), (b) general arrangement of prototype trap. 1 — beam path; 2 — coil; 3 — soft iron magnetic circuit; 4 — polepieces; 5 — anode tip; 6 — anode cup; 7 — coil assembly body; 8 — mounting plate; 9 — drift tube channel blocks

6 Prototype double bend magnetic trap: (a) exploded view, (b) complete assembly. 1 — coils; 2 — copper drift tube block; 3 — anode; 4 — mounting plate

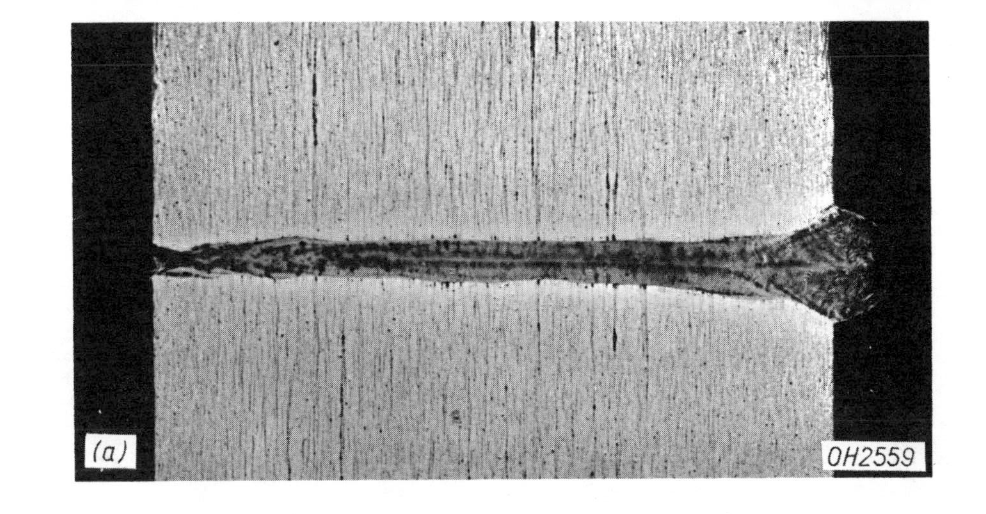

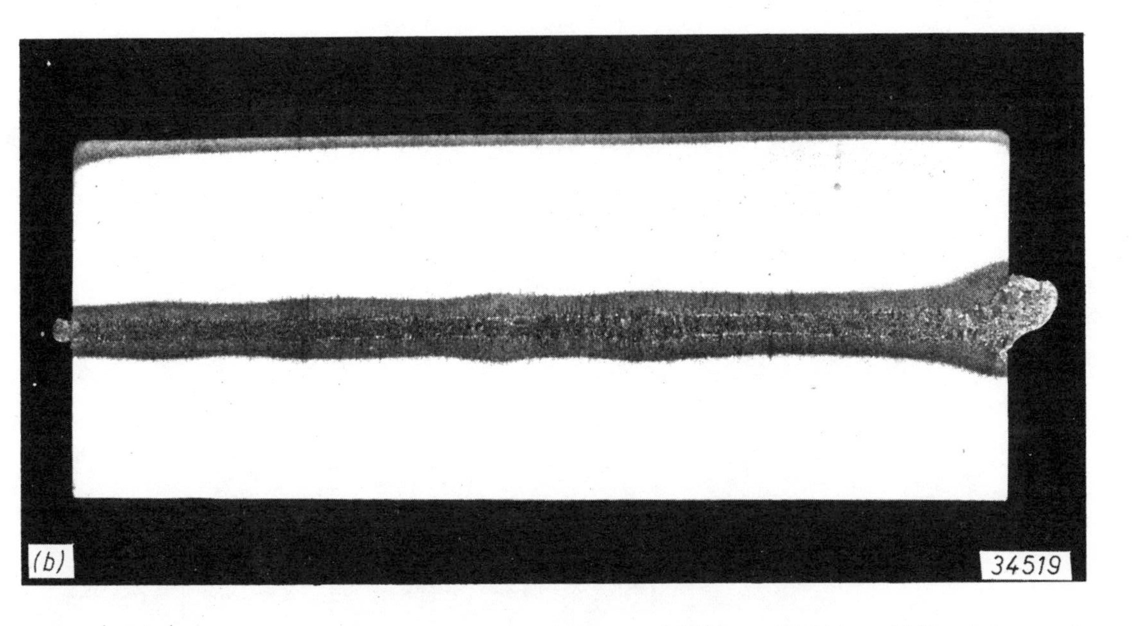

7 Melt runs made with double bend trap at working distance of 150mm: (a) 9.5mm 18/8 stainless steel; 150kV x 30mA (4.5kW), travel speed 2m/min (x 10), (b) 130mm CMn steel; 150kV x 250mA (37.3kW), travel speed 150mm/min

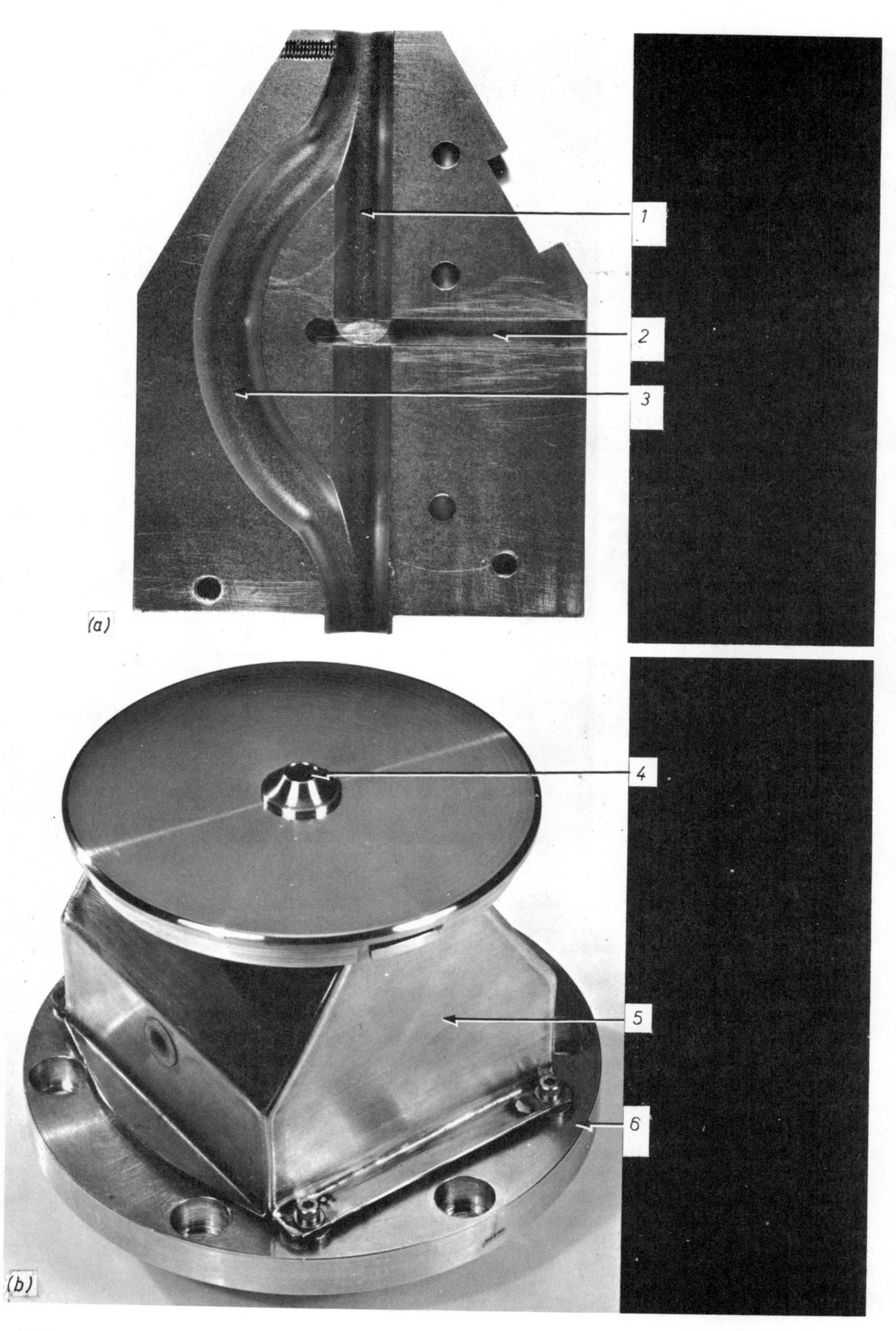

8 Triple bend magnetic trap: (a) copper drift tube split half component, (b) assembled prototype. 1 — undeflected beam channel; 2 — particle stop guide channel; 3 — deflected beam channel; 4 — anode tip; 5 — trap cover; 6 — mounting plate

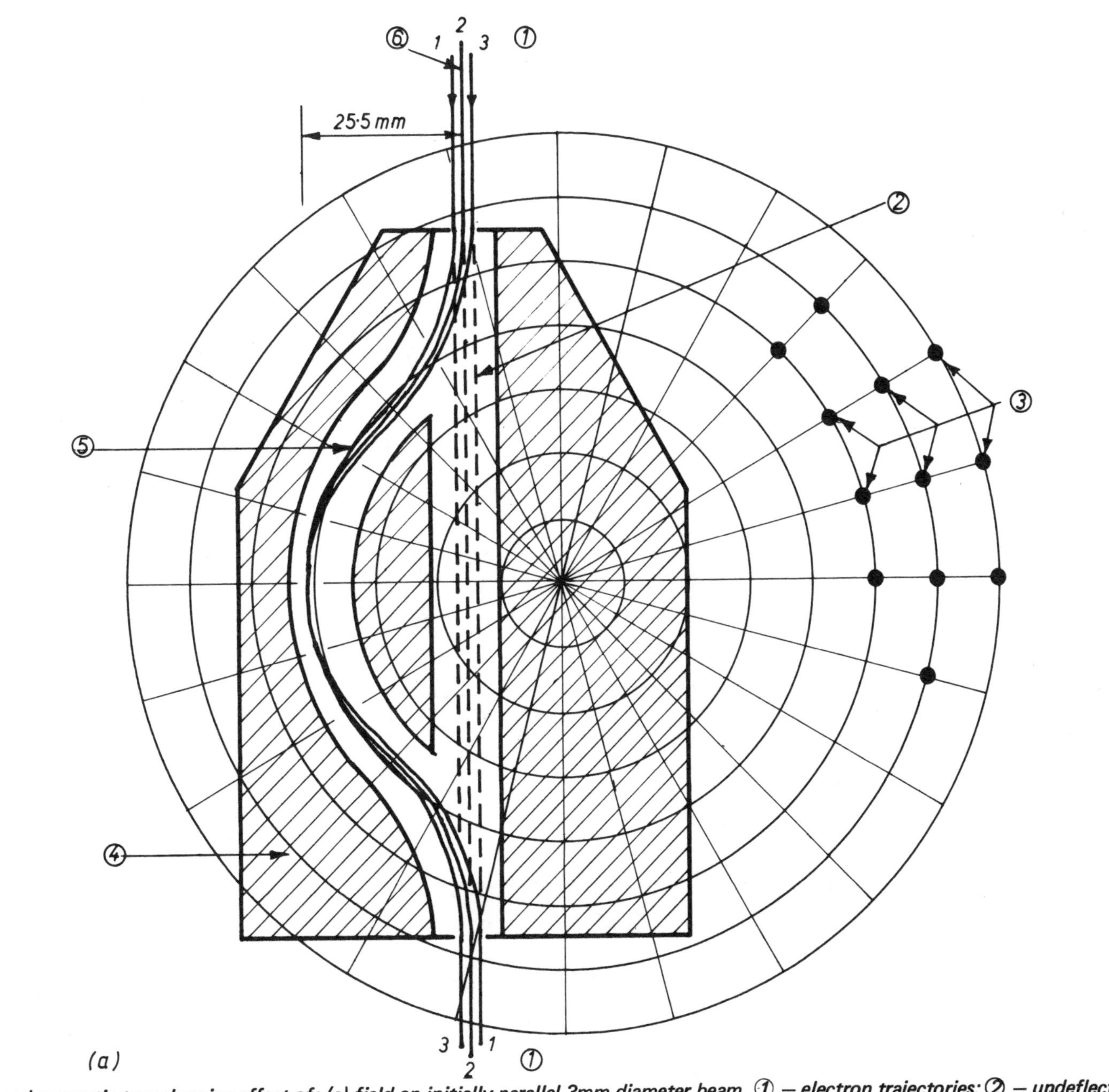

9 *Triple bend magnetic trap showing effect of: (a) field on initially parallel 3mm diameter beam, ① – electron trajectories; ② – undeflected trajectory; ③ – magnetic field measurement sites; ④ – copper drift tube component; ⑤ – deflected beam; ⑥ – central trajectory; ⑦ – electron beam; ⑧ – accelerating voltage, kV*

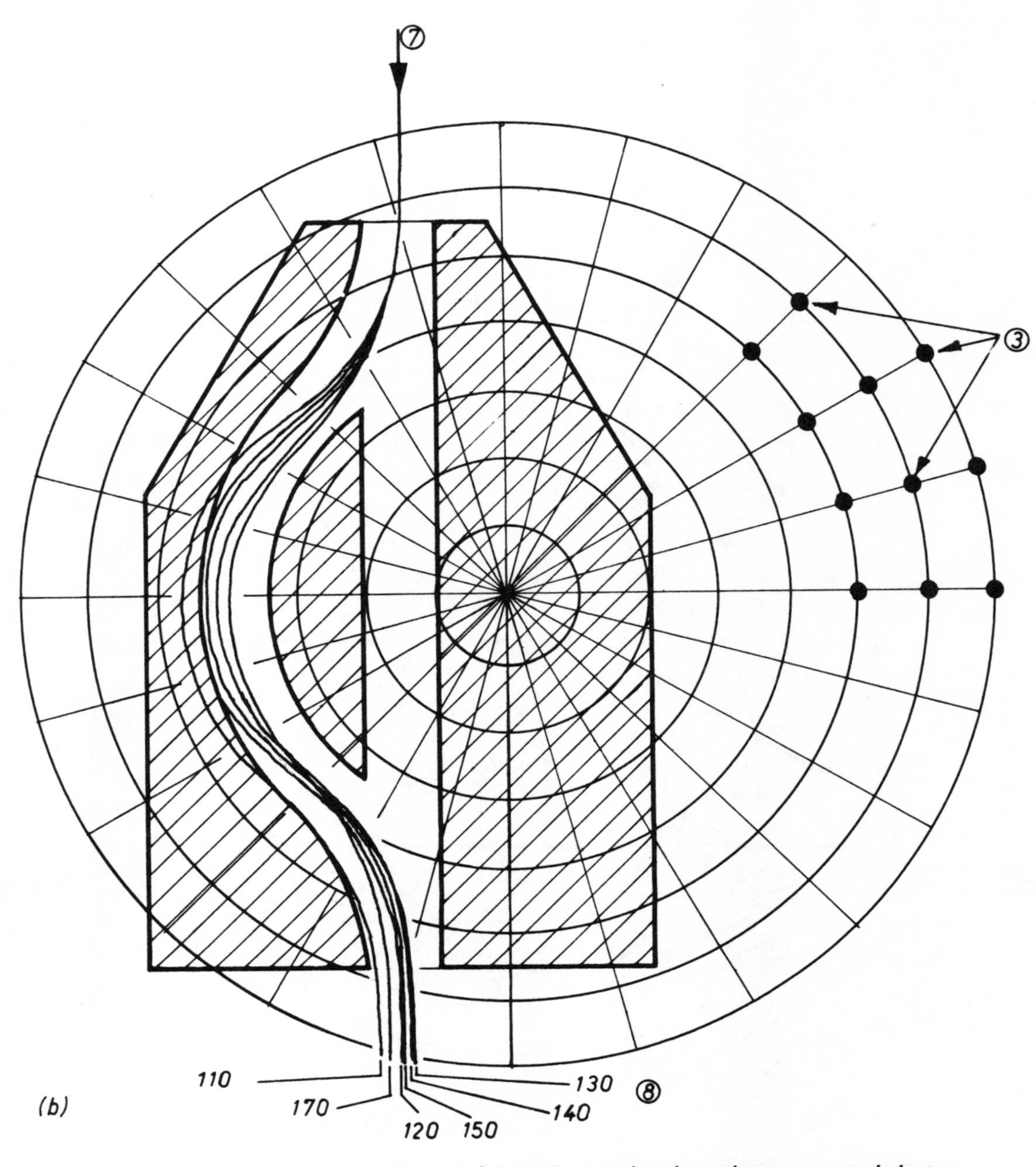

9 contd Triple bend magnetic trap showing effect of: (b) varying accelerating voltage on central electron trajectory

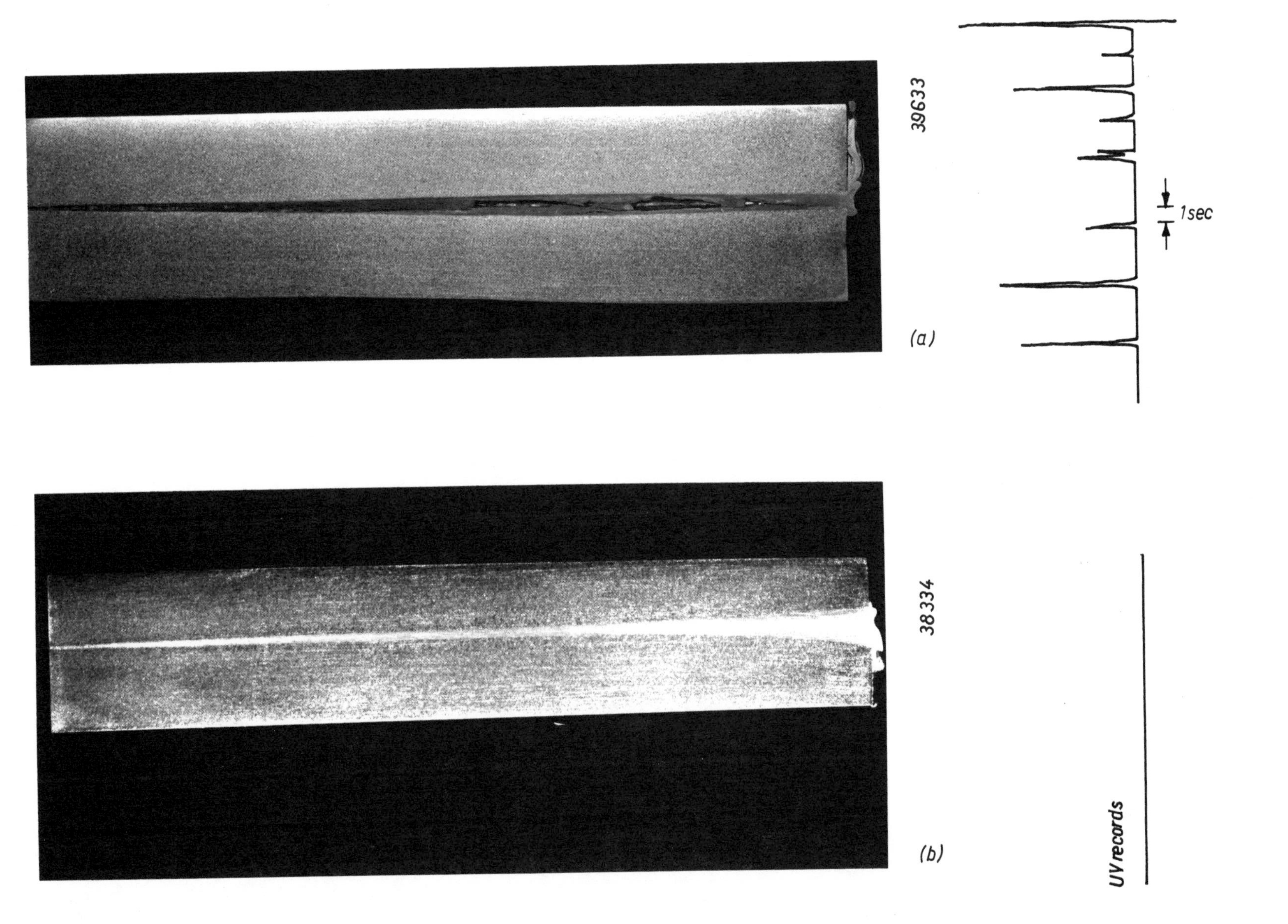

10 Welds made in 470mm thick aluminium alloy and corresponding UV beam current records. Welding conditions: 150kV x 520mA; 125mm/min; 125mm/min; working distance 150mm: (a) without trap showing gross cavitation defects, (b) with trap showing discharge-free current record and defect-free fusion zone

The design and realisation of a 100kV, 1A, electron gun

R.Cazes, Engineer ESE, and G.Sayegh, DSc, DEng

Theoretical methods, based on the resolution of the equations which define the electron-beam characteristics in an electron gun, were used to optimise the different components of the gun. Taking into consideration certain industrial restraints such as gun volume, cathode life, a 100kV, 1A, electron-beam column was designed and manufactured to weld up to 250mm thick steel.

Very good agreement between theory and experimental results was obtained for the electrostatic part of the gun (accelerating region) when space charge and relativistic effects were taken into account; whereas correction for partial neutralisation of space charge was needed to describe the actual behaviour of the electron beam in the electromagnetic part of the gun.

Experimental results on several materials with thicknesses up to 200mm are presented, and special attention is drawn to the importance of specific welding parameters (shape of the beam, oscillation, focus) on the weld quality.

To investigate further the industrial possibilities of the equipment, an international research group has been set up whose final goal is to butt weld in situ stainless and 533, Grade B, steel shells of about 3m diameter, taking into account the actual industrial conditions for joint preparation and gaps. Finally, some potential applications of the 100kV, 1A, electron-beam welding in heavy industry are enumerated.

INTRODUCTION

Electron-beam (EB) welding of heavy sections (>100mm) ceased to be a laboratory curiosity and became a reality after the first results obtained at The Welding Institute[1] in 1972 on the 150kV, 500mA, electron gun. Equipment manufacturers started to develop such high power electron guns to satisfy the needs of the heavy engineering industry.

Mr Cazes, R and D Manager, and Dr Sayegh, Manager — Scientific Department, are both with Sciaky SA, Vitry-sur-Seine, France.

This Paper briefly describes the methods used in the design of a 100kV, 1A, EB gun, taking into account the various restraints in the production of an industrial tool. The experimental results obtained on the gun are discussed in relation to the validity and limits of the theoretical methods used to determine its characteristics.

Finally, the gun performance is illustrated by laboratory results on the penetration (up to 250mm) of EB welding in various materials. The influence of the different welding parameters is discussed and the future of this high power EB welding technique analysed. Some indications are also given on the progress of an international research group investigating the possibilities of this high power EB technique for industrial butt welding of shells.

DESIGN OF THE GUN

The EB gun was designed bearing in mind the following guidelines which are believed to be indispensable to industrial use:

1. Cathode life greater than 20hr
2. Smallest possible gun for maximum mobility
3. Beam diameter less than 2mm at 600mm working distance so that 200-250mm of steel can be penetrated
4. Small diameter anode hole to decrease the inter-electrode flashover, by reducing the passage of metal vapour from the work-piece to the electrode region

The theoretical method used to optimise the different elements of the gun consists in resolving the various equations which define the beam characteristics[2] in a given configuration, followed by modifying this configuration to obtain the desired beam. The equations which define the electron beam comprise:

1. Electric potential distribution (Poisson's equation)
2. Magnetic vector potential (the Laplace equation)
3. Electron trajectories taking into account relativistic effects
4. Electron emission of the cathode

To reduce the design time the electron gun is divided into two parts which are computed successively, Fig.1, an electrostatic part and an electromagnetic part.

Definition of the electrostatic part

A computer program is used to resolve numerically the various equations in the gun. Finite difference, iterative methods combined with the relaxation technique permit rapid and accurate solutions of the equations to be obtained.

The optimisation of the gun consists in defining electrode configurations which are capable of producing a 100kW beam having a crossover of about $0.5mm^2$ with a divergence less than 20 milliradians. The emitting surface, which is bombarded indirectly with an auxiliary gun, has a 5mm diameter and operates with an acceptable emission density.

In choosing the accelerating voltage the problem was to find a compromise between two conflicting factors:

1. Use of high accelerating voltage, therefore a low current, thus reducing the space charge effect, but increasing the volume of the gun because of isolating distances needed for the high voltage

2. Using a low accelerating voltage, therefore high current, which would produce a relatively large beam but a reduced gun volume

The optimal solution is to choose the lowest accelerating voltage which is capable of producing the beam power in the desired cross-over section. Accelerating voltages ranging from 60-150kV have been explored and optimal electrodes capable of producing 100kW EB sought for each high voltage.[3] The gun operating at 100kV, 1A, was finally selected which fitted the technological guidelines believed to be necessary for an industrial gun.

Figure 2 represents the electron trajectories in the electrostatic part of the gun for varying bias ranging from 0 (diode) to the critical value, which results in the greatest beam diameter at the anode exit. The hole in the anode through which the electron beam passes should have a small diameter and long length to reduce the conductance between the electrostatic part of the gun and the chamber, thereby reducing the flow of metal vapour to the electrodes and avoiding flashover. Furthermore, the hole diameter should be defined in such a way that the envelope of the electron beams corresponding to various operating bias does not intersect the anode. This results in a diameter of 2.5mm and a length of 20mm for this gun.

Characteristics of the beam in the electromagnetic part

Electron trajectory in the electromagnetic part of the gun is given by

$$\frac{d^2r}{dz^2} = f(V) \left[\frac{1 + r'^2}{2P} \left(\frac{\delta P}{\delta r} - r' \frac{\delta P}{\delta z} \right) \right] \qquad [1]$$

where z = axis of the beam
r = radius perpendicular to the beam axis
$r' = \frac{dr}{dz}$
$P = V - kA^2$ (reduced potential)
V = accelerating voltage

k = constant depending on the particles
A = magnetic vector potential

To determine the characteristics of the beam in the electromagnetic part of the gun, eq.1 must be combined with Poisson's equation in the domain of integration, so it is necessary to know the value of the magnetic vector potential in this domain. Magnetic vector potential, A, is related to the magnetic induction, B, by the fundamental equations

$$\vec{B} = \text{Curl}\ \vec{A} \qquad [2]$$

$$\text{Div}\ \vec{A} = 0 \qquad [3]$$

In an axially symmetric system, $\vec{A}(z, r)$ can be theoretically expressed as a function of the induction, B(z), and its derivatives, in a series expansion.[4] Practically, the accuracy obtained by the successive derivation of B(z) is poor. It is, therefore, preferable to compute $\vec{A}$ directly by resolving eqs 2 and 3.

Using a finite element method, Fig.3, and taking into account the boundary conditions, the magnetic vector potential, A, can be calculated for all the points of interest in the magnetic part.[5] Vector A then is combined with V to resolve the electron trajectory equations.

Figure 4 shows the electron trajectories plotted by computer both when electron space charge is neglected and when it is taken into account. The following should be noted:

1 Scales on the r and z coordinates are very different
2 Electron trajectories, when space charge is not considered, are rectilinear outside the centre of the coil and focus on a spot located on the axis (spherical aberration)
3 Space charge considerably modifies the beam characteristics:

(a) rectilinear trajectories become curved and divergent
(b) cross-section of the beam increases at the centre of the coil
(c) focal point of the beam moves away from the coil
(d) diameter of the focal point increases

When comparing these results with those obtained by simplified methods (Gaussian optics) very significant differences are found. Thus it is not correct to apply the simplified methods to high power electron beams.

GUN CHARACTERISTICS

Figure 5 shows a general view of the 100kW gun in the horizontal position on a $5m^3$ vacuum chamber. The total length of the gun, including focus and deflection coils, is about 800mm, it weighs about 200kg, and includes a sliding table which allows displacement.

Figure 6 shows the comparison between the theoretical values of the emitted current and the experimental results measured on the gun. The following points are noted:

1 Agreement between theoretical values and experimental results is excellent
2 The gun is geometrically stable over the whole power range for which it was designed. Indeed, if the gun geometry became modified, i.e. if the position of the cathode changed while being heated to obtain the desired power, the perveance of the gun would change and the experimental points would not be located on the theoretical curve
3 The effect of positive ions on the beam current[3] in the electrostatic part of the gun is negligible

Figure 7 shows the power distribution in the beam at the focus point measured on a Faraday cup; the measured diameter is 2.2mm. When comparing this value to the theoretical one computed by neglecting or taking into account the space charge effect it is deduced, Fig.8, that the:

1 Actual beam diameter is in between the theoretical values of the beam diameters computed with and without space charge effects
2 Position of the focal point on the z axis for the theoretical excitation of the coil is located in between the theoretical focal points determined with and without space charge

The difference between theory and experimental results is attributed to the presence of a positive space charge (slow ion velocity) in the electromagnetic part of the gun which neutralises the electron space charge of the beam.[2] However good agreement between theoretical and experimental results is obtained when complete neutralisation of electron space charge is introduced in the programme in the volume confined between the entrance of the focusing magnetic coil and the target.

WELDING PERFORMANCE

From laboratory tests conducted on melt runs in the flat and horizontal positions the following conclusions are drawn:

1. Welding in the flat position is not favourable for deep penetration: in a given material there is a penetration limit beyond which the beam does not reach. For steel this limit seems to be around 130-150mm for the optimal welding conditions (power, welding speed, focus), Fig. 9. If the power is increased or the welding speed decreased no gain in beam penetration is obtained, only the molten zone increases in width. It seems that beyond a given penetration (which depends on the material and the thermal characteristics of the material) the bottom of the capillary under the beam impact cannot stand the weight of the molten metal located above, which flows down filling the capillary and limiting the penetration

2. In the horizontal position beam penetration seems unlimited. Different shapes of bead can be obtained by a suitable choice of welding parameters

3. One of the greatest difficulties in heavy section welding is obtaining sound welds, free of large porosity. Because of the importance of the molten liquid which is displaced during welding, any instability in welding conditions, stemming from beam characteristics, or the composition of the material can result in flow of the molten zone out of the workpiece.

4. High stability in welding parameters is required; this can be achieved industrially by a careful choice of gun components

5. The shape of the beam, and more specifically its divergence at the workpiece, can affect the quality of the bead: the thicker the section to be welded the smaller should be the beam divergence. A semi-divergence angle of 0.025 radians is sufficient to weld up to 150mm of steel

6. With suitable choice of welding parameters it is possible to obtain sound welds with different bead shapes. Figure 10 shows examples of several welds obtained in different materials with beads of various geometries

7. It seems that the most stable beads are those which are V-shaped rather than parallel-sided

8. When comparing the bead shape in heavy section, i.e. 200mm, to the natural shape of the beam the shape of the beam is much larger than the molten zone and *a fortiori* larger than the capillary hole created under the beam. This means that, during welding, the plasma contained in the capillary, which can be under high pressure, completely modifies the characteristics and the shape of the beam which passes through it. The effect of the plasma on the beam is to pinch it in the cavity throughout its path in the capillary. An important investigation to be undertaken is the better understanding and explanation of this interaction quantitatively

FUTURE POSSIBILITIES FOR HIGH POWER ELECTRON BEAMS

To explore the possibilities of the 100kV, 1A, EB gun under actual industrial conditions, a research programme (internationally supported with the participation of French, German, Italian, and Japanese companies) was set up at the beginning of 1977 for a period of two years. This programme consists of studying and resolving the industrial problems which are raised in butt welding thick shells (up to 130mm) of 3m diameter in stainless and 533, Grade B, steel.

Preliminary tests were conducted in the laboratory to define the best welding parameters with respect to the influence of joint geometry (gap, mismatch). Also, the overlap conditions of the weld were investigated and limiting values defined.

The particular feature of this research programme is that the working chamber is composed from the shells themselves. Figure 11 shows the schematic of the equipment which is now entirely manufactured. It comprises the following main components:

1. Mobile mechanical part on the top of the shells which can be adapted to several different diameters. The gun is fixed to this part and can rotate around the shells' axis. At the bottom of the shell another mechanical part is fixed through which the inner volume is pumped down
2. A counter chamber placed around the joint to be welded and pumped to the desired vacuum (primary vacuum)
3. Auxiliary elements: pumps, power supply, seam tracking

The programme should provide technical and economic information to heavy industry which will permit better evaluation of the industrial possibilities and the interest in using this high power EB in production.

Associated with a suction cup,[6] this gun can be used for longitudinal welds of shells or any other linear weld of heavy section in cumbersome components.

Among the possible applications of the 100kV, 1A, gun in industry are:

1 Nuclear reactor internal components: this is an interesting application because the stainless steel used can be EB welded without major metallurgical difficulties
2 Steam turbine diaphragms which already employ EB welding and which can take advantage of the available power to weld heavier sections
3 Steam turbine rotor discs: with increasing power in turbo-alternators, a welded rotor appears to be inevitable. Thicknesses of 200-250mm should then be achieved
4 Steam generators, hydraulic turbine components: solutions should still be found to obtain good metallurgical qualities in the low alloy steels which compose these parts, especially where very heavy sections (>150mm) are considered

CONCLUSIONS

Theoretical methods are now available to optimise the various elements of EB welding guns. Applied to a 100kV, 1A, EB gun, the method appears to be very efficient in characterising the electron beam in the electrostatic part of the gun (power, diameter of crossover). In the electromagnetic part of the gun, positive ions slightly modify the theoretical behaviour of the beam by neutralising some of the electron space charge.

Preliminary welding tests have revealed the limitation of penetration of EB in the flat position and the possibility of penetrating steels up to 250mm in the horizontal position. Sound welds in steel 150mm thick were obtained at only 50% of the available gun power, thus maintaining sufficient power in reserve to ensure that the equipment is not operating at its maximum limit. Some possible industrial applications for high power EB equipment are given.

The main objectives of an international research programme, which should be terminated by the end of 1978, are to establish industrial butt welding conditions of heavy section shells (up to 130mm).

REFERENCES

1 RUSSELL, J.D. 'Penetration in depth'. Welding Inst. Research Bull., 13 (11), 1972, 327-8.
2 SAYEGH, G. 'Détermination des caractéristiques des faisceaux dans un canon à électrons'. Thèse de Docteur es Sciences, Université de Paris Sud No.1723, 23 September 1976
3 SAYEGH, G. and DUMONTE, P. 'Theoretical and experimental techniques for design of EB welding guns'. 3rd Seminar on EB Processing, Stratford (UK), March 1974, Paper 3a, 88pp.
4 GRIVET, P. 'Electron optics', revised by A. Septier. Pergamon Press, 1965.
5 BONJOUR, P. 'Un nouveau type de lentille magnétique supra-conductrice à pole d'Holmium pour très haute tension'. Thèse de Docteur es Sciences, Université de Paris Sud No. 1973, CNRS-A09167.
6 SAYEGH, G. and DUMONTE, P. 'An electron-beam suction cup for linear welding large sheets'. Int'l Conference 'Structural Design and Fabrication in Shipbuilding', London, 18-20 November 1975, Paper 21, 109-16. Abington, Welding Inst., 1976.

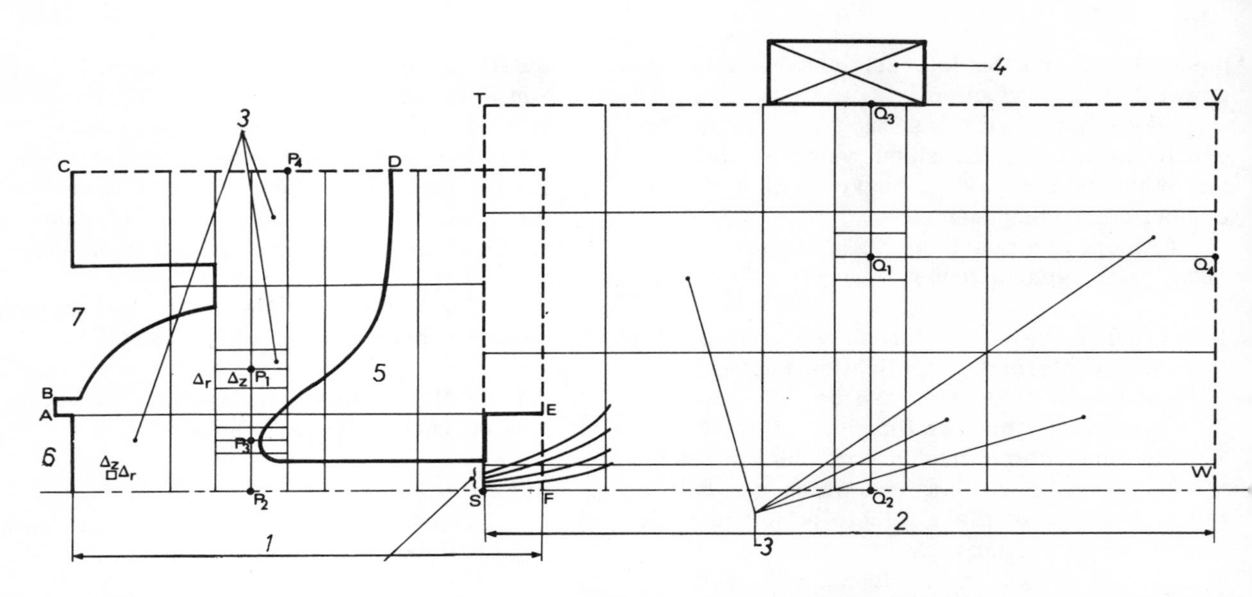

1 Domain of integration in electron gun: 1 — electrostatic; 2 — electromagnetic; 3 — different regions of domain; 4 — magnet coil; 5 — anode; 6 — cathode; 7 — bias cup

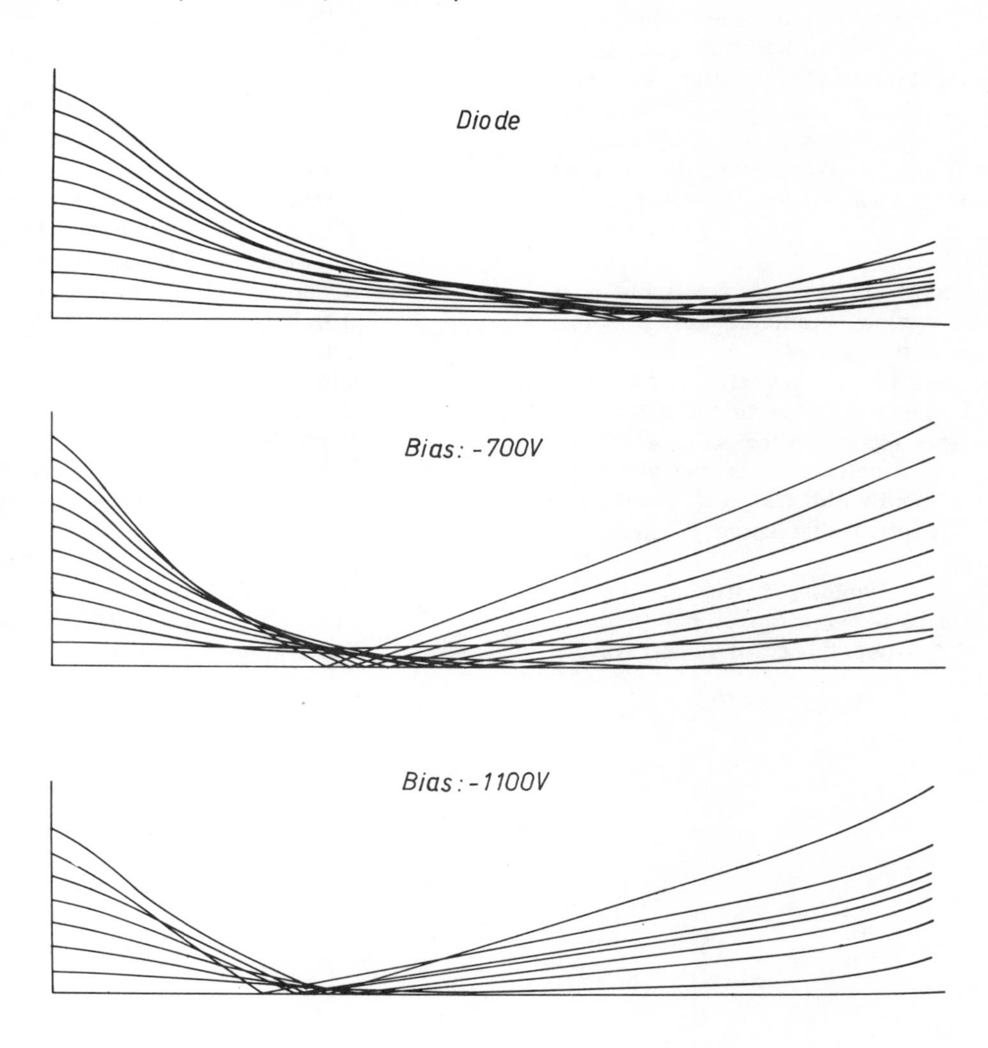

2 Electron trajectories in electrostatic part for different bias

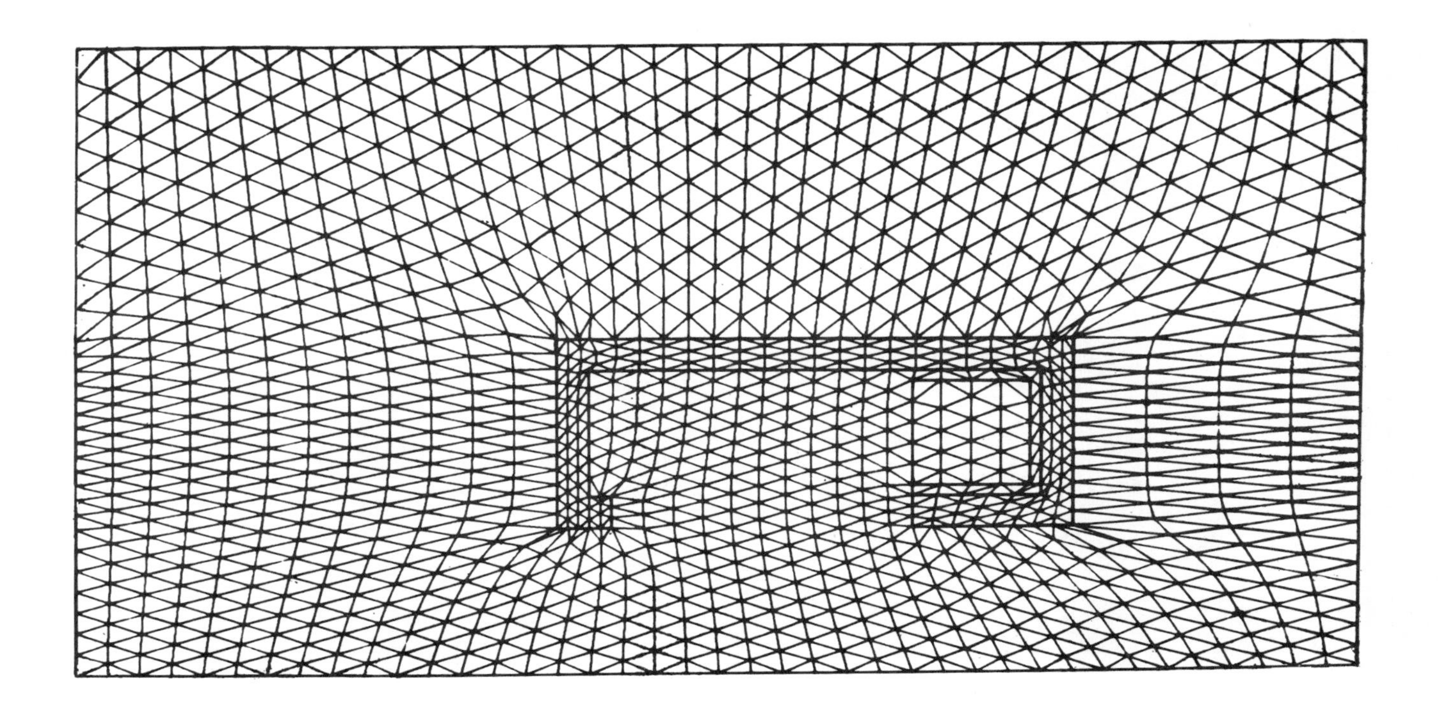

3 Finite elements used in computing magnetic vector potential $\vec{A}$ in electromagnetic part of gun

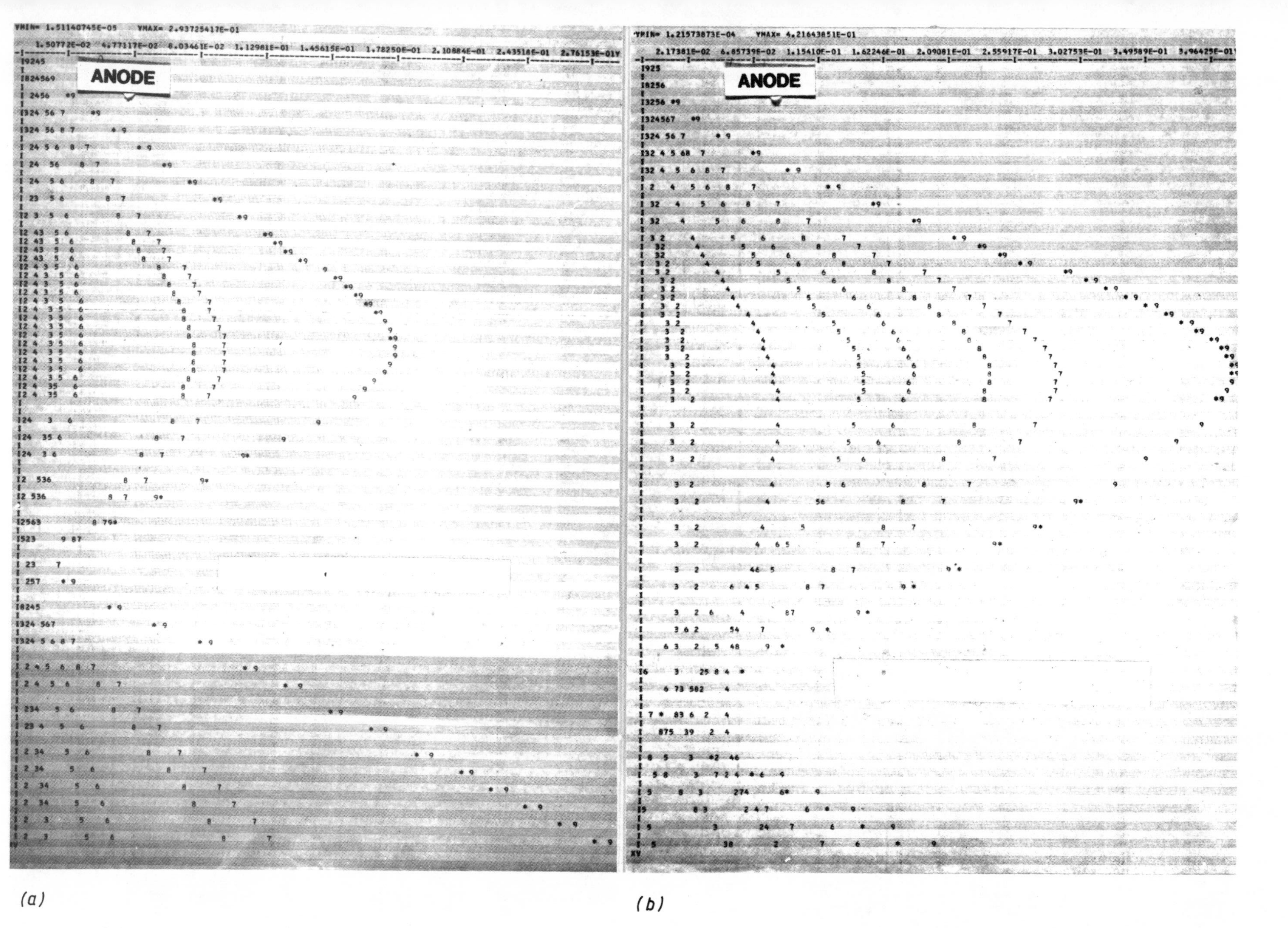

4 Electron trajectories plotted by computer in electromagnetic part of gun: *(a)* no space charge taken into account, *(b)* space charge taken into account

5 Electron gun, 100kV, 1A, in horizontal position. Total length of gun – 750mm; OD – 260mm

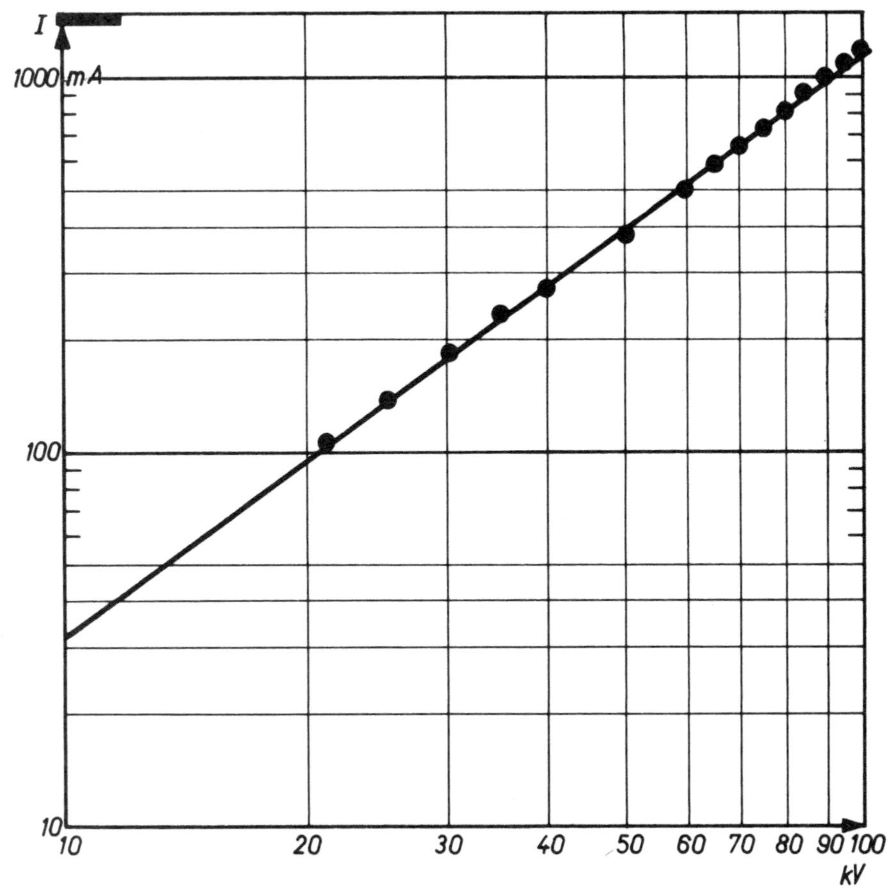

6 Beam current, I, as function of accelerating voltage. —— theory; ● – experimental results

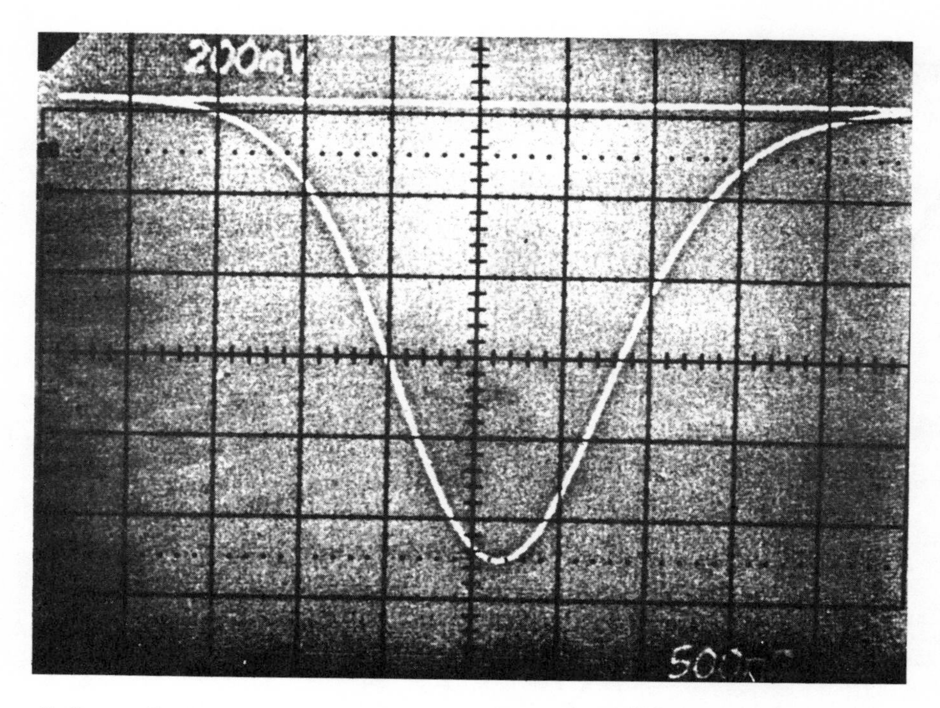

7 Power distribution at focus point. Diameter of beam containing 90% of beam power: 1.2mm

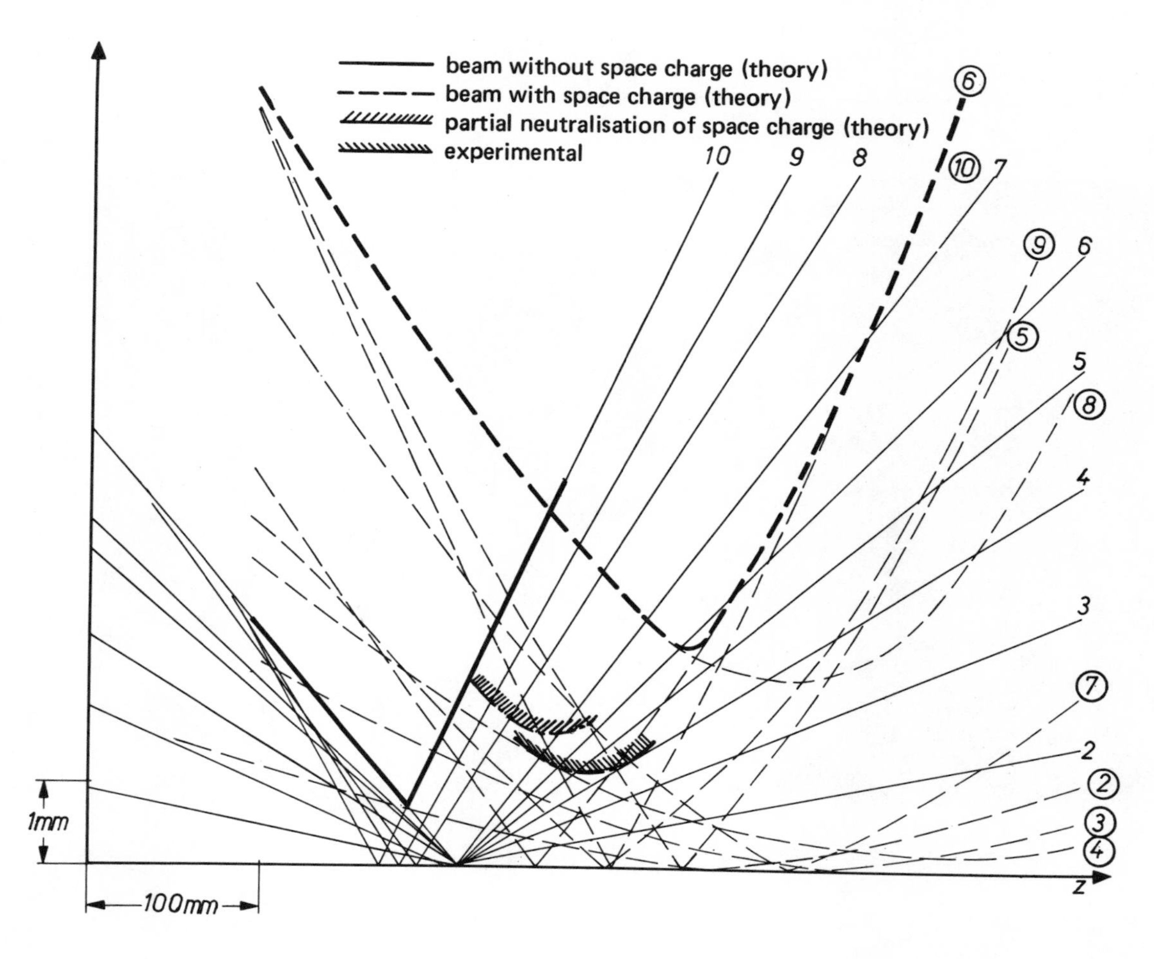

8 Comparison of theory with experiment at focus point

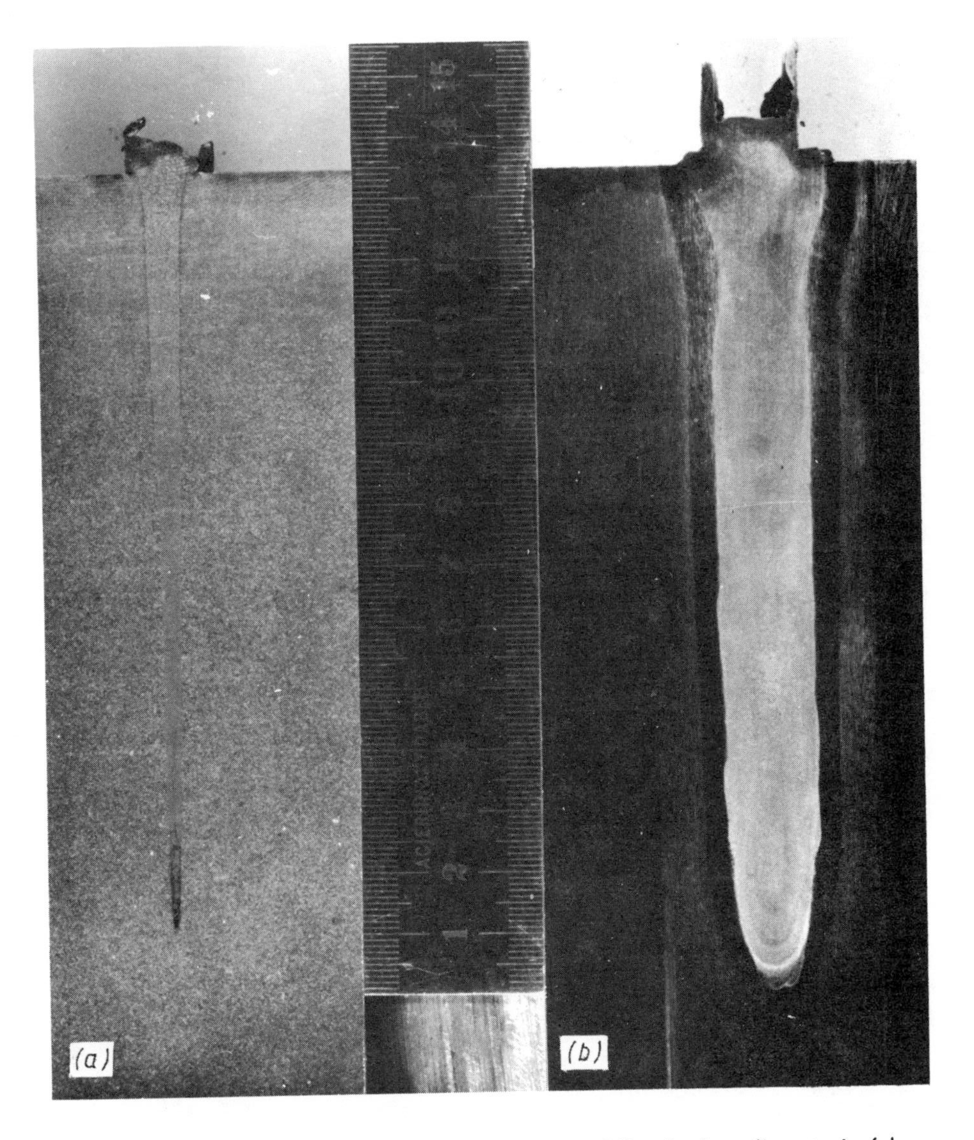

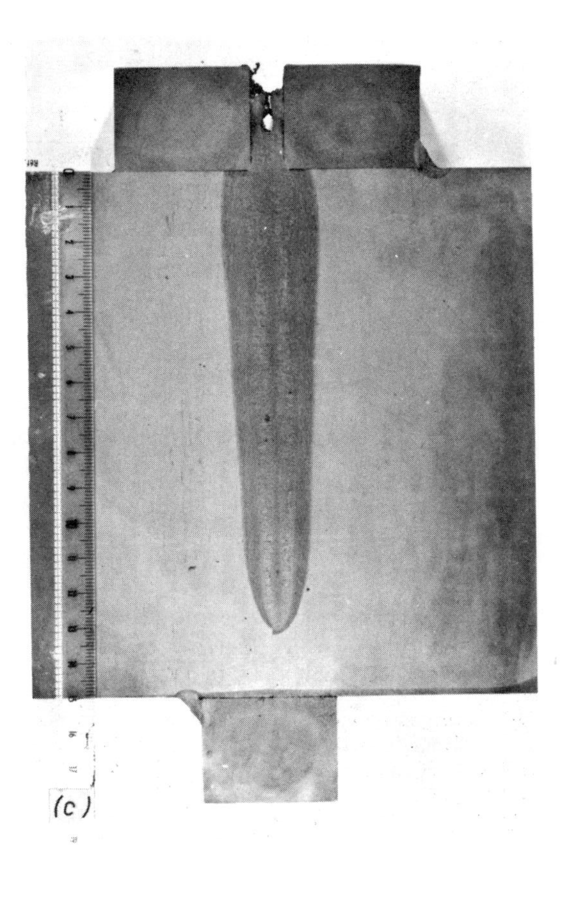

9 Limitation of beam penetration in flat position welding for low alloy steel: (a) power = 35kW, welding speed = 300mm/min, (b) power = 85kW, welding speed = 120mm/min, and (c) power = 85kW, welding speed = 120mm/min

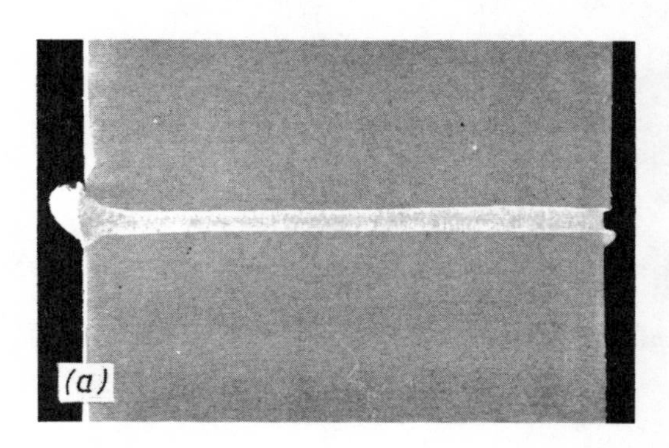

10 Macrographs of heavy section melt runs: (a) stainless steel 304L, 80mm, (b) 533, Grade B, steel, 127mm, (c) 387, Grade 22, steel, 150mm

10 cont. Macrographs of heavy section melt runs: (d) 200mm, 533, Grade B, steel, and (e) 250mm, 533, Grade 22, steel

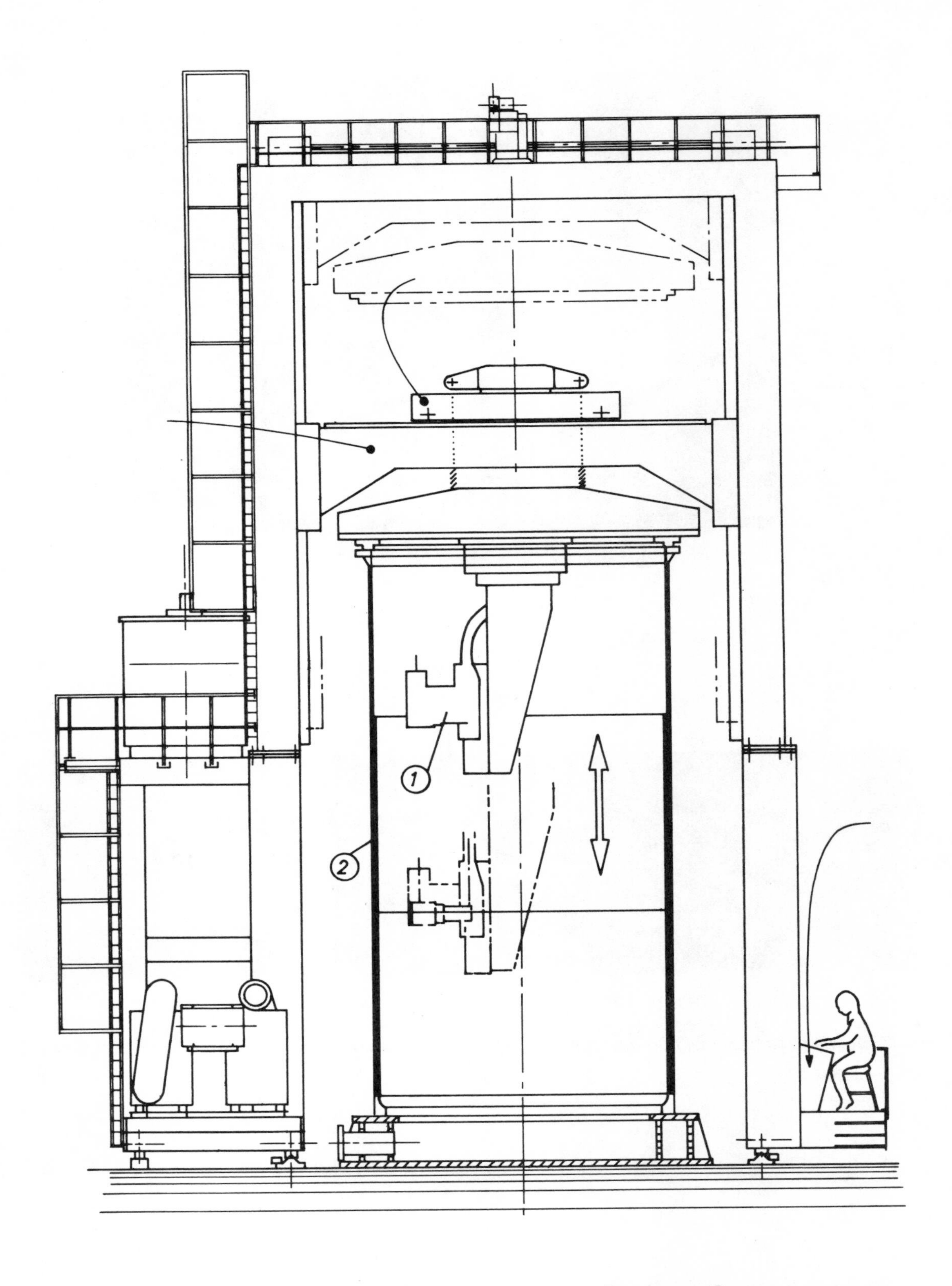

11 EB equipment for butt welding shells (under construction). ①– EB gun; ②– shells to be butt welded

High speed TIG-welding of stainless steel sheet

V.J.Hammond, BSc, and G.R.Salter, BSc, PhD, FWeldI

INTRODUCTION

There is a general and increasing interest throughout industry in methods for the improvement of productivity and process reliability. This is particularly true in arc welding, where production at maximum speed with the minimum of rectification work is desirable, especially in mechanised welding. Unfortunately, the degree of understanding of the factors limiting welding speed is often too restricted for maximum productivity to be reached. This situation can be improved only if the factors which limit productivity are defined, together with the procedural methods necessary to avoid defects.

The formation of defects in welds made at high production rates is most easily examined in autogenous welding, and several investigations have been made as to the cause of defects when using the TIG process. An examination of the formation of defects at high speeds, particularly undercut, was made by Bradstreet[1] who considered the significance of surface tension upon undercutting for mild steel bead-on-plate MIG welds. Attempts to control weld bead profile, and hence improve productivity in TIG-welding, have often used magnetic control of the arc column, e.g. Hicken and Jackson.[2] These investigations indicated that forward arc deflection was beneficial in raising the welding speed at which the onset of undercut occurred. The formation of undercut is usually attributed to the hydrostatic pressure caused by the excess height of the weld metal required to oppose the increased arc forces produced at high currents. To overcome these arc forces while maintaining sufficient heat input, several workers have used a multi-arc TIG system. Anderson and Yenni[3] showed that major improvements could be made in the maximum welding speed, which was proportional to the number of electrodes used for up to four electrodes. Wealleans and Adams[4] investigated the effects of welding current, welding speed, and arc length on the formation of undercut for TIG bead-on-plate welds. They concluded that arc force was the dominant factor in the formation of undercut.

The purpose of the present study was to examine the effects of different process variables on the incidence of defects in fully penetrating butt welds in 0.9mm stainless steel sheet using mechanised TIG-welding. The improvement in welding speed achieved by changes in electrode inclination was also examined for 1.6mm stainless steel to assess whether the effect of sheet thickness was significant.

Welding trials were made for a variety of current/speed combinations with all other factors remaining constant. Individual parameters were then varied around a standard experimental procedure and again welds of various current/speed combinations were made. This allowed operating tolerances to be established, so that improved tolerance and reliability at higher welding speeds were recorded. The techniques of producing operating fields is described by Doherty, James, and Spiller.[5] A study of weld pool fluid motion was also made in conjunction with process parameters to examine the dynamic conditions which produce different weld profiles.

Mr Hammond, formerly at The Welding Institute, is now a Metallurgist at the GKN Group Technological Centre and Dr Salter is Head of Production and Economics Research at The Welding Institute.

EXPERIMENTAL APPROACH

Butt welds were made in the flat position in 0.9 and 1.6mm stainless steel sheet to the AISI 316 specification, their compositions are given in Table 1. The 300mm long sheets were guillotine cut out, the edges rubbed down with 400 grade emery paper and degreased before being loaded into the welding jig. This consisted of a pneumatic finger clamp which held the sheets 10mm from the joint line on to a copper bar which contained a 5mm wide slot to carry the backing gas, Fig.1.

The power sources used throughout the work were transformer rectifiers of 400 or 600A capacity with a drooping output characteristic, which fed a 400A commercial welding torch mounted above the traverse carrying the

Table 1 Chemical analysis of materials used, wt%

	C	S	P	Si	Mn	Ni	Cr	Mo
AISI 316 specification	0.08max.	0.03max.	0.045max.	1.0max.	2.0max.	10-14	16-18	2-3
0.9mm sheet*	0.06	0.013	0.030	0.41	1.70	12.2	16.9	2.8
0.9mm sheet*	0.06	0.011	0.035	0.32	1.77	12.2	17.0	2.7
1.6mm sheet	0.07	0.013	0.027	0.45	1.72	11.8	16.9	2.8

* For the welding trials using 0.9mm sheet two separate casts were used; no difference in the welding properties was observed between the two compositions

testpieces. At the higher welding speeds used, a trailing gas shield was attached to the welding torch to improve weld surface appearance. In all instances the gas shield flow rates and composition were:

Shielding gas 10 litre/min Ar + 5%H_2
Backing gas 1 litre/min Ar + 5%H_2

The electrodes used were 2% thoriated tungsten, diameter 3.2mm below 360A and 4.8mm for greater currents.

Throughout the study the standard of acceptance used specified that the welds should not exhibit any of the following features:

Incomplete penetration
Centre line cracking
Burnthrough or drop through
Excessive undercut (greater than 10% of sheet thickness)
Excessive underbead width (greater than 4mm)

The effects of the process parameters were deduced by varying them individually from a standard procedure which was known to give good welds. For each welding parameter a region was mapped out on a current/speed diagram so that changes in tolerance could be seen. The method used was to set a welding speed above 20mm/sec and make welds at progressively increasing current levels until an acceptable weld was produced and subsequently an unacceptable weld. This procedure defined both the upper and lower limits of welding current for each welding speed examined.

PROCEDURE DEVELOPMENT – 0.9mm SHEET

The combinations of welding speed and current which could satisfactorily be welded using a set procedure were first established. For these tests a 3.2mm diameter electrode with a 60° vertex was mounted 2mm vertically above the weld joint. It was established that a maximum welding speed of 40mm/sec was obtained at 145A. When using currents above or below this level the achievable welding speed was significantly reduced. The factors limiting welding speed were either poor surface appearance, Fig.2, or burnthrough. Undercut, generally considered to be the main limitation on welding speed, was not encountered.

Poor surface appearance, in the form of discrete areas of lack of wetting, apparently resulted from insufficient heat input. This phenomenon decreased as current increased, but was sometimes present along the weld edges even when full penetration was achieved. Maximum welding speeds were also limited by:

(a) weld pool drop through under its own weight where excessively large (usually at low speeds and high current), and
(b) burnthrough immediately penetration was achieved

Following the development of 'standard' procedural limits for 0.9mm sheet, variations were made to some of the secondary welding parameters to determine their effect on maximum welding speed, i.e. arc length, electrode tip geometry and angle, and joint fitup.

Arc length

The procedural pattern followed with a 2mm arc length was repeated at 1 and 3mm. The results obtained at 1mm showed virtually no change from those with a 2mm arc, but increasing the arc length to 3mm made a substantial difference, Fig.3. With the longer arc a considerably greater current was necessary to produce an acceptable bead than was required for shorter arc lengths because of uneven weld edges, even though full penetration was easily achieved. Over the region where satisfactory welds could be made with the longer arc length of 3mm there was a slight improvement in tolerance, at the higher welding speeds, over that which existed for the smaller arc lengths.

Electrode tip geometry

The effect of electrode tip geometry on procedural tolerance was established for a 2mm length and vertical electrode, with the following electrode geometries; 30° vertex, 120° vertex, and 60° vertex with a 1mm diameter truncated tip. The data for a 30° tip was very similar to that found previously for 60°, Fig.4. When using the 120° tip an unstable arc resulted at lower currents, but at higher currents the maximum speed could be raised marginally. The truncated 60° vertex electrode had very poor operating characteristics at lower currents owing to an unstable arc, and poor weld quality resulted which restricted the speed range.

Electrode inclination

Beginning with the standard welding parameters of a 60° electrode tip and 2mm arc length, the range of acceptable welding procedures was developed for electrode inclinations from -10°, i.e. 10° trailing (with the arc deflected backwards), progressively through +20°, +30° to +45°, the maximum forward inclination achievable with the equipment.

The results obtained, Fig.5, indicate that increasing leading angle improved tolerance for both welding speed and current. An increase in maximum welding speed from 36mm/sec for a vertical electrode to 90mm/sec for an electrode angle of +45° was made possible, an increase of 150%. The corresponding macrosections are shown in Fig.6. However, when using an electrode with a trailing angle acceptable welds were difficult to produce because of rough surface appearance or burnthrough. Figure 7 shows macrosections of welds made with an electrode inclination of +45° for maximum and minimum current levels at a welding speed of 26mm/sec to demonstrate the tolerance achieved with a leading electrode. Because of this increased weld pool stability for +30° and +45° electrode inclinations, the limitation on current at speeds of less than 40mm/sec was the excessive weld underbead width, Fig.7b.

To confirm that it was the effect of electrode tip angle rather than that of the welding torch which influenced welding speed, a 6.4mm diameter was ground to the dimensions shown in Fig.8, with the tip axis at 45° to the electrode axis. It was readily determined that, with this configuration, a similar improvement in speed to that obtained by inclining the torch body could be achieved.

Joint fitup

Welding trials were carried out with 0.9mm sheet for a range of speeds and joint gaps, and electrode inclinations of 0° and 30° to the vertical, until the maximum gap had been found which would support a stable weld pool for a particular welding speed/current combination. The current used was always that which would just penetrate a close butt joint. Gaps were produced and controlled by surface grinding the two edges square to the sheet surface with one of the edges having a recessed space ground into it. For each weld the gap was measured with a feeler gauge.

For an electrode at +30° at welding speeds of 45 and 30mm/sec the maximum gaps that could be tolerated were 0.35 and 0.45mm respectively. A vertical electrode at 30mm/sec allowed a maximum gap of 0.30mm. Welds were made for the largest gap that could be tolerated, and in each instance the arc was extinguished over the recessed region. In all instances the weld pool remained stable and the region in front of the frozen weld pool always had a joint gap reduced to 0.05-0.10mm.

EFFECT OF ELECTRODE INCLINATION FOR 1.6mm SHEET

Following the success achieved in increasing welding speed for 0.9mm sheet by increasing the electrode inclination angle, trials were carried out to establish whether a similar improvement could be obtained when welding 1.6mm sheet to a similar composition, see Table 1.

The angles examined were 0, +30°, and +45°, and the study was limited to determining the maximum speed achievable rather than the full procedural range. For the three electrode inclinations examined, the maximum welding speeds achievable were: 0° – 22mm/sec, +30° – 39mm/sec, and +45° – 42mm/sec.

Thus, for an electrode of +30° an increase in welding speed of 77% and for +45° 91% over that of a vertical electrode was made possible. When using +45° at high welding speeds complete penetration was difficult to achieve, necessitating very high currents. Despite this, however, lack of penetration still occurred when the top bead was up to 7mm wide. The shape of the weld bead for maximum welding speed of 0.9mm sheet, Fig.6b, shows that, for this thickness also, a relatively wide top bead is necessary to produce full penetration.

WELD POOL BEHAVIOUR

High speed ciné films (1630fps) were made of the weld pool for fully penetrating bead-on-plate welds with various combinations of electrode inclination, speed, and current. The stainless steel sheet was first electron-beam welded on to a mild steel grid to give improved access to the weld pool and provide an efficient clamping system. For these trials liquid motion was traced by the movement of fine alumina particles sprinkled on to the sheet through a grade 200 sieve (0.075mm between parallel wires), ASTM E11-70. For these bead-on-plate runs no burnthrough was observed, even though currents of 50A more than would be required for a joint were used. Observation of burnthrough was made possible by welding a joint instead of using bead-on-plate. The welding of the joint was filmed using a split-image technique so that both the top bead and underbead could be simultaneously observed on the same film. Alignment of the two images was achieved by drilling 1.5mm diameter reference holes at 10mm intervals through the sheet and scribing lines at 5mm intervals on each side of the specimens to record scale.

For an electrode inclination of 0° the alumina tracers followed a straight path across the weld pool beneath the arc and parallel to the welding direction, stopping in a region immediately behind the arc, Fig.9a. At this point, about 3.5mm from the front of the pool, the particles accumulated in an irregularly moving mass. From movements of this impurity globule the weld pool motion appeared to be violent and without a regular flow pattern parallel to the surface. This would indicate that in this region behind the arc the molten metal travels vertically down into the pool where it changes direction and travels along the bottom surface towards the rear of the pool. Occasionally the larger globule behind the arc broke up because of its rapid irregular movements but soon re-formed again.

Welding with an electrode inclination of $+30^o$ gave a more ordered flow pattern. Initially alumina particles floated upon a 'bow' wave, which was always present at the front of the weld pool. Particles were then drawn around the edge of the weld pool into one of the two regions situated either side of the weld pool centre line behind the arc zone, Fig.9b. The speed of the alumina particles through the arc region for different electrode orientations was 570 ± 30 and 470-600mm/sec for $+0^o$ and $+30^o$ electrodes respectively. For the inclined electrode a wide range of speeds was observed which were dependent upon particle size, the larger moving substantially slower than the smaller. When using an inclined electrode two vortex regions were present behind the arc which rotated at speeds in excess of 50rev/sec in cycles of 1.0mm diameter.

Other measurements from these films included the distance from weld pool front to that point at which the weld pool achieved its maximum width, which was usually in the range 2.7-3.0mm and showed no dependence on electrode inclination. The horizontal distance from weld pool front to electrode tip was found to be in the range 0.8-1.2, 1.4-1.6, and 1.7-2.0mm for electrode angles of 0°, $+20^o$, and $+30^o$ respectively. If the arc continued parallel to the electrode, for a 2mm arc length these measurements show that the same distance exists between the weld pool front and the point of incidence of the arc axis on the weld pool for all inclinations. Using split-image photography for electrode inclinations of 0° and $+20^o$ it was shown that, when small gaps were present (of the order of 0.1mm), molten metal on the underside was able to advance along the joint as a very narrow underbead, burnthrough occurring when the leading edge of the top bead and underbead were coincident. During one of the welds the traverse was stopped and the growth of a stationary weld pool was observed. Since the joint fitup was good there was no point at which burnthrough could be initiated and the weld pool appeared very stable until it grew to a width of 5.3mm, about three times the size of a normal underbead. (In this situation burnthrough finally occurred when the weld pool encountered a reference hole drilled in the testplate to align the two images.)

DISCUSSION

The major defect limiting welding speed in the TIG process was found to be either the occurrence of burnthrough or the formation of an uneven weld surface along the edges. Undercut was produced only at high travel speeds on 1.6mm bead-on-plate welds used as a guide to establish working conditions. Of the four references cited[1-4] all consider bead-on-plate runs, though Anderson and Yenni[3] produced fully penetrating weld beads. Thus it is realistic to expect that the nature of the defects produced in fully penetrating butt welds may be different. When examining high speed TIG-welding several workers have used magnetic control of the arc to improve welding performance.[2] Although many different arrangements of applying a magnetic field to the welding arc have been attempted, all improvements have been brought about by applying a transverse magnetic field to deflect the arc in the direction of welding. This result has been confirmed in the present work by inclining the electrode up to 45°, or by using an electrode tip geometry which deflects the arc forward. Welding with an electrode inclined forward eliminates all those defects commonly associated with high speed welding, and both weld pool stability and resistance to surface defects are increased.

Changes in electrode inclination produced by far the most dramatic increase in welding speed for any of the variables considered, with an increase of up to 150% made possible with an electrode inclined +45°. The effects of changes in electrode inclination are not as strong as those achieved by Anderson and Yenni[3] who produced a 300% increase in welding speed with the use of four electrodes for fully penetrating bead-on-plate welds. The use of a magnetic field would probably produce comparable increases in welding speed as is achieved by changes in electrode inclination. However, compared with these latter two methods, the use of an inclined electrode has the advantage of cost and simplicity when being used in industry. On an industrial level the technique is also important because of the large increase in tolerance to changes in current and travel speed which is offered. (The tolerance to changes in current is demonstrated in Fig. 7.) A major limitation would appear to be present when welding material of over 1.6mm thickness because of the difficulty in achieving penetration. However welds of excellent surface appearance were produced with high currents when penetration was a problem, and so this same procedural change may be beneficial in the MIG process. In this the penetration requirement can be controlled by joint design and the surface of the weld bead could possibly be improved by electrode inclination.

To relate this improvement to some physical process within the weld pool, weld pool motion was observed. This was found to be quite different between a vertical torch and one inclined +30° in the direction of welding, Fig. 9. For a vertical torch the alumina tracer particles were directed to a region behind the arc where they would remain, indicating that there is no mass flow parallel to the surface from this point and that liquid flows to this region and then vertically down. When welding with an electrode inclination of +30° two regions of swirling motion were in evidence. An electrode of +20° produced a pool motion which was mixed, at times swirling in a well-ordered pattern and occasionally becoming less ordered with small erratic movements of particles on the pool surface. The relevance of flow pattern to weld pool burnthrough is not clear since only surface movements could be traced. However, as is shown in Fig. 9a, the molten metal is forced down beneath the arc for a vertical electrode, and this flow, when mixed with the metal flowing vertically down behind the arc, may produce an unstable turbulent region within the pool. Although forward electrode inclination is influential in avoiding burnthrough, the quality of the jigging is also of importance. It is because of this that the nature of the limiting defects produced within this work are different from those produced in bead-on-plate experiments. As mentioned earlier the filming of the underbeads demonstrated the importance of good joint fitup in overcoming burnthrough. At a constant welding speed of 30mm/sec a 50% increase in the maximum permissible gap size was possible for an electrode of +30° inclination. Thus poor weld pool stability caused by joint gaps can be partly alleviated by the use of an inclined electrode. It was shown when producing the ciné films that surface tension forces are capable of supporting large volumes of metal despite the incidence of arc forces normal to the unsupported weld pool. Any factors which disturb the continuity of a weld pool surface or link them by a free surface of molten metal (as when a joint gap is present) increase the possibility of burnthrough. The advantageous effects of a leading electrode were also produced on 1.6mm thick material. In this however the possible increase in welding speed was considerably less than for 0.9mm materials. A factor which limited speed was the difficulty in achieving full penetration for an electrode of +45° even though high welding currents were used which resulted in a top bead over 7mm wide. Similar increases in welding speed on the 0.9mm sheet were brought about by using the electrode tip shown in Fig. 8 which simulated an inclined electrode.

Increasing arc length above the standard 2mm of the main series of trials enabled successful welding over a wide operating range with a minor increase in maximum speed. The increase in procedural tolerance found when the arc length was increased from 2 to 3mm may be attributed to the diffuse nature of the longer arc. This produced a wide molten zone which could more easily bridge the joint line than could a smaller weld pool. If a small gap is present a good weld could be produced using an increase of current and longer arc length. If a low current was used for the same joint fitup, a narrow molten zone would be unable to bridge the gap and produced a stable bead.

Electrode tip geometries showed effects which were similar to those produced by varying arc lengths. The 30° vertex showed little variation on the operating conditions produced with a 60° vertex. This similarity in welding tolerance is attributed to the maintenance of a highly directional arc caused by the narrow vertex angle, as would be the position with a short length. With a vertex angle of 120° improved operating conditions were brought about and an improvement in maximum welding speed from 36 to 42mm/sec was made possible. At low currents the arc produced a series of molten spots along the plate because of arc instability. This necessitated much higher currents than when using narrower vertex angles. Using an electrode with a truncated tip produced strong deleterious effects. Thus, though a fully penetrating joint was easily achieved, it was difficult to avoid the formation of an uneven surface along the edges. This can be attributed to an ill-defined anode spot on the stainless sheet which will show changes in the localised current density and consequently possibly induce turbulence in the weld pool.

The increase in tolerance bought about by forward inclination of the electrode allows welds to be made over a wide range of current settings. This results in weld profiles which can vary anywhere from that shown in Fig. 7b a conventionally shaped TIG weld in sheet material, to that shown in Fig. 7a, a wide weld with parallel sides. How these changes in weld profile may affect mechanical properties has not been ascertained and thus further work will be necessary to examine if the weld profile has any detrimental effect on mechanical properties. From the macrosections it is evident that drop through has not occurred on any of the specimens resulting in reduced cross-sectional area nor were notches formed at the toe of the weld.

The effect of the thermal cycle associated with high speed welding on metallurgical and mechanical properties will also be of significance for certain alloys.

CONCLUSIONS

The present work has shown that the maximum butt welding speed of 0.9 and 1.6mm AISI 316 stainless steel sheet using the flat position autogenous TIG process was limited by weld pool instability or bad surface appearance. Procedural changes produced the following effects:

1. A leading electrode inclination was found to be beneficial in eliminating all the different types of defect at a given welding speed. For 0.9mm sheet an increase in the maximum travel speed from 36 to 90mm/sec was produced by using a leading angle of $+45^{o}$. On 1.6mm sheet a 45° leading angle produced an increase in welding speed from 22 to 42mm/sec, 91%. The inclination of the electrode also greatly enhanced speed and current tolerance

2. Arc lengths of 1 and 2mm showed very little difference in maximum welding speed on 0.9mm sheet. An arc length of 3mm allowed a slight increase in welding speed from 36 to 42mm/sec. Longer arcs were unstable and so could not be used

3. A small increase in welding speed over that obtained with a 60° vertex electrode was produced with a 120° vertex electrode on 0.9mm sheet. A special electrode tip has been designed which offers on a vertical electrode speed increases comparable to changes in electrode inclination

4. The use of alumina tracers has shown that different weld pool flow patterns exist for different electrode inclinations

5. The presence of a joint gap has strong deleterious effects on process tolerance. A 0.45mm gap reduces the maximum welding speed of 0.9mm sheet from 76 to 30mm/sec for an electrode inclination of $+30^{o}$.

REFERENCES

1. BRADSTREET, B.J. Welding J., 47 (7), 1968, 314s-22s.
2. HICKEN, G.K. and JACKSON, C.E. Welding J., 45 (11), 1966, 515s-24s.
3. ANDERSON, J.E. and YENNI, D.M. Welding J., 44 (7), 1965, 327s-31s.
4. WEALLEANS, J.W. and ADAMS, B. Welding and Metal Fab., 37 (5) and (6), 1969, 210-13, 255-7.
5. DOHERTY, J., JAMES, R., and SPILLER, K.R. Welding Inst. Members Report PE12/74.

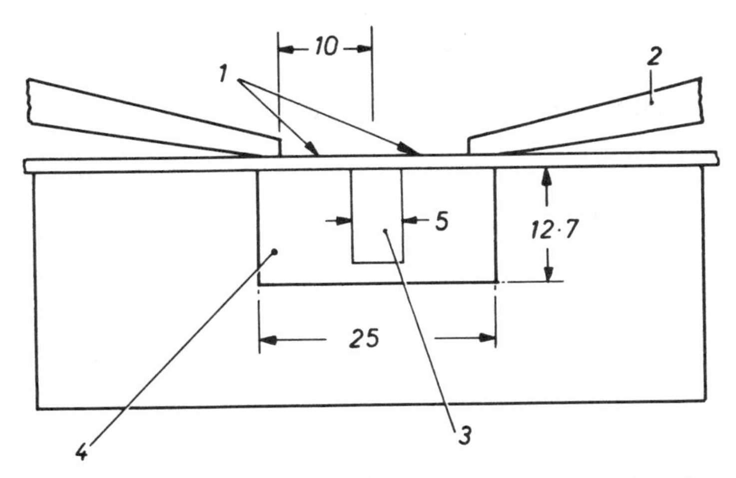

1 Cross-section through jig to show backing system and clamping arrangement; dimensions in millimetres. 1 — specimens; 2 — pneumatically controlled finger clamp; 3 — backing gas channel; 4 — copper backing bar

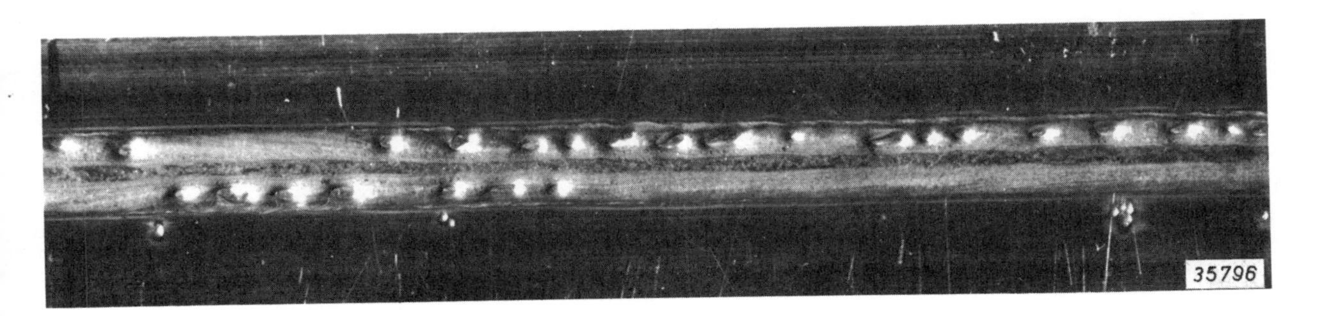

2 Example of unacceptable weld surface caused by poor wetting of weld metal. x2

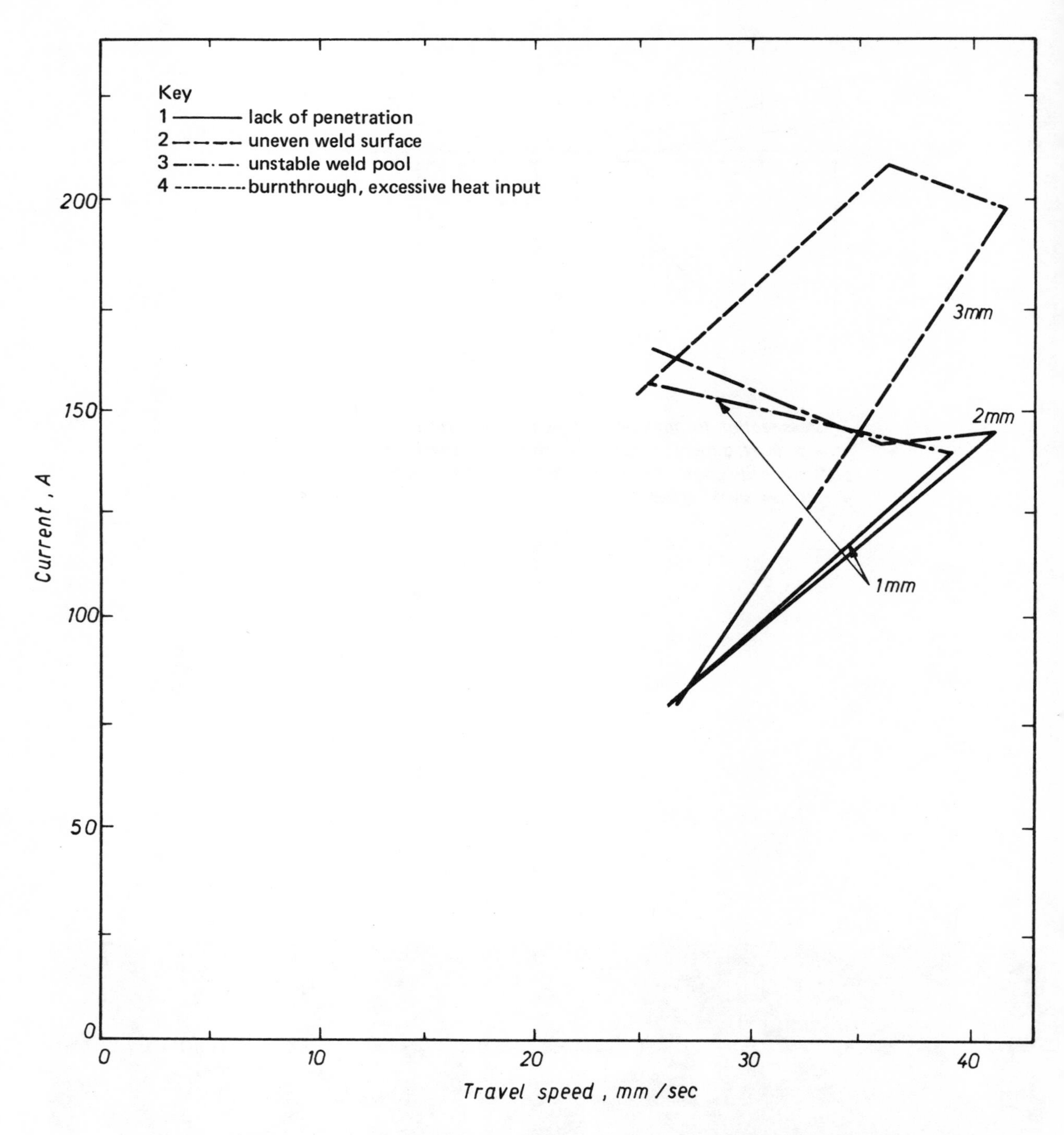

3 Effect of arc length on current/speed relationship for butt welds in 0.9mm stainless steel sheet

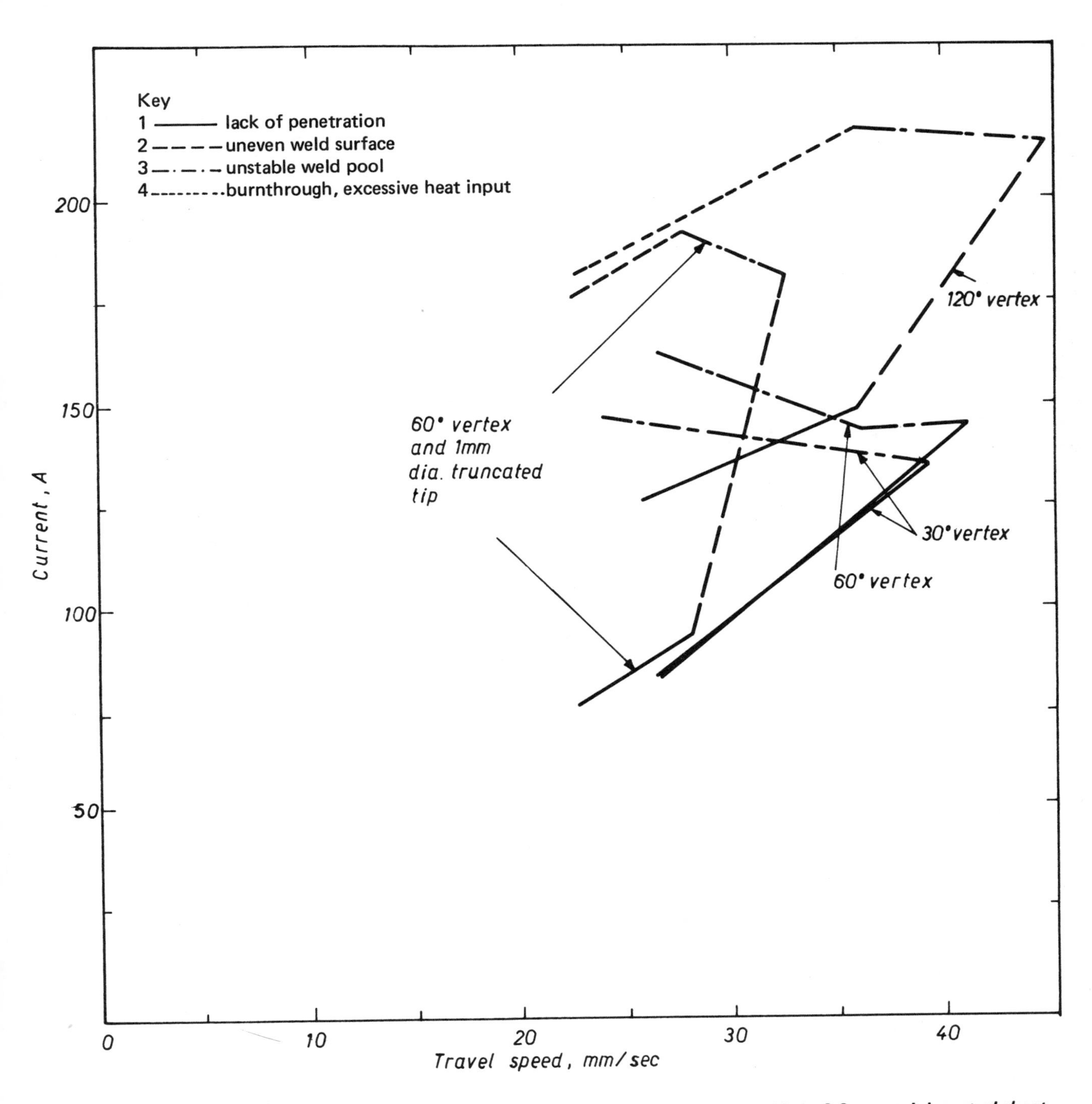

4 Effect of electrode tip geometry on current/speed relationship for butt welds in 0.9mm stainless steel sheet

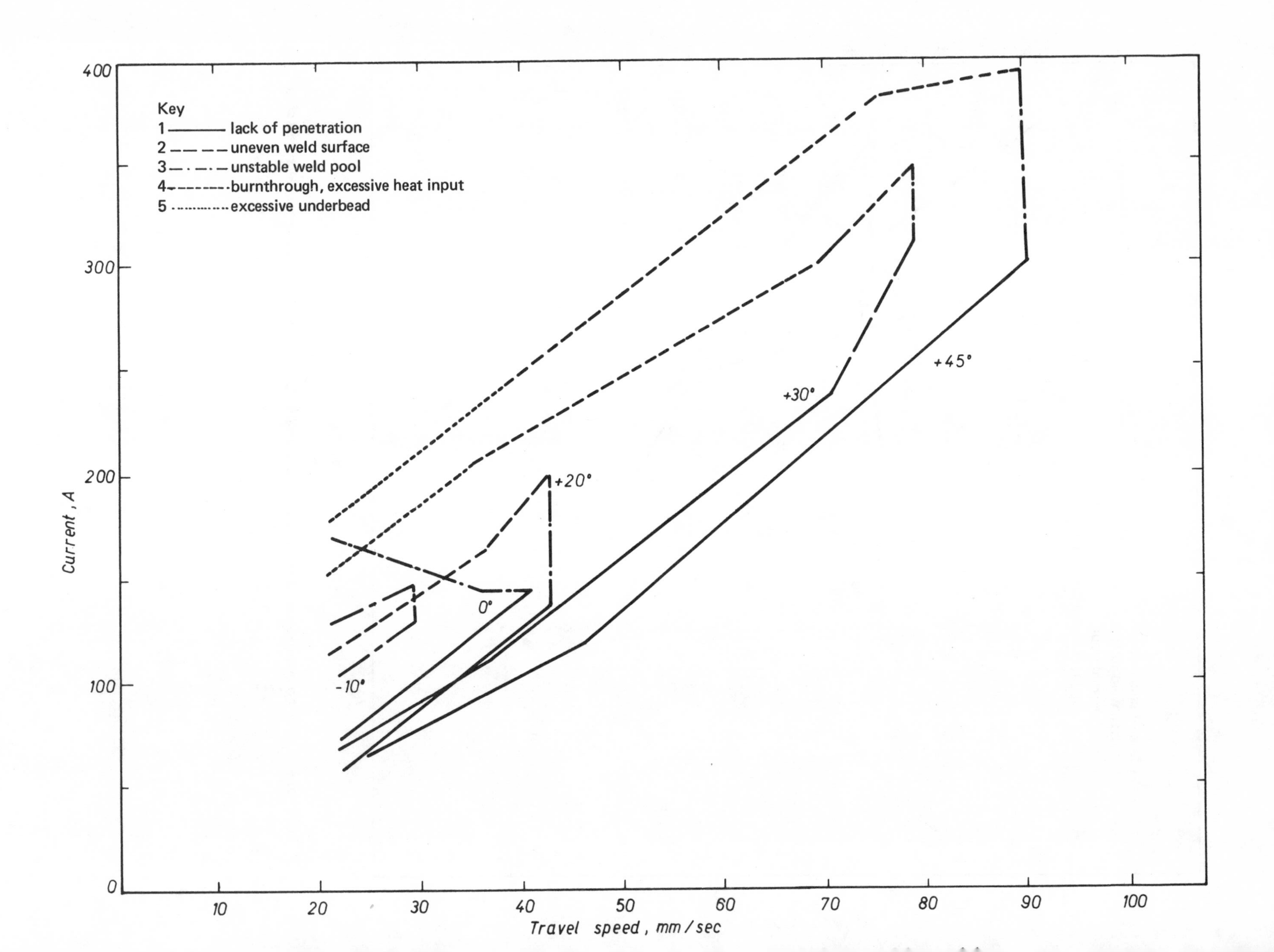

Key
1 —— lack of penetration
2 — — — uneven weld surface
3 —·—·— unstable weld pool
4 ------- burnthrough, excessive heat input
5 .... excessive underbead
+45°
+30°
+20°
0°
-10°
Current, A
400
300
200
100
0
Travel speed, mm/sec
10
20
30
40
50
60
70
80
90
100

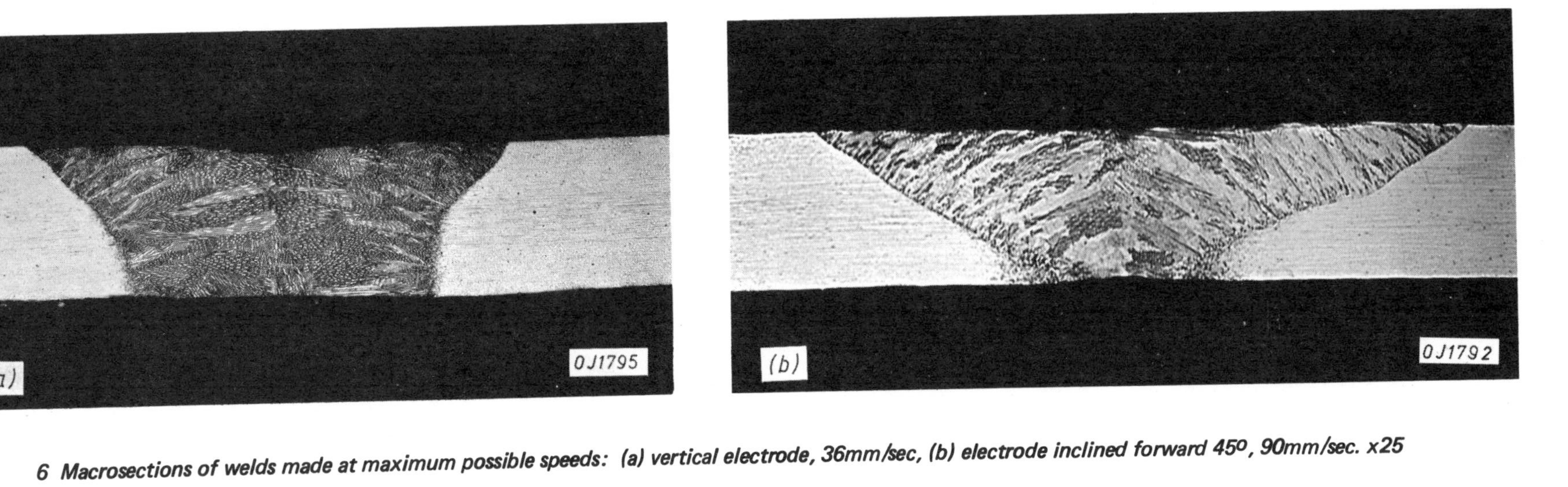

6 Macrosections of welds made at maximum possible speeds: (a) vertical electrode, 36mm/sec, (b) electrode inclined forward 45°, 90mm/sec. x25

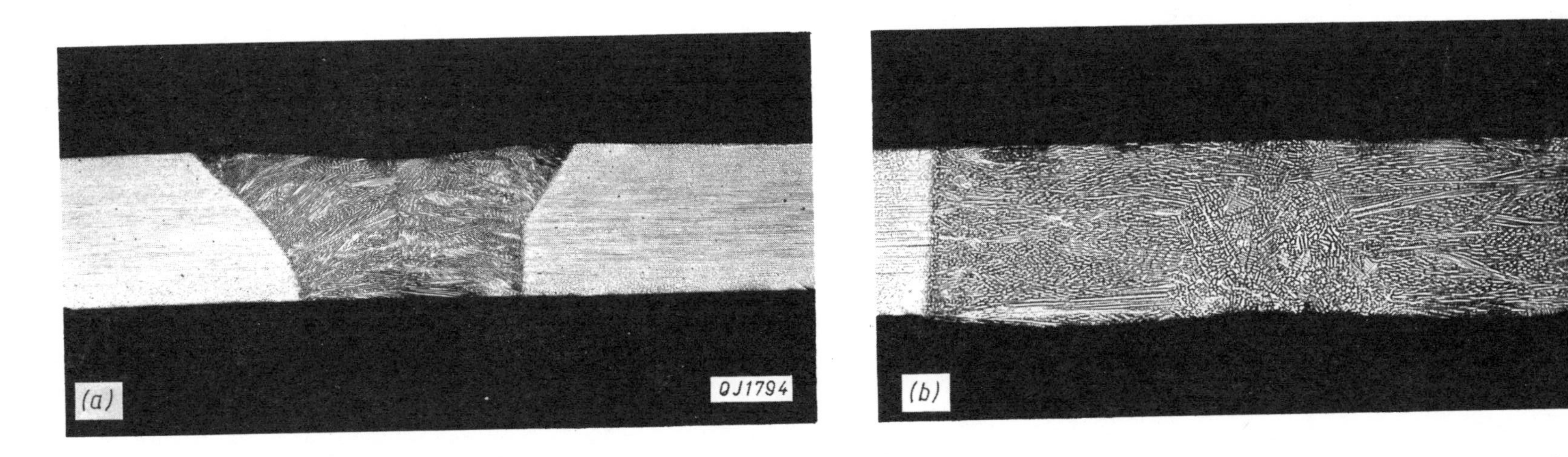

7 Macrosections of welds made with electrode inclined 45° forward, 26mm/sec: (a) minimum current, 80A, (b) maximum current, 190A. x25

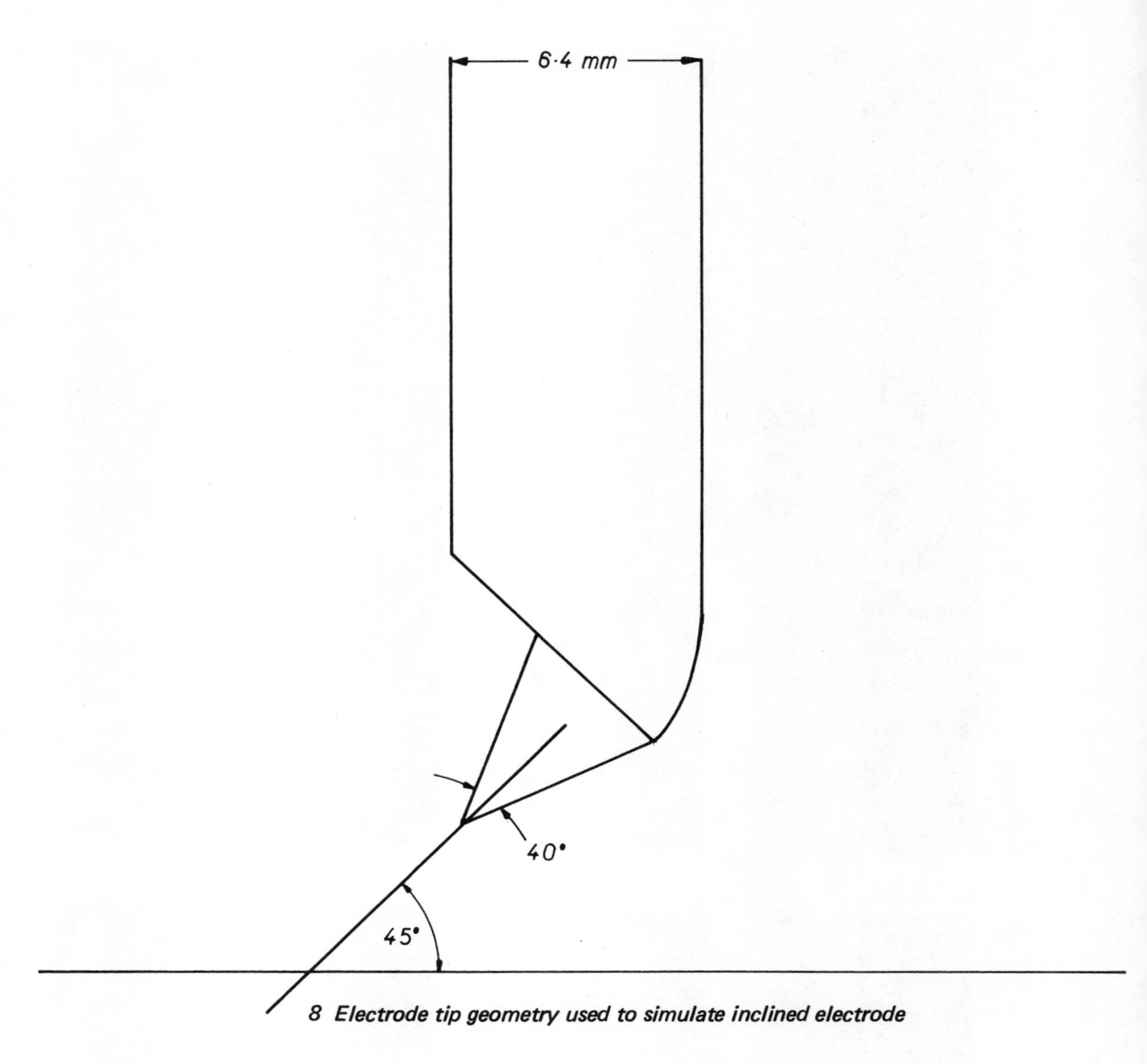

8 Electrode tip geometry used to simulate inclined electrode

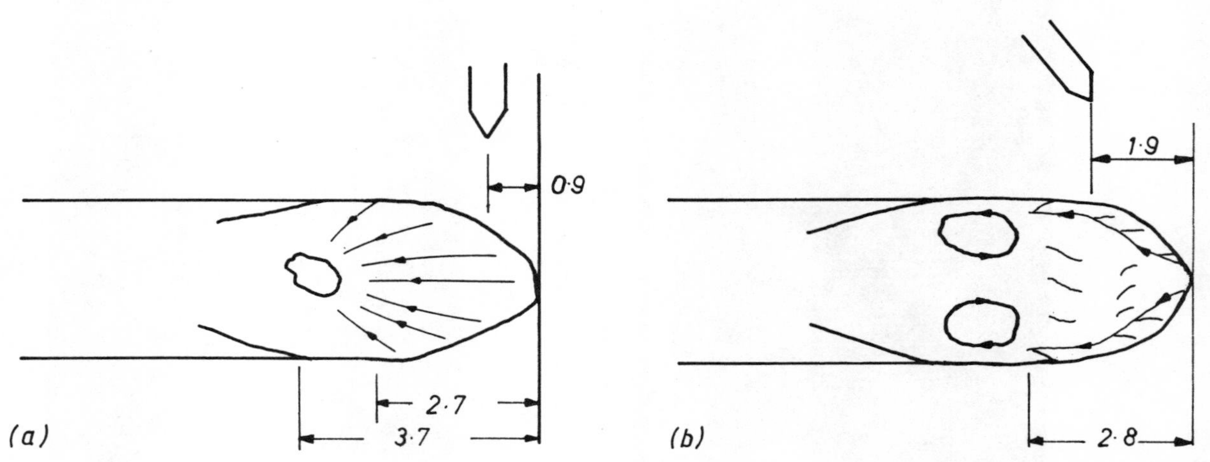

9 Weld pool circulation for TIG arcs (dimensions in millimetres): (a) vertical torch, (b) torch with 30° leading angle

Atmosphere control criteria for welding titanium

E.J.Morgan-Warren

Titanium is increasingly being used in the chemical and power industries for its outstanding corrosion resistance in certain chemical and marine environments. Although readily weldable, titanium reacts strongly with oxygen and nitrogen at elevated temperatures, leading to severe hardening and embrittlement. The use of automatic welding techniques enhances the necessity for a clear specification of welding conditions, including the quality of the shielding atmosphere. The application of gas analysis to the monitoring and control of shielding atmosphere during welding enables the atmosphere purity requirement to be specified. It is shown that for tube-to-tubeplate welding in commercially pure titanium, levels of atmosphere contamination below 0.1% are required to ensure that normal weld quality criteria are met. For the highest quality welds less than 0.02% air contamination would be required, which could not be assured by the absence of surface discoloration alone.

The effect of gas purity on the process operation is described, and the economic advantages of atmosphere control and the development potential are discussed.

INTRODUCTION

The use of titanium has broadened during recent years into the fields of chemical and power plant engineering. Its high resistance to oxidation, which results from the formation of a tenacious surface oxide film, renders titanium inert to many chamical and marine environments. In the power industry, titanium's resistance to marine corrosion has given rise to specification of the metal for condenser tubes in coastal installations.

Although titanium is not a difficult material to weld, its strong affinity for oxygen and nitrogen places very stringent requirements on the prevention of contamination during welding. Failure to maintain an efficient inert gas shield around the heated and molten material results in absorption of these two atmospheric gases and consequent weld embrittlement.

With manual welding the necessary inert gas coverage is usually achieved by means of trailing gas shields or glove boxes, and quality is monitored by checking discoloration on test runs.

In this Paper a technique to control shielding gas purity based on continuous analysis of the shielding gas is described. The application of this technique to a specific tube-to-tubeplate welding development programme is outlined and the potential of the method for further development is discussed.

Mr E. J. Morgan-Warren is at the Marchwood Engineering Laboratories of the Central Electricity Generating Board.

ABSORPTION OF ATMOSPHERIC GASES BY TITANIUM

The atmospheric gases — oxygen and nitrogen — dissolve interstitially in alpha titanium to the extent of 14.5 and 9% respectively, and 0.5% is sufficient to cause severe (up to 95%) loss of ductility because of solid solution hardening.[1] In solid titanium, absorption takes place initially by surface reaction to form the oxide or nitride and is linearly time-dependent. A film soon forms whereafter diffusion-controlled absorption is observed. This is governed by the characteristic parabolic rate law

$$W^2 = Kt + c \qquad [1]$$

where W = weight of contaminant absorbed
t = time
K, c = constants

The reaction rate is exponentially dependent on temperature, so that, below 300°C, absorption is minimal but above this temperature it rises steeply. For example, experimental measurements have shown the absorption rate to rise fifteenfold when the temperature is increased from 300° to 600°C.[2] In the weld heat-affected zone (HAZ) absorption would occur to a depth dependent on the thermal cycle, but in the fused zone, absorption of impurities would be aided by the movement of the molten metal so that the whole cross-section of the weld bead would be equally contaminated. The degree of contamination would then depend not only on the atmosphere purity, absorption rise, and thermal cycle, but also on the surface-to-volume ratio of the weld pool.

Surface oxidation which gives rise to progressive discoloration is much used as an indicator of weld contamination. However the discoloration shows only surface oxide which has formed in the solid state, and does not relate directly to the more important bulk contamination which is chiefly responsible for poor weld properties. It is apparent, therefore, that the theoretical analysis which may predict gas absorption behaviour in simple systems is insufficient for the complex welding situtation. The level of weld metal contamination arising from a given atmosphere impurity level, and the effect of this on mechanical properties, must therefore be determined by experiment. In practice it is a knowledge of the direct effect of atmosphere contamination on weld properties that is required, so that a shielding gas purity requirement can be established.

EQUIPMENT AND EXPERIMENTAL TECHNIQUE

The aim of the experimental programme was to determine the required purity of shielding atmosphere for the welding of tube-to-tubeplate joints in titanium. Furthermore, the sensitivity of surface discoloration, as a check on weld quality, was to be examined. The equipment consisted of a divided chamber, Fig. 1, in which controlled atmospheres could be applied separately to both sides of the tube-to-tubeplate assembly. It was thus possible to simulate a complete in-chamber situation with gas shielding all round the joint, or a procedure in which shielding is applied to the face side only. All-round gas coverage was used throughout the present series of experiments. Welding was carried out by the TIG process using an orbital head driven via a shaft through the lid of the chamber.

The required atmosphere was obtained by partially evacuating the chamber, refilling, and purging with the appropriate gas mixture set up using gas flow meters. The atmosphere in the chamber was continuously checked by sampling through a capillary tube and analysing by means of a quadrupole mass spectrometer. An oxygen analysis was used as the basis for the air content of the gas. The equipment, comprising the chamber, flow meters, and mass spectrometer is shown in Fig. 2.

Test specimens consisted of commercially pure titanium tube 25.4mm OD, 0.71mm wall, face welded into commercially pure titanium bushes 25mm thick, to simulate tube-to-tubeplate welds. Typical welding conditions were: current, 25A; travel speed, 75mm/min; electrode, 1.6mm diameter, 2% thoriated tungsten. Accordingly welds were produced under shielding gases consisting of high purity argon, commercial purity argon, and argon-air mixtures in the range 0.05 to 1% air. The specifications for commercial purity argon and super purity argon are shown in Table 1. The influence of air contamination on the welding process was observed, and the effects on the appearance and properties of the welded joints were measured.

RESULTS

Welding process effects

Noticeable effects of gas shielding on the welding process characteristics were confined to the very pure and very impure atmospheres. It was found that, under high purity argon shielding, arc initiation was not readily achieved by means of the high frequency spark starter. Several attempts were often required

Table 1 Specifications for commercial and high purity argon

	O_2	N_2	H_2	H_2O	Hydrocarbons	Argon
High purity argon	<2vpm	<10vpm	<1vpm	<2vpm	<1vpm	>99.998%
Commercial purity argon	<5vpm	<20vpm	<1vpm	<4vpm	<1vpm	>99.996%

to start the arc. High purity argon contains reduced levels of oxygen and nitrogen (see Table 1), these gases having a somewhat lower ionisation potential than argon. The values are: oxygen, 13.62eV; nitrogen, 14.53eV; argon, 15.75eV. The shortage of the more easily ionised gases may have been sufficient to hinder the establishment of the arc column. Once initiated, however, no lack of arc stability was observed. At the other end of the impurity scale, when 1% air was present, some evidence of oxidation of the tungsten electrode was found. At other purity levels, i.e. in the range from commercially pure argon to argon + 0.5% air, no abnormal arc behaviour was observed.

Metallurgical effects

The degree of contamination exerted a far greater influence on the properties of the welds. The weld bead hardness commonly forms the basis of acceptance standards and surface discoloration is much used as an indicator of inadequate protection. Thus the effects of contamination on these properties was measured. The relationship between the contamination and weld bead hardness is shown in Fig. 3. Theoretical considerations lead to the deduction that the degree of hardening is proportional to the square root of the impurity level. A regression analysis shows that this relationship is obeyed very closely in the present situtation, the equation being:

$$H = A + B\sqrt{C}$$

where H is the Vickers hardness HV and C is the air concentration in the shielding gas (vol. %) for the present data, A = 213, B = 97, and the correlation coefficient, r = 0.997, indicating a statistical significance level better than 99.9%. The values of A and B are specific to the joint geometry and conditions used. Acceptance standards for weld bead hardness vary, but a common criterion for commercial purity titanium welds is a maximum hardness of 250HV. On this basis a maximum value for the air content of the shielding gas is 0.1%. On the more rigorous but nevertheless often-used standard of a maximum hardness increase of 30HV over the parent material, a maximum impurity level of 0.02% air would be required.

The degree of weld surface discoloration was found to increase systematically as the air content of the shielding gas was raised, Table 2. With both super purity argon and commercial purity argon no noticeable discoloration occurred. Although 0.05% air gave a pale straw colour, with lower levels however, discoloration would hardly be discernible. Thus colour alone would not be sensitive enough as an indicator for the highest quality welds.

DISCUSSION

The sensitivity of titanium to atmosphere purity during welding is well known; however the extent of this sensitivity from a technological point of view is less well appreciated. In this work the technique of atmosphere control has been used to quantify the effect of air contamination in a specific instance of tube-to-tubeplate welding.

Gas shielding requirements for titanium welding

The experiments described have confirmed the need for stringent precautions when welding titanium and quantified the purity required to achieve a given level of metallurgical quality. Thus, for example with the welding geometry and materials used in this investigation, a

Table 2 The effect of air contamination in shielding gas on surface discoloration

Contamination, vol. % air	Surface colour
0	Bright silver
0.05	Pale straw
0.1	Straw
0.2	Deep straw/some blue
0.5	Deep straw/blue HAZ
1.0	Deep grey blue

level of air contamination less than 0.1% is required to keep the bead hardness below 250HV. This is a lower level than has been suggested in some previous investigations; the reason for this is almost certainly the high surface-to-volume ratio associated with tube-to-tubeplate face seal welds. It is noteworthy that, in this investigation, high purity argon gave no noticeable advantage over the commercial grade, which reinforces the view that the source of gas shielding problems is more likely to lie in the system than in the gas cylinder itself.

From a welding process point of view high purity argon was found to give rise to arc starting difficulties, but purity levels satisfying both process and metallurgical requirements were not found to be difficult to achieve.

Implementation of gas shielding requirements

In the present investigation welding was carried out in a chamber. The attainment of the required atmosphere and the maintenance of its quality during welding was verified by gas analysis. For production welding, the use of a chamber is to be recommended where possible; a number of techniques may be employed to maintain the atmosphere quality depending on the degree of control required. Most simply the system may be purged for a predetermined length of time before welding. A first order estimate of the required purging time may be made using the following equation

$$t = \frac{V}{Q} \log_e \frac{Co}{C_1}$$

where t = purging time
V = volume of welding chamber
Q = rate of gas flow
Co = initial concentration of impurity
C_1 = maximum permitted concentration of impurity

Reliance on purging time alone, however, has a number of disadvantages. Firstly, the calculated time must be regarded as approximate and a substantial safety factor (up to 100%) would be required. Furthermore this technique gives no assurance that the required atmosphere is maintained during welding.

The technique of continuous sampling and analysis as used in the present experimental programme is probably the simplest which provides an assurance of quality. The technique could be developed to incorporate automatic comparison of the required and actual gas analyses, and coupled to a switch which initiates the welding operation when required conditions are obtained. The basic technique might be further developed to encompass such features as a reserve gas line which is automatically applied if failure of the gas supply is detected.

The investigation was carried out using a mass spectrometer as the analysing facility. This instrument is capable of high sensitivity (of the order of a few parts per million) and fast response (~15msec). The response of the analysing system as a whole is limited by the sampling tube, and for a tube 2m in length it is of the order of 0.1sec. The mass spectrometer is a versatile instrument and has the advantage of being applicable to a wide range of gases. The benefits must be set against the cost, and to monitor only one or two gases, e.g. oxygen, less expensive instruments such as the zirconium oxide cell may suffice.

Advantages of controlled and monitored shielding atmospheres

Among the advantages to be gained from controlled and monitored atmosphere welding are:

(a) monitoring shielding gas in the area of welding provides a continuous check on the gas shielding system and enables faults to be investigated at an early stage

(b) in titanium welding, gas analysis is a more sensitive quality indicator than weld surface discoloration

(c) controlled atmosphere welding offers the potential economic advantage of reducing the excessive purging times and gas flow rates which would otherwise be needed to ensure that the required atmosphere purity had been achieved and maintained

CONCLUSIONS

1 Atmosphere control enables gas shielding requirements to be specified quantitatively. For tube-to-tubeplate welding in titanium with 0.71mm thick wall tube, a shielding atmosphere containing less than 0.1% air was found to be necessary to ensure a weld bead hardness below 250HV. To ensure the highest standard (hardness increase of less than 30HV), less than 0.02% air would be required

2 Shielding gas analysis has been found to be more sensitive than surface discoloration as an indicator of weld contamination

3 The use of a high purity argon atmosphere was found to cause difficulty with arc

striking, and in the present example offered no noticeable metallurgical advantages over commercially pure argon.

4 Atmosphere monitoring obviates the need for prolonged gas purging with consequent economic advantages in terms of welding time and gas consumption

5 The technique is amenable to incorporation into automated welding systems to increase the quality level and consistency of welds.

ACKNOWLEDGEMENT
This work is published by permission of the Central Electricity Generating Board.

REFERENCES

1 JAFFEE, R.I. 'General physical metallurgy titanium reviewed'. J. Metals, February 1955, 247-52.

2 GULBRANSEN, E.A. and ANDREW, K.F. 'Kinetics of the reactions of titanium with O_2, N_2, and H_2'. J. Metals, 1 (8), 1949, Sect. 3 (transactions), 515-25.

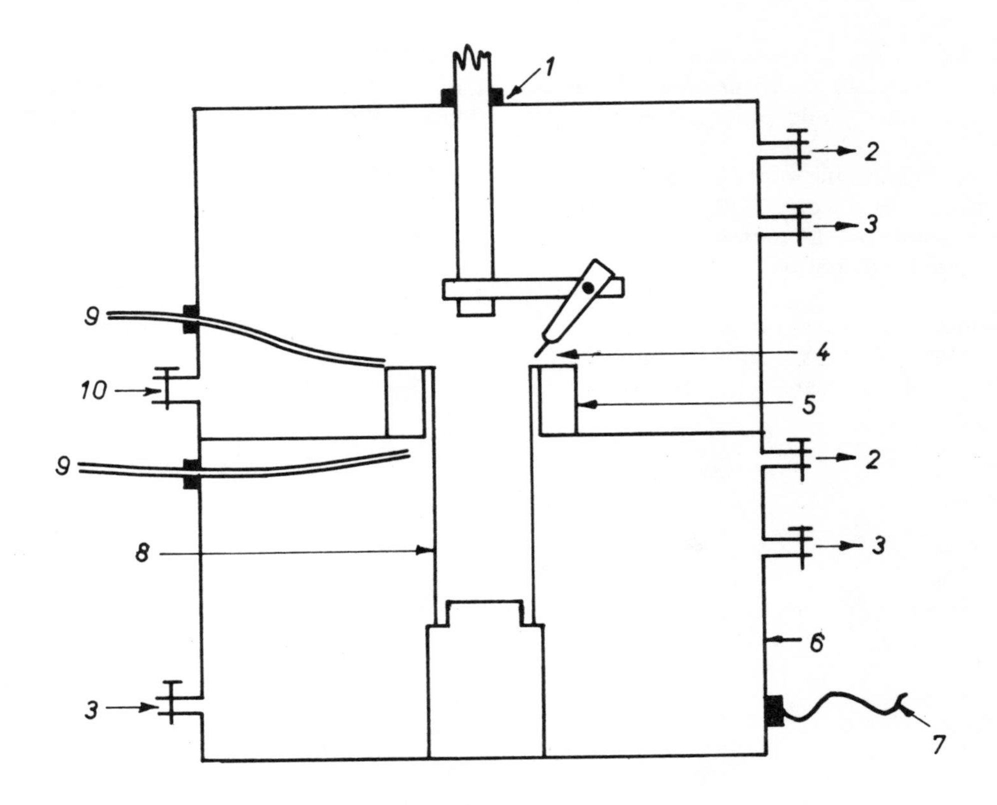

1 Schematic diagram of controlled atmosphere welding chamber. 1 — rotating seal; 2 — to vacuum pump; 3 — gas outlets; 4 — welding electrode; 5 — tubeplate sample; 6 — steel chamber; 7 — welding return; 8 — tube sample; 9 — to mass spectrometer; 10 — gas inlet

2 Experimental equipment comprising welding chamber, gas-mixing gauges, and mass spectrometer

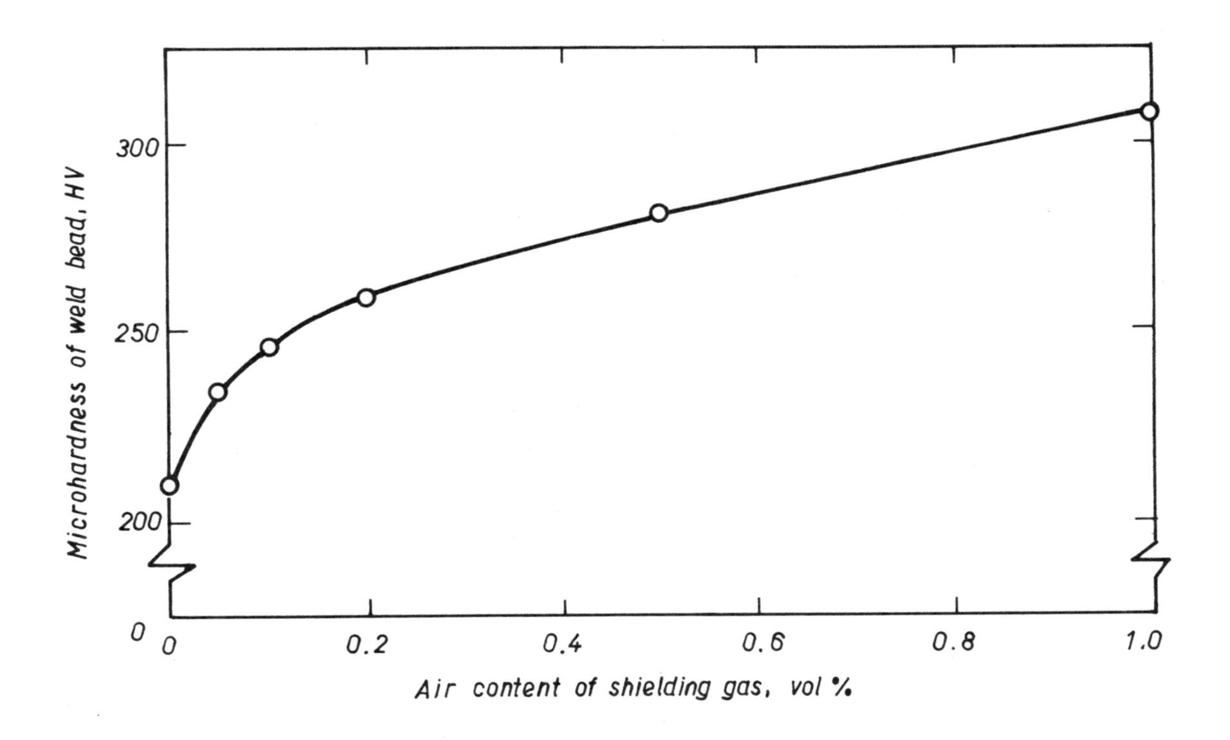

3 Dependence of weld bead hardness on shielding gas air contamination

PAPER 2

Narrow gap orbital welding

R.Hill, BSc, and M.R.Graham, MMet, FIM

A welding technique has been developed for thick-walled pipes using the pulsed TIG process and a deep narrow weld preparation. Welds have been made up to 110mm deep both downhand and in all positions, a single weld bead being deposited for each weld layer. A development programme to optimise weld preparation dimensions, welding parameters, and production procedures has been completed. Metallurgical evaluation and mechanical testing showed both the weld and heat-affected zones to have superior properties to 2Cr-1Mo/½CrMoV weldments made by other processes.

The prototype equipment, developed to demonstrate the feasibility of the process, was primarily suited to downhand welding. All-positional equipment is essential for production welding, and a commercial orbital welding system was modified to take the special narrow gap welding torch for full-scale laboratory tests. Following successful tests with this, a production welding head was designed and manufactured. Test welds made in ½CrMov high pressure pipes using this equipment proved satisfactory and it is now being used for production pipework welds.

INTRODUCTION

During the early 1970s all high pressure pipe welds at C.A.Parsons were made by the manual metal-arc (MMA) welding process. This relies totally on the skill of the welding operators and makes control of quality and productivity extremely difficult. To obtain welds of consistently high quality a portable automatic welding equipment, capable of operating in all positions, was considered essential and the development or availability of such systems was investigated.

The all-positional requirement eliminated both submerged-arc and flux-cored electrode processes as these can be used effectively only in the downhand position. This narrowed the possibilities to the MIG (metal inert gas), MAG (metal active gas), and TIG (tungsten inert gas) processes which have proven all-positional capability.

A market survey of commercial orbital TIG- and MIG-welding systems indicated that twelve were then available (1970) and that five of these could weld pipes greater than about 120mm diameter. Only one system used MIG-welding and the technical information indicated that it would be unsuitable as it could only weld pipes greater than 350mm diameter and, moreover, alteration of welding parameters was necessary during welding. The TIG process was considered to be most capable of producing mechanised all-positional welds of the quality required and this was used in the remaining

Mr Hill, Group Leader — Production Welding Group, and Mr Graham, Chief Metallurgist — Applications, are both with C.A.Parsons and Co. Limited.

four equipments. None were considered suitable in their existing form for operation at the high preheat and interpass temperatures required for welding 1/2CrMoV pipes, but were believed to be capable of further development.

NARROW GAP WELDING

The major drawback to TIG-welding is the relatively low deposition rate, and with standard weld preparations the process is uneconomic compared with MMA welding. To overcome this it was proposed to use a narrow, practically parallel-sided, weld groove. Such grooves had been used[1,2] with the MIG- and TIG-welding processes. The volume of such a narrow gap preparation in 60mm walled pipes would be only about 30% that for submerged-arc and MMA welding, Fig.1 and Table 1, and this reduction more than compensated for the lower weld metal deposition rate of the TIG process.

Table 1 Area of MMA, submerged-arc, and narrow gap weld preparations

Thickness, mm	Cross-sectional area, mm^2		
	MMA	Submerged-arc	Narrow gap
25	415	415	185
50	1290	1290	440
75	2410	1970	735
100	3750	3000	1080

It was proposed that metal deposition for narrow gap TIG-welding would be by multilayer single-pass beads. Control of the molten pool for all-positional operation was expected to be achieved by a combination of capillary surface tension between the sidewalls and the liquid weld pool, and the controlled pool solidification pattern obtained by the use of pulsed TIG-welding.

Further anticipated advantages of the process were lower distortion and residual stress levels as a result of the reduced weld volume, greater support of the weld by the parent material, and the ability to use high purity filler wires to give improved weld metal properties.

Process development

Before starting on a costly equipment and process development programme, tests were made to prove the feasibility of narrow gap welding.

Adequate gas coverage of the weld pool for TIG bead-on-plate tests was achieved with the tungsten electrode protruding 19mm beyond the gas cup, providing that a gas lens is used in the standard welding torch to produce lamellar flow of the shielding gas. Further extension of the electrode beyond the gas cup was considered possible for narrow groove welding, as the sidewalls, being close together, would channel the gas to the weld area.

Test welds in a 7mm wide groove between two 25mm square mild steel bars previously tack welded to a backing plate were unsatisfactory, however, because of porosity, even though the end of the gas cup was immediately above the upper surface of the bars and the argon gas flow rate more than 19 litre/min. As the groove was bridged satisfactorily by the weld deposit, further tests were carried out in an effort to overcome the porosity problem, which was attributed to inadequate gas coverage. The porosity was found to be caused by air entrainment arising from turbulence and the level was reduced in subsequent welds by lowering the shielding gas flow rate to 14-16 litre/min. In these tests, however, extensive distortion, caused by contraction during weld solidification and the low restraint of the test assemblies, closed the groove and prevented full depth welding. To increase the restraint and obtain more relevant information on the feasibility of the process, groove-in-pipe test welds were made. Grooves 7mm wide by 16mm deep were machined in 19mm wall x 250mm OD steel pipe, and a series of welds made by pulsed TIG-welding using mild steel (1.2mm diameter) filler wire. The welds were made in the downhand position by rotating the pipe below a fixed tungsten arc torch. The welding parameters investigated included peak and background current, pulsation frequency, voltage, welding speed, filler wire feed speed, tungsten electrode tip geometry, shielding gas composition, and weld preparation dimensions.

Lack of sidewall fusion and porosity in these welds were both attributed to low heat input and inadequate shielding gas coverage. The lack of sidewall fusion was overcome by changing the shielding gas from argon (99.99%) to a 99% argon-1% hydrogen mixture to increase the arc voltage for a given length and hence to increase the heat input for given welding conditions. This mixture was adopted for all subsequent work. Porosity was reduced to an acceptable level in shallow (13-19mm) weld grooves when this gas mixture was used but increased to an unacceptable level as the groove depth was increased.

To improve the efficiency of the gas shielding, alternative gas cup designs were investigated. Ceramic designs completely surrounding the electrode to within the weld pool area were unsuccessful because of limited access and cup failures arising from high temperature gradients during the welding cycle. Auxiliary gas flow external to the gas cup was also unsuccessful, as turbulence occurred. A method which did ensure adequate gas shielding was to fit thin-walled (0.8mm) stainless steel U channel side legs at the exit from the gas cup. The legs, which fitted into the weld groove, were 1.6mm narrower than the groove and their length varied with the weld depth. To maintain effective gas coverage the end of the gas cup had to be kept within 5mm of the weld surface; this necessitated frequently stopping welding to fit progressively shorter legs. The final solution to the problem of producing adequate gas cover during continuous welding was to use long side legs surrounded by a stainless steel telescopic shielding arrangement, Fig.2.

With the optimised welding parameters, a series of welds was made in ½CrMoV pipe of 63mm wall thickness and 230mm bore, using 2Cr-1Mo filler wire. Initially a filler wire designed for MAG-welding was used but, after temper embrittlement of the weld deposit was observed, an alternative filler wire, Oerlikon SD2 2½Cr-1Mo, was chosen for subsequent test welds. Three of the welds, one made with the MAG wire and two with the SD2, were subjected to metallurgical evaluation, mechanical testing, and residual stress measurement. Non-destructive testing showed some of the welds met BS 2533 : 1973 requirements.

Having proved the feasibility of the narrow gap welding process for downhand welding, further tests were made to establish that welding could be done in all positions. Welds up to 60mm deep were made in the vertical and horizontal positions by attaching the weld torch and wire feed to a rotator and orbiting this around fixed pipes. The quality did not meet BS 2633 : 1973 requirements because of porosity and sidewall fusion defects resulting from difficulties in operating the equipment, and from not having an automatic arc length control system. The technique (Patent applied for)[3] was however considered capable of producing acceptable welds, and improved equipment was therefore sought.

Equipment development

The equipment developed for the process feasibility tests was unsuitable for detailed evaluation tests and production welding trials because of its bulk, lack of an arc length control system, and inability to rotate around a fixed pipe under its own power. The design and manufacture of a suitable welding system in-house was considered but it was decided instead to purchase and modify an existing system. An orbital welding head, control system, and a 400A pulsed TIG-welding rectifier were purchased together with a welding torch to fit into narrow groove weld preparations, so that welds up to 150mm deep could be continuously executed.

The equipment package, Fig.3a, consisted of a welding head with a TIG torch attached to a control box containing an electronic arc voltage control unit and a wire feed system. These were attached to a four-wheeled carriage which fitted astride a segmented metal link guide belt previously located at a fixed position from the weld to ensure accurate rotation. The assembly was driven around the pipe by a spring-tension belt and roller-drive system powered by an electric motor fixed to the carriage.

The equipment reliability was poor, primarily because of the high equipment temperatures attained during welding at up to 350°C preheat. Insulated aluminium heat shields were fitted to reduce heat transfer by radiation from the hot pipe and a water-cooling coil was fitted round the carriage drive motor. The fabric drive belt supplied was replaced by a stainless steel mesh belt which worked satisfactorily for a short period. Further modifications were found essential to the drive gearing system to give consistent rotation of the equipment around the pipe and, most importantly, the belt drive system was replaced by a chain and sprocket arrangement, Fig.3b. Subsequent testing showed the equipment to operate satisfactorily but the design was not sufficiently robust for production welding.

PRODUCTION WELDING

A production welding system, of simple, robust construction, Fig.4, was designed and manufactured in collaboration with Weldcontrol (Devon) Limited. The segmental guide belt of the development equipment was replaced by split, fixed-diameter, machined rings having close tolerance machined grooves into which the guide wheels on the welding head and drive carriages were located. The rings fit clear of the pipe to minimise heat transfer from the hot pipe, being located by spring-loaded adjustable screws. The welding head and drive carriages are separate and located diametrically opposite each other by quickly detachable link arms. This separation reduces the out-of-balance loading on the drive motor. The welding torch, the arc length control system, and wire feed arrangements were also modified and both carriages were fully insulated to minimise heat buildup. The fixed-diameter guide ring system was considered the best as it eliminated difficult setting up operations. To cope with pipe diameters from 180 to 650mm, three guide rings and associated link arms are necessary each having a range of 150mm below the fixed diameter. Setting up the rings in relation to the weld preparation is of prime importance and must be accurate within ±0.8mm of the vertical plane through the weld. To achieve this accuracy easily, three detachable locating arms have been fitted to the rings at 120° intervals. Concentricity should be within ±1.6mm to ensure that the gas surface cover and arc radiation protection shield remain close to the bead surface throughout welding, minimising the possibility of air entrainment into the argon gas stream and eye damage caused by the high intensity arc.

Preproduction tests were satisfactorily completed and a procedure test was made in ½ CrMoV pipework using 2Cr-1Mo welding wire and qualified to BS 4870 : 1974. The procedure was approved by the CEGB for welding turbine steam high pressure pipework and the process was used to weld two pipework loops, four welds per loop, in 325mm OD x 40mm wall ½CrMoV material. All welds were examined by radiography and ultrasonics prior to and after furnace stress relief.

The narrow gap process was thus proven to be capable of operation in a production environment, Fig.5. Average 'arc' time for these welds was 6.5hr with an average through-put time, i.e. setup, preheat, and weld, of 12hr. The process is now in normal production use the Company's pipe shop and an agreement has been signed with Clarke Chapman - John Thompson Limited to market the process through their Weldcontrol subsidiary. Further development work is in hand to investigate the use of photodiode feedback control for root fusion and the welding of stainless steel and other alloys.

ADVANTAGES OF NARROW GAP WELDS

Metallurgical

Metallurgical examination and mechanical testing of welds made in ½CrMoV pipe material using the correct type of 2Cr-1Mo filler wire showed that the ambient and elevated properties were at least equal and in most instances superior to welds made using the MMA and submerged-arc welding processes. Figure 6 shows the as-welded surface weld bead appearance and a typical macrosection of a narrow gap weld. The regular appearance of the weld and weld heat-affected zones (HAZs) is readily apparent.

During the development programme the effect of preheat temperature on the dimensions and microstructure of narrow gap welds was investigated using standardised welding parameters and preheat and interpass temperature ranges of 250° to 300°C, 100° to 150°C, and 15° to 50°C. Increasing the preheat temperature lowered the hardness of the weld and HAZs. The weld metal and HAZs of the unpreheated weld had mean hardnesses of 370 and 285HV, reducing to 345 and 270HV respectively for the 100° to 150°C preheat and to 320 and 250HV for the 250° to 300°C preheat.

Increasing the preheat temperature increased the width of the HAZs and the amount of grain refinement in them, except in the capping pass region. None of the sidewall zones had any totally unrefined, coarse-grained, bainite of the type under the capping pass, but the welds made without preheat and at the lower preheat temperature contained partially transformed material. The weld made at the highest preheat had an almost homogeneous fine-grained HAZ. The increase in preheat had a similarly beneficial effect of the weld metal refinement, the unpreheated weld metal containing very little refined material but the amount increasing with increasing preheat until, at 250° to 300°C, a very high proportion was refined, Fig.7.

Test welds were made using standard welding conditions and the capping pass was remelted up to three times using standard and reduced heat input conditions, to determine the effects of such remelting on the refinement of the weld metal structure and associated HAZs. Remelting using standard conditions had no effect on the grain size, as the same amount of melting occurred each time and the same HAZ was regenerated. Remelting with 30% reduction in heat input with the same pulsing conditions reduced the amount of remelting and improved the degree of refinement. With steady current TIG-welding no refinement occurred at heat input levels approximately 75% of the original, but progressive reductions to 60% and 40% on subsequent remelts gave increasing grain refinement.

Residual weld stresses

Surface residual weld stresses were determined on a weld made using the SD2 2½Cr-1Mo filler wire and ½CrMoV pipe. The stresses in the as-welded and stress-relieved (700°C for 12hr) welds were measured using stress relaxation by spark erosion. In the as-welded condition the axial and circumferential surface stresses were both compressive, Fig.8, unlike those for a 50mm thick-walled ½CrMoV weld made using MMA 2Cr-1Mo electrodes. Such compressive stresses will be beneficial in preventing the initiation and propagation of cracks during stress-relief heat treatments. After stress relief the axial stresses remained slightly compressive and were similar in magnitude to those for the MMA weld; the circumferential stresses became slightly tensile.

Nondestructive testing

The radiographic inspection of narrow gap welds is straightforward as the defects in the weld will be either lack of sidewall fusion or porosity. The alignment of the weld preparation is ideal for revealing lack of sidewall fusion on the radiograph.

To check that the narrow gap welds can be adequately inspected by ultrasonic methods a test weld was made having deliberately introduced lack of root fusion with associated extension cracking and lack of sidewall fusion defects. All the features gave good ultrasonic responses and were readily detected by this inspection technique.

CONCLUSIONS

The pulsed TIG narrow gap welding process has been developed and is now used as a production welding technique to weld carbon and low alloy steel pipework, and the equipment is being marketed.

REFERENCES

1 VAGNER, F.A. and STEPANOV, V.V. 'An automatic pulsating arc welding machine for unrotated butt joints in thick-walled tubes'. Welding Production, 14 (8), 1967, 55-60.

2 US Patent no. 3 328 556, 1967. Battelle Development Corporation 'Process for narrow gap welding'.

3 British Patent no. 1 476 321, 1977. Reyrolle Parsons Limited 'Welding by fusion'.

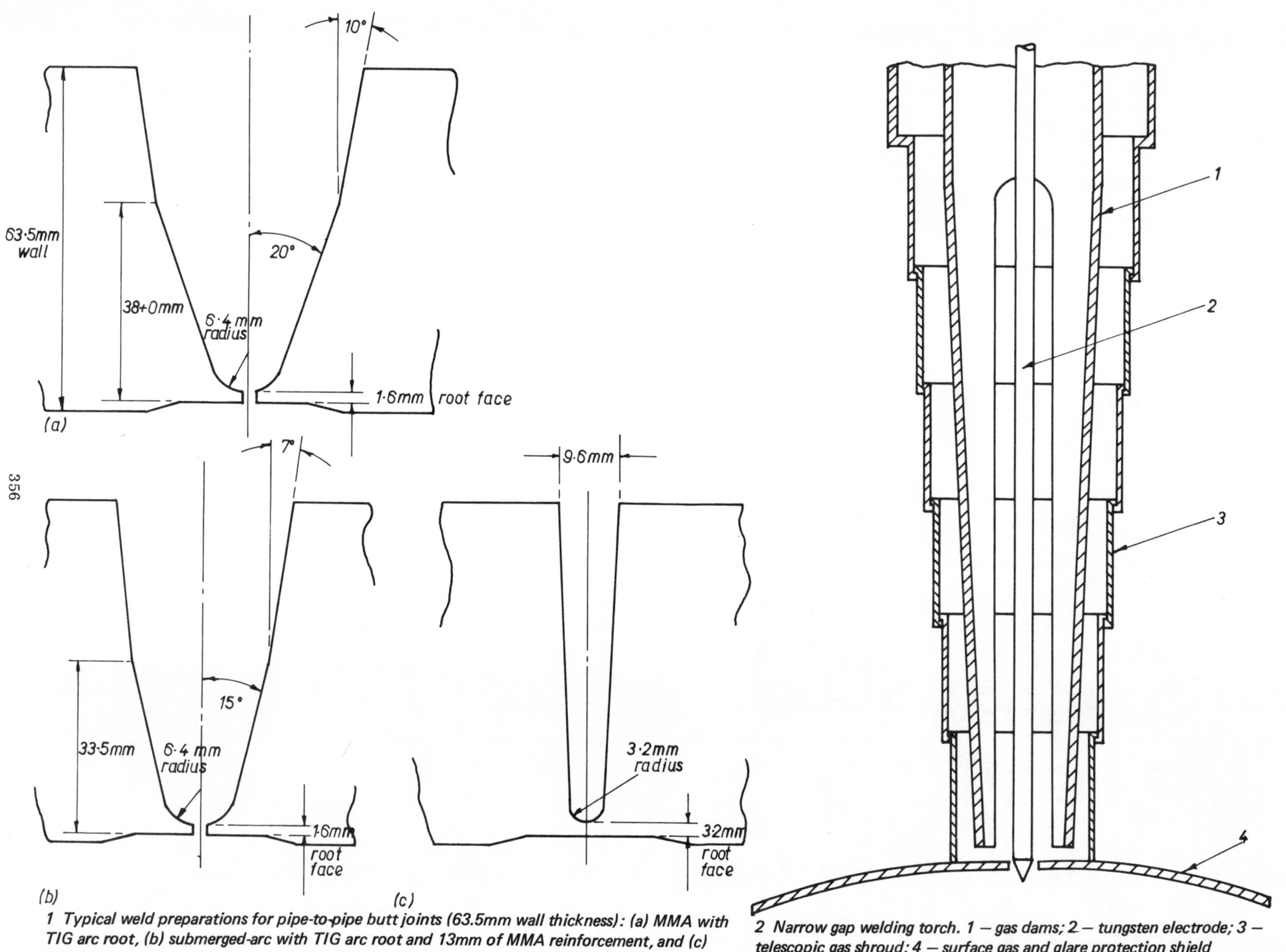

1 Typical weld preparations for pipe-to-pipe butt joints (63.5mm wall thickness): (a) MMA with TIG arc root, (b) submerged-arc with TIG arc root and 13mm of MMA reinforcement, and (c) narrow gap

2 Narrow gap welding torch. 1 – gas dams; 2 – tungsten electrode; 3 – telescopic gas shroud; 4 – surface gas and glare protection shield

3 Prototype narrow gap orbital welding equipment: (a) belt-driven weld head, (b) modified chain-driven weld head

4 Production welding equipment

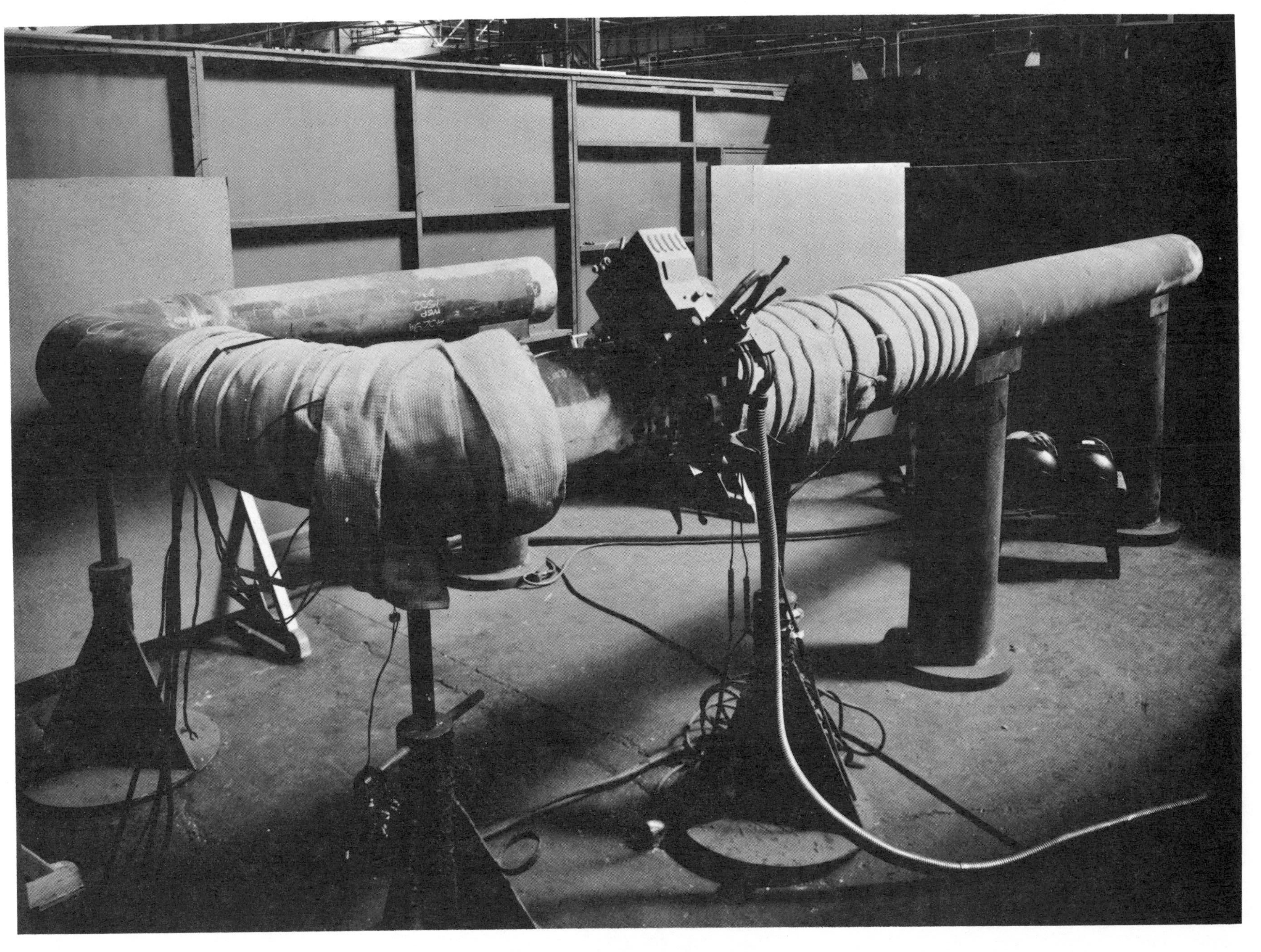

5 Narrow gap production welding area

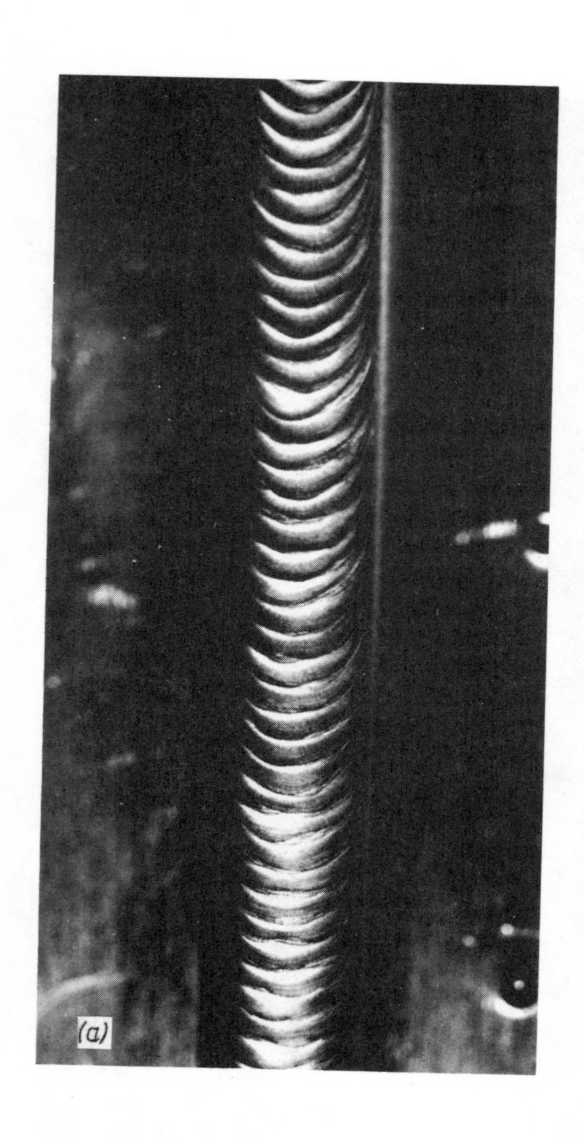

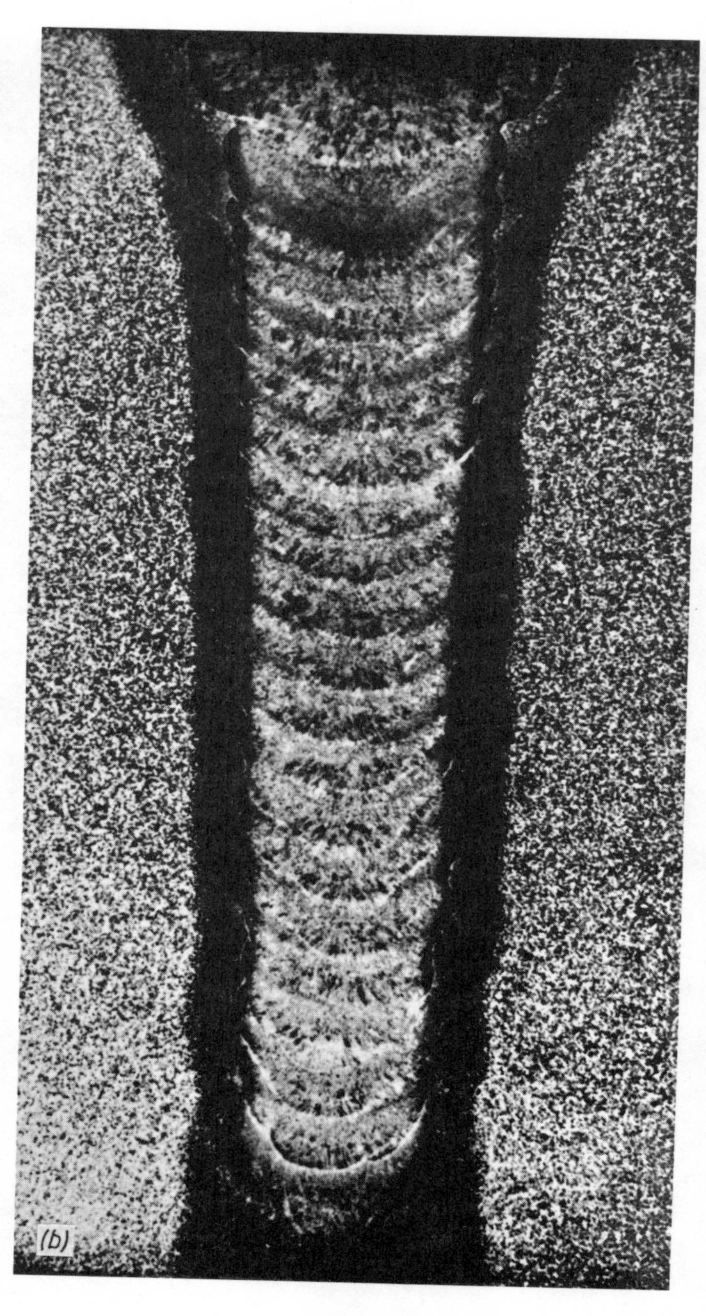

6 Surface and cross-section of typical narrow gap weld: (a) surface weld bead (x 2), (b) macrosection (x 4)

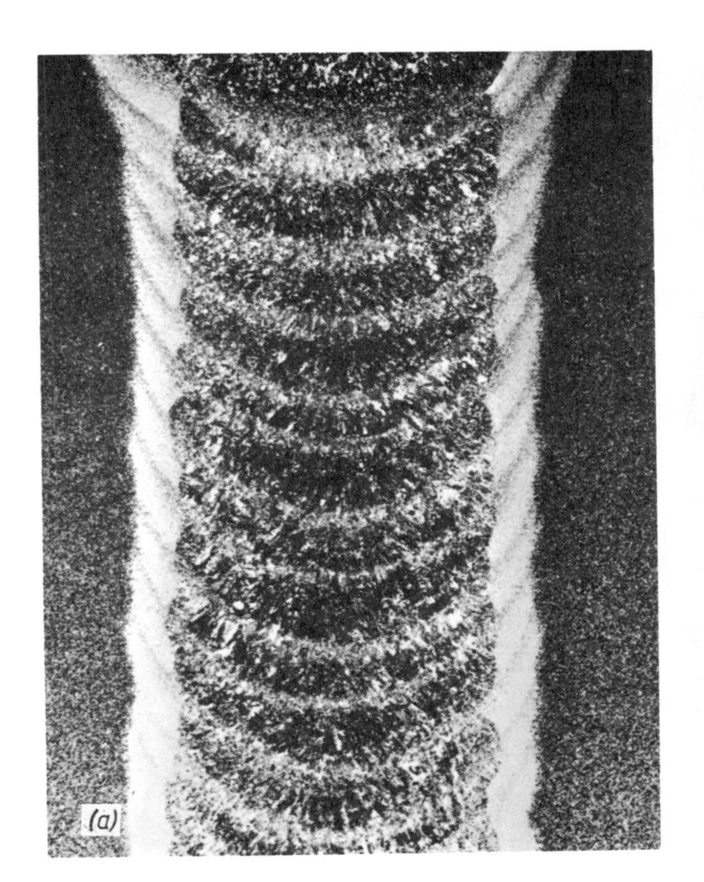

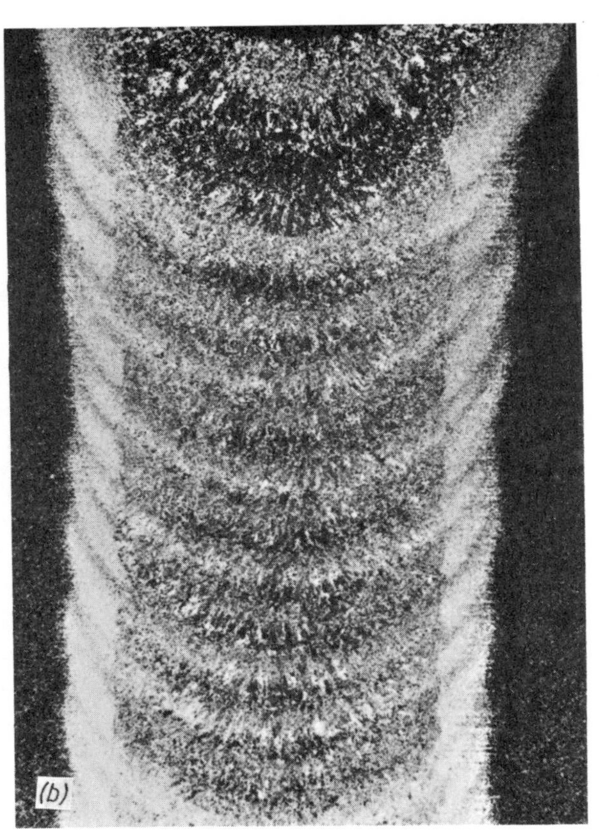

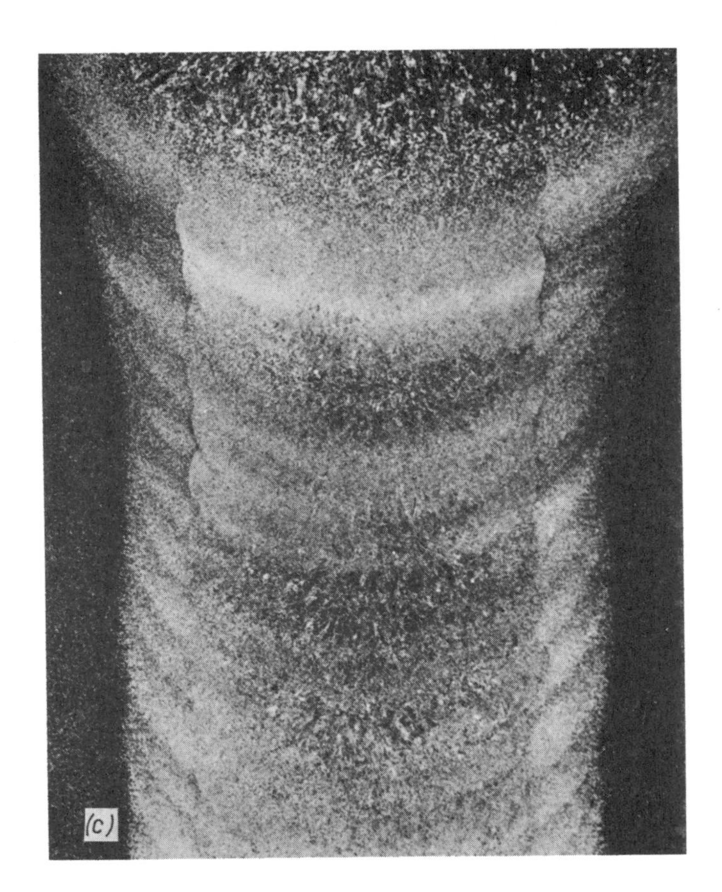

7 Effect of preheat temperature on weld and HAZ grain refinement: (a) no preheat, (b) 100° to 150°C preheat, and (c) 250° to 300°C preheat × 7

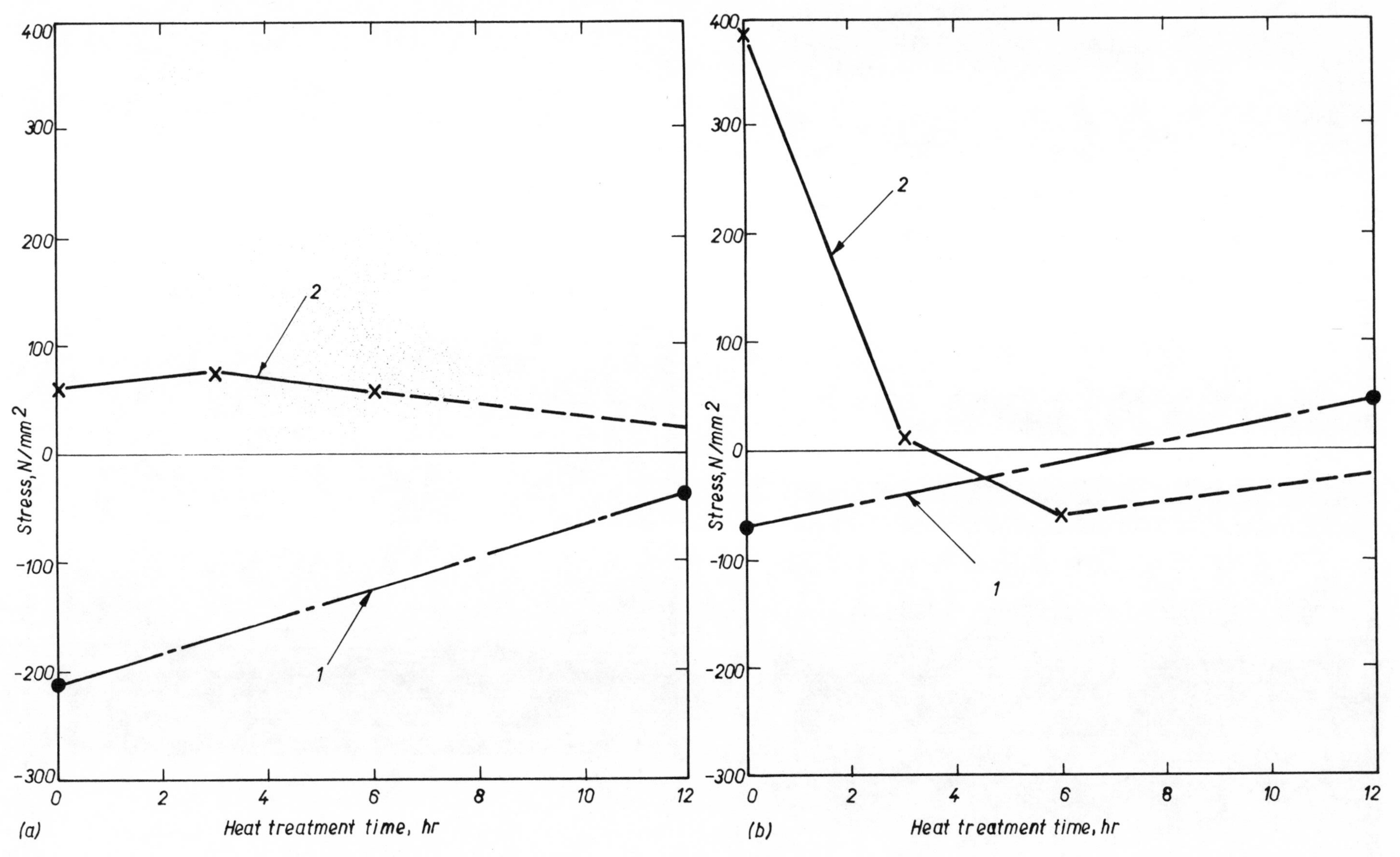

8 Surface residual stress in narrow gap (1) and MMA (2) welds: (a) axial, (b) circumferential

New trends in narrow gap welding

A.Ducrot, Engineer CNAM, M.Koffel, and G.Sayegh, DSc, DEng

Developments in narrow gap welding techniques were recently undertaken to improve the industrial possibilities for this process.

These developments were oriented towards rendering the process more adapted for certain practical applications where welding in all positions is not indispensable. This was obtained by increasing the gap to 14-20mm, thus rendering the welding head sturdier and the equipment more reliable. Even though the volume of the deposited metal is larger than in the first narrow gap version, which uses a gap of 8mm, it is shown that by using a third torch in the welding head the total welding time is not increased dramatically.

Oscillating the welding arc, with controlled amplitude and frequency, appeared to be very useful to obtain good weld quality. This can be achieved on one or two torches. When two oscillating torches are used, tolerances on the gap can be as high as $^{+7mm}_{-0mm}$.

Finally, comparison between the different narrow gap welding techniques is presented where it is shown that the choice of narrow gap equipment, for a given application, should be a judicious compromise between the advantages of the different new versions and their capabilities.

INTRODUCTION

One of the objectives of narrow gap welding was to accomplish welds in all positions. This necessitated a small molten pool and consequently the narrowest gap between the workpieces. A gap of 6 to 8mm was chosen which implied the use of small, tiny welding heads. The first results obtained on such equipment created an undeniable interest in the process, and samples with excellent metallurgical qualities were obtained.

Looking at other potential applications of the process it is apparent that many do not need all-position welding. So for these applications use of a very narrow gap, which is indispensable for welding in all positions, is not a necessity. Conversely, the use of a very narrow gap for heavy gauges introduces the disadvantages of relatively fragile welding heads, which are unnecessary for this specific application, and presents difficulties of access for eventual repair.

After a review of the main characteristics of the first generation narrow gap machines, this Paper describes recent technological investigations to improve the sturdiness of the equipment. These investigations were directed towards increasing the width of the gap and using an oscillating arc.

It can be said that there should not be a universal narrow gap machine for all application: according to the industrial application considered, appropriate equipment can be designed

M. Ducrot, Metallurgical Laboratory, M. Koffel, Arc Welding Laboratory, and Dr Sayegh, Scientific Manager, are all with Sciaky SA, Vitry-sur-Seine, France.

with optimised gaps to take full advantage of the process and to alleviate the constraints imposed by a universal machine.

FIRST GENERATION NARROW GAP WELDING EQUIPMENT

Designed for welding in all positions, narrow gap welding can produce a molten pool which adheres to the previously deposited layers, even upside down as in the overhead position. That is why the gap between the pieces to be welded and the different elements of the welding head (torch, gas supply, wire feed) have been designed to produce the desired fusion. [1] The consequences are:

(a) small gap of 8mm necessary for a stable deposit of the molten wire
(b) two torches operating simultaneously and symmetrically to reduce the welding time
(b) feed wire prestressed and formed outside the workpiece

First results obtained by the narrow gap process were apparently, because of its numerous advantages, of great interest to heavy industry. Many laboratories were equipped to explore the possibilities of the process and to define the conditions of its exploitation in an industrial environment. [2] Results have shown that this welding process can respond favourably to the industrial constraints of welding in all positions and thicknesses as great as 200mm were welded, Fig. 1. However, this welding process designed to operate in all positions has in consequence the following requirements:

1 A tight restraint on the width of the gap, resulting from the need to operate in all positions. This means the employment of small, tiny, and fragile welding heads (torch, sensor, gas nozzle)
2 Difficulty of repairing because of the difficulty of introducing tools into such a very narrow gap

On closely examining the possible industrial applications, Fig. 2, it can be appreciated that welding in all positions is not always necessary (welding in flat position, welding pieces on a positioner beneath or in front of a torch). It is not, therefore, indispensable to operate with the very narrow gap (8mm) which was necessary for welding in all positions with the corresponding constraints which are implied.

To satisfy the specific needs for these last-mentioned applications other versions of the narrow gap process have been developed which can be adapted particularly to welding in the flat position, which latter covers a large domain of industrial application.

MAIN OBJECTIVES OF THE NEW DEVELOPMENTS

The most critical point in the industrial usage of narrow gap welding for thicknesses exceeding 100mm was found to be related to the width of the gap. So the first objective of the new development was to increase the gap to 16 and even 20mm which produced the following consequences on the operating conditions:

(a) relaxation of the tight tolerance on the gap; tolerances of $^{+3mm}_{-0mm}$ became acceptable
(b) possibility of using sturdier welding head (torch, gas supply)
(c) possibility of stressing the wire near the weld, thus eliminating the difficulty met on the first machine in obtaining regular shape of the wire when it was prestressed outside the workpiece
(d) possibility of introducing a tool into the joint for machining in the event of a weld repair, and when for one reason or another a defect is present in the weld
(e) possibility of depositing the wire symmetrically in the gap

On the other hand, increasing the gap involves an increase in the quantity of molten metal required to fill it.

These conditions led the new development to be oriented in several directions:

1 The use of a welding head with three torches, permitting operation in all positions as in the previous condition but with a wider gap
2 Use of a welding head with one or two oscillating torches for applications where welding in all positions is not required (welding in the flat position, welding workpieces fixed on a positioner which rotates beneath the torch)
3 Oscillating the arc and using CO_2 gas shielding appeared to be very beneficial in improving the operating conditions of the equipment

RESULTS AND DEVELOPMENT TRENDS

Welding head with three stationary torches

This is a similar solution to the first narrow gap equipment, with an additional axial torch which deposits the feed wire between the lateral beads. The arrangement for shielding gas protection is identical to that used in the first version.

The operating conditions in this version are similar to those obtained on first-generation equipment. Nevertheless, advantage can be taken of the relatively wide gap between the workpieces to introduce sturdy torches with curved extremities enabling the feed wire to be directed towards the deposit point. Such equipment can comprise:

One orientable torch welding successively the three passes in single layers
Three torches welding simultaneously and oriented one to the right, another to the left, and the third to the middle

It should be noted that, when operating with three heads, it is possible to weld in all positions because the molten zone is very small. Figure 3 shows a macrograph of a weld in a low alloy steel accomplished with a three-torch machine.

Welding head with one torch, with wire bending and arc oscillation

Increasing the gap has permitted a system to be employed which strains the feeding wire in the gap towards the deposit point. It is thus possible to maintain with high precision the orientation of the wire near the deposit point, Fig. 4. The orientation can attain 20°, thus ensuring very good fusion between the parent metal and the deposited wire.

The addition of a motorised device which controls the rotation of the torch about the axis of the straining system permits arc oscillation. This possibility has been tested by Ito et al.[3] In the system currently developed the amplitude and frequency of oscillation can be controlled appropriately according to the desired application.

The welding cycle can comprise either a constant welding speed with an adjustable dwell time for the wire which is a function of the arc position on each lateral surface, or a variable welding speed. For the latter, incremental forward displacement of the torch or of the workpiece occurs only when the arc is directed towards the lateral surface of the gap; conversely, the welding speed is zero during rotation of the torch. This combination provides a very flexible oscillating method for general-purpose application.

It should be noted that the tolerances on the gap in this version are normally very difficult to obtain in heavy industry. Indeed this tolerance should be no more than $^{+1mm}_{-0mm}$, thus limiting its use to tight tolerance gaps like those which can be met in certain mechanical applications. Figure 5 shows a macrograph obtained with one oscillating torch.

Welding head with two oscillating torches

The previous version, already tested by some equipment manufacturers,[4,5] has the following drawbacks:

(a) long welding time

(b) tight tolerances needed for the gap $^{+1mm}_{-0mm}$

To reduce the welding time a second torch can be added thus halving the time. The use of the oscillating arc permits the second point to be satisfied which otherwise remains a major obstacle to the use of this process in heavy industry. As a matter of fact, an oscillating movement of the torches can be used which covers two-thirds to three-quarters of the gap with each torch, while aligning each welding torch to the gap by means of a seam-tracking device. The regulation of the amplitude of oscillation depends upon the value of the maximum gap tolerance measured. This system enables gap variations of $^{+7mm}_{-0mm}$ to be absorbed without major difficulties. This tolerance is compatible with those met in heavy industry. Moreover this solution maintains the advantages brought by the oscillation of the torch.

The shielding gas normally used in narrow gap welding is a mixture of 80% argon and 20% carbon dioxide. This proportion has been chosen to reduce the spatter from the molten wire, which depends upon the nature of the gas (impurities) and the welding parameters.

Pure carbon dioxide has been previously used for shielding with one torch.[4] This shielding gas associated with the narrow gap version using two oscillating torches appears to promote high efficiency, after the optimal welding parameters have been chosen to reduce spatter from the wire. Figure 6 shows a macrograph of a weld with excellent compactness and fusion between the deposited layers and between the fused zone and parent metal. The improvement in the fusion between deposited bead and parent metal permits this process to be applied (under certain conditions and depending on the metallurgical qualities required of the joint) to weld pieces which have been flame cut.

DISCUSSION AND COMPARISON OF THE DIFFERENT VERSIONS OF NARROW GAP WELDING

The increased gap in the new versions of the

Table 1 Comparison of the different narrow gap processes (see also Fig. 7)

Reference	Width of the gap, mm	Tolerances on gap geometry, mm	Thickness of the weld, mm	Number of torches	Prestraining the feed wire	Oscillating the arc
A	6 to 12	+1.5 -0.0	50 to 200	2	Outside the gap	No
B	12 to 16	+3.0 -0.0	50 to 200	3	Inside the gap	No
C	16 to 18	+1.0 -0.0	50 to 200	1	Inside the gap	Yes
D	16 to 20	+7.0 -0.0	50 to 300	2	Inside the gap	Yes

narrow gap processes involves unavoidably an increase in the volume of the deposited feed wire. This implies a longer welding time than in the original version. Practically, increasing the gap from 8 to 16mm does not double the welding time, because the diameter of the wire employed in the new version is larger and the number of torches can be increased to three.

Figure 7 shows diagrammatically a comparison between the different versions: the welding times for 1m long and 100mm thick are noted for different gaps as a function of the weight of the molten wire deposited and as a function of gap.

It is noted that in the three torches employed on a 16mm gap the welding time is 1.3x that necessary for a two-torch version, with an 8mm gap. Furthermore, the addition of a supplementary torch to a machine already equipped with two torches is not an expensive operation, because the main elements of the machine remain unchanged. It is thus possible to reduce the welding time without dramatically increasing the cost of the welding machine.

On the other hand, the welding time with one torch and a 16mm gap is about 3.5x that using three torches for the same gap. It should be noted, incidentally, that the welding time measured for one torch is comparable to that obtained by Yabuki et al.[5] Indeed the latter obtained a welding time of 110min for a 10mm gap and 100mm thickness, which is to be compared with 104min obtained on the same thickness by the present authors.

Also, the welding time with only one torch is of the same order as that obtained with automatic submerged-arc welding using 5.6mm diameter wire. Nevertheless the following remarks concerning the drawbacks of the submerged-arc process can be made:

1 A large metal deposit (twice that obtained with the narrow gap process) which, by its weight, tends towards lack of fusion between layers
2 Existence of porosity essentially caused by the use of flux
3 The energy input per unit length for submerged-arc welding is much higher than that employed in the narrow gap process (by 3 to 4 times). This will increase the HAZ with the corresponding consequence on the metallurgical quality

In laboratory tests it is important to note the angular distortion of the workpieces during the successive passes. It has been found that a chamfer of 2° from the bottom is necessary finally to obtain parallel welded surfaces.

Concerning the metallurgical characteristics of the weld, excellent quality is obtained by the narrow gap process. In addition to the favourable results obtained in the classical tensile and bend tests, the Charpy tests data can be better in the weld zone than in the parent metal. Even in the HAZ, where results are usually poor, the Charpy tests give good values. It is interesting also to note that very good explosion bulge test results have been obtained on 40mm thick samples which indicate excellent welded joint behaviour.[6]

Among the different versions, what is the best equipment to choose? It seems obvious that the choice should be made with respect to the considered application. This choice is a compromise between the different advantages of the several versions and their various drawbacks. When an application necessitates welding in all positions the first-generation equipment with two torches can be used, or alternatively the version with a larger gap and three torches. Where the workpiece can be rotated on a positioner, or the weld achieved in the flat position, it is possible to use the version with 16 to 20mm gap equipped with one, two, or three oscillating torches according to the production rate required. It should be pointed out that the narrow gap process can be favourably adapted with respect to the last aspect. As a matter of fact, multiplication of the welding torches affects the total cost of the equipment only very slightly. The different characteristics and performances of the various versions described above are summarised in Table 1 (cf Fig. 7).

CONCLUSION

New developments have been brought about in the narrow gap welding process to render it more industrial by eliminating or reducing the drawbacks met in the operation of the first-generation equipment, designed to operate in all positions, and characterised by the narrowness of the gap.

In the newer versions the gap is increased from 8 to 16mm and even more, and the welding head composed of one, two, or three torches becomes sturdier. In addition, bending the wire is achieved near the impact point of the arc, thus eliminating the cause of lack of reproducibility of curvature found in the first machines.

Increasing the gap from 8 to 16mm, which as a consequence improves the practical conditions for using the process, does not increase the welding time considerably when the three-torch version is employed.

When the application necessitates operation in all positions the first version can be used with an 8mm gap, or that with a 16mm gap equipped with three torches. Where the workpiece can be rotated on a positioner or where it can be welded in the flat position the version with larger gaps (16 to 20mm) can be used with two torches associated with an arc-oscillating device. It is therefore important to choose equipment according to the application needs. Finally, it is claimed that the last versions are particularly adapted to such eventual modification.

These latest developments of the narrow gap welding process render it sturdier, more reliable, and consequently more industrially justifiable.

REFERENCES

1. BATTELLE MEMORIAL INSTITUTE. 'All-position method welds thick plates'. Iron Age, (12), 1963, 102.
2. HENDERSON, I.D. and DUCROT, A. 'Narrow gap welding of heavy section pressure vessel'. IIW 1976 Public Session and Metals Technology Conference, Sydney, 1976, Vol.B, Session 16-6-1—16-6-18.
3. ITO, T. et al. 'Development of arc-oscillating narrow gap welding process'. Symposium 'Advanced Welding Technology', 25-27 August 1975, Osaka. Japan Welding Soc., Vol.II, Session 2, 391-5.
4. NAKAYAMA, H. et al. 'Application of narrow gap CO_2 arc weaving'. Symposium 'Advanced Welding Technology', 25-27 August 1975, Osaka. Japan Welding Soc., Vol.II, Session 2, 403-408.
5. YABUKI, Y. et al. 'A new narrow gap welding process with oscillating arc'. IIW Doc-XII-212-77.
6. SCIAKY SA. 'Narrow gap welding'. Cahier technique, (10), May 1977.

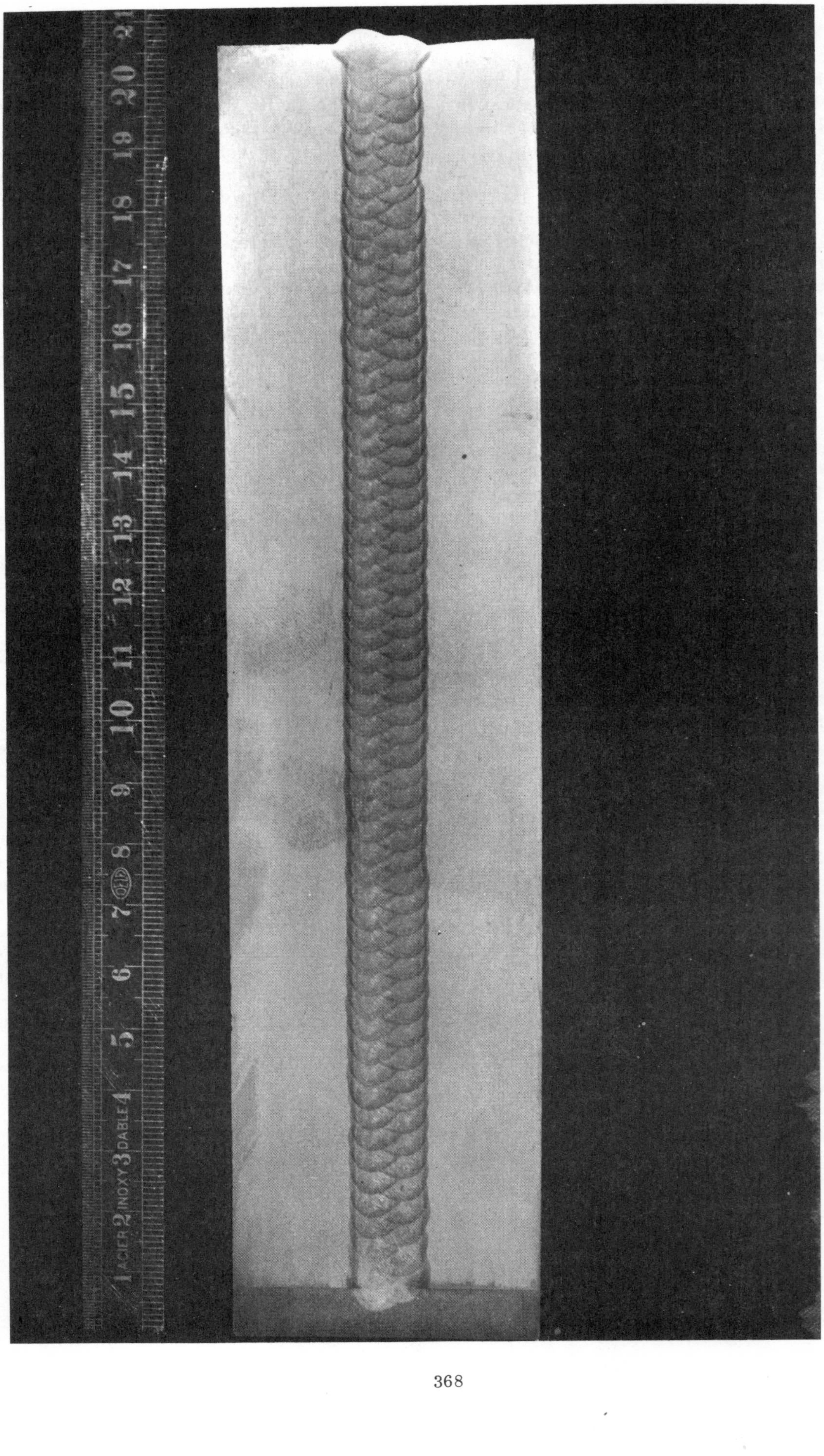

1 Macrograph of weld in low alloy steel, 200mm thick, by narrow gap process. First version: A

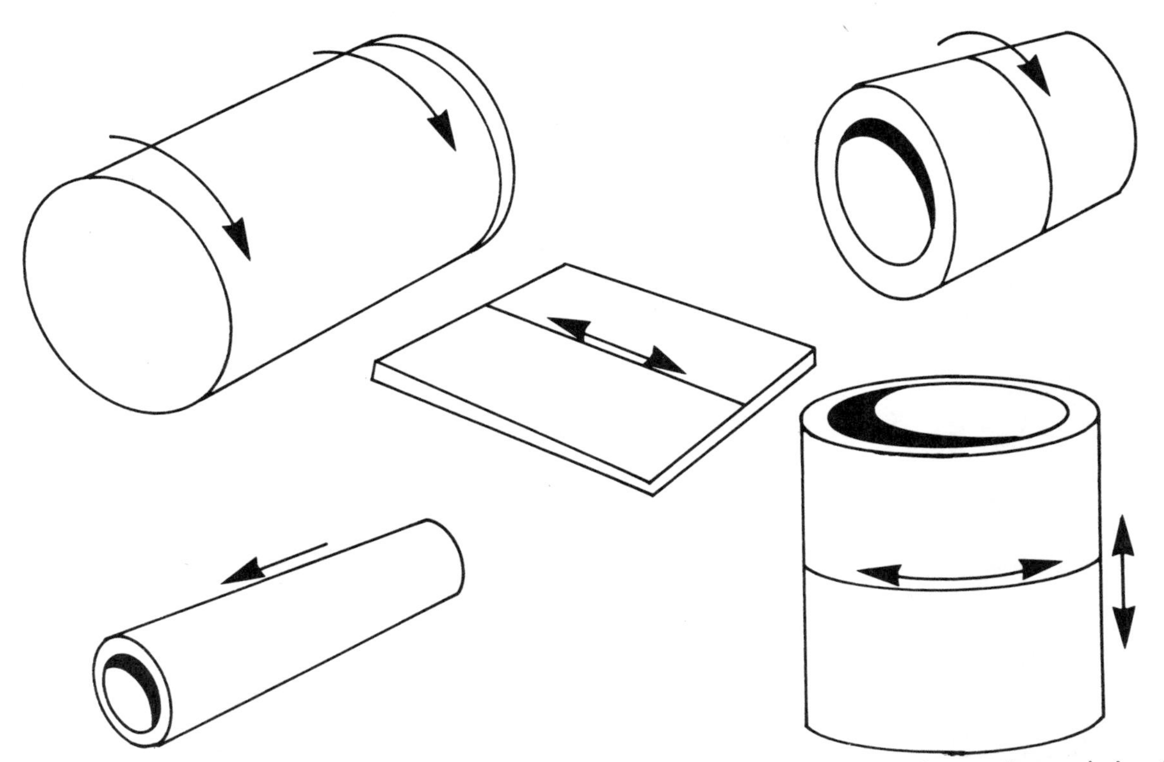

2 *Potential applications for narrow gap process which can be accomplished in flat position or by rotating workpiece in front of torch*

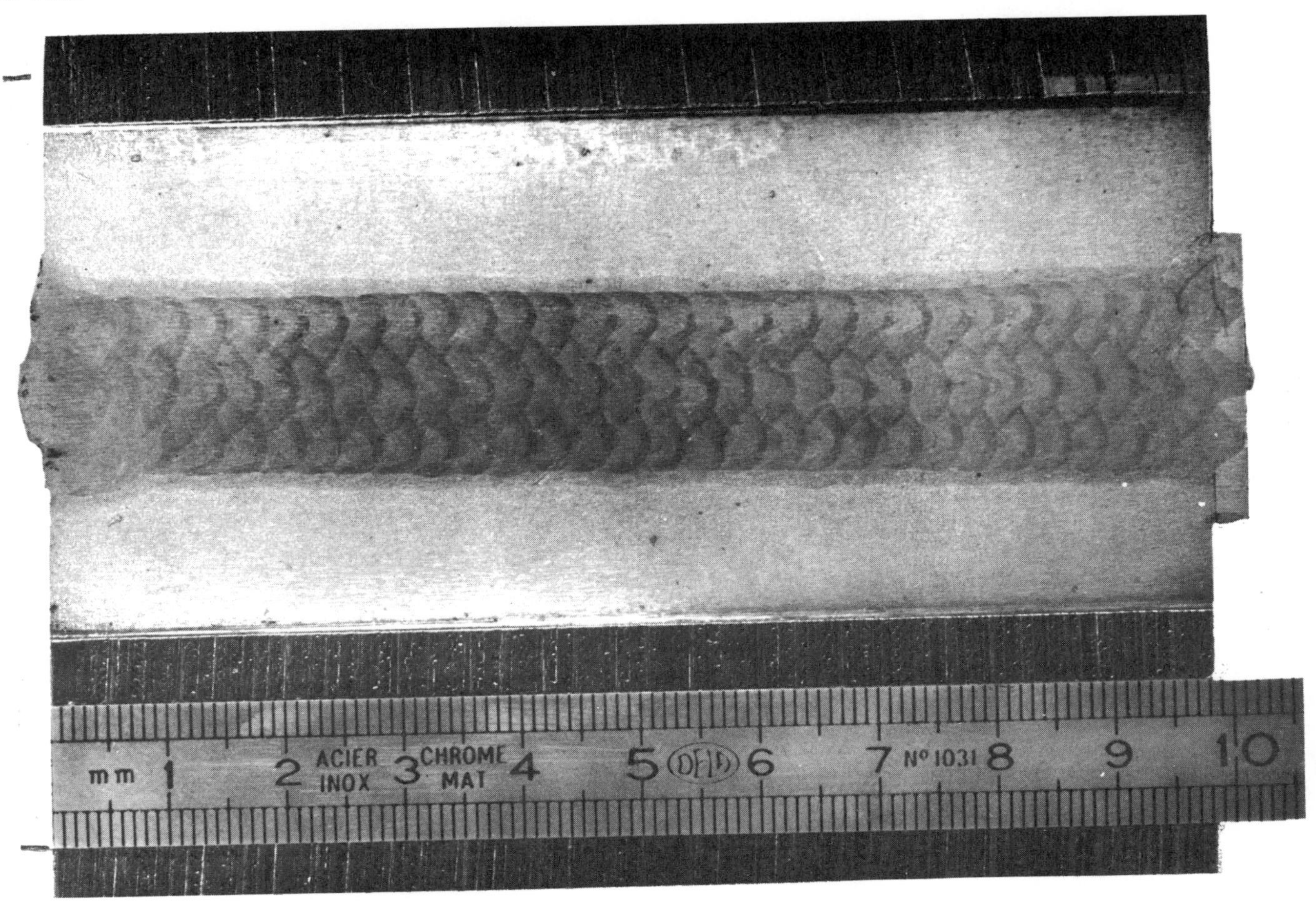

3 *Macrograph of weld in 100mm thick low alloy steel using narrow gap machine with three torches*

4 Straining feed wire in gap by special device. Tilting up to 20° off axis can be obtained

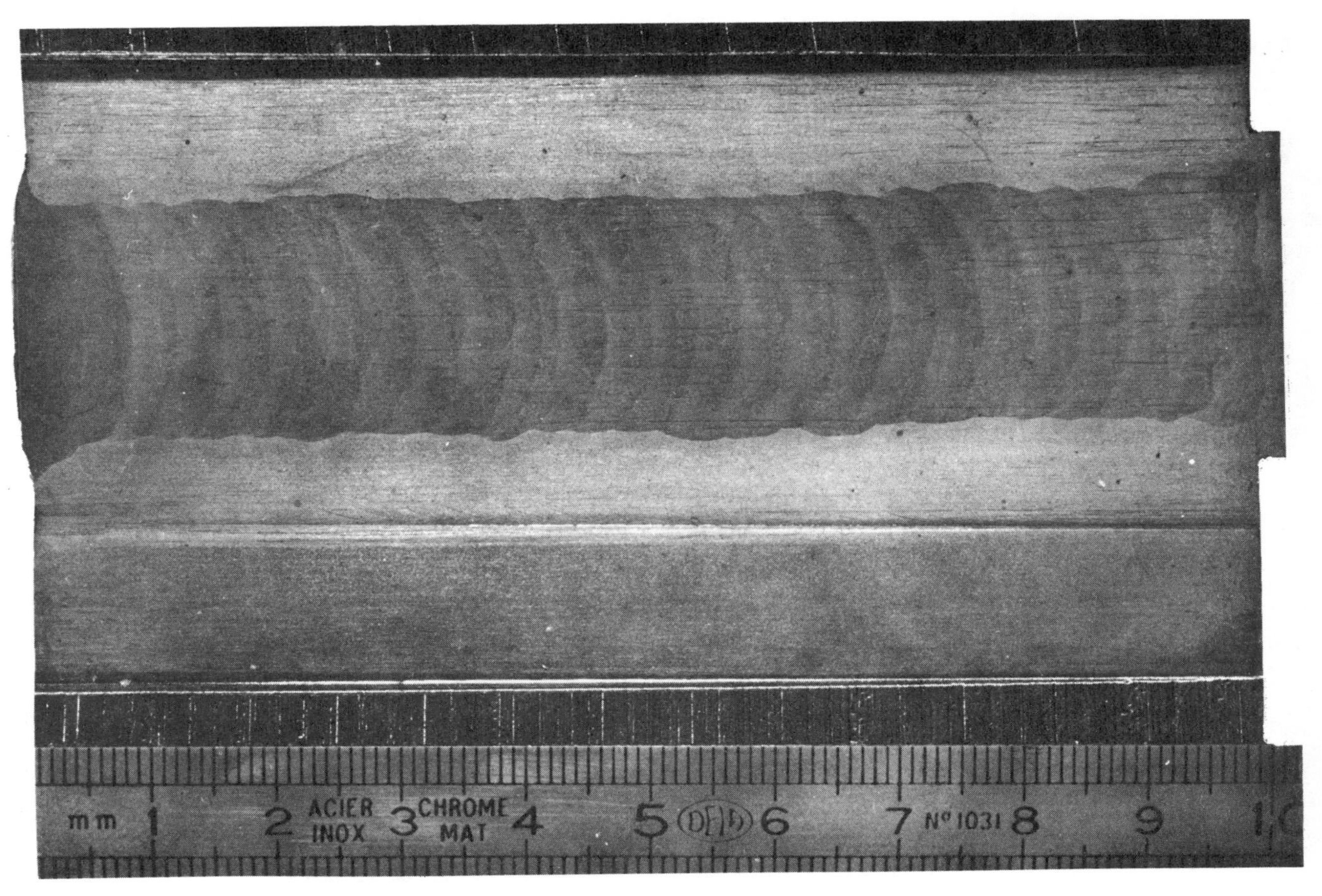

5 Macrograph of weld in low alloy steel obtained with one oscillating torch

6 Weld in low alloy steel using two torches with CO_2 gas shielding

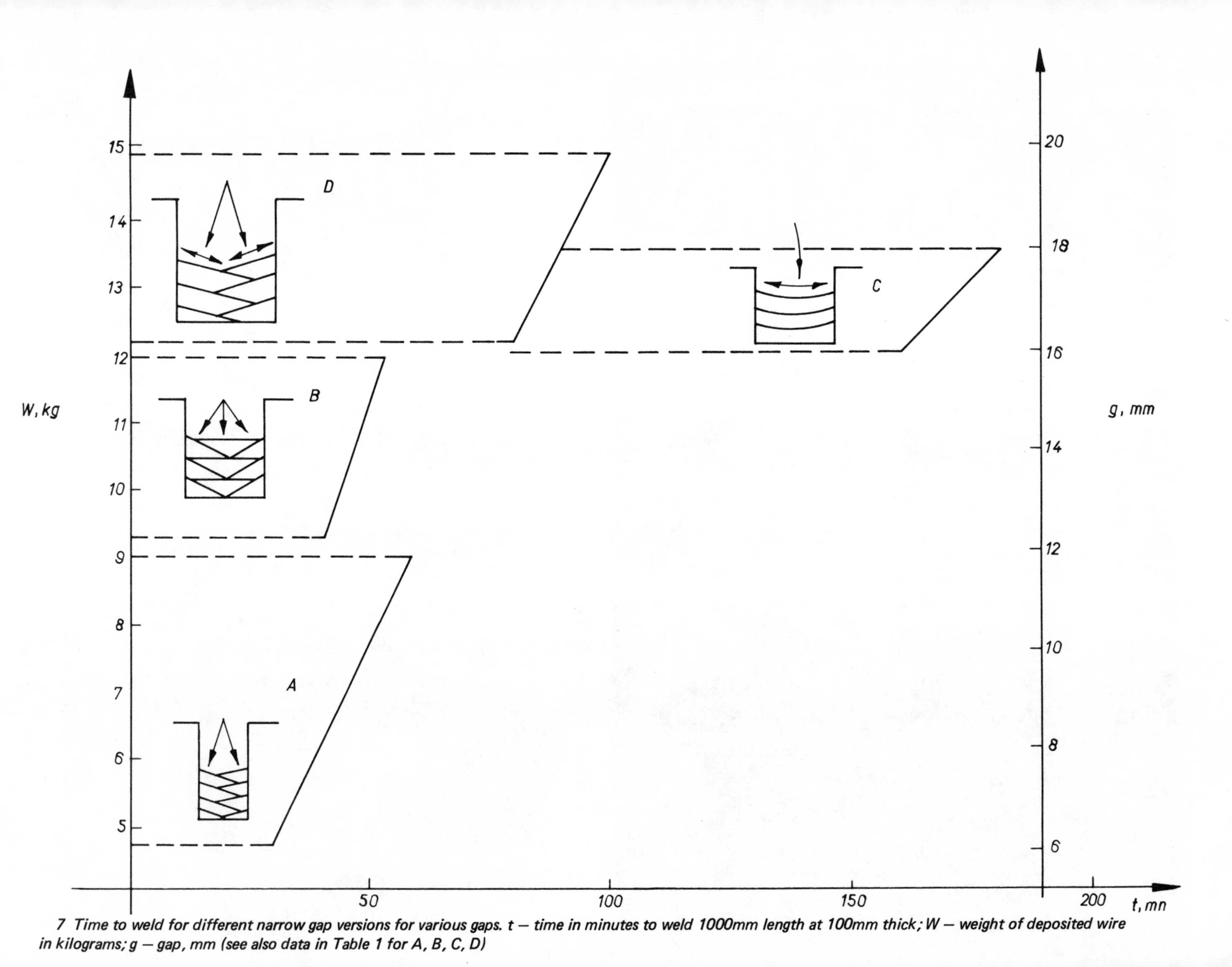

7 Time to weld for different narrow gap versions for various gaps. t – time in minutes to weld 1000mm length at 100mm thick; W – weight of deposited wire in kilograms; g – gap, mm (see also data in Table 1 for A, B, C, D)

PAPER 3

Welding the core support plate for the Clinch River breeder reactor plant, USA

W.W.Canary, BS, MS, and E.A.Franco-Ferreira, BEngPhys

The core support structure for the USA's Clinch River breeder reactor plant consists of a 51mm thick cylindrical barrel welded to a 610mm thick core support plate, both in Type 304 stainless steel. The reactor plant will operate at temperatures in excess of 427°C and hence the design of the core support structure is not covered by the ASME Code.

One of the more difficult weldments in the support structure was associated with the fabrication of the 610mm thick core support plate. It was made from two forgings comprising a 3.5m diameter forged disc placed inside a 4.5m diameter forged ring which were joined by a circular, 530mm thick, full penetration weld. A narrow gap submerged-arc welding process was successfully developed for this application.

INTRODUCTION

The USA electric power industry and the Department of Energy are working jointly to build the Clinch River breeder reactor plant. This plant will be the country's first large-scale demonstration of the liquid metal fast breeder reactor, and is an integral part of the national programme to achieve energy independence.

The reactor plant will operate at temperatures in excess of 427°C; hence the design of the core support structure is not covered by the ASME Code. However, fabrication and inspection were performed to the requirements of ASME, Section III, subsection NB as supplemented by NG, for Class 1 components, and modified by Reactor Development and Technology (RDT) Standards. The RDT Standards supplement the ASME Code and are prepared by the US Department of Energy. The one exception to this is the nondestructive examination of the core support plate weldment which is too thick to be conveniently radiographed. As an alternative, the liquid penetrant inspection of each 12mm or so of weld thickness was permitted.

The 610mm thick Type 304 stainless steel core support plate, Fig.1, was made from two forgings because it was too large to be forged in one piece. The assembly was made by placing a 3.5m diameter forged disc inside a 4.5m diameter forged ring and joining the two with a circular, 530mm thick, full penetration weld.

A survey of the welding literature and discussions with pressure vessel manufacturers in the USA established that stainless steel had not commonly been welded in thicknesses much above 150mm. Some of the problems to be faced at the outset were thus:

(a) qualification of a nominally 600mm thick weldment
(b) selection of a weld groove geometry which would minimise welding time and distortion
(c) obtaining tooling for slag removal, grinding, and wire brushing in a deep, narrow

Mr Canary, Welding Engineer, Nuclear Components Division, and Mr Franco-Ferreira, Manager — Projects Hydraulic Turbine Division, are both with the Allis-Chalmers Corporation, York, Pa, USA.

groove, and tooling for backmachining (for a two-sided joint geometry)

(d) obtaining a weld wire/flux combination which would meet the specification requirements, and

(e) development of a method for dye penetrant inspection in a deep groove

The overall solution, which was successfully accomplished, is summarised below.

DEVELOPMENT OF WELDING PROCEDURE

Consumable electrode process

The submerged-arc welding process was selected for the core support plate weldment because of the high deposition rate, low propensity for defects associated with this process, and because of the ability to adapt the equipment for welding in a deep circular groove, as indicated in Fig.1. Details of the groove selected are described below under 'Weld preparation'.

Flux was dispensed into the groove using a standard flux hopper and flexible tubing which was attached to the wire guide tube. The position of the end of the tubing was adjusted above the bottom of the groove to control the depth of the flux during welding.

Welding consumables posed something of a problem because of RDT Standards chemistry limitations. The weld deposit was required to have a silicon content of 0.25-1.00%, and a ferrite content of 5.9%, as determined by chemical analysis and use of the Schaeffler diagram. The compositions of both the bare filler wire and the submerged-arc flux were carefully chosen to achieve these weld deposit requirements. Type ER308 wire was used with a modified Arcos S-4 flux. The actual silicon content of the weld deposit produced by this combination was 0.96% and the ferrite level was 9%.

With the narrow groove finally selected the top face of each layer of deposited weld metal was concave which reduced the probability of slag entrapment at the sidewalls in the succeeding pass. The concave bead also aided the removal of slag. The welding parameters selected were current: 440A, voltage: 32V, and speed: 230mm/min.

Additional provisions were made to allow the use of the narrow weld joint and to meet all the operating criteria desired (see below). A specially made needle-shaped descaler, Fig.2 (centre), with extra-long needles and a guide tube, was used to break the slag into small pieces which allowed their removal with a standard vacuum flux recovery system. Both the needles and guide tube were made from stainless steel to avoid contamination of the workpiece. A long, small-diameter brass guide tube, Fig.2 (bottom), was made to conduct the filler wire from the wire feeder to the welding contact tip at the bottom of the joint. This guide was electrically insulated on its outer surface by wrapping with a low contamination tape to prevent accidental short-circuiting to the groove sidewalls. An air-operated hand grinder with a long, narrow guided belt, Fig.2 (top), was used to grind starts and stops where necessary. Similarly, the backmachining was accomplished using a specially made reinforced, extended-length grooving tool with a high speed cutting insert.

Weld preparation

Selection of the weld groove geometry was considered to be crucial because of its major impact on fabrication time and costs. The fundamental approach was that the volume of weld metal required to fill the joint should be minimised, together with the ability to statisfy the following criteria:

1. To make the weld without extensive repairs
2. To backmachine the groove
3. To produce a desirable weld bead contour
4. To remove the slag easily
5. To make meaningful liquid penetrant examinations

Normal practice by the Company for conventional weldments (25 to 100mm thick) had been to use a joint with a 9.5mm groove radius and a 15° included angle, Fig.3a. If this geometry had been used for the support plate weldment it would have taken an estimated 360hr of arc time to deposit the required 2165kg of weld metal. Several mock-ups were welded to determine the narrowest groove opening which could be satisfactorily welded and still meet all other selection criteria. The joint which was finally chosen, Fig.3b, had a 9.5mm groove radius and a 2° included angle. This permitted the use of a deposition technique which resulted in a single bead per layer of weld.

For the full penetration weld a double U groove was used with the first side to be welded having a depth of 300mm, and the second side a depth of 230mm with slightly greater width. The second — or reverse — side groove was designed to accommodate the tooling requirements for the backmachining operation which was carried out on a vertical boring mill. This joint design resulted in the actual deposition of 810kg of weld metal in an arc time of 135hr (6kg/hr average). This represents a saving of 62.5% in time and material when compared with conventional practice.

PRACTICAL APPLICATION

Welding quality assessment

Procedure qualification

Prior to fabrication the welding procedure was qualified to the requirements of ASME, Section IX. To do this in an expeditious manner three 200mm thick rolled plates were stacked and edge-welded together to provide the 600mm thick test assembly shown in Fig.4. This testpiece, which weighed 800kg, was then prepared with the proposed joint design. To duplicate the inherent restraint of the production assembly, C-shaped blocks were welded to each end of the qualification test assembly to reduce the closure of the joint. Welding of the test assembly was completed in eight days.

The ASME destructive testing requirements called for two transverse tensile tests and four side bend tests on the full weld thickness. The test specimens were split into a number of pieces, as permitted by ASME, to accommodate the limitations in testing equipment capacity In this instance, a total of eighteen tensile specimens and seventy-two bend specimens were required to accomplish testing of the full thickness. An array of the tested specimens, along with a macro-etched cross-section of the qualification weld, is shown in Fig.5. All ASME and RDT requirements were fully met.

Nondestructive examination

A postemulsified visible dye penetrant procedure was used to make the nondestructive examination of each 12mm or so of weld thickness with the weld surface in the as-deposited condition. This penetrant method was selected for its inherent adaptability to areas of limited access. The inspection of a single layer was completed in less than 4hr. The dye penetrant examination was accomplished by dispensing the dye from a standard pressurised can with a long, small-diameter tube attached to the spray can nozzle. After emulsification the penetrant was removed using a wet pickup vacuum cleaner. The developer was applied using a standard pressurised can with a spray nozzle modification similar to that used for the application of the penetrant.

Fabrication results

During production welding the entire 300mm deep groove was welded full prior to inverting the core support plate for backmachining. Joint closure was monitored after each dye penetrant examination. Total closure at the top surface was only 3.2mm when the 300mm deep groove was full. The plate was then inverted and backmachined to sound metal prior to welding the 230mm deep groove. Total closure for this side was only 2.4mm.

Welding progressed with absolutely no difficulties. In the forty-four dye penetrant examinations of completed weld layers no defects were found. The support plate assembly was dimensionally inspected before and after welding. Bowing in a dished fashion across the face of the plate was only 0.94mm, well within tolerance. A photograph of the completed weldment is shown in Fig.6. Original schedule planning allowed four months for the welding operation; actually it took only six weeks.

To finish the assembly of the reactor plant core support structure the support plate was welded into the bottom of a cylindrical fabrication, also in Type 304 stainless steel, of 3.4m OD, 51mm wall thickness. which was 4.8m high. The completed core support structure was delivered to the pressure vessel manufacturer early in 1978 for installation into the reactor pressure vessel.

CONCLUSIONS

The following steps were realised in the fabrication of the core support plate weldment:

1. A submerged-arc, single weld bead per layer, technique in a deep, narrow groove was successfully developed with no requirement for interpass grinding, resulting in a minimum amount of welding consumables and arc time being needed to fabricate the core support plate

2. Tools were obtained which successfully removed slag and ground starts and stops, and dye penetrant examinations were satisfactorily and routinely performed in the deep, narrow weld groove

3. A 600mm thick, Type 304 stainless steel, weld was qualified and a 530mm thick weldment was successfully made in production using the particular submerged-arc multipass process in a narrow groove

4. The welding of the core support plate was completed without requiring any rework, and 2½ months ahead of schedule

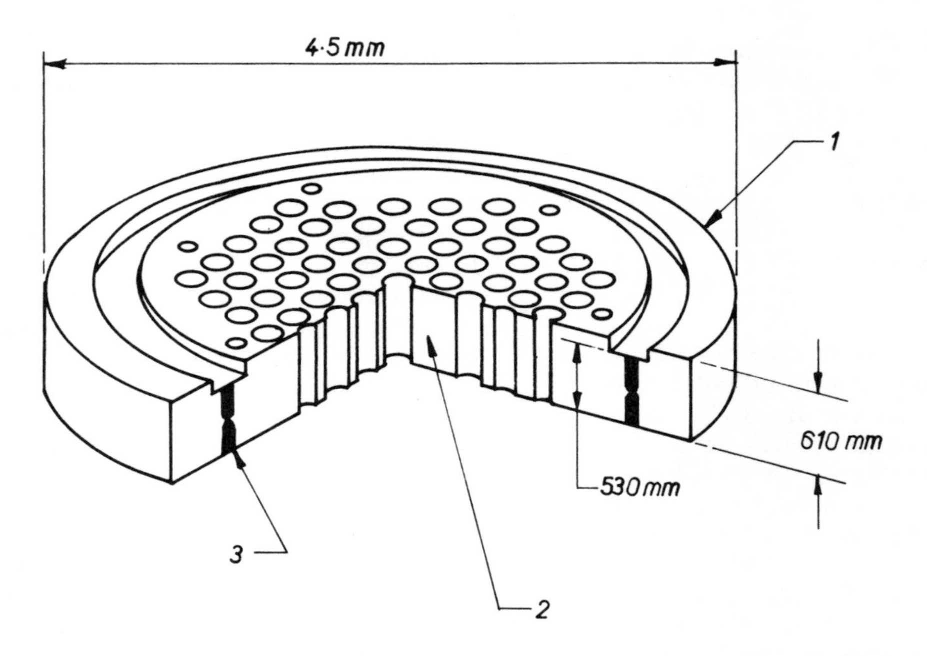

1 Core support plate assembly showing circumferential full penetration weld required. 1 – ring forging; 2 – plate forging; 3 – weld

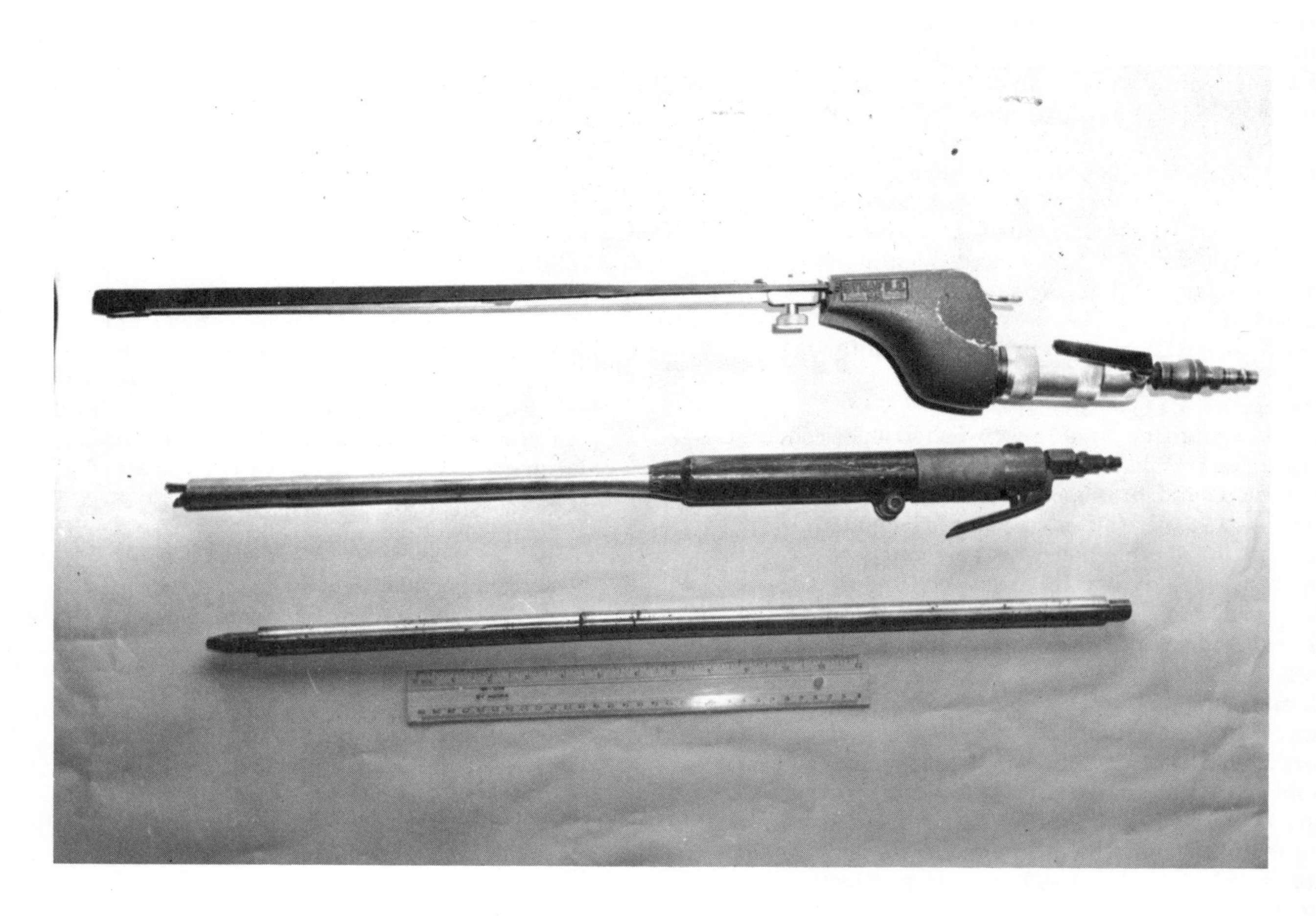

2 Tooling for deep groove operation (from top): belt sander, needle scaler, wire guide and contact tip

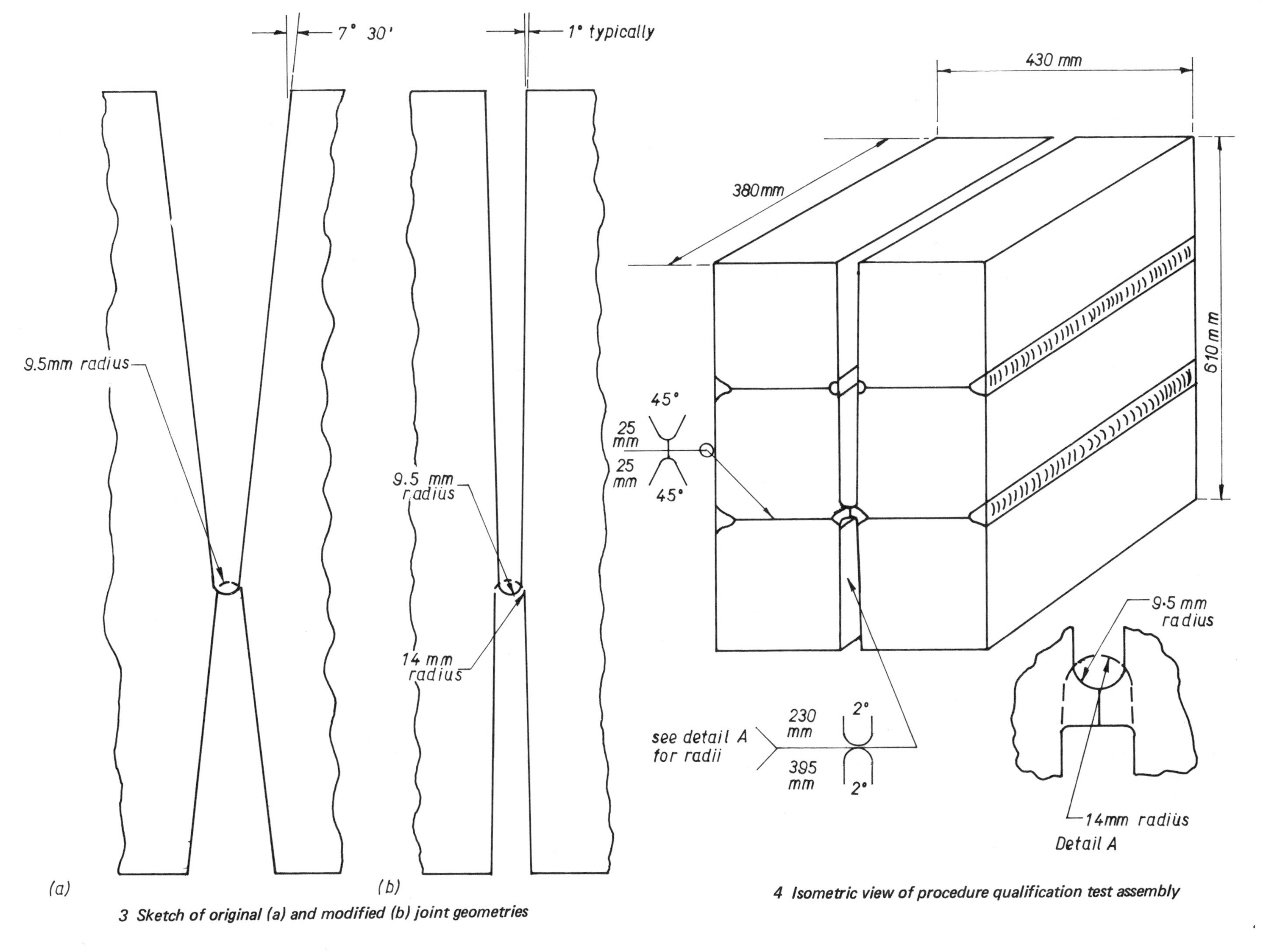

3 Sketch of original (a) and modified (b) joint geometries

4 Isometric view of procedure qualification test assembly

5 Macro-etched cross-section and tested specimens from 610mm thick procedure qualification

6 Completed core support plate weldment

Welding problems in the construction of tube bundles for heat exchangers

R.Torrani, Ing, and A.Paolocci, PerInd

INTRODUCTION

In the manufacture of tube bundles for heat exchangers the joint between tubes and tubesheet is most crucial. However, in detail, tube welding and the choice of system utilised are based on manufacturing and operating experience rather than on first principles.

Among the most frequent examples where welding is essential are:

(a) where the circulating fluids are dangerous because of their high toxicity (halogens, sulphur, and ammonia products), or because mixing with the atmosphere may cause an explosion, e.g. hydrogen, or when the presence of radioactive products (as in nuclear plant) is feared

(b) where it is imperative to avoid intermixing the circulating fluids, either because of the risk of reaction between the fluids themselves or when contamination of one of them must be avoided absolutely, as in food products, drinking water, closed cycle fluids, etc.

(c) for very thin tubesheets where the tubes reinforce the plate itself

(d) where there is a possibility of corrosion stemming from the circulating fluids

(e) where service conditions are extremely severe, such as high pressure or temperature, or fatigue caused by pulsating or alternating loads

The choice of system for welding the tubes to the tubesheet is determined by many factors such as: service pressure and temperature, tube pitch, material to be welded, thickness of the tubes and tubesheet, conditions under which welding has to be carried out, and last, but not least, the question of cost.

The joint designs are generally determined by the welding process adopted, but they fall into two basic classes, i.e. tube welding on the front face of the tubesheet, and on the backface.

Many procedures have been developed for the first category, but only some two or three for the second. Of the former arrangements those which best allow automatic welding are those in which the tube simply protrudes from the tubesheet without any preparation on it. Here welding can be compared with a fillet joint, where the root penetration is accomplished relatively easily. Therefore this latter configuration was adopted for the applications described here.

WELDING EQUIPMENT

General

A careful consideration of GTAW equipment, for tubes of small to medium diameter, at present on the market laid stress on finding a type which would satisfy all the basic requirements in view of the type of production envisaged. In particular these include:

1. Manageable guns with completely automatic operation so that two welding equipments utilised on the same tubesheet could be controlled by one operator only
2. Full mechanical accuracy in torch positioning with a wide range of adjustment
3. Constant filler wire feed with the possibility for full three-axis attitude with respect to the weld pool
4. Fixed wiring and cable booms to the welding head, and not rotating or winding around the tube
5. Welding to be continuously observable during running
6. Since the tubes are arranged horizontally, a wide range of programming for the welding cycle with orbital variation of basic parameters (current, rotational speed, wire feed)
7. Programming and its orbital variations to be based on the actual torch position during welding for rapid setting up of basic parameters and welding equipment
8. Square wave pulsed current operation,

Mr Torrani is Manager, RTA, Rome, and Mr Paolocci is Manager, Welding Office, Nuovo Pignone, Massa, both in Italy.

with compensation for line voltage variations up to ±10%

9 Easy and rapid maintenance of mechanical, electronic, and electrical components with replacement spare parts

10 Utilisation of standard 100mm diameter spools

From the applications standpoint the equipment should also provide for:

1 Welding tubes protruding up to 50mm from the tubesheet

2 Correct centring of the gun in the tube easily and quickly

3 Control of the distance between the tungsten electrode and the workpiece during welding

4 Ease of operation and positioning of the torch

5 Possibility of carrying out an orbital weld in tube of 32mm OD with very limited space between tubes where the spool of filler wire cannot rotate around the tube but has to remain stationary in space. Furthermore, the torch should lock on both tubes to be welded to be sure that it rotates coaxially, with provision for axial positioning with the machine mounted on the tube to centre the welding plane exactly

Programme control

After choosing several types of welding equipment and testing each for the degree of repeatability of the programmed procedure, the TIG-A-MATIC system was finally selected, Fig.1. This comprises a welding head, electronic programmer, remote control, and welding power supply, and in its latest form interesting additional features, in particular a standard programming approach for all requirements which, at the same time, is easy to use both for programming and to control the welding parameters. Moreover, the same programmer could operate both welding heads on the tubesheet and other torches to join elbows on to the tubes. However the welding heads themselves required some adaptation to suit the geometry and dimensions required. This was relatively simple and allowed the bodies of the standard heads to be used. The welding power source was a 'Miller Analog' modified for connection to the programmers provided, complete with safety stop controls in the event of failure of the cooling water or shielding gas. Figure 1 shows the whole TIG-A-MATIC equipment comprising the welding gun (model 112), welding torch (model 139), and the electronic programmer (model 152).

The operating principle of the electronic programmer is of particular interest as it controls the torch (or gun) accurately and with ease. Inside the welding head there is a precision potentiometer, properly protected against the harmful effects of high frequency arc striking, rotating synchronously with the welding gun. The resistance values of this potentiometer are recorded in the programmer. Here they are read and, by integrated circuit comparators, checked against the values set on the programmer itself. That is to say, they are compared with the front face potentiometers for programming, or with the internal potentiometers if preset. In this way it is possible both to programme and control the welding cycle as a function of the torch position.

Since a precise resistance value, of the potentiometer inside the welding head, always corresponds to a specific angular position of the torch, there is the possibility to establish a certain number of coincidence points with the corresponding values in the programmer. Thus, with preset torch positions, it is possible to check the events taking place such as changes in rotational speed in an orbital programme. Equally, it is easy to establish that a full revolution has been achieved independently of the time required.

Programmer module

A brief description is given below of the overall programmer, Fig.1c, consisting of three modules for: basic programming, wire feed, and orbital programming (called VARIOTIG).

The programmer, complete with a remote control and a separate panel with instrumentation for welding current and voltage, caters for three welding stages: preheat, welding, stress relief, each with its own independent parameters which are stored in the basic programming module. A switch selects the number of stages required, and in addition there are also the torch position indicator, an on/off key, an emergency pushbutton stop, and warning lights for water or gas failure.

In this module it is possible to select the:

1 Current level for arc striking up-slope duration

2 Current level for forming the weld pool and its duration

3 Welding current (steady or pulsed)

4 Synchronisation of head rotation with current pulsation (if any)

5 Torch rotation speed and lag (if any) between starting the wire feed with respect to starting torch rotation
6 Wire feed speed and synchronisation with current pulsation (if any)
7 Current down-slope time and final current level and duration

Where welding can be influenced by gravity, as here, the orbital module permits the rotational speed, wire feed rate, and current (whether pulsed or not) to be changed automatically every 60°, i.e. on six sectors. There is a seventh, adjustable-amplitude, sector which establishes the welding parameters during the overlap stage.

The programmer is locked via a transparent front panel but can feed a six-channel recorder and has provision for pressurising by air or inert gas. The internal construction is based on extractable printed circuit cards, using a strip wiring system which has proved to be extremely reliable.

Torch head

The modifications to the standard welding heads did not present serious difficulties as the standard bodies could be used. The welding gun, Fig.1a (model 112), was adapted to allow it to centre inside the tube and to weld with the tube protruding 50mm from the tubesheet. The centring system is important and is based on expanding spherical plugs which not only position the welding head but can also cater for possible deformation of the tube. By counterbalancing the weight of the guns, and without using expensive positioners or manipulators, one operator could handle two machines by simply inserting and extracting the centring device from the bore of the tube to be welded.

The tungsten electrode is free to be swivelled with respect to the tube axis, and the inlet guide nozzle for the filler wire is likewise free. This allows rapid optimisation of the electrode angle and of the relevant inlet position for the wire. The welding head rotates without limit since the current, cooling water, and gas shield are fed to the torch through a special rotating gland, leaving the supply cables stationary. Also, to avoid any kinking, the wire spool rotates coaxially with the machine.

All running operations and normal maintenance (such as spool and electrode change, replacement of gas cups and wire nozzles) are easy and can be directly carried out by the operator.

The more sophisticated torch, Fig.1b (model 139), also distributes the current, water, and gas via a rotating gland, so that the external cables are stationary. The machine is very compact and is constructed like a long box with a U opening, where the body encloses all the mechanism, cables, and sheaths, so that it is easy to use among the tubes in situ, Fig.2. The filler wire is also completely contained within the machine body and the wire spool turns on its own axis but does not orbit around the tube.

The rotating part of the machine is supported on two open ball bearings set well apart to guarantee a long operating life without requiring clearances. Two clamps, placed at the end of the body, are elastically mounted to exert a predetermined load on the tube, sufficient to support the machine in any welding position.

The torch itself is miniaturised, water-cooled, and insulated in heat-resisting resin. During welding it is mechanically guided to follow the tube profile, and hence to complete joints of imperfect preparation.

Because of its elastic support the torch can be adapted to other welding diameters, the welding head being locked on to the tubes and the axial position of the electrode adjusted by a knurled roll.

On both machines the filler wire feed is insulated to avoid damaging the motor and electronic controls if the spark starting accidentally strikes on to the wire.

EVALUATION IN APPLICATION

The equipments described above were used in welding heat exchangers for uranium hexafluoride and ammonium carbamate. For the former, carbon steel tubes, 32mm OD and 2mm thick, were welded to a carbon steel tubesheet, 40mm thick, Fig.3a. The tubes projecting 50mm from the tubesheet also had to be butt welded to similar steel tubes to make the bend connections, Fig.3b. For both these welds very simple joint preparations were chosen with no groove in the tubesheet and no chamfer on the tube-tube joint.

Regarding preparation, it was easy enough to obtain constant joint dimensions on the tubesheet before welding, but, for the joint between the elbows and the tubes, wide tolerances had to be accepted with misalignment or gaps ranging from 0 to 1mm, Fig.3b. Sometimes both maximum mismatch and gap occur together and yet should not cause weld defects such as lack of penetration or undercutting.

Tables 1 and 2 give the main welding parameters chosen together with the results of nondestructive tests for the tube-to-tubesheet and tube-to-tube joints respectively.

Table 1 Tube-to-tubesheet welds, see Fig.3a

Welding parameters		
Tube material	:	carbon steel
Tubesheet material	:	carbon steel
Current (DC electrode -ve)	:	75-200A (pulsed)
Voltage	:	10-20V
Speed	:	110sec rotation each tube
Position	:	tubesheet vertical
Filler	:	0.8mm diameter, 1.2m per tube
No. of passes	:	1
Welds carried out	:	62 000
Leakages on helium test	:	6

Table 2 Tube-to-tube welds, see Fig.3b

Welding parameters		
Materials	:	carbon steel
Current (DC electrode -ve)	:	30-80A (pulsed and orbital moduled)
Voltage	:	10-20V
Speed	:	140sec rotation each tube
Position	:	tube with axis horizontal (nonrotating)
Filler	:	0.8mm diameter, 0.5m per tube
No. of passes	:	1
Welds carried out	:	45 000
Leakages on helium test	:	4

Table 3 Procedure for ammonium carbamate heat exchanger

Welding parameters		
Tube material	:	AISI 316L (urea grade)
Tubesheet material	:	overlay AISI 316L (urea grade)
Current (DC electrode -ve)	:	75-150A (pulsed)
Voltage	:	10-20V
Speed	:	40sec, 1st pass (without filler) 45sec, 2nd pass (with filler)
Position	:	tubesheet vertical
Filler	:	0.8mm diameter, 0.65m per tube
No. of passes	:	2 (1st without filler metal)
Welds carried out	:	7500
Leakages to the air test	:	12 (after 1st pass)

Figure 4a shows a typical section through a tube-to-tubesheet weld from the 112 type torch, and Fig.4b shows the internal and external appearance of the tube-to-tube weld made with the 139 type gun for the uranium hexafluoride heat exchanger.

Finally, for the ammonium carbamate condenser, the tube-to-tubesheet weld preparation and typical sections are shown in Fig.5, using the welding conditions specified in Table 3.

In conclusion it can be stated that the welding equipment chosen has proved very satisfactory in operation and a high degree of repeatability has been achieved with a combination of welding procedures. Moreover, in spite of the apparent sophistication, the equipment is essentially simple to operate and set up, and does not require highly skilled welding operatives. Furthermore, the programming has been found to be particularly suited to these applications.

BIBLIOGRAPHY

SQUADRELLI SARACENO, F. and PALAZZI, S. 'Welding tubes to the tubesheet in heat exchangers'. Quaderni Pignone, 8, 1967, 17-27.

LEVENE, L.H. 'Performance of tube-to-tubeplate welded joints in high pressure feed heaters'. Welding Inst. Conference 'Fabrication and Reliability of Welded Process Plant', London, 16-18 November 1976. Abington, Welding Inst., 1977, 21-7.

HICKIN, P. and BULLOCK, J. 'The automatic welding of tube-to-tubeplate joints in heat exchangers'. Welding Inst. Conference 'Fabrication and Reliability of Welded Process Plant', London, 16-18 November 1976. Abington, Welding Inst., 1977, 1-11.

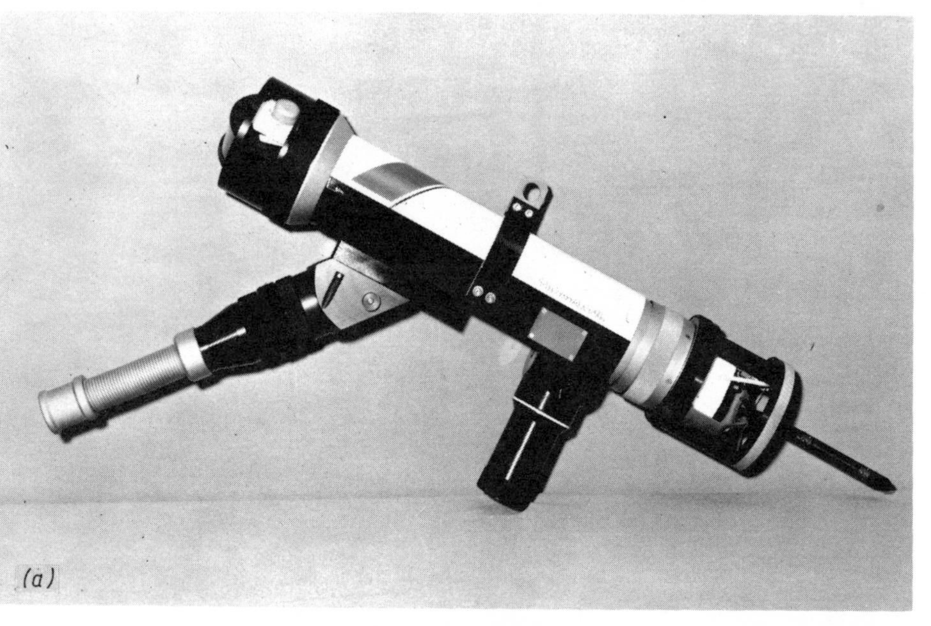

1 Automatic welding equipment for heat exchangers: (a) gun head model 112, (b) gun head model 139, and (c) programmer console (interior view)

2 Compact torch head for tube-to-tube welding

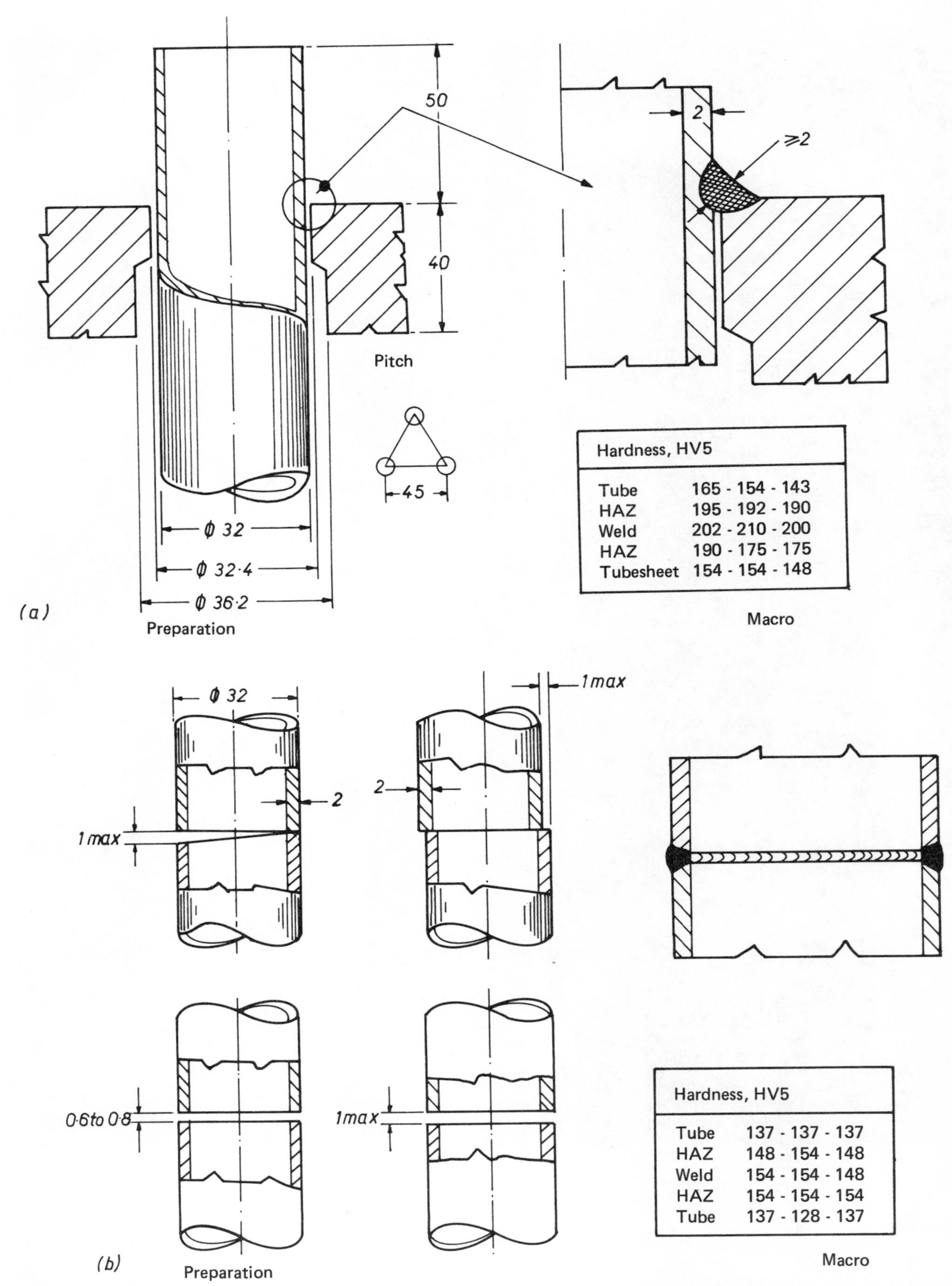

Hardness, HV5	
Tube	165 - 154 - 143
HAZ	195 - 192 - 190
Weld	202 - 210 - 200
HAZ	190 - 175 - 175
Tubesheet	154 - 154 - 148

Hardness, HV5	
Tube	137 - 137 - 137
HAZ	148 - 154 - 148
Weld	154 - 154 - 148
HAZ	154 - 154 - 154
Tube	137 - 128 - 137

3 Weld preparation and fitup for uranium hexafluoride heat exchanger: (a) tube-to-tubesheet joint, (b) tube-to-tube conne joint. Dimensions in millimetres

(a)

(b)

4 Weld appearance for uranium hexafluoride plant: (a) macrosection of tubesheet joint, (b) internal and external view of tube-to-tube joint × 2

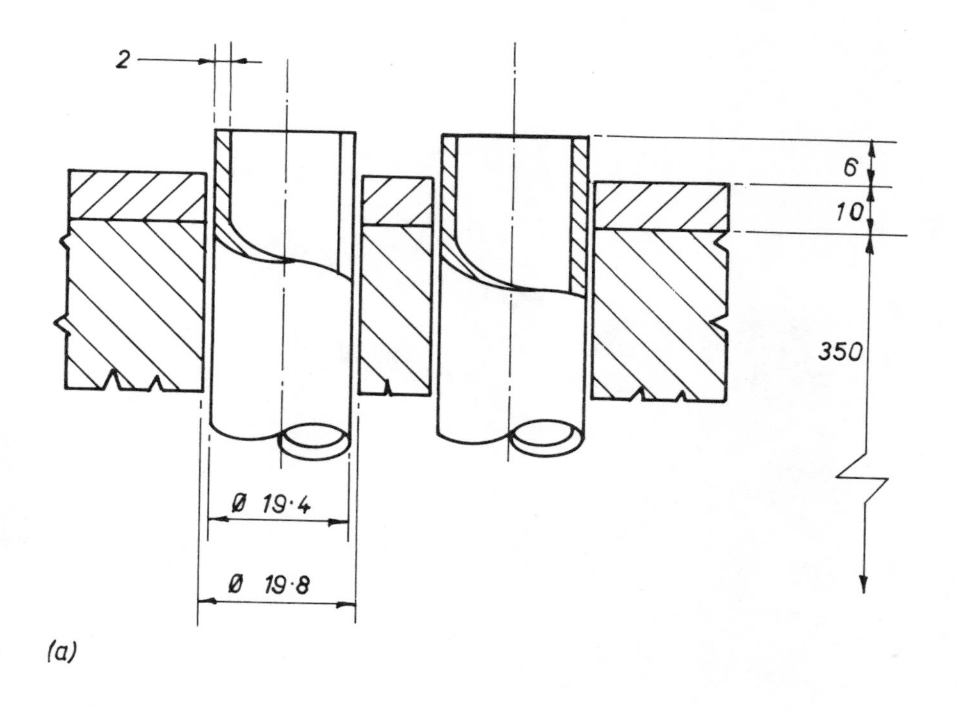

5 Tube-to-tubesheet welds for ammonium carbamate heat exchangers: (a) joint dimensions (in mm), (b) macrosection of tubesheet weld (x 2)

'he implications of reducing diffusion bonding temperatures

1. Bartle, MIM, MWeldI

iffusion bonding mechanisms are briefly reviewed, especially regarding e reduction of residual porosity to a size where it can be eliminated ' vacancy diffusion within required time/temperature limits. The effects reducing temperature in relation to surface mating, and the problems residual oxide and contamination, are discussed, and the value of using termediary materials and the significance of their form are emphasised. ie practical difficulties associated with producing ultrafine surface finishes, with advanced cleaning techniques, are considered to restrict their tential as aids to reduction in bonding temperature.

TRODUCTION

ffusion bonding is gradually gaining recogni-n as a process with potential for industrial ploitation, especially for large area joints. 'pically temperatures of 0.7Tm (where Tm the melting point in degrees absolute), and essures of <7Nmm^{-2} are employed to pro-ce sound bonds between similar materials thin a few minutes at temperature. Moderate 'els of macroscopic deformation do little to omote bonding and therefore pressures are equently chosen to minimise or avoid defor-tion. An area of particular interest is nding at lower temperatures to avoid damage heat treated material, or unwanted phase insformation or grain growth.

The practical options for achieving this e essentially to:

Increase bonding time and pressure
Use an intermediary material, probably with increase in time and pressure
Use a liquid phase technique

tailed data on low temperature bonding are irse, but the value of these approaches varies h the applications and materials involved. is Paper reviews the present understanding bonding mechanisms, and notes that a key ture is the reduction of residual porosity to a level where it can be eliminated with the time/temperature limits chosen. It is usually assumed that residual porosity needs to be entirely eliminated before joints attain both the strength and ductility of the parent metal, but this may not be so.

In discussing the probably implications of the various techniques the emphasis is on the practical viability of the approaches for large joints, e.g. >10 000mm^2. Consequently the points made relate to bonding surfaces machined to 0.4μm CLA and degreased only, except where otherwise noted. Reference is restricted to metal/metal joints, but the approaches can generally be applied to joints involving inorganic nonmetals.

Bartle is Head of Diffusion Bonding and ess Wave Studies at The Welding Institute.

BONDING MECHANISMS

The present understanding of the mechanisms[1] involved in diffusion bonding have been developed from the studies by Roberts-Austen[2] and early industrial considerations.[3] In essence, two mechanisms need to occur for the formation of a sound bond, i.e. the bringing together of the surfaces on an atomic scale, and the elimination of residual oxide and contaminant. For solid state bonding the former is a three-stage process of initial yielding of surface asperities, followed by interfacial creep giving an increase in the size and number of areas of true contact until these are interspersed only with lenticular pores sufficiently small to be eliminated by vacancy diffusion mechanisms. The time required to complete the joint, and the significance of each of the three stages, is dependent on the material, the pressurisation cycle, and the thermal cycle. The key to solid state bonding, however, is the reduction of residual pore sizes to a level where they can be eliminated by vacancy diffusion mechanisms within acceptable time/temperature limits. The residual oxide and contaminant are eliminated by solution in the parent material, since normally there is very little, if any, scope for atmospheric reduction in oxides.

In special circumstances, solid state interdiffusion between suitable dissimilar materials (occurring even through residual oxide and contaminant) results in a change of composition at the interface and the formation of a liquid film. This is the basis of liquid phase diffusion bonding, where the liquid wets the faying surfaces, bridging them to form the joint. Macroscopic deformation is not essential to bond formation with either solid state or liquid phase diffusion bonding; consequently, although it can sometimes be used to advantage, it is frequently minimised or avoided altogether. It is standard practice to use intermediary materials (foils or coatings) either to aid solid state bonding or permit liquid phase bonding to occur.

SURFACE MATING

Similar parent materials alone

The practical consequence of reducing temperature is to increase the time required for bond formation. If pressure is set just to avoid macroscopic deformation it is increased as bonding temperature is decreased. To some extent this ameliorates the effect of temperature reduction on yielding and the time required for creep, but not on the time required for vacancy elimination of voids of constant size. The net increase in total bonding time may, however, restrict the increase in pressure because of long-term bulk creep considerations.

The separate development with time of different joint properties is illustrated in Fig. 1. The more rapid development of strength than ductility is a result of the formation of large numbers of small area bonds which develop high strengths when triaxially stressed, but ductility requires the elimination of at least some of the microvoids from the interface.

To obtain an indication of the effect of temperatures on bonding time, mild steel, nickel, and titanium are taken as materials of similar melting point and bonding propensity. Time v. bonding temperature data are plotted for these from various experimental results for adequate bonds, Fig. 2, and show a trend towards a logarithmic increase in time with lin decrease in temperature. The results for bo ing mild steel at 710°C illustrate the extent t which bonding temperature can practicably be reduced, although success here, Fig. 3, may a consequence of the high diffusion rate in α iron than in γ iron. The marked significanc of pressure on the viability of reducing bondi temperature is also apparent. Thus where th applied pressure is kept constant, Fig. 2, indicates that time needs to be increased about te fold per 100degC drop in temperature, but onl threefold if pressure can be suitably increase These trends are compatible with the formatic of sound solid phase resistance welds, or 'Quanta' spot welds[8] in times of 10 to 50mse at temperatures close to the melting point, bu at much higher pressures, e.g. $\sim$50Nmm^{-2} fo resistance welding.

It is important to promote sufficient yiel ing and creep to reduce the pore size to a le that can be eliminated by vacancy diffusion wi in a realistic time. Once sufficient creep has been obtained, however, which occurs relative early in the overall process, continued applica tion of pressure is not essential to the impro ment of the bond, as illustrated by the results in Table 1.

Table 1 Development of bond properties by pos bond heating without load (Ti, 6Al, 4V)

Condition	Properties	
	UTS, N/mm^2	E on 25m %
As bonded 20min, 850°C, 5N/mm^2	820	0
As above, plus 6hr, 850°C	960	10

Thus mating difficulties provide one practical limit to the extent to which bonding temperature can be lowered without the use of an intermediary material, the limit being dependent upon the maximum acceptable bondi time and the viability of increasing pressure.

Addition of intermediary materials

Where it is not possible, within acceptable time/temperature/pressure combinations, to bring the whole of the parent metal faces directly into atomic contact, the options for improving surface mating (and hence reducing bonding temperature) are to:

Table 2 Properties of joints made with and without interlayers

Materials		Bonding conditions			Joint properties		
Parent	Interlayer	Temp., ^oC	Pressure, Nmm^{-2}	Time, ksec	UTS, Nmm^{-2}	E, %	RA
Stellite	None	1150	5	1.2	80-230	-	-
Stellite	Nickel	1100	7	1.2	600-920	-	-
Nimonic	None	1100	5	1.2	695	1	3
Nimonic	None	1150	7	1.2	765	0	-
Nimonic	Nickel	1050	7	1.2	920	8	10
10Ni, 8Co, 2Cr 1Mo steel	None	1150	4	1.8	825	-	1
10Ni, 8Co, 2Cr 1Mo steel	Same composition as parent	1150	4	1.8	1170	-	1
En3B	None	900	7	1.2	250	-	-
En3B	Mild steel	900	7	1.2	330	-	-

(a) add a foil interlayer
(b) coat the parent metal surfaces
(c) use coatings and a foil together

The classic examples of the use of foils are with strong creep-resistant materials such as Stellite or Nimonic, Table 2. Here abnormally high temperatures have to be used to bond in the absence of interlayers, but, by embedding the hard parent metal surfaces into 'soft' foils, bonding temperatures can be brought down to more acceptable levels. Additionally, however, improvements in properties have been found when foils of parent metal compositions are used, Table 2.

Coatings ought perhaps to allow bonding to be carried out at temperatures appropriate to the coating material rather than the parent metal. This approach is successful in degree, Table 3 and Fig.4, but appears not to be as satisfactory as the use of a foil (presuming it bonds to the parent metal) unless very high pressures are used. The use of foil and coatings together may be essential if a foil has to be used which itself does not bond readily to the parent metal.

The alternative diffusion bonding approach is the liquid phase technique. The combination of materials from which the liquid is generated may be the parts themselves, or the parts and a foil or a coating, or intermediary materials only. The bonding temperature is approximately that at which the liquid forms and thus, to reduce bonding temperature, a change in material combination is necessary. One very successful use of this approach is in the bonding of tungsten to other materials, where the tungsten is copper-coated and a silver foil is used as the second material to give a bonding temperature of 780°C (0.3Tm of tungsten).

Table 3 Ultimate tensile strengths of joints in tantalum made at 800°C for 3.6ksec at $26Nmm^{-2}$ with various copper intermediaries

Form of intermediary	UTS, Nmm^{-2}
Ion plating	154
Foil 0.0125mm thick	246
Foil 0.0062mm thick	289

Phenomenological considerations

In bringing the surfaces into atomic contact, both long-range waviness, Fig.5, and short-range roughness should be considered, the former possibly being of greater practical concern. The benefits which might be expected from the use of a foil interlayer are most relevant to long-range waviness and the statistical occurrence of indentations on opposite faces. The advantages are a potential halving of the gap to be closed, a change in pore geometry, and displacement of material into the voids. On loading in the absence of a foil some fraction of all the peaks coincide and many more make an 'angled' approach, Fig.6a. The angled contacts can 'bind' and support a considerable proportion of the total load, greatly reducing the stress on the 'in-line' peaks thus retarding the closure of the faying surfaces. If the peaks that otherwise would make angled contact are separated by a foil, Fig.6b, the latter would be expected to bend, stretch, and shear as the surfaces are brought together. The coincident peaks will move together mainly by indentation, especially where soft foils are used.

Thus the benefits to be obtained from the use of a foil relate mainly to its effect on the residual pore size and geometry. The more readily the final pores can be diffused away the lower the temperature that can be used to effect a bond in an acceptable time. The mere inclusion of a foil can reduce the 'thickness' dimension of troughs by about half, increase the aspect ratios (flatten the pores) and the surface to volume ratios, and the foil acts as an additional vacancy sink. These all speed the final elimination of the pores by vacancy diffusion mechanisms (the slowest part of the bonding process). The increase in number of pores to be eliminated has no detrimental effect in itself, especially since the total volume is more likely to decrease than increase.

The presence of the foil should result in greater movement together of the parent metal faces, under the same applied load, than occurs in the absence of a foil. This will further reduce the size of the pores to be eliminated. Even small additional movements, e.g. $\sim 1\mu m$, could significantly reduce, Fig.6c, the volume of the pore to be eliminated by vacancy diffusion.

If a parent metal were coated with a material having, for example, a 20% lower melting point but with otherwise similar bonding propensities in relation to its melting point as the parent, it might be expected that if the parent metal could be bonded readily at 0.7Tm the coated parent metal could be bonded at 0.7 x 0.8Tm of the parent metal. In practice, however, coatings are frequently sufficiently thin that restraint by the parent metal is a serious consideration. Hence coated surfaces may behave (mechanically) more like the parent metal than the coating. This may explain why O'Brien et al[9] needed to use very high pressures, Fig.4 (more than twenty times those normally employed), to bond silver coated steel at 0.57Tm of silver, even though the steel faces had been smoothed by lapping.

Thus from the mechanical aspects of mating alone there is reason to believe that foil interlayers should offer advantages over coatings. The restraint argument applied to coatings can however be extended to foils,[10] although the effect of creep in the two examples has yet to be quantified, and, for both, thicknesses will be an important consideration.

The thicker a coating the more likely it is to behave like itself and not its backing, and the thinner a foil the less should be the effect of its bulk restraint. Effectively therefore (as the various examples show), advantage can be gained from the use of an intermediary, but whether it is of practical significance depends upon both the parent material and the form of the intermediary, with respect to the bonding conditions employed.

SURFACE OXIDE AND CONTAMINATION

There are only a few materials where the oxide or contaminant give rise to problems at bonding temperatures of about 0.7Tm and, with these, increases in bonding temperature lead only to a marginal improvement until close to the melting point. Here use of suitable intermediary materials can be beneficial in allowing bonding temperatures to be reduced to more desirable levels. For example, a steel or nickel foil will absorb carbon contamination from the surface of graphitic cast irons, and suitable coatings on tough pitch copper allow the oxide problem to be overcome with solid state diffusion bonding.

Apart from these special situations, oxide and contaminants introduce limitations only when bonding temperatures are reduced and their solution becomes too slow to permit bonding. To overcome this, clean adherent coatings of noble metal need to be used, since these do not oxidise. Such coatings can be most effective especially on finely finished surfaces bonded using high pressures, Fig.4. The use of foil interlayers alone can be virtually ineffective under solid state conditions. For example, no bonding was observed when joining was attempted with a silver foil between mild steel pieces held for 20min at 520°C under a pressure of $5Nmm^{-2}$ (cf Fig.4). Alternatively, liquid phase techniques can be used where the liquid may dissolve or erode the oxide.

DISCUSSION

Temperature reduction using conventional techniques

Temperatures of about 0.7Tm are typically used to achieve bonding within a few minutes or less, with pressure close to the bulk yield stress. The yielding of surface asperities and microcreep result in residual pore sizes sufficiently small to allow their rapid elimination. Without resorting to the use of intermediary materials, there is scope for a reduction in bonding temperature, but probably only to about 0.5Tm in the best examples. This is achieved by greatly increasing bonding time to compensate for reduction in diffusion rates, and probably increasing pressure to compensate for increases in yield stress and creep resistance. Practical considerations limit the temperature reduction which can be obtained in this manner. Twenty-four hours (86ksec) is possibly the practical time limit, and buckling problems or equipment capacity (for large joint area, e.g. plate cladding) restrict pressure increases.

The other factor to be considered is the elimination of residual oxide and contaminants by solution in the parent material. As bonding temperature is reduced solution rates also fall rapidly and may become the limiting factor.

Intermediary materials can be used as aids in the reduction of temperatures. In solid state bonding, foils promote mating by modifying residual pore size and geometry so that they can be diffused away within set time/temperature limits, but rarely provide advantage with regard to surface oxide or contamination. On the other hand coatings (particularly noble metal) can provide surfaces where the oxide and contaminant is sufficiently controlled to permit bonding within the limits required, but usually are of limited benefit regarding mating itself. In practice, at lower temperatures, e.g. below about 600°C, both approaches will probably be required if unduly high pressures are to be avoided. Both coating and foil will probably need to be noble metal to avoid further oxide and contaminant problems. From cost and melting point considerations silver would be favoured, but in certain circumstances other noble metals, e.g. gold, may be preferred.

Thus by using silver coatings together with a silver interlayer it would be expected that virtually any material with a melting point above about 400°C could be bonded at temperatures down to ~350°C (0.5Tm for Ag). In fact bonding with silver at even lower temperatures (and without a foil) has been achieved[9] by using very high pressures. On the other hand the pressures required may be prohibited with some parent materials or assembly geometries. In this context it should be understood that the limiting pressure for a given material in bulk form at a particular temperature can be greatly exceeded, with benefit, when the same material is used as a thin intermediary.

The use of silver, therefore, can be regarded as a key feature in the future development of low temperature bonding techniques. Where silver or a similar noble metal is not acceptable the choice of an alternative for solid state bonding is currently very limited. Tin may be usable at very low temperatures, but there is a temperature gap for which there is no suitable pure metal and for which an alloy may need to be developed.

If liquid phase techniques are employed the mating problem is obviated, because the liquid can bridge any small gaps between the parent metal surfaces which would normally result in porosity. Moreover, oxide problems can be ameliorated if the oxide is dissolved or eroded by the liquid. Although there is no specific need for noble metals, experience suggests that here silver may again be preferred. The scope for the parent metal to serve as one of the materials forming the liquid is often very limited. If, however, all the materials that have to be reacted together are introduced between the parent metals, the scope for variation is quite large. Ideally, to avoid wetting problems in such situations, one or more of the materials introduced should be in the form of coatings on the parent metals.

Alternative approaches for temperature reduction

Improvement in surface finish, to reduce the extent of yielding, creep, and possibly vacancy diffusion necessary to bring the surfaces into atomic contact, might appear to be a reasonable route to pursue. This would be practicable, however, only with small assemblies, e.g. up to 100mm across. Even at this size it is unlikely to be economic and with larger joints adequate control of long-range features would be difficult to achieve. Thus for most applications the surface finish of 0.4 µm CLA currently recommended can be taken as a practical limit.

Similarly advanced cleaning techniques such as ion bombardment can be used to remove residual oxide and contaminant. Their value, however, as a direct replacement for conventional cleaning has yet to be established. In this respect the major drawbacks are twofold: firstly, cleanliness rather than surface finish aspects is normally improved, so that intermediary materials, which may themselves need cleaning, are still required for mating; secondly, the cleaned surfaces should not be exposed to air before bonding. Preferably therefore the cleaned surfaces have to be transferred to a bonding machine in a vacuum, or cleaning and bonding need to be carried out in the same equipment, both approaches posing further practical constraints. Thus advanced cleaning techniques as a direct substitute for conventional cleaning may be restricted to permitting temperature reduction for small but otherwise important assemblies.

Indirectly, however, the advanced cleaning techniques may prove exceptionally valuable. All present indications point to the use of coatings and foils in combination as the key to low temperature bonding, but if adequate joint strengths are to be obtained high quality, adherent coatings are required. With low bonding temperatures, any improvement in coating adhesion during the bonding cycle may be minimal, and therefore the adhesion must be good, 'as-deposited'. Here parent metal cleanliness is an important factor. Taking this into account, together with the very rapid general development of vacuum deposition techniques and an increasing interest in their use on ecological grounds, the wide use of combined vacuum cleaning and deposition techniques as a preparation for low temperature bonding should be expected.

Joint properties

The extent to which bonding temperature can be reduced without resort to an intermediary material will depend upon the properties required. Once another material has been introduced, joint properties and behaviour are likely to be modified, and care is required to avoid the formation of excessive amounts of brittle intermetallic compounds. With axially stressed joints, tensile strengths several times those of the bulk intermediary material can be achieved, Fig.4, but actual values will be influenced by interlayer thickness and joint geometry. Other properties, particularly ductility, and shear, torsional, and fatigue strength, may be poor in relation to those of the parent metal if this is far stronger than the interlayer material in bulk form. Also, electrochemical problems may be introduced in relation to corrosion, stress corrosion, etc. Thus, although intermediary materials can be used to produce sound strong joints at low temperatures, extra caution will be required regarding their application.

CONCLUSIONS

1 Where sound joints can be effected without the use of an interlayer within a few hundred seconds at temperatures around 0.7Tm, there is scope for bonding at lower temperatures, but probably only down to 0.5Tm at best, by using much longer times at temperature and increasing bonding pressure as far as possible

2 Alternatively, reduction of solid state bonding temperature, to less than 0.5Tm, can be achieved with the aid of intermediary materials particularly noble metals. Again high pressures and long bonding times will normally be required

3 Foils are preferred to aid the mating of surfaces, but coatings are preferred to avoid oxide and contaminant problems. Both coatings and foils will probably need to be used together for bonding below about 600°C

4 Vacuum cleaning and coating techniques used together should play a major role in the future development of low temperature bonding procedures.

5 Liquid phase techniques can also be used to reduce bonding temperature, and, when both the required dissimilar materials have to be introduced into the joint, coatings are preferred to minimise wetting problems

ACKNOWLEDGEMENT

The author would like to thank several colleagues, particularly Dr I.A.Bucklow, for fruitful discussion during the preparation of this Paper.

REFERENCES

1 BARTLE, P.M. 'Basic features of diffusion bonding', in 'Diffusion bonding as a production process' to be published by The Welding Institute.

2 ROBERTS-AUSTEN, Sir W. 'On the diffusion of gold in solid lead at ordinary temperature'. Procs Roy. Soc., 67, 1900, 101-105.

3 KINZEL, A.B. Adams Lecture 'Solid phase welding'. Welding J., 23 (12), 1944, 1124-44.

4 STOCKHAM, N.R. and WESTGATE, S.A. 'Variation of tensile properties with bonding parameters for diffusion bonded En3B steel'. Welding Inst. Report 7225/7/75, 1975.

5 OWCZARSKI, W.A., KING, W.H., and O'CONNOR, J.W. 'The tensile properties and fracture characteristics of titanium diffusion welds'. Welding J., 48 (9), 1969, 377s-83s.

6 KAZAKOV, N.F., SHISKOVA, A.P., and CHARUKHINA, K.E. 'Vacuum diffusion bonding of titanium'. Automatic Welding, 16 (10), 1963, 71-4.

7 SHORSHOROV, M.Kh et al. 'The problem of quantitative estimation of pressure welding conditions'. Welding Production, 14 (7), 1967, 24-31.

8 'Diffusion bonding: off the shelf?'. Iron Age, 202 (7), 15 August 1968, 85-6.

9 O'BRIEN, M., RICE, R.C., and OLSON, D.L. 'High strength diffusion welding of silver coated base metals'. Welding J., 55 (1), 1976, 25-7.

10 BUCKLOW, I.A. Private communication

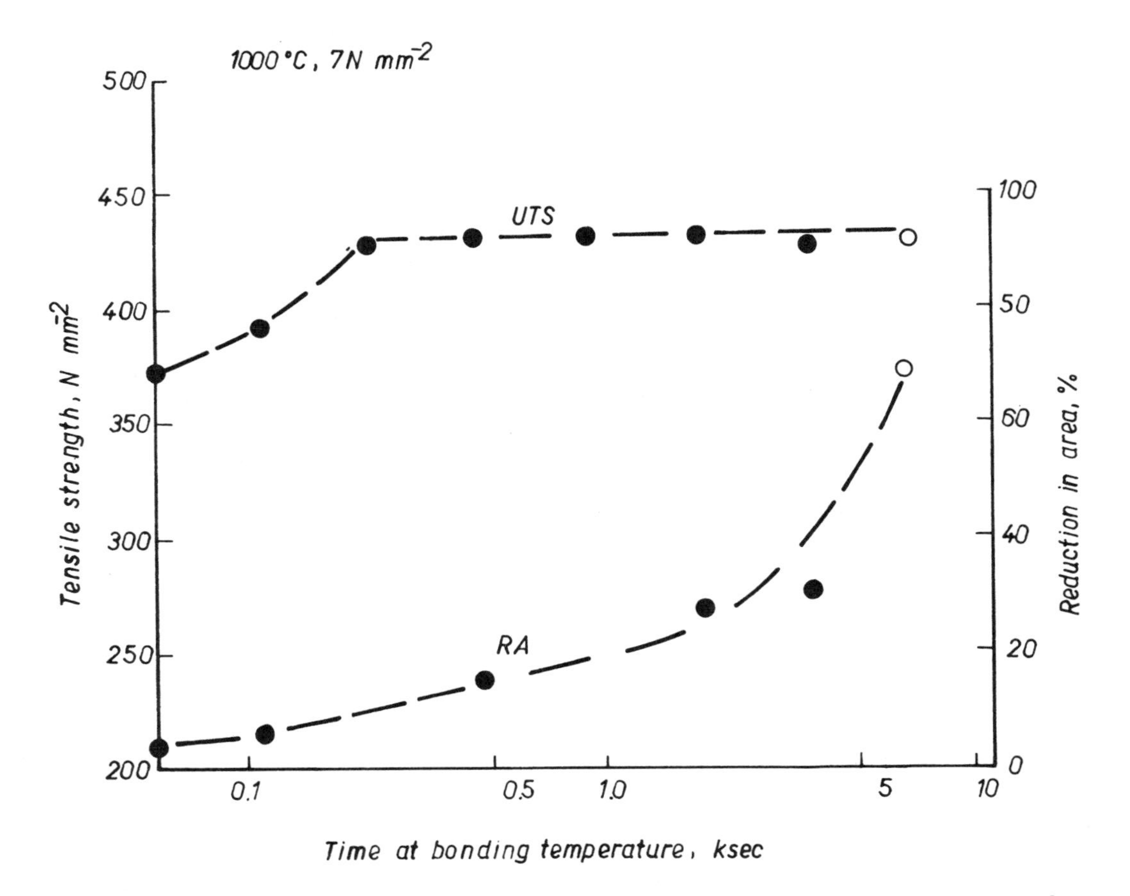

1 Separate development with time of joint properties in En3B (after Stockham and Westgate[4]). Failure: ● — at joint; ○ — away from joint

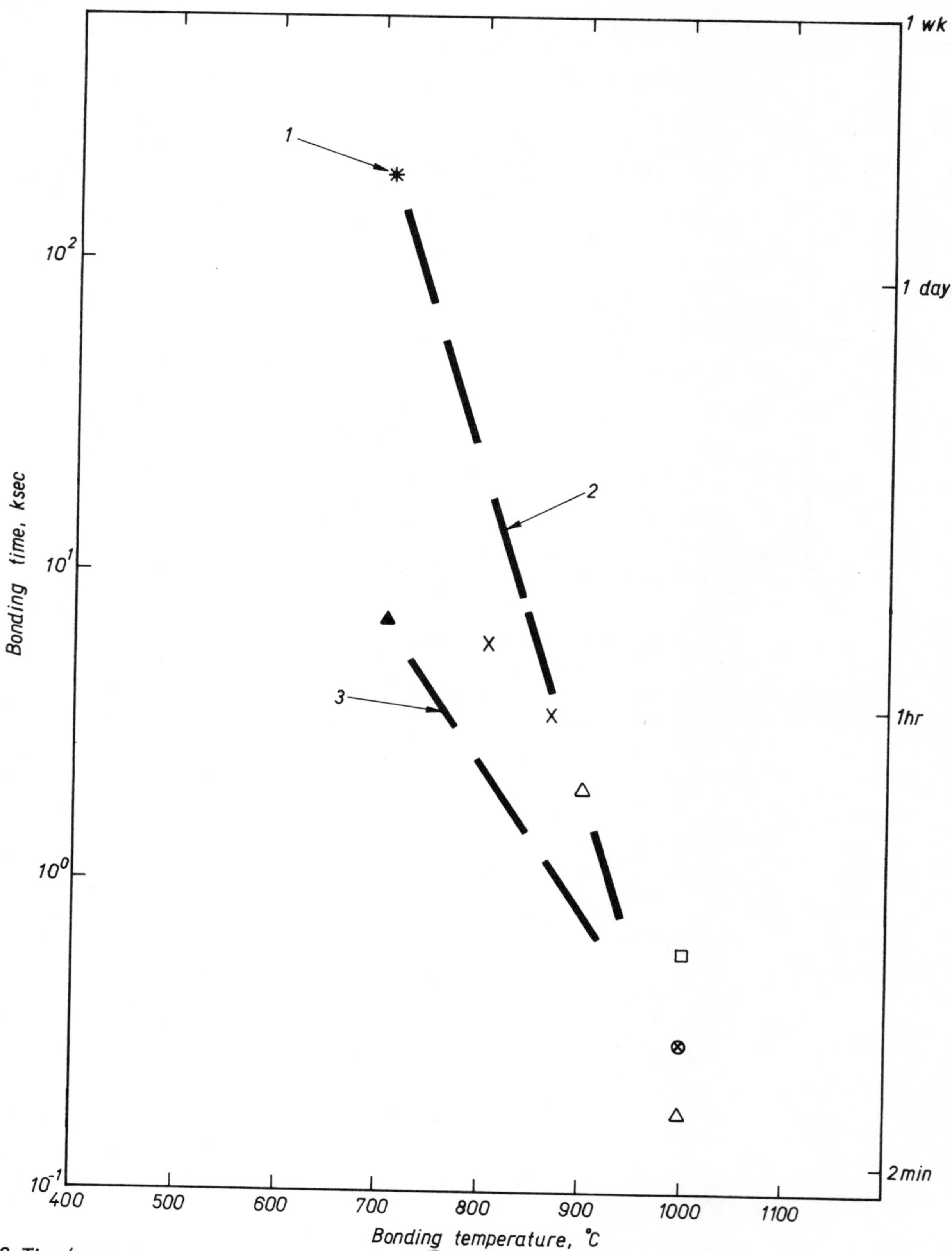

2 Time/temperature combinations used to produce 'sound' joints in three materials of similar melting point: △ – mild steel 7N mm^{-2}(4); ▲ – mild steel 20N mm^{-2}(4); X – titanium 3N mm^{-2}(5); ○ – titanium alloy 2N mm^{-2}(6); □ – nickel 6N mm^{-2}(7). 1 – bonding time probably required for mild steel under 7N mm^{-2} (extrapolated from data for shorter times at 710°C); 2 – slope tenfold per 100degC (appropriate to bonding at constant pressure) (note: intended only to indicate general trend for all results plotted for pressures of 7N mm^{-2}); 3 – slope threefold per 100degC (appropriate to bonding at increased pressure at lower temperatures) (note: indicates lessening of increase in time required for bonding at lower temperatures, occasioned by increase in pressure)

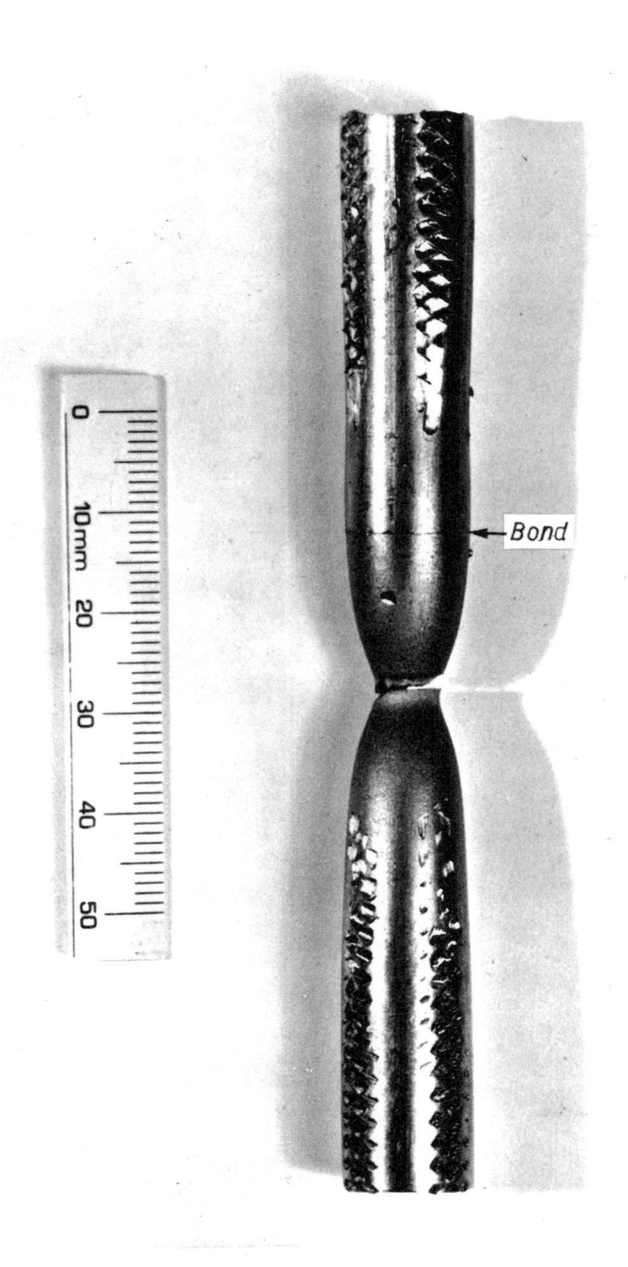

3 Tensile failure of mild steel specimen bonded for 7ksec (2hr) at 710°C under pressure of 20N mm^{-2} without interlayer (after Stockham and Westgate[4]

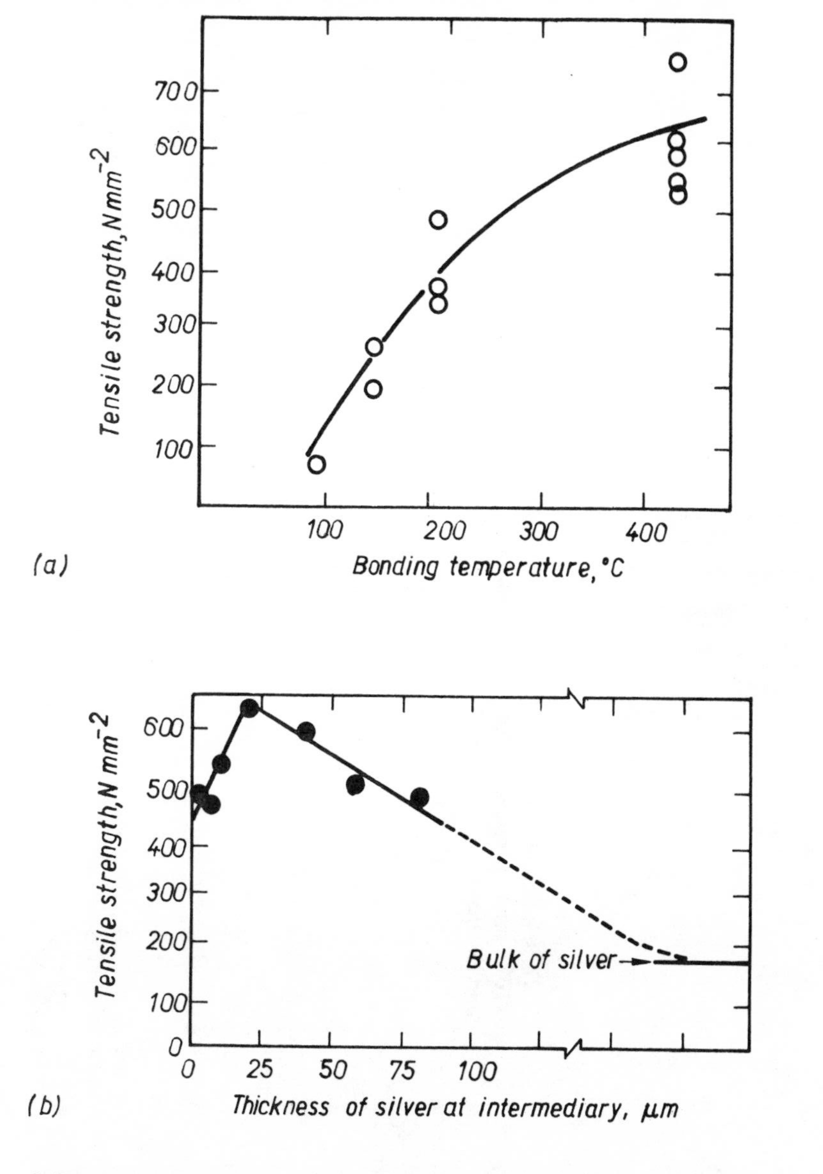

4 Properties of joints in lapped and silver plated maraging steel: (a) variation with bonding temperature, (b) variation with silver thickness; bonding temperature 425°C. Bonding: pressure – 140N mm^{-2}; time – 0.6ksec (after O'Brien et al[9])

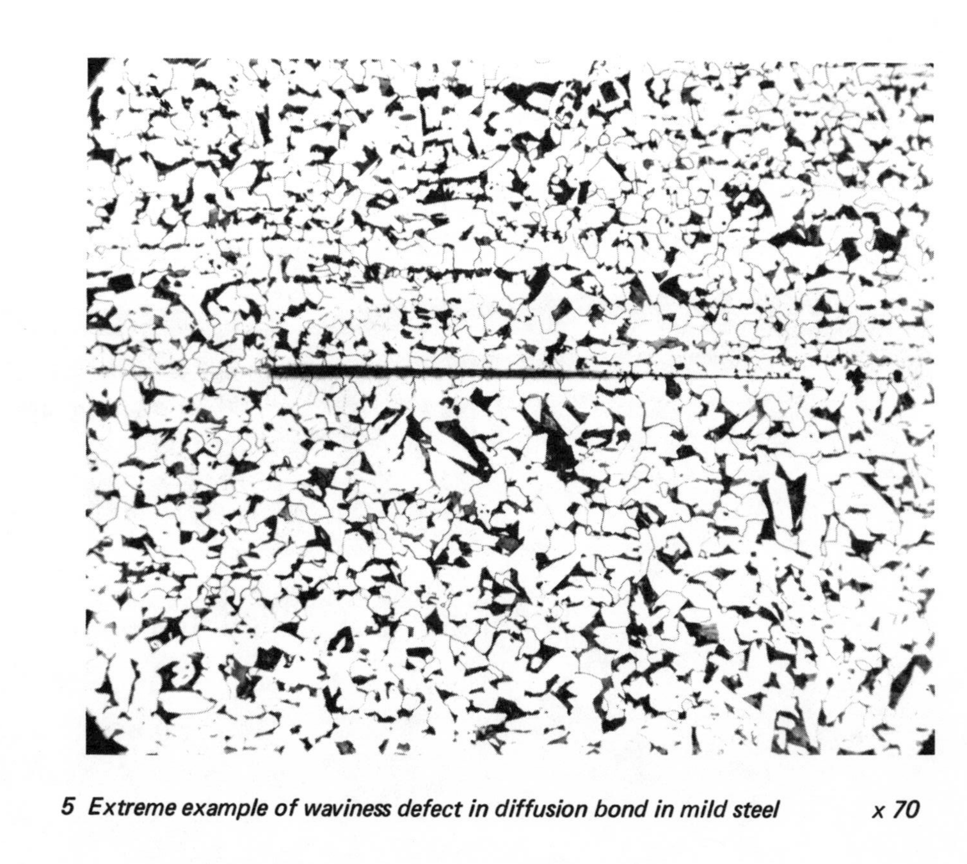

5 Extreme example of waviness defect in diffusion bond in mild steel x 70

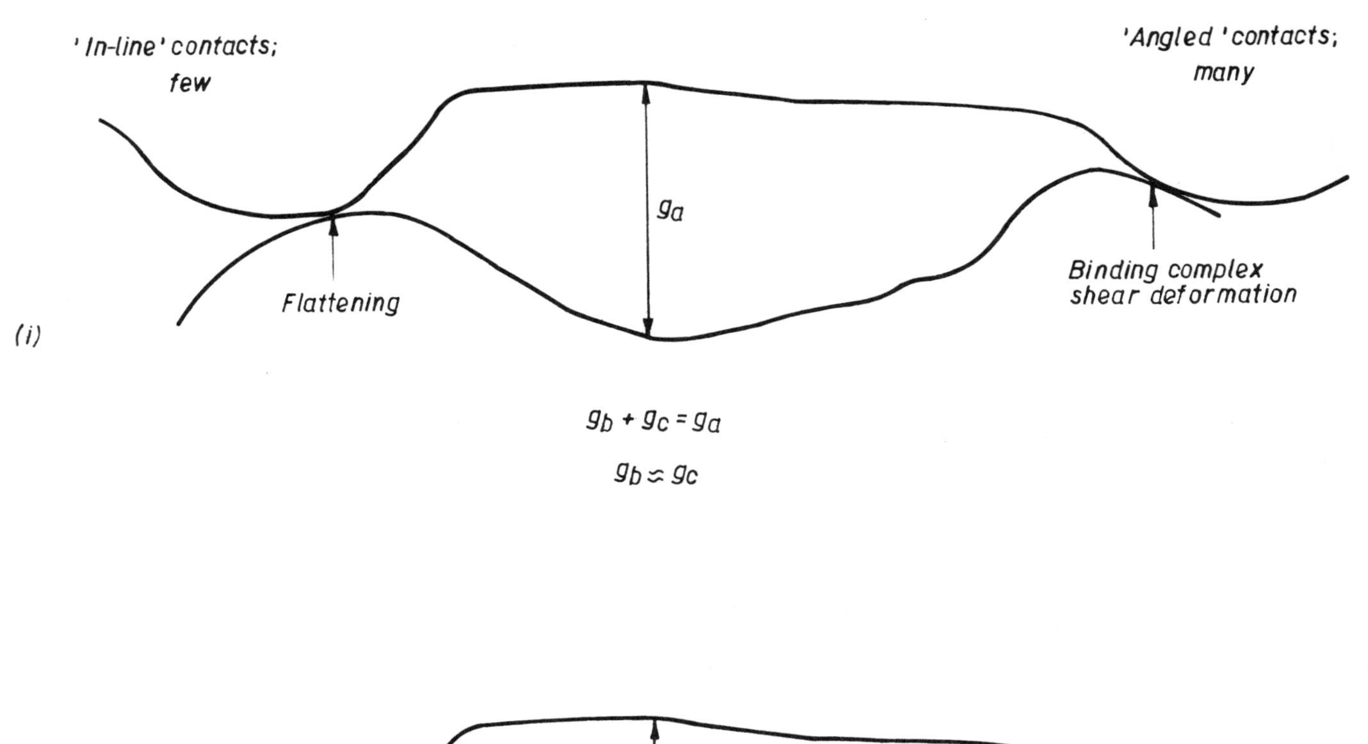

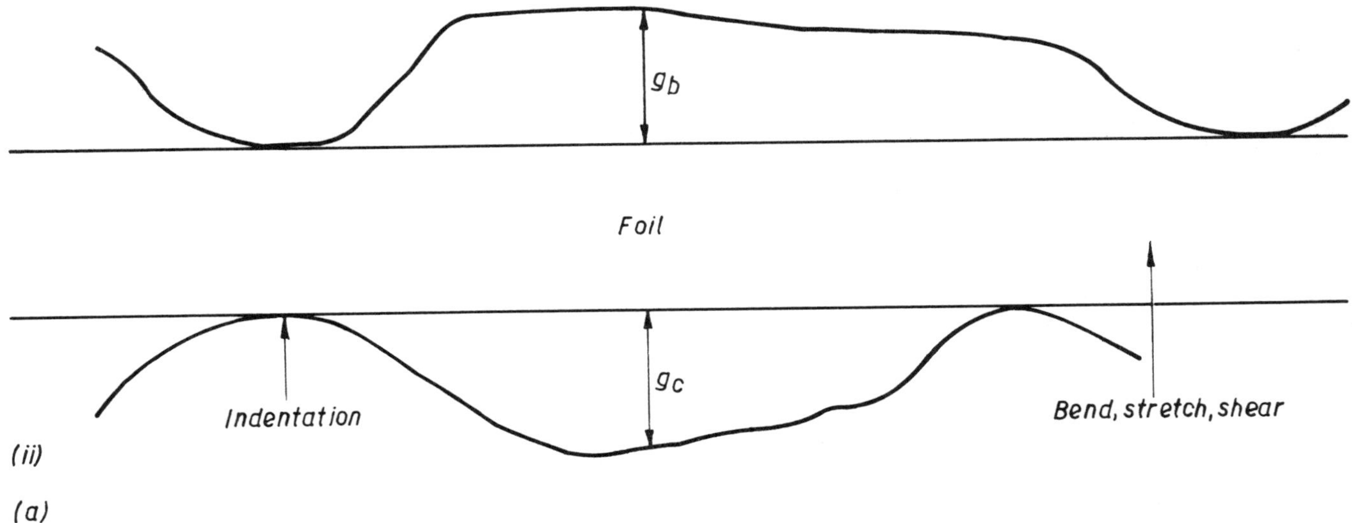

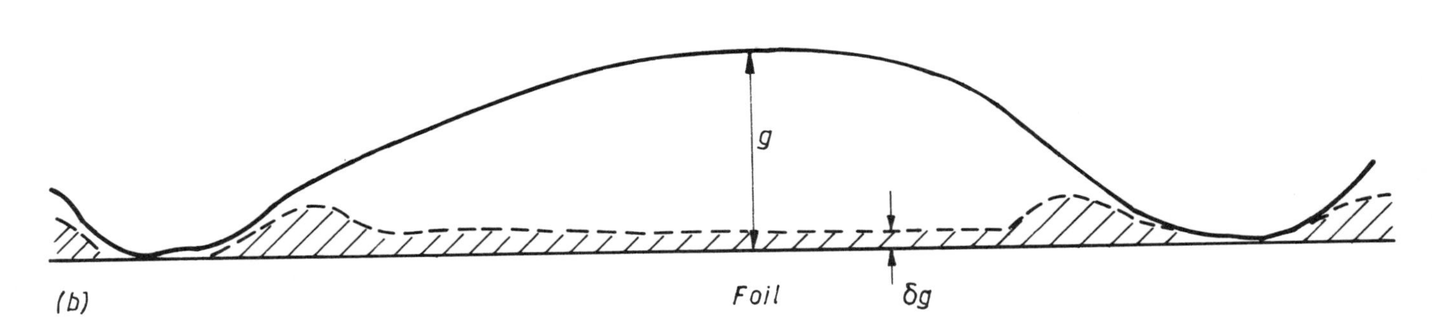

6 Schematic representation of mating of surfaces: (a) comparison of (i) without foil and (ii) with foil situations, (b) reduction in pore volume associated with small indentation, δg (shaded area represents significant proportion of original pore volume)

Recent developments and applications of the gas-metal-plasma-arc process

M. Areskoug and H. Widgren

The gas-metal-plasma-arc (GMPA) process was introduced for the first time at The Welding Institute Conference 'Exploiting Welding in Production Technology' in London in April 1975.[1] A continuous development in the improvement of the equipment and finding new applications has taken place ever since. Careful studies of metal transfer and the mode of solidification have served as the basis. The analysis of the process characteristics gives an idea of the possibility to yield a uniform and reproducible welding result. Today this is one of the most important factors when calculating the total economy of a welding procedure. Another important factor is the increasing materials costs. All possibilities of saving on filler materials in weld surfacing are of importance. This Paper gives results from studies of the fundamental characteristics of the GMPA process and its correlation to practical results. New developments based on this are described and applications presented.

PROCESS FUNDAMENTALS: RESULTS OF STUDIES

The GMPA process is based on advanced plasma technology and electronics. It has previously been described[1] and the principles will be repeated here only briefly. Figure 1, a sketch of the process, shows the two separated current circuits, one for transferred plasma arc and another for the filler wires. The current from power source no. 1 (current-controlled) is fed through the plasma torch and the transferred arc to the workpiece. Power source no. 2 (voltage-controlled) feeds current over sliding contacts to the filler wires and through the upper part of the plasma arc to the torch. By this system it is possible to control the amount of deposited filler metal independently of the penetration into the base metal. The filler metal is transferred to the weld puddle as droplets in the plasma arc jet. This will be discussed below. To have a wide weld bead the head is oscillated. The welding sequence is started by preheating the base metal with the plasma torch until a weld puddle is formed. Then the filler metal addition starts and when the desired chemical composition is reached the travel of the workpiece starts. By this sequence exactly the same properties are ensured in the start point as in the rest of the clad layer. When stopping, a slope-down sequence for both the filler wires and the plasma arc eliminates the risk of defects as pipes in the stopping area.

The droplet formation and transport to the weld puddle are very important for weld metal quality. The drops are formed from a preheated wire tip in the plasma arc. Current densities within the working range of the equipment — 400A maximum, 1.6mm diameter wire — give a very stable and regular drop formation. Low current densities give larger drops, but this creates problems only at extremely low

Mr Areskoug and Mr Widgren are with AGA Welding, Lidingö, Sweden.

deposition rates from the normal working range of the equipment. It is possible to adjust the stickouts and power source to give the optimum current densities for filler wires with different electrical properties. This is an important development to enable the use of certain material. The droplet formation can be disturbed by the formation of oxide slag films on the drops. This leads to larger drops and irregular pinching of them. An oxidising atmosphere and slag-forming elements in the filler metal may cause these disturbances. By using only inert or reducing shielding gases and ensuring good gas protection this problem is solved. Slag-forming elements create problems only when using certain flux-cored wires.

The weld bead geometry is determined by the surface tension and viscosity of the weld puddle only. The shielding gas influences especially the edge angle between the base and weld metals. The addition of hydrogen to the shielding gas gives a larger angle than pure argon. The absence of slag on the surface gives a very flat surface without sink in the middle of the bead, Fig. 2. By adjusting the dwell times for the oscillating movement of the welding head the heat balance can be controlled to ensure complete penetration into previous beads and yield a flat overlap surface.

The wetting characteristics for the weld puddle are very important for a good bead geometry. When using filler metals containing high amounts of easily oxidising elements, such as titanium and aluminium, the formation of surface oxides must be prevented by ensuring an extremely good gas shield. This can easily be done on the welding head.

The transfer of alloy elements from the filler wires to the weld metal happens in an inert or slightly reducing atmosphere. Numerous investigations have shown that the droplets carry almost completely the alloy elements from the wire tips to the weld puddle. The only exception is the decarburising effect found especially when using reducing shielding gases. This effect is very useful when surfacing with corrosion-resistant materials; Table 1 shows an example.

The oscillating movement affects the solidification both by sometimes breaking the columnar grain growth and forming a ripple pattern. This pattern is influenced by the choice of shielding gas, filler metals, and the welding parameters. By using shielding gases with additions of hydrogen (5-15%) the surface smoothness is improved.

Table 1 Chemical composition of clad layers made with argon +15% hydrogen shielding gas

Material	Alloy element, %				
	C	Si	Mn	Cr	Ni
Base metal	0.13	0.34	1.30	0.22	0.21
Filler metal	0.014	0.90	1.75	20.1	9.7
Weld metal	0.020	0.90	1.73	18.3	9.0

WELDING WITH TUBULAR WIRES

A wide range of special alloys is very difficult or impossible to draw as wire. This is particularly true for hardfacing alloys. Equipment has been developed to solve the problems involved in extending the working range of the GMPA process also to this extensive field. It involves the possibilities mentioned above to adjust the stickouts and power source for the difference of solid wires in electrical properties. It is also essential to ensure that the wires used contain only metallic components and no fluxing agents. This means that, for example, self-shielding wires cannot be used as the drop transfer and weld pool formation is disturbed, as mentioned above.

As an example of welding with tubular wires, a test weld with a cobalt-base alloy (Stellite) is reported.

When corrosion- and abrasion-resistance is required cobalt-base alloys are often used. Surfacing with these alloys has been a very time-consuming process. With the introduction of cobalt-base alloy cored wires it has been possible to use automatic welding methods. Weld tests have been performed with the GMPA process using cored cobalt-base alloy filler wire. The wire was of the 1%C-26%Cr-5%W type, 1.6mm diameter. The weld metal composition is shown in Table 2 for the first and second layers. The dilution ratio is about 12%. The microstructure over the fusion line parent metal/weld metal is shown in Fig. 3. The zone near the fusion line is free of carbides and represents the gradient in chemical composition. The microhardness figures are 155HV in the base metal, 302HV in the gradient zone, and 397HV in the weld metal. There was no sign

Table 2 Chemical composition of clad layer made with Co-base (Stellite) cored filler wire

Material	C	Cr	W	Co
Filler wire	1.0	26	5	Balance
First layer	0.75	22.0	3.8	51.2
Second layer	0.81	25.5	4.4	58.1

of cracks in the samples examined. The weld metal surface was very smooth and the beads were approximately 5mm thick and 70mm wide. The deposition rate in this test was approximately 16kg/hr. The welding performance was very smooth and stable throughout the test with 180A current through the plasma torch and 340A through the wires.

WELDING WITH ALLOYS HAVING HIGH CONTENT OF EASILY OXIDISED ELEMENTS

As described above, high contents of aluminium or titanium, for example, may cause problems because of the formation of oxide films on the droplets and the weld puddle. A test weld has been performed showing that by improving the gas shield and using high purity shielding gases it is possible to have good results. By using a curtain of fireproof material the gas shield will be sufficient for most purposes. As an example test welds with aluminium bronze are reported.

To minimise metal-to-metal wear aluminium bronzes are often used. Test welds with an alloy of 85%Cu-11.5%Al and 3.5%Fe have been performed. The filler metal was in the form of 1.6mm diameter spooled wire. The wire guides had to be extended to give the correct preheat to the filler wires. Because of the high content of aluminium it was necessary to improve the gas shielding by using curtains around the weld bead. The first layer was welded with 160A to the plasma torch and 290A to the wires, and the second with 120A and 300A respectively. The first layer had a thickness of approximately 3mm and a width of 70mm. The deposition rate was approximately 10kg/hr for the first layer and 13kg/hr for the second.

Figure 4 shows the microstructure of the area adjacent to the fusion line. There was no sign of cracks in the samples investigated. It should be observed that the parent metal is an austenitic-martensitic stainless steel. The microstructure is normal, with the stainless steel represented as the round grey phase in the matrix. The welding performance was very stable and the surface of the weld beads smooth. The chemical composition of the first layer was 9.4%Al-7.3%Fe and Cu balance which corresponds to a dilution of approximately 10%. The hardness of both the first and second layers was 160HV, Fig. 5.

In Table 3 some results from the test welds are summarised. It illustrates the flexibility of the process. Each material has its own characteristic which must be considered when welding. When welding Ni, Ni alloys, NiCu

Table 3 Chemical composition of clad layers made with the GMPA process

Alloy type		Chemical composition, %								
		C	Si	Mn	Cr	Ni	Mo	Nb	Cu	Remarks
70Ni20Cr	FM	0.021	0.22	3.00	19.8	72.8	-	2.75	-	Fe 0.28
Second layer	WM	0.015	0.19	3.04	19.2	71.8	-	2.6	-	Fe 2.88
65Ni35Cu	FM	<0.15	<1.25	3.5	-	65	-	-	Bal.	<Fe 2.5
First layer	WM	0.069	0.71	3.04	-	55.6	-	-	23.0	Fe 15.2
Second layer	WM	0.063	0.74	3.36	-	62.8	-	-	25.7	Fe 4.75
25Cr22Ni2Mo	FM	<0.020	≤0.20	4.5	25	22	2.1	-	-	N 0.1
One layer	WM	0.017	0.12	4.42	24.2	21.6	1.95	-	-	N 0.108
Two layer	WM	0.020	0.14	4.67	24.9	22.1	2.09	-	-	N 0.105
19Cr12Ni2Mo	FM	0.020	0.90	1.8	19	12	2.7	-	-	-
One layer	WM	0.026	0.94	1.54	18.2	10.9	2.52	-	-	-

WM = weld metal
FM = filler metal

alloys, there is a risk of end crater cracks. This can be avoided by a slope-down sequence both for the plasma arc and filler wires, which can be programmed on the GMPA unit. Other materials such as cored wires have completely different physical properties which need some correction of the characteristics of the process which also can be programmed. Materials such as nickel need higher energy to ensure complete penetration because of differences in the melting point; this can be obtained by adjusting the plasma torch.

AUTOMATION

The trend towards more mechanisation and automatic systems within welding technology puts new requirements on the equipment. Stability in the performance and fully controllable parameters are necessary. The GMPA process belongs to the new generation of welding equipment which uses fully electronically controlled power sources and all parameters are of the feedback type, not only the stability of the equipment itself but also its tolerance to other variables such as variation in surface-to-welding head distance and bead overlapping. Tests have shown that variations of ±6mm surface-to-welding head distance do not affect the penetration or other properties of the weld metal. This is very important in production welding where relative movements of the workpiece, e.g. on turn rolls, and welding head, e.g. welding column, occur. The overlapping can be controlled by a sensor system. No other variable than the penetration and filler metal addition affects the chemical composition of the bead; compare this with the submerged-arc processes where variations in flux and flux consumption influence the weld metal chemistry. The properties of both the start and stop areas can be controlled by easily programmable sequences to ensure consistent chemistry and microstructure. All these characteristics of the GMPA equipment make it extremely suitable for computer control and also to work in combination with robots.

BUTT-WELDING UNIT

Butt welding with high deposition rates, high speed, is usually performed with the submerged-arc process. These high speed, high deposition, rates are accompanied by relatively high heat input to parent metal. For many types of material high heat input is disastrous for the properties of the heat-affected zone (HAZ) and also increases the risk of dimensional distortion. Examples of materials are high alloyed stainless steels and LPG steels. By using the GMPA concept for butt welding it is possible to considerably reduce the heat input at the same deposition rates and welding speeds, and at the same time to have improved control over the bead geometry.

The equipment works on the same basic principle as the GMPA weld surfacing equipment. In butt welding, however, there is no oscillating unit, and thus the gas shield of the head can be smaller. For further control of the heat balance a preheat plasma torch can be added. This ensures a good penetration to the edges. An example of a GMPA butt welding head is shown in Fig. 6. In this design there are three plasma torches, one for preheat, one for the root run, and the third for filling up the groove. By this combination of torches it is possible to weld up to 20mm stainless steel in one pass with full control over both the root and the reinforcement on the top. The root is welded with the keyhole technique with backing gas. Figure 7 shows a cross-section of a stainless steel weld made with a GMPA head of this design. The most remarkable properties are the absence of reinforcement both at the top and the root and the favourable penetration geometry. In addition, very low dimensional distortion was experienced. The solidification pattern shows an equiaxed zone in the middle of the weld metal which reduces the risk of cracking. Until now the maximum deposition rate has been 22kg/hr and the maximum welding speed 0.7m/min. The heat input to the parent metal is about half that of the comparable submerged-arc welding procedure. This is because of the high efficiency of the plasma process and the separated preheated filler metal addition. With about 200A through the plasma torch and 350A through the wires, the same speed and deposition rates can be obtained as with 1000A submerged-arc welding, with half the heat input per unit length.

The application of the GMPA butt welding unit is most favourable when the benefits of the low heat input can be utilised, as in high alloyed steels and low alloyed steels with high requirements on the HAZ impact strength, e.g. LPG steels.

APPLICATIONS OF THE GMPA PROCESS

Weld surfacing of cylindrical workpieces with high speeds and a smooth surface is a difficult problem, especially at small diameters. One problem is the heat balance: too high a heat input gives overheating of the workpiece. Another problem is to control the weld pool from flowing away. This was encountered in the manufacture of bottom nozzles for a boiling water nuclear reactor at the Uddcomb Sweden AB plant. The diameters of the smallest nozzles were only 90mm. The requirements for the weld surface were high because of the inability to carry out a nondestructive test. To manage the surfacing problem the GMPA process was used. This gave possibilities to keep the weld pool under control despite the small radius of the workpiece. Although the GMPA process gives a comparatively low heat input for the deposition rates used, it must be balanced. In heavy pieces it is done by maintaining the preheat with additional heating. In small workpieces, a 90mm diameter nozzle for example, cooling must be applied to avoid excessive temperature. In this instance it was arranged by a water-cooling unit put in the central hole of the nozzle, which kept the temperature to a steady 60°C. Figure 8 shows the setup used in production.

The welding was performed with stainless steel, type 20%Cr-10%Ni extra low carbon, and a deposition rate of 12kg/hr; the result can be seen in Fig. 9. The bead overlaps are very smooth and the chemical composition of the one-layer surface was 0.020%C-0.90%Si-1.73%Mn-18.3%Cr-9.0%Ni and ferrite content of 9%. The dilution of the weld metal was approximately 8%. This procedure was approved for welding nuclear nozzles and two complete sets have so far been manufactured by this procedure. Side bend, intergranular corrosion, and hardness tests of the procedure qualification testpiece showed good results. The extensive PT and UT test of the production welded nozzles gave extremely good results, and only single defects were found.

This application example demonstrates how high productivity can be combined with high weld metal quality on strongly curved surfaces. When welding with nickel-base and other nonferrous alloys the dilution from the base metal must be minimised to obtain properties comparable with solid material. This means that for most welding methods a multilayer technique must be used. Furthermore, the difference in melting point between the alloys gives a risk of lack of fusion when trying to weld with low dilution. To solve these problems, and others encountered when weld cladding tubesheets for nuclear steam generators, the GMPA process has been used.

By using the possibility to control the transferred plasma arc independently from the filler metal addition and the control of the oscillation mode, safe and regular penetration can be ensured even at a dilution of 8 to 10%. In this way it is also possible to reduce the number of layers necessary to obtain the required chemical composition. It is possible to obtain a ferrous content of 10-12% in the first layer with a 5mm thickness and a deposition rate of 16kg/hr. The second layer can then be welded with 8mm thickness if necessary and below 3% iron content and a deposition rate of 22kg/hr.

It is sometimes required to normalise the HAZ of the base metal. By reducing the thickness of the first layer and keeping the dilution to approximately 10%, the second layer can be used as a normalising pass. The thickness of this pass can be chosen by the filler metal addition circuit and the required heat penetration can be set by the transferred plasma arc.

A two-layer welding procedure has been developed for weld surfacing with nickel-chrome alloy (AWS NiFe-3) for nuclear steam generator tubesheets. The procedure qualification tests have shown very good results and the procedure has been approved. The production of the first set of tubesheets has started.

CURRENT DEVELOPMENTS

The further development work with the GMPA process is concentrated on three areas:

1. Equipment for the inside cladding of pipes and nozzles: the aim is to be able to clad the internal diameter with very high deposition rates and smooth surfaces and high weld metal quality
2. Further developments of the multihead butt welding unit: the possibility to combine the high deposition rates with low heat input gives very promising results for, especially, alloyed steels
3. Adapting the process for new types of advanced material, e.g. in hardfacing: very promising results have been obtained with complicated high alloyed hardfacing materials

CONCLUSIONS

The GMPA process has a unique set of useful properties. These are used for continuous development work to solve welding problems both as regards production economics and weld metal quality. With the GMPA process it is possible to have a more consistent quality in the welding result than before, and this is combined with high deposition rates and economical filler metals. As the production continues to show excellent results, further applications within the industry will be made.

ACKNOWLEDGEMENTS

The authors wish to thank Uddcomb Sweden AB for giving permission to publish results of production welding.

REFERENCE

1 SMARS, E. and BACKSTROM, G. 'Gas-metal-plasma-arc welding, a new method for cladding'. Welding Inst. Conference 'Exploiting Welding in Production Technology', London, 22-24 April 1975. Abington, Welding Inst., 1975, Paper 9, 179-87.

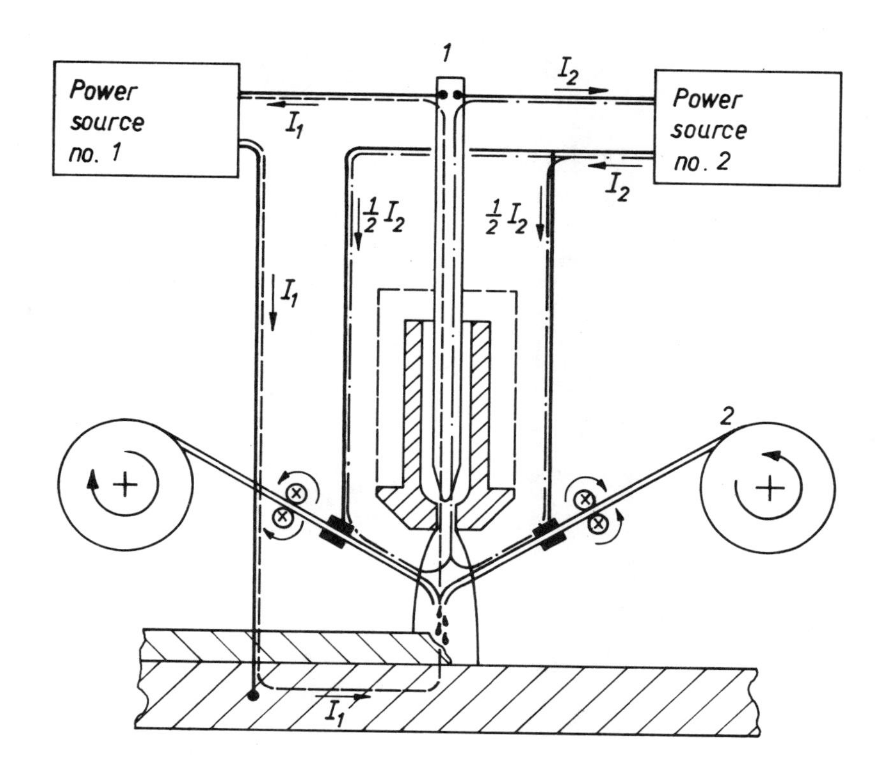

1 Explanatory sketch of GMPA process. 1 – plasma torch; 2 – wire feed

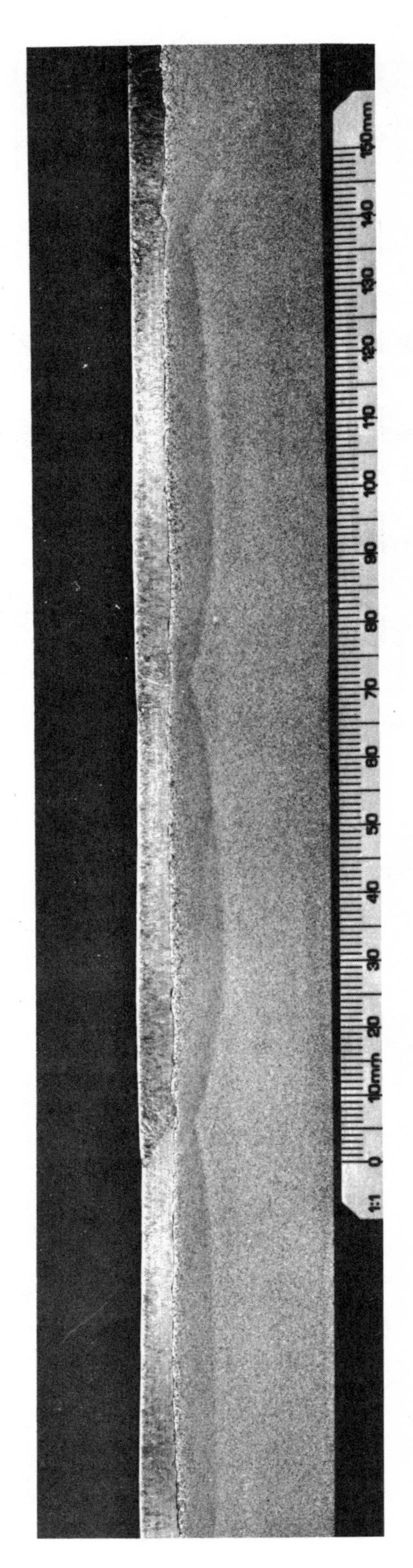

2 Cross-section of one-layer surfacing with austenitic stainless steel

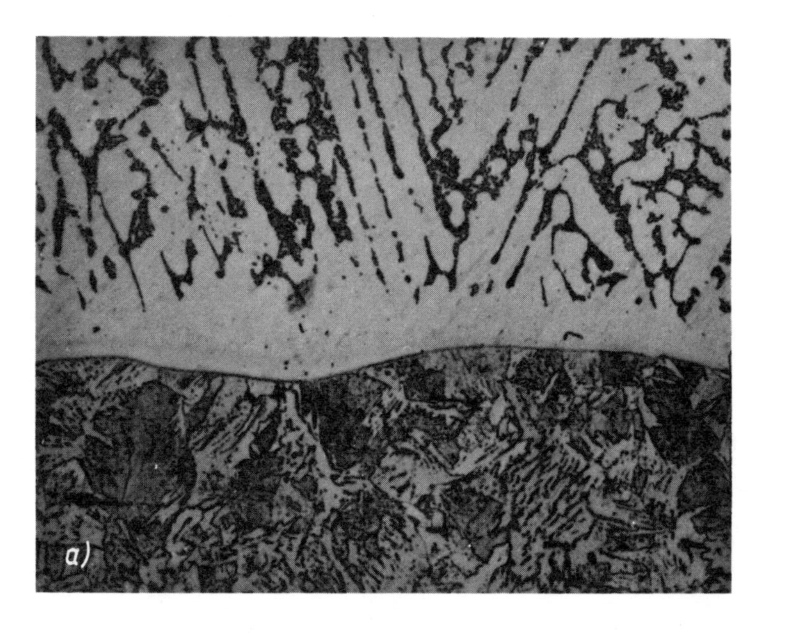

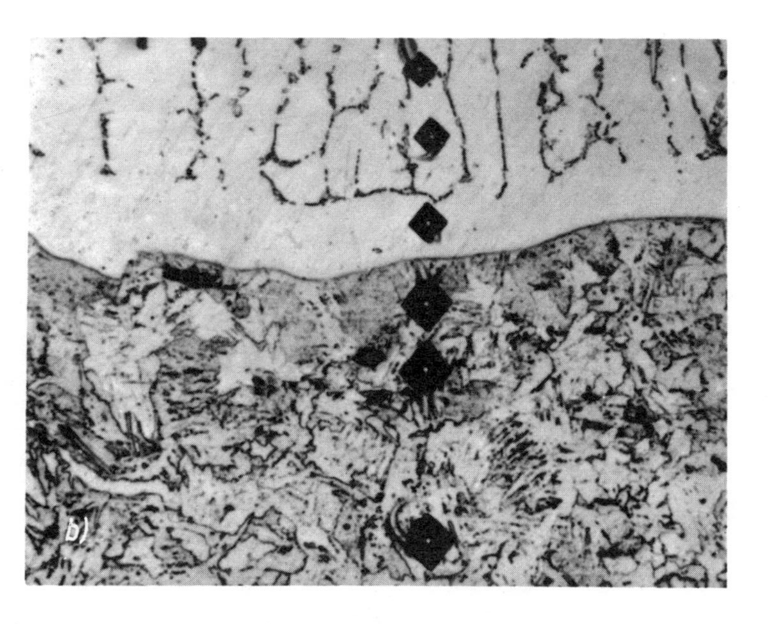

3 (a) fusion line with CMn steel base metal and cobalt-base weld metal (x 250), (b) as (a) but with microhardness prints showing hardness distribution (x 100)

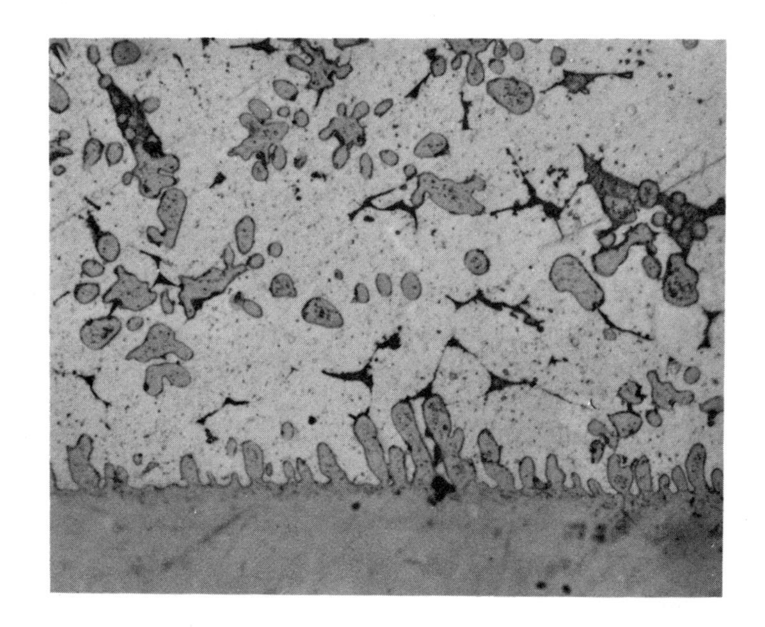

4 Fusion line between CMn steel base metal and aluminium bronze clad layer x 640

5 Two-layer cladding with aluminium bronze

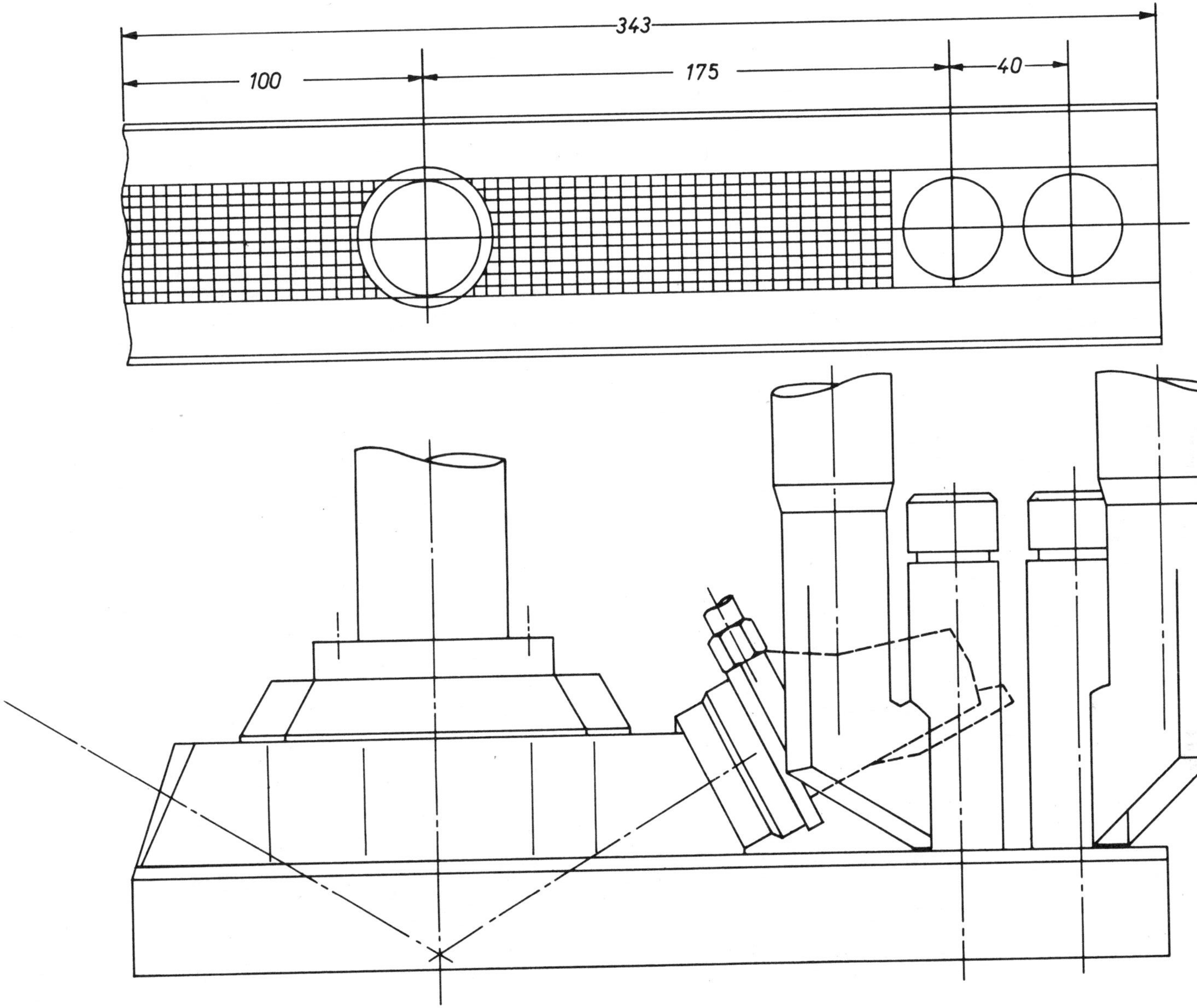

343

100

175

40

6 Multitorch head for butt welding with two 300A plasma torches and one GMPA head for welding 20mm thick stainless steel in one pass

7 Cross-section of stainless steel weld made in 15mm plate with GMPA butt welding equipment

8 Production setup for GMPA weld surfacing 90mm nozzles at Uddcomb Sweden AB

9 Bottom nozzles for boiling water nuclear reactor vessel made at Uddcomb Sweden AB